SIXTH EDITION

HANDBOOK
OF TECHNICAL
WRITING

Gerald J. Alred
University of Wisconsin-Milwaukee

Charles T. Brusaw
NCR Corporation (retired)

Walter E. Oliu
U.S. Nuclear Regulatory Commission

ST. MARTIN'S PRESS   New York

FOR BEDFORD/ST.MARTIN'S

Senior Editor: Michelle McSweeney
Associate Editor: Ellen Thibault
Developmental Editor: Mimi Melek
Senior Editor, Publishing Services: Douglas Bell
Production Supervisor: Joe Ford
Project Management: Books By Design, Inc.
Marketing Manager: Karen Melton
Text Design: Claire Seng-Niemoeller
Cover Design: Diana Coe/ko Design Studio
Cover Art: Sunset Fog and Highway 101 Bridge © Neil Gilchrist/Panoramic Images, Chicago, 1999
Composition: Pine Tree Composition, Inc.
Printing and Binding: Haddon Craftsman, an R. R. Donnelley & Sons Company

President: Charles H. Christensen
Editorial Director: Joan E. Feinberg
Editor in Chief: Karen S. Henry
Director of Editing, Design, and Production: Marcia Cohen
Manager, Publishing Services: Emily Berleth

Library of Congress Catalog Card Number: 99-62313

5   4   3   2   1   0

f   e   d   c   b   a

For information, write: Bedford/St. Martin's, 75 Arlington Street, Boston, MA 02116   (617-399-4000)

ISBN: 0-312-19804-3 (paperback)
         0-312-25496-2 (hardcover)

ACKNOWLEDGMENTS

Acknowledgments and copyrights appear on page 687, which constitutes an extension of the copyright page.

Contents

Preface v

Five Steps to Successful Writing x

Checklist of the Writing Process xviii

Topical Key to the Alphabetical Entries xx

Handbook of Technical Writing:
Alphabetical Entries 1– 686

Index 689

Preface

The sixth edition of the *Handbook of Technical Writing*, like previous editions, aims to serve as a comprehensive resource for both academic and professional audiences. The more than 500 alphabetically arranged entries in the *Handbook* provide practical guidance on the technical writing process, abundant real-world examples of technical writing—including models of effective reports, proposals, visuals, and correspondence—and thorough coverage of grammar and usage.

For this edition of the *Handbook,* we have updated each entry in the book to reflect the demands of an increasingly technological, global, and cross-cultural workplace. In addition to revising and expanding our coverage of technology and global cross-cultural material, we have improved the help offered to nonnative speakers of English and provided a fresh design that makes information even easier to access. The sixth edition of the *Handbook,* though contemporary in its scope, continues to provide time-honored strategies to help writers in the classroom and the workplace communicate correctly, strategically, and effectively.

How to Use This Book

The four-way access system of the *Handbook of Technical Writing* provides readers with multiple ways of retrieving information.

- The **alphabetically organized entries** with color tabs enable readers to find information quickly. Within the entries, terms in **boldface type** provide links to other entries that contain key definitions of concepts, further information on topics, and additional entries on related subjects.
- The **Topical Key to the Alphabetical Entries,** pages xx–xxii, groups the entries into categories and serves as a table of contents to all subjects covered in the book. The key can help a

writer focusing on a specific task or problem locate helpful entries; it is also useful for instructors who want to correlate the *Handbook* with standard textbooks or their own course materials.
- The **Checklist of the Writing Process,** pages xviii–xix, allows readers to reference all writing-related entries.
- The **comprehensive index** lists all the topics covered in the book, including those topics that are not main entries in the alphabetical arrangement.

In the classroom, the *Handbook* can be used in conjunction with a standard textbook, but instructors who want to avoid the constraints imposed by a textbook could easily use the *Handbook* as a core text, supplementing it as necessary. For that purpose, we offer an instructor's manual that provides 16 suggested lessons based on a survey of the most popular topics in technical communication texts.

New to This Edition

Updates, additions, and a new design help make the sixth edition of the *Handbook of Technical Writing* an accurate reflection of contemporary workplace communication.

Updated and Expanded Coverage.   In response to reviewer and reader requests, we have thoroughly updated and expanded our coverage of technology in the workplace, including new and revised entries on the **World Wide Web, Web page design, search engines, selecting the medium,** and **email.**

We have added coverage of **Internet research** and offer solid advice on finding, evaluating, and documenting Internet sources; we also have revised our general **research** and **documenting sources** entries.

To help readers meet the challenges of communicating and functioning effectively in the business environment, we have significantly revised a number of major entries, including **meetings, presentations, proposals, job search, résumés,** and **layout and design,** and have added a new entry for **usability testing.**

To address the cross-cultural, global nature of technical writing today, we have added the new entries **ethics in writing, biased language, global communications,** and **global graphics.** We have also addressed the issue of how audience and purpose affect the selection of a communication medium in the updated "Five Steps to Successful Writing."

Our expanded ESL coverage includes 22 new ESL boxes, provided by an ESL specialist, that offer writing advice to nonnative speakers of English. We have also revised the **English as a second language** and **varieties of English** entries.

Numerous Writer's Checklists offer the reader quick reference tips on the writing process, including advice on revising, formatting, and using visuals. Throughout the book, the checklists reinforce the "Five Steps to Successful Writing."

A Bolder, Clearer, Two-Color Design.   New visual elements such as the Writer's Checklists and ESL boxes highlight key information and useful advice in a visually appealing format. A new ESL symbol ⓔⓢⓛ identifies entries helpful to speakers of English as a second language.

Alphabetical tabs printed on the outside margins and clear, accessible main headings make it easier for readers to find what they are looking for. Hand-edited corrections allow readers to see at a glance how a passage can be corrected or improved.

Acknowledgments

For sharing their valuable advice for the sixth edition of the *Handbook,* we wish to express our appreciation to the following reviewers:

Ralph Batie, Oregon Institute of Technology
Shane Borrowman, University of Arizona
Patricia C. Click, University of Virginia
Elizabeth Coughlin, DePaul University
Evan Davis, Southwestern Oregon Community College
Pat Dorazio, SUNY Institute of Technology at Utica
James P. Farrelly, University of Dayton
M. L. Flynn, South Dakota State University
Heather Bradie Graves, DePaul University
David Hatfield, Marshall University
Susan Tyler Hitchcock, formerly University of Michigan
Suzanne Karberg, Purdue University
Kristie Kemper, Floyd College
Nancy Pointer, Normandale Community College
Alan Rauch, Georgia Institute of Technology
Margaret B. Walters, Kennesaw State University

Special thanks are due to Bob Hemmer, Editorial Director of Nonce Publishing Consultants, for reviewing each entry to make it accessible to speakers of English as a second language and for writing the ESL Tips boxes. We also wish to acknowledge and express our appreciation to Rachel Spilka, University of Wisconsin-Milwaukee, who wrote the entry on usability testing; to TyAnna K. Herrington, Georgia Institute of Technology, for her expert contributions to the copyright and plagiarism entries; and to Daniel J. Cleary, University of Wisconsin-Milwaukee, who contributed to the email entry.

We are deeply indebted to the many professionals, instructors, students, reviewers, and others who helped us prepare and who made important contributions to the first five editions of the *Handbook of Technical Writing:*

Sandra Balzo, First Wisconsin National Bank
Deborah Barrett, Houston Baptist University-Houston
Eleanor Berry, University of Wisconsin-Milwaukee
Beatrice Christiana Birchac, University of Houston-Downtown
James H. Chaffee, Normandale Community College
Darlene Chang, Freelance Graphics Designer
Mary Coney, University of Washington
Edmund Dandridge, North Carolina State University
Donna Lee Dowdney, De Anza College
Paul Falon, University of Michigan
James Farrelly, University of Dayton
Pat Goldstein, University of Wisconsin-Milwaukee
Susan Hitchcock, University of Virginia
William H. Holahan, University of Wisconsin-Milwaukee
Patrick Kelley, New Mexico State University
Wayne Losano, Rensselaer Polytechnic Institute
Carol Mablekos, Temple University
David Mair, University of Oklahoma-Norman
Roger Masse, New Mexico State University
Judy McInish, Hornbake Library, University of Maryland
Susan K. McLaughlin, Business Communications Consultants
Mimi Mejac, Nuclear Regulatory Commission
Mary Mullins, Delux Data Corporation
Brian Murphy, NCR Corporation
Lee Newcomer, University of Wisconsin-Milwaukee
Wendy Osborne, Society of American Foresters
Keith Palmer, Harnischfeger Corporation
Thomas Pearsall, University of Minnesota
Alan Rauch, Georgia Institute of Technology
Diana Reep, University of Akron

2

Lois Rew, San Jose State University
Rayleona Sanders, Nuclear Regulatory Commission
Peter V. Sands, University of Wisconsin-Milwaukee
Elizabeth A. Schulz, Olympic College
Stuart A. Selber, Pennsylvania State University
Charles Stratton, University of Idaho
Donald R. Swanson, Wright State University
John Taylor, University of Wisconsin-Milwaukee
Erik Thelen, Marquette University
Ann Thomas, Nuclear Regulatory Commission
Mary Thompson, Purdue University
William Van Pelt, University of Wisconsin-Milwaukee
Thomas Warren, Oklahoma State University
Merrill Whitburn, Rensselaer Polytechnic Institute
Conrad R. Winterhalter, NCR Corporation
Kristin R. Woolever, Northeastern University
Arthur Young, Michigan Technological University

We also want to thank all of the organizations that granted us permission to use examples of their documents: Altavista; Ballantine Books; Biospherics, Inc.; Brenda Sims, the Association of Teachers of Technical Writing, and the Department of Rhetoric at the University of Minnesota; *Chemical Engineering;* Firststar Bank; Harnischfeger Corporation; Hewlett Packard Company; Houghton Mifflin Company; Johnson Service Company; Ken Cook Company; Macmillan Computer Publications; NCR Corporation; Topica, Inc.; the University of Wisconsin Press; the University of Wisconsin-Milwaukee; and YAHOO! Inc.

We most gratefully acknowledge the contributions of colleagues at St. Martin's Press over the years and, most recently, the outstanding editorial work of Mimi Melek, Michelle McSweeney, the staff at Bedford/St. Martin's, and Herb Nolan of Books By Design.

Finally, for her patience and the time spent preparing the manuscript for the first five editions, we offer our heartfelt thanks to Barbara Brusaw. For this edition, very special thanks go to Janice Alred, Elizabeth Ignatowski, Rosemary Lesnik, Andrea Deacon, Nikki Brazy, and Seon Ju Oh, who spent many hours working on various stages of this project.

G. J. A.
C. T. B.
W. E. O.

Five Steps to Successful Writing

Many people in today's fast-paced, technological workplaces have difficulty writing because they approach the task of writing haphazardly. They sit down at a computer, for example, and hope to fill the screen without realizing that successful writing, like the technical skills in which they are proficient, requires the thoughtful application of a systematic approach.

Successful writing on the job is not the product of inspiration, nor is it merely the spoken word converted to print; it is the result of knowing how to structure information using both words and design to achieve an intended purpose. The best way to ensure that a writing task will be successful—whether it is a memo, a résumé, a proposal, or a Web page—is to approach writing using the following steps:

1. Preparation
2. Research
3. Organization
4. Writing a draft
5. Revision

You must follow those five steps consciously—even self-consciously—at first. But the same is true the first time you use new software, interview a candidate for a job, or chair a committee meeting. With practice, the steps become nearly automatic. That is not to suggest that writing becomes easy. It does not. But the easiest and most efficient way to write is to do it systematically.

As you master the five steps, keep in mind that they are interrelated and sometimes overlap. For example, your reader's needs and your purpose, which you determine in step 1, affect decisions you make in subsequent steps. You may also find that you need to retrace steps. For example, when you begin organization, you may discover that you need to return to the research steps to gather more information.

The time required for each step varies with different writing tasks. For example, when writing an informal memo, you might follow the first three steps (preparation, research, and organization) by simply listing the points in the order you want to cover them. In other words, you gather and organize information mentally as you consider your purpose and your audience. For a formal report, on the other hand, the first three steps require well-organized research, careful note taking, and detailed outlining. And for a routine email message to a coworker, the first four steps merge as you type the information on the screen. In short, the five steps expand, contract, and at times must be repeated to fit the complexity or context of the writing task.

Dividing the writing process into steps is especially useful for **collaborative writing,** in which you must divide work among team members, keep track of a project, and avoid wasting time by duplicating effort. (In this discussion, as elsewhere throughout this book, words and phrases printed in **boldface type** refer to specific alphabetical entries.) When you collaborate, you can use email to share data, exchange texts, suggest improvements, and generally keep everyone informed of your progress as you follow the steps in the writing process.

Preparation

Writing, like most professional tasks, requires solid **preparation,** In fact, adequate preparation is as important as writing the draft. In preparation for writing, you must accomplish four major tasks:

- Establish your purpose (or objective).
- Assess your readers (or audience).
- Determine the scope of your coverage.
- Select the appropriate medium.

Purpose.   You establish your **purpose** simply by determining what you want your readers to know, believe, or be able to do when they have finished reading what you have written. Be precise: It is all too common for a writer to state a purpose in terms so broad that it is almost useless. A purpose such as "to report on possible locations for a new facility" is too general. But "to compare the relative advantages of Paris, Singapore, and San Francisco as possible locations for a new engineering facility so top management can choose the best location" is a purpose that can guide you throughout the writing process.

Reader.   The next task is to assess your **reader.** Again, be precise. What are your reader's needs in relation to your subject? What does your reader already know about your subject? You need to know, for example, whether you must define basic terminology or whether such definitions will merely bore or even impede your reader. Is your "reader" actually multiple readers with different interests and levels of specialized knowledge? Are you communicating with an international audience and therefore dealing with issues inherent in writing **international correspondence?** For the purpose stated in the preceding paragraph, the reader was described as "top management." But *who* is included in that category? Will one of the people evaluating the report be the Human Resources Manager? If so, that person likely would be interested in the availability of qualified professionals as well as the presence of training, housing, and perhaps even recreational facilities available to potential employees in each city. The Director of Engineering might be concerned about possible research and development partnerships with local universities. The Marketing Manager would give priority to the facility's proximity to the primary markets for its products and services and the transportation facilities that are available. The Chief Executive Officer would be interested in all those things and even more (for example, the convenience of personal travel between corporate headquarters and the new facility).

In addition to knowing the needs and interests of the readers, you should learn as much as you can about their background knowledge. For example, have they visited all three cities? Have they already seen other reports on the three cities? Is this the first new facility, or have they chosen locations for new facilities before? If you have many readers, one way to accommodate their needs is to combine all readers into a composite reader and write for *that* reader. Another way is to design different parts of a document to reach different readers: an **executive summary** for a top manager, an **appendix** with detailed data for technical specialists, and the body of the report with appropriate **headings** for middle managers.

ESL TIPS FOR WRITING TO INTERNATIONAL AUDIENCES

In many cultures, it is inappropriate to express strong conclusions too directly, so competent writers do not *state* their ideas—they hint at them and approach the topic indirectly in as many different ways as possible. Writing around an issue is highly valued in those cultures. In other cultures, competent writers are generally expected to write in an elaborate style, with references to broad philosophical issues, however minimally and regardless of how mundane the topic is.

In North American English, *conciseness, coherence, clarity,* and *cogency* are the four words most often used to express the values assigned to "good" writing. In other words, be brief, make sure your writing holds together, be clear, and say only what is necessary to communicate your message. Of course, no writing style is inherently better than another, but to be a successful writer in any language, you must understand the cultural values that underlie the language in which you are writing. As a guide for nonnative speakers of English, we have highlighted the North American English values with an icon ESL throughout this book. See ESL Advice listed in the Topical Key to the Alphabetical Entries; see also **awkwardness, biased language, clarity, coherence, conciseness/wordiness, copyright, documenting sources, plagiarism,** and **varieties of English.**

Scope.   Determining your purpose and assessing your readers will help you decide what to include and what not to include in your writing. Those decisions establish the **scope** of your writing project. If you do not clearly define the scope, you will spend needless hours on research because you will not be sure what kind of information you need or even how much. For example, given the purpose and the readers established for the report on facility locations, the scope would include such information as land and building costs, available labor force, cultural issues, transportation facilities, proximity to suppliers, and so forth. However, it probably would not include the early history of the cities being considered or their climate and geological features, unless those aspects were directly related to your particular organization.

Selecting the Medium.   Finally, you need to determine the most appropriate medium for communicating your message. Contemporary technical communicators face a wide array of options, from email, fax, voice mail, and video conferencing to more traditional means like letters and memos, telephone calls, and face-to-face meetings.

The most important considerations in selecting the appropriate medium are the audience and the purpose of the communication. For example, if you need to collaborate with someone to solve a problem or if you need to establish rapport with someone, written exchanges (even by **email**) could be far less efficient than a phone call or a face-to-face meeting. On the other hand, if you need precise wording or you need to provide a record of a complex message, communicate in writing. If you need to make information that is frequently revised accessible to employees at a large company, the best choice might be to place the information on an intranet "web" page. (See **Web page design.**) If reviewers need to make handwritten comments on a proposal, you may need to provide paper copies that can be faxed. The comparative advantages and primary characteristics of the most typical means of communication are discussed in the entry **selecting the medium.**

Research

The only way to be sure that you can write about a complex subject is to thoroughly understand it. To do that, you must conduct adequate **research,** whether that means extensive investigation and **note taking** for a major proposal or simply checking a company Web page and jotting down points before you send an email to a colleague.

Methods of Research.   Researchers frequently distinguish between primary and secondary research, depending on the types of sources consulted and the method of gathering information. *Primary research* refers to the gathering of raw data compiled from direct observations, surveys, experiments, **questionnaires,** and the like. Such information is frequently gathered through such means as interviews and audiotape or videotape recordings. In fact, direct observation and hands-on experience may be the only ways to obtain certain kinds of information, such as the behavior of people or animals, certain natural phenomena, mechanical processes, and the operation of tools and equipment. *Secondary research* refers to gathering information that has been analyzed, assessed, evaluated, compiled, or otherwise organized into accessible form. Forms or sources include books, articles, **reports,** Web pages, dissertations, operating and procedure manuals, brochures, and so forth. Use the methods most appropriate to your research needs, recognizing that some projects will require several types.

Sources of Information.   As you conduct research, numerous sources of information are available to you.

- Your own knowledge, which is often the most common and important source on the job and which can be extracted and cultivated through **brainstorming**
- Internet sources, as discussed in **Internet research**
- Printed sources in the workplace, such as **reports, memos,** and **email**
- Library resources, as described in **library research**
- People, through **interviewing for information** and **questionnaires**

Consider all sources of information when you begin your research and use those that fit your needs. The amount of research you will need to do depends on the scope of your project.

Organization

Without **organization,** the material you gather during research will be incoherent to your reader. To organize information effectively, you must determine the best way to structure your ideas; that is, you must choose a **method of development.**

Methods of Development.   An appropriate method of development is the writer's tool for keeping information under control and the reader's means of following the writer's presentation. Your subject may lend itself readily to a particular method of development. For example, if you were writing instructions for assembling office equipment, you would naturally present the steps of the process in the order they should be performed: the **sequential method of development.** If you were writing about the history of an organization, your account would go from the beginning to the present: the **chronological method of development.** If your subject naturally lends itself to a certain method of development, use it—do not attempt to impose another method on it. Choose the method that best suits your subject, your readers, and your purpose. The entry

methods of development describes the most typical ones used in on-the-job writing.

Outlining.   Once you have chosen a method of development, you are ready to prepare an outline. **Outlining** breaks large or complex subjects into manageable parts. It also enables you to emphasize key points by placing them in the positions of greatest importance. Finally, by structuring your thinking at an early stage, a well-developed outline ensures that your document will be complete and logically organized, allowing you to focus exclusively on writing when you begin the rough draft. Even in a short letter or memo, successful writing needs the logic and structure that an outline and an appropriate method of development provide, even if the method of development and outline exist in your mind rather than on screen or paper.

Make sure at this point that you begin to consider a **layout and design** that will be helpful to your reader and appropriate to your subject and your purpose. If you intend to include **illustrations, photographs,** or **tables,** a good time to think about them is when you are outlining, especially if they need to be prepared by someone else while you are writing and revising the draft.

Writing a Draft

When you have established your purpose, your reader's needs, and your scope and have completed your research and your outline, you will be well prepared to write a first draft. Expand your outline into **paragraphs,** without worrying about **grammar,** refinements of language, or **punctuation.** Writing and revising are different activities; refinements come with revision.

Write the rough draft quickly, concentrating entirely on converting your outline into sentences and paragraphs. Write as though you were explaining your subject to someone sitting across the desk from you. Do not worry about a good opening. Just start. There is no need in the rough draft to be concerned about exact **word choice** unless it comes quickly and easily—concentrate instead on ideas.

Even with good preparation, writing the draft remains a chore for many writers. The most effective way to get started and keep going is to use your outline as a map for your first draft. Do not wait for inspiration—you need to treat writing a draft as you would any on-the-job task. The entry **writing a draft** describes tactics used by experienced writers — discover which ones are best suited to you and your task.

Consider writing the **opening** or **introduction** last because you will know more precisely what is in the body. Your opening should announce the subject and give the reader essential background information. For longer documents, an introduction should serve as a frame into which readers can fit the detailed information that follows.

Finally, you need to write a **conclusion** that ties the main ideas together and emphatically makes a final significant point. The final point may be to recommend a course of action, make a prediction or a judgment, or merely summarize your main points — the way you conclude depends on the purpose of your writing and your readers' needs.

Revision

It is likely that the clearer a finished piece of writing seems to the reader, the more effort the writer has put into the **revision.** If you have followed the steps of the writing process to this point, you will have a very rough draft. The next step, revision, requires a different frame of mind than does writing the draft. Read and evaluate the draft from the reader's point of view. Be eager to find and correct faults and be honest. Be hard on yourself for the reader's convenience rather than easy on yourself at the reader's expense. (See also **ethics in writing.**)

Do not try to do all your revising at once. Read your rough draft several times, each time looking for and correcting a different set of problems or errors. Concentrate first on larger issues, such as **unity** and **coherence;** save mechanical ones, like **spelling,** until later.

Check your draft for accuracy and completeness. Trim extraneous information that does little to accomplish your purpose. Remember that your writing should give readers exactly what they need, but it should not burden them with unnecessary information or sidetrack them into loosely related subjects. Use the Checklist of the Writing Process on the inside front cover to guide you as you revise. The checklist refers you to specific entries grouped according to the five steps and can help you diagnose writing problems and show you how to solve them.

CHECKLIST OF THE WRITING PROCESS

This checklist arranges key entries of the *Handbook of Technical Writing* according to the sequence presented in "Five Steps to Successful Writing" on pages x–xvii. The checklist is useful both for following the steps and for diagnosing writing problems. The titles of entries are in **boldface type** for quick reference, followed by the page numbers. When you turn to the entries themselves, you will find boldface cross-references to other entries that may be helpful.

You may also wish to refer to the Topical Key to the Alphabetical Entries on pages xx–xxii as well as the index, which begins on page 689.

PREPARATION   462

- ☑ Establish your **purpose**   527
- ☑ Identify your **readers**   542
- ☑ Determine your **scope** of coverage   575
- ☑ **Select the medium** appropriate to your readers, purpose, and scope   580

RESEARCH   551

- ☑ **Brainstorm** to explore what you know   75
- ☑ Conduct **Internet research**   309
- ☑ Conduct **library research**   351
- ☑ Take notes (**note taking**)   406
- ☑ **Interview for information**   314
- ☑ Create and use **questionnaires**   530

- ☑ Avoid **plagiarism**   455
- ☑ **Document sources**   172

ORGANIZATION   428

- ☑ Choose the best **method of development**   383
- ☑ **Outline** your ideas   430
- ☑ Create and integrate **illustrations**   277
- ☑ Consider **layout and design**   343

WRITING A DRAFT   682

- ☑ Select an appropriate **point of view**   456
- ☑ Adopt an appropriate **style** and **tone**   607, 633
- ☑ Use effective **sentence construction**   585
- ☑ Construct effective **paragraphs**   434

☑ Use **quotations** and
  paraphrasing   538, 441
☑ Write **introduction**
  or **opening**   322, 423
☑ Write **conclusion**   123
☑ Choose a **title**   631

REVISION   571

☑ Check for
  completeness and
  accuracy   572
☑ Check for
  unity and
  coherence   648, 96
  conciseness   119
  pace   434
  transition   640
☑ Check for **sentence
  variety**   594
  emphasis   203
  parallel structure   438
  subordination   611
☑ Check for **clarity**   93
  ambiguity   50
  awkwardness   66
  logic errors   364
  positive writing   458
  voice   664
☑ Check for **ethics in
  writing**   214
  biased language   70
  copyright   128

plagiarism   455
sexism in language   598
☑ Check for appropriate
  word choice   679
  abstract words/
  concrete words   9
  affectation and
  jargon   33, 330
  clichés and trite
  language   95, 644
  connotation   127
  defining terms   153
☑ Eliminate problems
  with **grammar**   256
  agreement   37
  case   84
  modifiers   390
  pronoun reference   487
  sentence faults   590
☑ Review mechanics
  and **punctuation**   526
  abbreviations   2
  acronyms and initialisms 17
  capital letters   79
  contractions   128
  dates   150
  indentation   283
  italics   327
  numbers   413
  proofreading   494
  spelling   606
  symbols   613

TOPICAL KEY TO THE ALPHABETICAL ENTRIES

This list arranges the alphabetical entries into subject categories. See the Checklist of the Writing Process (pages xviii–xix) for a list of the key entries in a recommended sequence of steps for any writing project.

TECHNICAL WRITING FORMS AND ELEMENTS

Correspondence 130
  Acknowledgment Letters 16
  Adjustment Letters 25
  Complaint Letters 114
  Cover Letters 144
  Inquiries and Responses 287
  International Correspondence 298
  Memos 378
  Reference Letters 544
  Refusal Letters 546
Description 158
Global Communication 251
Headers and Footers 265
Indexing 283
Instructions 292
Job Descriptions 330
Literature Reviews 362
Mathematical Equations 370
Meetings 372
Minutes of Meetings 385
Newsletter Articles 399
Presentations 466
Press Releases 479
Process Explanation 482
Proposals 496
Questionnaires 530
Reports and Components
  Abstracts 11
  Appendixes 56
  Executive Summaries 219
  Feasibility Reports 225
  Formal Reports 239
  Investigative Reports 325
  Laboratory Reports 339
  Progress and Activity Reports 482
  Table of Contents 616
  Test Reports 628
  Trip Reports 643
  Trouble Reports 644
Specifications 604
Technical Manuals 619
Trade Journal Articles 635
Usability Testing 649

FINDING A JOB

Acceptance Letters 14
Application Letters 56
Interviewing for a Job 316
Job Search 333

Resignation Letters or Memos 553
Résumés 557

WORKPLACE TECHNOLOGY

Email 197
Fax 224
Hypertext 270
Internet 305
Internet Research 309
Web Page Design 669
World Wide Web 680

PLANNING AND RESEARCH

Preparation 462
  Purpose 527
  Readers 542
  Scope 575
  Selecting the Medium 580
  Topics 634
Research 551
  Bibliographies 73
  Brainstorming 75
  Copyright 128
  Documenting Sources 172
  Internet Research 309
  Interviewing for Information 314
  Library Research 351
  Listening 358
  Note Taking 406
  Paraphrasing 441
  Plagiarism 455
  Questionnaires 530
  Quotations 538
  Search Engines 577

ORGANIZATION, WRITING, AND REVISION

Collaborative Writing 97
Ethics in Writing 214
  Biased Language 70
  Sexism in Language 598
Forms of Discourse 245
  Description 158
  Exposition 223
  Narration 395
  Persuasion 448
Methods of Development 383
  Cause and Effect 88
  Chronological 91
  Comparison 112

Decreasing Order of
    Importance   151
Definition   155
Division and Classification   169
General to Specific   250
Increasing Order of Importance   281
Sequential   596
Spatial   602
Specific to General   604
Organization   428
Outlining   430
Proofreading   494
Revision   571
    Clarity   93
    Coherence   96
    Conciseness/Wordiness   119
    Defining Terms   153
    Logic Errors   364
    Transition   640
    Unity   648
Writing a Draft   682
    Conclusions   123
    Introductions   322
    Openings   423
    Point of View   456
    Titles   631

DESIGN AND ILLUSTRATIONS

Headers and Footers   265
Headings   267
Illustrations   277
    Drawings   188
    Flowcharts   234
    Global Graphics   252
    Graphs   257
    Maps   368
    Mathematical Equations   370
    Organizational Charts   429
    Photographs   450
    Schematic Diagrams   575
    Tables   616
Indentation   283
Layout and Design   343
    Lists   361
    Quotations   538

ESL ADVICE

Agreement of Pronouns and
    Antecedents   38
English as a Second Language
    (ESL)   205
Idioms   275
Verbs   656

ESL TIPS BOXES

Adjectives   24
Agreement of Pronouns and
    Antecedents   42
Articles   62
Dictionaries   167

Gender   249
International Audiences (see "Five Steps
    to Successful Writing"   xiii)
Mood   393
Numbers   416
Possessive Case   460, 461
Prepositions   465
Pronouns   493
Quotation Marks   538
Sentence Construction   586, 590
Syntax   614
Tense   627
That/Which/Who   630
Understanding Native Speakers (see
    English as a Second Language   206)
Verbs   659, 661
Voice   668

STYLE AND LANGUAGE

Style   607
    Allusion   48
    Ambiguity   50
    Analogy (see Figures of Speech   231)
    Awkwardness   66
    Clarity   93
    Coherence   96
    Comparison   112
    Conciseness/Wordiness   119
    Emphasis   203
    Figures of Speech   231
    Nominalizations   404
    Pace   434
    Parallel Structure   438
    Paraphrasing   441
    Point of View   456
    Positive Writing   458
    Repetition   549
    Rhetorical Questions   573
    Subordination   611
    Technical Writing Style   621
    Telegraphic Style   623
    Tone   633
    Unity   648
    "You" Viewpoint   685
Language
    Absolute Words   9
    Abstract Words/Concrete
        Words   9
    Affectation   33
    Antonyms   53
    Blend Words   74
    Clichés   95
    Clipped Forms of Words   96
    Compound Words   118
    Connotation/Denotation   127
    Contractions   128
    Defining Terms   153
    Diction   162
    Dictionaries   163
    Direct Address   168
    Double Negatives   186

Elegant Variation   194
English, Varieties of   211
Euphemisms   217
Foreign Words in English   237
Functional Shift   246
Glossaries   254
Gobbledygook   255
Idioms   275
Intensifiers   296
Jargon   330
Long Variants   366
Malapropisms   368
New Words   399
Prefixes   462
Suffixes   613
Synonyms   613
Thesaurus   631
Trite Language   644
Usage   651
Vague Words   653
Vogue Words   664
Word Choice   679

SENTENCES AND PARAGRAPHS

Sentence Components
    Clauses   94
    Phrases   451
Sentence Construction   585
    Complements   117
    Expletives   222
    Garbled Sentences   248
    Mixed Constructions   389
    Modifiers   390
    Objects   418
    Restrictive and Nonrestrictive
        Elements   555
Sentence Errors
    Dangling Modifiers   148
    Misplaced Modifiers (see
        Modifiers   390)
    Sentence Faults   590
    Sentence Fragments   592
Sentence Variety   594
    Emphasis   203
    Parallel Structure   438
    Subordination   611
Paragraphs   434
    Coherence   96
    Topic Sentences (see
        Paragraphs   434)
    Transition   640
    Unity   648

PARTS OF SPEECH AND GRAMMAR

Agreement   37
    Agreement of Pronouns and
        Antecedents   38

Agreement of Subjects and
    Verbs   43
Pronoun Reference   487
Grammar   256
    Case   84
    Gender   249
    Mood   392
    Number   413
    Person   447
    Tense   624
Parts of Speech   443
    Adjectives   19
    Adverbs   30
    Conjunctions   125
    Functional Shift   246
    Interjections   297
    Nouns   409
        Appositives   61
        Possessive Case   459
        Substantives   612
    Prepositions   463
    Pronouns   488
    Verbs   656
        Mood   392
        Number   413
        Person   447
        Tense   624
        Verbals   653
        Voice   664

PUNCTUATION AND MECHANICS

Punctuation   526
    Apostrophes   53
    Brackets   75
    Colons   100
    Commas   103
    Dashes   149
    Ellipses   195
    Exclamation Marks   218
    Hyphens   271
    Parentheses   442
    Periods   445
    Question Marks   529
    Quotation Marks   536
    Semicolons   583
    Slashes   600
Mechanics
    Abbreviations   2
    Acronyms and Initialisms   17
    Ampersands   52
    Capital Letters   79
    Contractions   128
    Dates   150
    Diacritical Marks   161
    Indentation   283
    Italics   327
    Numbers   413
    Proofreaders' Marks   494
    Proofreading   494
    Spelling   606
    Symbols   613

A

a / an

A and *an* are indefinite articles; "indefinite" implies that the **noun** designated by the article is not a specific person, place, or thing, but is one of a group.

- She installed *a* program.
  [not a specific program but an unnamed program]

Use *a* before words beginning with a consonant or consonant sound (including *y* or *w* sounds).

- The year's activities are summarized in *a* one-page report.
  [Although the *o* in *one* is a vowel, the first sound is *w*, a consonant sound. Hence, the article *a* precedes the word.]

- *A* manual has been written on that subject.

- It was *a* historic event for the laboratory.

- The office manager felt that it was *a* difficult situation.

- He took *a* TWA flight.

Use *an* before words beginning with a vowel or vowel sound.

- The report is *an* overview of the year's activities.

- He seems *an* unlikely candidate for the job.

- The interviewer arrived *an* hour early.
  [Although the *h* in *hour* is a consonant, the word begins on a vowel sound; hence, it is preceded by *an*.]

- He bought *an* SLR camera.

Be careful not to use unnecessary indefinite articles in a sentence.

- I will meet you in *a* half *an* hour.

(See also **adjectives** and **articles**.)

a lot / alot

A lot is often incorrectly written as one word (*alot*). Write the phrase as two words: *a lot*. It is, however, an informal phrase that normally should not be used in technical writing. Use *many* or *numerous* instead or give a specific number or amount.

- The peer review group had ~~a lot of~~ *numerous* objections.

abbreviations   ESL

Abbreviations are shortened versions of words or combinations of the first letters of words. (See also **acronyms and initialisms.**)

- Avenue/Ave.
- Corporation/Corp.
- disk operating system/DOS
- dozen/doz
- Hypertext Markup Language/HTML

Abbreviations, like symbols, can be important space savers in technical writing because it is often necessary to provide the maximum amount of information in a limited amount of space.

Using Abbreviations

Use abbreviations only if you are certain that your readers will understand them as readily as they would the terms for which the abbreviations stand. Remember that memos or reports addressed to specific people may be read by other people; you must consider those secondary **readers** as well. A good rule of thumb: When in doubt, spell it out.

Except for commonly used abbreviations (U.S.A., a.m.), a term to be abbreviated should be spelled out the first time it is used, followed by the abbreviation in parentheses. Thereafter, the abbreviation may be used alone, as in the following example.

- The National Aeronautics and Space Administration (NASA) has accomplished much in its short history. NASA will now develop a permanent orbiting space laboratory.

Do not add an additional period at the end of a sentence that ends with an abbreviation.

- The official name of the company is Data Base, Inc.

For abbreviations specific to your profession, use a style guide provided by your professional organization or company. (A listing of style guides appears at the end of **documenting sources**.) Normally you should not make up your own abbreviations because they will probably confuse your readers.

Forming Abbreviations

Measurements.   The following list contains some common abbreviations used with units of measurement. Notice that, except for abbreviations that form words (fig., gal., in.), abbreviations of measurements do not require periods.

amp, ampere	Hz, hertz
atm, atmosphere	in., inch
bbl, barrel	kc, kilocycle
bhp, brake horsepower	kg, kilogram
Btu, British Thermal Unit	km, kilometer
bu, bushel	lb, pound
cal, calorie	lm, lumen
cd, candela	min, minute
cm, centimeter	oz, ounce
cos, cosine	pa, pascal
ctn, cotangent	ppm, parts per million
doz or dz, dozen	pt, pint
emf or EMF, electromotive	qt, quart
force	rad, radian
F, Fahrenheit	rev, revolution
fig., figure (illustration)	sec, second or secant
ft, foot (or feet)	tan., tangent
gal., gallon	yd, yard
hp, horsepower	yr, year
hr, hour	

Abbreviations of units of measure are identical in the singular and plural: 1 *cm* and 15 *cm* (not 15 *cms*).

Personal Names and Titles.   Personal names generally should not be abbreviated.

- Charles [*not* Chas.], Thomas [*not* Thos.], William [*not* Wm.]

An academic, civil, religious, or military title should be spelled out and in lowercase when it does not precede a name.

- The C̶a̶p̶t̶. *captain* wanted to check the orders.

When they precede names, some titles are customarily abbreviated. (See also **Ms./Miss/Mrs.**)

- Dr. Smith, Mr. Mills, Ms. Katz

An abbreviation of a title may follow the name; however, be certain that it does not duplicate a title before the name.

- D̶r̶. Angeline Martinez, Ph.D.
- Dr. Angeline Martinez, ̶P̶h̶.̶D̶.

When addressing **correspondence** and including names in other documents, you normally should spell out titles.

- The Honorable Mary J. Holt
- Professor Charles Matlin
- Captain Juan Ramirez
- The Reverend James MacIntosh

The following is a list of common abbreviations for personal and professional titles.

Atty.	Attorney
B.A.	Bachelor of Arts
B.S.	Bachelor of Science
B.S.E.E.	Bachelor of Science in Electrical Engineering
Dr.	Doctor [used for anyone with a doctorate]
Drs.	Plural of Dr.
Ed.D.	Doctor of Education
Hon.	Honorable
	[used with various political and judicial titles]
Jr.	Junior
	[used when a father with the same name is living]
M.A.	Master of Arts
M.B.A.	Master of Business Administration

M.D.	Doctor of Medicine
Messrs.	Plural of Mr.
Mr.	Mister
	[spelled out only in the most formal contexts]
Mrs.	Married woman
Ms.	Female equivalent of Mr.
M.S.	Master of Science
Ph.D.	Doctor of Philosophy [for many disciplines]
Rev.	Reverend
Sr.	Senior
	[used when a son with the same name is living]

Names of Organizations. A company may include in its name a term such as *Brothers, Incorporated, Corporation,* or *Company.* If the term appears as an abbreviation in the official company name, use it in its abbreviated form: *Bros., Inc., Corp.,* or *Co.* If not abbreviated in the official name, the term should be spelled out in writing other than addresses, footnotes, bibliographies, and **lists,** where abbreviations may be used. A similar guideline applies for use of an **ampersand** (&); that symbol should be used only if it appears in the official company name. For names of divisions within organizations, terms such as *Department* and *Division* should be abbreviated (*Dept.* and *Div.*) only when space is limited.

Dates and Time. The following are common abbreviations used with dates and times.

A.D.	*anno Domini,* "in the year of the Lord" [placed before the year: A.D. 1790]	Jan.	January
		Feb.	February
		Mar.	March
B.C.	before Christ [placed after the year: 647 B.C.]	Apr.	April
		Aug.	August
B.C.E.	before the common era [corresponds to B.C.]	Sept.	September
		Oct.	October
		Nov.	November
C.E.	common era [corresponds to A.D. but placed after the year]	Dec.	December
a.m.	*ante meridiem* ["before noon"]		
p.m.	*post meridiem* ["after noon"]		

Months should be spelled out when only the month and the year are given. Standard abbreviations for dates and times appear in current dictionaries, either in regular alphabetical order (by the letters of the abbreviation) or in a separate index.

Common Scholarly Abbreviations.   The following is a partial list of abbreviations commonly used in reference books and for documenting sources in research papers and reports. Latin abbreviations generally should be avoided in all but formal scholarly work.

anon.	anonymous
assn.	association
bibliog.	**bibliography,** bibliographer, bibliographic, bibliographical
c., ca.	*circa,* "about" [used with approximate dates: c. 1756]
cf.	*confer,* "compare"
ch., chs.	chapter, chapters
cit.	citation, cited
diss.	dissertation
ed., eds.	edited by, editor(s), edition(s)
e.g.	*exempli gratia,* "for example" [see **e.g./i.e.**]
enl.	enlarged [as in "rev. and enl. ed."]
esp.	especially
et al.	*et alii,* "and others"
etc.	*et cetera,* "and so forth" [see **etc.**]
f., ff.	and the following page(s) or line(s)
fig.	figure
fwd.	foreword or foreword by
GPO	Government Printing Office, Washington, D.C.
i.e.	*id est,* "that is" [see **e.g./i.e.**]
l., ll.	line, lines
ms, mss	manuscript, manuscripts
n., nn.	note, notes [used immediately after page number: 56n., 56n.3, 56nn.3–5]
N.B.	*nota bene,* "take notice, mark well"
n.d.	no date [of publication]
n.p.	no place [of publication]; no publisher
n. pag.	no pagination
p., pp.	page, pages
pref.	preface or preface by
proc.	proceedings
pseud.	pseudonym
pub (publ.)	published by, publisher, publication

rev.	revised by, revised, revision; review, reviewed by [spell out "review" where "rev." might be ambiguous]
rpt.	reprinted by, reprint
sec., secs.	section, sections
supp.	supplement
trans.	translated by, translator, translation
UP	University Press [used in MLA style of **documenting sources,** as in Oxford UP]
viz.	*videlicet,* "namely"
vol., vols.	volume, volumes
vs., v.	*versus,* "against" [v. preferred in titles of legal cases]

Postal Abbreviations.   The U.S. Postal Service specifies the following abbreviations for states and protectorates in postal addresses.

Alabama	AL	Montana	MT
Alaska	AK	Nebraska	NE
Arizona	AZ	Nevada	NV
Arkansas	AR	New Hampshire	NH
California	CA	New Jersey	NJ
Colorado	CO	New Mexico	NM
Connecticut	CT	New York	NY
Delaware	DE	North Carolina	NC
District of Columbia	DC	North Dakota	ND
Florida	FL	Ohio	OH
Georgia	GA	Oklahoma	OK
Guam	GU	Oregon	OR
Hawaii	HI	Pennsylvania	PA
Idaho	ID	Puerto Rico	PR
Illinois	IL	Rhode Island	RI
Indiana	IN	South Carolina	SC
Iowa	IA	South Dakota	SD
Kansas	KS	Tennessee	TN
Kentucky	KY	Texas	TX
Louisiana	LA	Utah	UT
Maine	ME	Vermont	VT
Maryland	MD	Virginia	VA
Massachusetts	MA	Virgin Islands	VI
Michigan	MI	Washington	WA
Minnesota	MN	West Virginia	WV
Mississippi	MS	Wisconsin	WI
Missouri	MO	Wyoming	WY

The names of the Canadian provinces and territories are abbreviated as follows. (The first abbreviation given for each is the one specified by the U.S. Postal Service.)

Alberta	AB or Alta.
British Columbia	BC or B.C.
Manitoba	MB or Man.
New Brunswick	NB or N.B.
Newfoundland	NF or Nfld.
Northwest Territories	NT or NWT
Nova Scotia	NS or N.S.
Ontario	ON or Ont.
Prince Edward Island	PE or PEI
Quebec	PQ or Que
Saskatchewan	SA or Sask.
Yukon Territory	YT

Note that when *United States* is used in its noun form it is spelled out, but its abbreviation, U.S., may be used as an adjective.

* New York will be our first stop in the United States. [noun form]

* The U.S. economic policy has a profound effect on the global economy. [adjective form]

above

Avoid using *above* to refer to a preceding passage, illustration, or table in your writing. Its reference is often vague, and if the passage referred to is far from the current one, the reader must go back to the earlier passage to understand the reference. The same is true of *aforesaid, aforementioned, above mentioned, the former,* and *the latter.* In addition to distracting the reader, those words contribute to a heavy, wooden **style.** If you must refer to something previously mentioned, either repeat the **noun** or **pronoun** or construct your paragraph so the reference is obvious (see also **former/latter** and **aforesaid**).

* Please fill out and submit ~~the above~~ *your travel voucher* by March 1.

absolute words

Absolute words (such as *round, unique, exact,* and *perfect*) cannot be used logically in **comparison** (*rounder, roundest*), especially in technical writing, where accuracy and precision are crucial. (See also **adjectives.**)

- We modified our technical manual to more ~~exactly~~ *closely* reflect existing specifications.

absolutely

Absolutely means "definitely," "entirely," "completely," or "unquestionably." It is not an **intensifier** and should not be used to mean "very" or "much."

- A new analysis is ~~absolutely~~ impossible.

abstract words / concrete words   (ESL)

Abstract words refer to general ideas, qualities, conditions, acts, or relationships—intangible things that cannot be detected by the five senses.

- work, freedom, courage, crime, kindness, idealism, pride, loyalty

Abstract words must frequently be qualified by other words.

- What the members of the Research and Development team need is freedom *to explore the problem further*.

Concrete words identify things that can be perceived by the five senses (sight, hearing, touch, taste, and smell).

- monitor, telephone, keyboard, coffee, gasoline

Abstract and concrete words are usually best used together, in support of each other.

- A well-designed office *chair* [concrete] helps reduce *stress* [abstract].

As another example, the abstract idea of *transportation* is made clearer with the use of specific concrete words, such as *subway* and *buses.*

- *Transportation* was limited to the *subway* and *buses.*

A word that is concrete in one context can be less concrete in another. The same word may even be abstract in one context and concrete in another. The choice between abstract and concrete terms always depends on the context. Just how concrete a particular context might require you to be is shown in Figure A–1, which goes from the most abstract, on the left, to the most concrete, on the right.

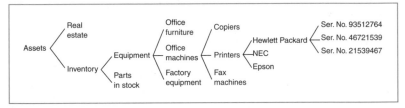

FIGURE A–1. Example of Abstract-to-Concrete Words

    The example in Figure A–1 represents seven levels of abstraction; the appropriate level depends on your **purpose** in writing and on the context in which you are using the word. For example, a company's annual report might logically use the most abstract term, *assets,* to refer to all the property and goods the firm owns; shareholders probably would not require a less abstract term. Interoffice **memos** between the company's accounting and legal departments would call the firm's holdings *real estate* and *inventory.* To the company's inventory control department, however, the word *inventory* is much too abstract to be useful, and a report on inventory might contain the more concrete terms *equipment* and *parts in stock.* But to the assistant inventory control manager in charge of equipment, that term is still too abstract; he or she would speak of particular kinds of equipment: *office furniture, factory equipment,* and *office machines.* The breakdown of the types of office machines for which the inventory control assistant is responsible might include *copiers, fax machines,* and *printers.* However, even that classification would not be concrete enough for the company's purchasing department to obtain service contracts for the maintenance of its printers. Because the department must deal with different printer manufacturers, printers would have to be listed by brand name: *Hewlett Packard, NEC,* and *Epson.* And the Epson technician who performs the maintenance must go one step further and identify each Epson printer by serial number. As the example in Figure A–1 shows, then, a term that is sufficiently concrete in one context may be too abstract in another. (See also **word choice.**)

abstracts

An *abstract* summarizes and highlights the major points of a document. Abstracts are written for many **formal reports,** many **trade journal articles,** and most dissertations, as well as for many other works. Their primary purpose is to enable readers to decide whether to read the work in full. (For a discussion of how summaries differ from abstracts, see **executive summaries.**)

Although abstracts are published with the longer works they condense, they also are often published independently in periodical indexes and abstracting services (see **library research**). Abstracting services are devoted exclusively to abstracting information in specific fields of study. They enable researchers to review a great deal of **literature** in a short time. (See also **documenting sources.**)

Usually 200–250 words long, an abstract must be able to stand on its own because abstracts may be published independently of the main document. Depending on the kind of information they contain, abstracts are often classified as descriptive or informative.

A *descriptive abstract* includes information about the **purpose, scope,** and methods used to arrive at the reported findings. It is a slightly expanded table of contents in paragraph form. Provided that it adequately summarizes the information, a descriptive abstract need not be longer than several sentences. A typical descriptive abstract is shown in the following example.

ABSTRACT

This report investigates the long-term effects of long-distance running on the bones, joints, and general health of runners aged 50 to 72. The Sports Medicine Institute of Columbia Hospital sponsored this investigation, first to decide whether to add a geriatric unit to the Institute, and second to determine whether physicians should recommend long-distance running for their older patients. The investigation is based on recent studies conducted at Stanford University and the University of Florida. The Stanford study tested and compared male and female long-distance runners between 50 and 72 years of age with a control group of runners and nonrunners. The groups were also matched by sex, race, education, and occupation. The Florida study used only male runners who had run at least 20 miles a week for five years and compared them with a group of runners and nonrunners. Both studies based findings on medical histories and on detailed physical and X-ray examinations.

An *informative abstract* is an expanded version of the descriptive abstract. In addition to information about the purpose, scope, and methods of the original document, the informative abstract includes the results, **conclusions,** and any recommendations. The informative abstract retains the **tone** and essential scope of the report while omitting its details, as shown in the following example.

ABSTRACT

This report investigates the long-term effects of long-distance running on the bones, joints, and general health of runners aged 50 to 72. The Sports Medicine Institute of Columbia Hospital sponsored this investigation, first to decide whether to add a geriatric unit to the Institute, and second to determine whether physicians should recommend long-distance running for their older patients. The investigation is based on recent studies conducted at Stanford University and the University of Florida. The Stanford study tested and compared male and female long-distance runners between 50 and 72 years of age with a control group of runners and nonrunners. The groups were also matched by sex, race, education, and occupation. The Florida study used only male runners who had run at least 20 miles a week for five years and compared them with a group of runners and nonrunners. Both studies based findings on medical histories and on physical and X-ray examinations.

Both studies conclude that long-distance running is not associated with increased degenerative joint disease. Control groups were more prone to spur formation, sclerosis, and joint-space narrowing and showed more joint degeneration than runners. Female long-distance runners exhibited somewhat more sclerosis in knee joints and the lumbar spine area than matched control subjects.

Both studies support the role of exercise in retarding bone loss with aging. The investigation concludes that the health risk factors are fewer for long-distance runners than for those less active between the ages of 50 and 72.

The investigation recommends that the Sports Medicine Institute of Columbia Hospital consider the development of a geriatric unit a priority and that it inform physicians that an exercise program that includes long-distance running can be beneficial to their patients' health.

The organization for which you work or the publication for which you are writing determines the type of abstract you should write. Otherwise, aim to satisfy the needs of your primary **readers.** Informative abstracts satisfy the needs of the widest possible readership, but descriptive abstracts are preferable for any document that compiles a variety of information. Typically, an abstract follows the title page and is numbered page iii.

Writing Style

Write the abstract after you have finished the report or document. Otherwise, the abstract may not accurately reflect the longer work. Begin with a topic sentence that announces the subject and the scope of the original document. Then, using the major and minor headings of your outline to distinguish primary ideas from secondary ones, decide what material is relevant to your abstract. Write with **clarity** and **conciseness,** eliminating unnecessary words and ideas. Do not, however, become so terse that you omit articles (*a, an, the*) and important transitional words and phrases (*however, therefore, but, in summary*). Write in complete sentences but avoid stringing a group of short sentences end to end; instead, combine ideas by using **subordination** and **parallel structure.** As a rule, spell out most acronyms and all but the most common **abbreviations** (e.g., CD-ROM, RAM).

Writer's Checklist: Abstracts

Because readers of abstracts do not yet know anything about your document, except what its title announces, you should include the following information:

☑ The subject

☑ The scope

Writer's Checklist: Abstracts (continued)

- ☑ The purpose
- ☑ The methods used
- ☑ The results obtained [informative abstract only]
- ☑ The recommendations made, if any [informative abstract only]

Do not include the following kinds of information:

- ☑ A detailed discussion or explanation of the methods used
- ☑ Administrative details about how the research was undertaken, who funded it, who worked on it, and the like, unless such details have a bearing on the document's purpose
- ☑ Illustrations, tables, charts, maps, and bibliographic references
- ☑ Any information that does not appear in the original document

For guidance about the placement of abstracts in **reports,** see **formal reports.**

accept / except

Accept is a **verb** meaning "consent to," "agree to take," or "admit willingly."

- I *accept* the responsibility that goes with the appointment.

Except is normally used as a **preposition** meaning "other than" or "excluding."

- We agreed on everything *except* the schedule.

acceptance letters

When you receive an offer of a job that you want to accept, begin your letter by accepting the job with pleasure. Your letter should identify the job you are accepting and state the salary you have been offered to avoid confusion on those two important points.

The second paragraph might go into detail about your plans for moving and reporting to work—the details will vary depending on what you discussed during the job interview. Conclude the letter by stating that you are looking forward to working for your new employer. (See also **correspondence** and **refusal letters**.)

Figure A–2 is an example of an acceptance letter written by a college student.

2647 Sitwell Road
Charlotte, NC 28210
March 26, 20--

Mr. F. E. Vallone
Manager of Human Resources
Calcutex Industries, Inc.
3275 Commercial Park Drive
Raleigh, NC 27609

Dear Mr. Vallone:

I am pleased to accept your offer of $30,500 per year as a junior ACR designer in the Calcutex Group.

After graduation, I plan to leave Charlotte on Tuesday, June 16. I should be able to find suitable living accommodations within a few days and be ready to report for work on the following Monday, June 22. Please let me know if this date is satisfactory to you.

I look forward to working with the design team at Calcutex.

Very truly yours,

Philip Ming

Philip Ming

FIGURE A–2. Acceptance Letter

accumulative / cumulative

Accumulative and *cumulative* are **synonyms** that mean "massed" or "added up over a period of time." *Accumulative* is rarely used except in reference to accumulated property or wealth.

• The Radiology Department reported that last month's exposures to X-rays were acceptable, but we must be careful so that *cumulative* doses do not exceed the hospital's guidelines.

• The corporation's *accumulative* holdings include 19 real-estate properties.

acknowledgment letters

One way to build goodwill with colleagues and clients is to let them know that something they sent arrived and to express thanks. A letter that serves that function is called an acknowledgment letter. It is

```
To: James G. Evans <jge.waterford@omni.com>
From: Roger Hammersmith <hammer@eci2.com>
Attchmnt:
Subject : Report Received
- - - - - Message Text - - - - -

Dear Mr. Evans:

I received your report today; it appears to be complete and well done.

When I finish studying it thoroughly, I'll send you our cost estimate for
the installation of the Mark II Energy Saving System.

Again, thanks for your effort.

Regards,

Roger

=====================================
Roger Hammersmith, Manager
Sales Division, Ecology Systems, Inc.
1015 Clarke Street, Chicago IL 60615
Off (312) 719-6620 Fax (312) 719-5500
http://www.ecology.sys/com
email: hammer@eci2.com
=====================================
```

FIGURE A–3. Email Version of Acknowledgment Letter

usually a short, polite note. If you have established a working relationship with someone, letting them know by **email** is appropriate. (See also **correspondence.**) The example shown in Figure A–3 on page 16 is typical and could be sent as a letter or an email.

acronyms and initialisms

An *acronym* is an **abbreviation** that is formed by combining the first letter or letters of several words. Acronyms are pronounced as words and are written without periods.

- *L*ocal *a*rea *n*etwork/LAN
- *a*lternating *g*radient *s*ynchrotron/AGS
- *r*andom-*a*ccess *m*emory/RAM
- *r*ecirculation *a*ctuation *s*ymbol/RAS
- *r*ead-*o*nly *m*emory/ROM

An *initialism* is an abbreviation that is formed by combining the initial letter of each word in a multiword term. Initialisms are pronounced as separate letters.

- *f*or *y*our *i*nformation/FYI
- *e*nd *o*f *m*onth/e.o.m.
- *S*ociety for *T*echnical *C*ommunication/STC
- *c*ash *o*n *d*elivery/c.o.d.
- *p*ost *m*eridiem/p.m.

In business, industry, and government, people working together on particular projects or having the same specialties—for example, financial analysts or accountants—often use acronyms and initialisms. As long as such specialists communicate with one another, shortened forms will be easily recognized and understood. If the same acronyms or initialisms were to be used in **correspondence** to someone outside the group, however, they might be incomprehensible to that reader and should be explained.

Acronyms and initialisms, if used appropriately, can be convenient for both the reader and the writer. Many people, however, overuse them, either as an **affectation** or in a misguided attempt to make their writing concise, especially in email.

Writer's Checklist: Acronyms and Initialisms

Follow these general guidelines to decide whether to use acronyms and initialisms.

☑ The first time an acronym or initialism appears in a written work, write the complete term, followed by the abbreviated form in parentheses. Thereafter, you may use the acronym or initialism alone.

- The Capital Appropriations Request (CAR) controls the spending of money. However, if the CAR is not approved by senior management, we must find . . .

☑ In a long document, help the readers by repeating the full term in parentheses after the abbreviation at regular intervals so they do not have to search back to the first time the acronym or initialism was used to find its meaning.

- Remember that the CAR (Capital Appropriations Request) controls the spending of money.

☑ If something is better known by its acronym or initialism than by its formal term, use the abbreviated form. The initialism *a.m.,* for example, is much more common than the formal *ante meridiem.*

☑ Write acronyms in capital letters without periods. The only exceptions are acronyms that have become accepted as common nouns, which are written in lowercase letters.

- laser [for *l*ight *a*mplification by *s*timulated *e*mission of *r*adiation]
- scuba [for *s*elf-*c*ontained *u*nderwater *b*reathing *a*pparatus]

☑ Initialisms may be written in either uppercase or lowercase. Generally, use periods for lowercase initialisms but not for uppercase ones. Two exceptions are geographic names and academic degrees.

- EDP/e.d.p., EOM/e.o.m., OD/o.d.
- U.S.A., U.K., E.U.
- B.A., M.B.A., Ph.D.

☑ Form the plural of an acronym or initialism by adding an *s.* Do not use an **apostrophe.**

- CARs, CRTs

activate / actuate

Both *activate* and *actuate* mean "make active," although *actuate* is usually applied only to mechanical processes.

- The relay *actuates* the trip hammer. [mechanical process]
- The electrolyte *activates* the battery. [chemical process]
- The governor *activated* the National Guard. [legal process]

active voice (*see* **voice**)

actually (*see* **really**)

ad hoc

Ad hoc is Latin for "for this" or "for this particular occasion." An ad hoc committee is one set up to consider a particular issue, as opposed to a permanent committee. The term has been fully assimilated into English and thus does not have to be italicized (or underlined). (See also **affectation** and **foreign words in English.**)

adapt / adept / adopt

Adapt is a **verb** meaning "adjust to a new situation." *Adept* is an **adjective** meaning "highly skilled." *Adopt* is a verb meaning "take or use as one's own."

- The company will *adopt* a policy of finding engineers who are *adept* managers and who can *adapt* to new situations.

adjectives (ESL)

DIRECTORY
Overview   20
Limiting Adjectives   20

A

20 adjectives

Comparison of Adjectives 22
Placement of Adjectives 23
Use of Adjectives 23
Tips for Using Adjectives (ESL) 24

Overview

An adjective makes the meaning of a **noun** or a **pronoun** more specific by highlighting one of its qualities (descriptive adjective) or by imposing boundaries on it (limiting adjective).

* a *hot* iron [descriptive]

* *ten* automobiles [limiting]

* *his* desk [limiting]

Limiting Adjectives

Limiting adjectives include these categories:

> Articles (*a, an, the*)
> Demonstrative adjectives (*this, that, these, those*)
> Possessive adjectives (*my, your, his, her, its, our, their*)
> Numeral adjectives (*two, first*)
> Indefinite adjectives (*all, none, some, any*)

Articles. Articles are considered to be adjectives because they modify the items they designate by either limiting them or making them more precise. (See also **articles.**) There are two kinds of articles, indefinite and definite.

The *indefinite* articles, *a* and *an,* denote an unspecified item.

* *A* program was run on our new computer.
  [Not a specific program but an unspecified program. Therefore, the article is indefinite.]

(See also **a/an** and **English as a second language.**)

The *definite* article, *the,* denotes a particular item.

* *The* program was run on the computer.
  [Not just any program but *the* specific program. Therefore, the article is definite.]

Do not omit all articles from your writing. Including articles costs nothing; eliminating them makes reading more difficult. (See also **telegraphic style.**) On the other hand, do not overdo it. An article can be superfluous.

- I'll meet you in *a* half *an* hour.
  [Choose one and eliminate the other.]

- Fill with *a* half *a* pint of fluid.
  [Choose one and eliminate the other.]

Do not capitalize articles when they appear in titles except as the first word of the title. (See also **capital letters.**)

- *Time* magazine reviewed *The Old Man and the Sea.*

Demonstrative Adjectives. A *demonstrative* adjective "points to" the thing it modifies, specifying the object's position in space or time. *This* and *these* specify a closer position; *that* and *those* specify a more remote position.

- *This* information we received today is more complete than *that* information we received earlier.

- *These* blood samples we tested today were easier to process than *those* samples tested last week.

Demonstrative adjectives often cause problems when they modify the nouns *kind, type,* and *sort.* Demonstrative adjectives used with those nouns should agree with them in **number.**

- this kind / these kinds
- that type / those types
- this sort / these sorts

Confusion often develops when the **preposition** *of* is added (*this kind of, these kinds of*) and the object of the preposition does not conform in number to the demonstrative adjective and its noun.

- *This kind* of hydraulic ~~cranes~~ crane is best.

- *These kinds* of hydraulic ~~crane~~ cranes are best.

Using demonstrative adjectives with words like *kind, type,* and *sort* can easily lead to vagueness. It is better to be more specific.

VAGUE   It was *kind of* a bad year for the firm.

SPECIFIC   The firm's profits were down 10 percent from last year.

Possessive Adjectives.   Because *possessive* adjectives (*my, your, his, her, its, our, their*) directly modify **nouns,** they function as adjectives, even though they are pronoun forms. Anything that modifies a noun functions as an adjective.

- The *proposal's* virtues outweighed *its* defects.

Numeral Adjectives.   *Numeral* adjectives (or **numbers**) identify quantity, degree, or place in a sequence. They always modify count nouns. Numeral adjectives are divided into two subclasses: cardinal and ordinal. A *cardinal* adjective expresses an exact quantity.

- *one* pencil, *two* computers, *three* airplanes

An *ordinal* adjective expresses degree or sequence.

- *first* quarter, *second* edition, *third* degree, *fourth* year

In most writing, an ordinal adjective should be spelled out if it is a single word (*tenth*) and written in figures if it is more than one word (*312th*). Ordinal numbers can also function as adverbs. (See also **first/firstly.**)

- John arrived *first.*

Indefinite Adjectives.   *Indefinite* adjectives are so called because they do not designate anything specific about the nouns they modify.

- *some* CD-ROMs, *any* branches, *all* aircraft engines

The articles *a* and *an* are included among the indefinite adjectives. (See also **a/an.**)

- It was *an* hour before the needle of the gauge finally reached the danger zone.

Comparison of Adjectives

Most adjectives in the "positive form" add the **suffix** *-er* to show **comparison** with one other item and the suffix *-est* to show comparison with two or more other items.

- The first ingot is *bright.* [positive form]
- The second ingot is *brighter.* [comparative form]
- The third ingot is *brightest.* [superlative form]

Many two-syllable adjectives and most three-syllable adjectives are preceded by the word *more* or *most* to form the comparative or the superlative.

- The new facility is *more* impressive than the old one.

- The new facility is the *most* impressive in the city.

A few adjectives have irregular forms of comparison (*much, more, most; little, less, least*).

Absolute words (*round, unique, exact, accurate*) are not logically subject to comparison; nevertheless, they are sometimes used comparatively.

- Phase-loop circuits make FM tuner performance more ~~exact~~ *accurate* by decreasing tuner distortion.

Placement of Adjectives

When limiting and descriptive adjectives appear together, the limiting adjectives precede the descriptive adjectives, with the articles usually in the first position.

- *The ten red* cars were parked in a row.
  [article, limiting adjective, descriptive adjective]

Within a sentence, adjectives may appear before the nouns they modify (the attributive position) or after the nouns they modify (the predicative position).

- *The small* jobs are given priority. [attributive position]

- Tests are taken even when exposure is *brief*. [predicative position]

Use of Adjectives

Because of the need for precise description, it is often necessary to use nouns as adjectives.

- The *test* conclusions led to a redesign of the system.

When adjectives modifying the same noun can be reversed and still make sense or when they can be separated by *and* or *or,* they should be separated by commas.

- The company is seeking a *young, energetic, creative* management team.

When an adjective modifies a phrase, no comma is required.

ESL TIPS FOR USING ADJECTIVES

Unlike many other languages, adjectives in English have only one form. Do not add -s or -es to an adjective to make it plural.

- the long trip
- the long letters

    Capitalize adjectives of origin (city, state, nation, continent).

- the Venetian canals
- the Texan hat
- the French government
- the African desert

    In English, verbs of feeling (for example, *bore, interest, surprise*) have two adjectival forms: the present participle (-*ing*) and the past participle (-*ed*). Use the present participle to describe what causes the feeling. Use the past participle to describe the person who experiences the feeling.

- We heard the surprising election results.
  [The *election results* cause the feeling.]

- Only the candidate was surprised by the election results.
  [The *candidate* experienced the feeling of surprise.]

    Adjectives follow the noun in English in only two cases: when the adjective functions as a subjective complement, as in

- That project is not *finished*.

and when an adjective phrase or clause modifies the noun, as in

- The project *that was suspended temporarily* . . .

In all other cases, adjectives are placed before the noun.

    When there are multiple adjectives, it is often difficult to know the right order. The guidelines illustrated in the following example would apply in most circumstances, but there are exceptions. (Normally do not use a phrase with so many stacked **modifiers.**) (See also **articles.**)

The	six	extra large	rectangular	brown	Chinese	cardboard	take-out	containers
	number	size		color		material		noun
determiner	comment		shape		origin		qualifier	

- He was wearing his *old cotton tennis hat.*
  [*Old* modifies the phrase *cotton tennis hat.*]

Never use a comma between a final adjective and the noun it modifies.

- He is a conscientious, honest, reliable, worker.

Writers sometimes string together a series of nouns to form a unit modifier, thereby creating jammed **modifiers.** Be aware of that problem when you use nouns as adjectives.

JAMMED	The test control group meeting was held on Wednesday.
IMPROVED	The meeting of the test control group was held on Wednesday.

As a rule of thumb, it is better to avoid general (*nice, fine, good*) and trite (a *fond* farewell) adjectives; in fact, it is good practice to question the need for most adjectives. Often, your writing will not only read as well without an adjective, but it may be better without it. If you must use an adjective, try to find one that is as precise as possible. (See also **articles.**)

adjustment letters

An adjustment letter is written in response to a complaint letter and tells the customer what your company intends to do about the complaint (see **complaint letters**). You should settle such matters quickly and courteously and always try to satisfy the customer at a reasonable cost to your company.

Although sent in response to a problem, an adjustment letter actually provides an excellent opportunity to build goodwill for your company. An effective adjustment letter both repairs the damage that has been done and restores the customer's confidence in your company.

Grant adjustments graciously; a settlement made grudgingly will do more harm than good. The **tone** of your **correspondence** is critical. No matter how unreasonable the complaint, your response should be positive and respectful. Avoid emphasizing the unfortunate situation; instead, focus on what you are doing to correct it. Not only must you be gracious, you must also acknowledge the error in such a way that the buyer does not lose confidence in your company.

Before granting an adjustment to a claim for which your company is at fault, you must determine what happened and what you can do to satisfy the customer. Be certain you are familiar with your company's adjustment policy. In addition, be careful about your wording; for example, "We have received your letter of May 7 about our *defective product*" could be ruled in a court of law as an admission that the product is, in fact, defective. Treat every claim individually and lean toward giving the customer the benefit of the doubt. An adjustment letter is shown in Figure A–4.

Computer Solutions, Inc.
521 West 23rd Street
New York, NY 10011

Customer Relations
Phone: (212) 574-3894
Fax: (212) 574-3899
email: sgtv@juno.com

September 28, 20--

Mr. Paul Denlinger, Manager
Diagnostic Services Dept.
Baker Memorial Hospital
501 Main Street
Springfield, OH 45321

Dear Mr. Denlinger:

Thank you for your letter regarding your order for nine HV3 monitors. We have shipped the correct units by UPS; you should receive them shortly after you receive this letter. I have also canceled your original order so you will not be sent overdue notices and so we can charge you at our preferred-customer rate.

Please accept our apologies. Evidently, Shipping and Receiving misread your order, and sent it to our Service Department. That is why your note did not come to the attention of our sales staff.

To prevent further inconvenience, please send any future inquiries directly to Mr. Gene Sanchez, Sales Manager, at our Newark facility (email: gs2.csi.exec.com).

We appreciate your business and look forward to working with you in the future.

Sincerely,

Susan Siegel

Susan Siegel
Assistant Director

SS/mr

FIGURE A–4. Adjustment Letter When the Company Is at Fault

Sometimes you may decide to grant a partial adjustment, even though the claim is not really justified, as shown in Figure A–5. You would do that only to regain the lost goodwill of the customer.

You may need to educate your reader about the use of your product or service. Customers sometimes submit claims that are not justified, even though the customers honestly believe them to

---

Computer Solutions, Inc.
521 West 23rd Street
New York, NY 10011

Customer Relations
Phone: (212) 574-3894
Fax: (212) 574-3899
email: sgtv@juno.com

September 28, 20--

Mr. Fred J. Swesky
7811 Ranchero Drive
Tucson, AZ 85761

Dear Mr. Swesky:

Thank you for your letter regarding the replacement of your CS5 Portable PC.

You said in your letter that you used the unit on an open deck. As our service representative pointed out, this model is not designed to operate in extreme heat. As the instruction manual accompanying your new CS5 states, such exposure can produce irreparable damage. Because your unit was used in such extreme heat conditions, we cannot honor the warranty.

However, we are enclosing a certificate entitling you to a trade-in allowance equal to your local CSI dealer's markup for the unit. This means you can purchase a new unit at wholesale, provided you return your original unit to your local dealer.

Sincerely yours,

Susan Siegel

Susan Siegel
Assistant Director

SS/mr
Enclosure

FIGURE A–5. Adjustment Letter When the Customer Is at Fault.

be (for example, a problem resulting from a customer not following prescribed maintenance instructions properly). You would grant such a claim only to build goodwill. When you write a letter of adjustment in such a situation, it is wise to give the explanation before granting the claim; otherwise, your reader may never get to the explanation. If your explanation establishes customer responsibility, be sure to do so tactfully. Figure A–6 on page 29 is an example of an educational adjustment letter.

Writer's Checklist: Adjustment Letters

Use the following guidelines to help you write adjustment letters.

☑ Open with what the **reader** will consider good news:
- *Grant the adjustment,* if appropriate, for uncomplicated situations ("Enclosed is a replacement for the damaged part").
- Reveal that you intend to grant the adjustment by admitting that the customer was right ("Yes, you were incorrectly billed for the delivery"). Then explain the specific details of the adjustment. This method is good for adjustments that require detailed explanations.
- Apologize for the error ("Please accept our apologies for not acting sooner to correct your account"). This method is effective when the customer's inconvenience is as much an issue as money.
- Use a combination of techniques. Often, situations that require an adjustment also require flexibility.

☑ If an explanation will help restore your reader's confidence or goodwill, explain what caused the problem.

☑ Explain specifically how you intend to make the adjustment, if it is not obvious in your opening.

☑ Express appreciation to the customer for calling your attention to the situation, explaining that this helps your firm keep the quality of its product or service high.

☑ Point out any steps your company is taking to prevent a recurrence of whatever went wrong, giving the customer as much credit as the facts allow.

☑ Close pleasantly, looking forward, not back. Avoid recalling the problem in your closing ("Again, we apologize . . .").

SWELCO Coffeemaker, Inc.

9025 North Main Street
Butte, MT 59702
Phone: 800-233-5656
Fax: 800-233-3010

August 26, 20--

Mr. Carlos Ortiz
638 McSwaney Drive
Butte, MT 59702

Dear Mr. Ortiz:

Enclosed is your SWELCO Coffeemaker, which you sent to us on August 17.

In various parts of the country, tap water may contain a high mineral content. If you fill your SWELCO Coffeemaker with water for breakfast coffee before going to bed, a mineral scale will build up on the inner wall of the water tube — as explained on page 2 of your SWELCO Instruction Booklet.

We have removed the mineral scale from the water tube of your coffeemaker and thoroughly cleaned the entire unit. To ensure the best service from your coffeemaker in the future, clean it once a month by operating it with four ounces of white vinegar and eight cups of water. To rinse out the vinegar taste, operate the unit twice with clear water.

With proper care, your SWELCO Coffeemaker will serve you faithfully and well for many years to come.

Sincerely,

Helen Upham

Helen Upham
Customer Services

HU/mo
Enclosure

Email: swelco@janus.com
Web: http://www.swelco.com

FIGURE A–6. Educational Adjustment Letter

adverbs　(ESL)

DIRECTORY
Overview　30
Types of Adverbs　30
Comparison of Adverbs　31
Placement of Adverbs　32

Overview

An adverb modifies the action or condition expressed by a **verb.**

- The wrecking ball hit the side of the building *hard.*
  [The adverb tells *how* the wrecking ball hit the building.]

An adverb also can modify an **adjective,** another adverb, or a **clause.**

- The graphics department used *extremely* bright colors.
  [modifying an adjective]

- The redesigned brake pad lasted *much* longer.
  [modifying another adverb]

- *Surprisingly,* the machine failed.
  [modifying a clause]

An adverb answers one of the following questions:

Where? (adverb of place)

- Move the throttle *forward* slightly.

When? (adverb of time)

- Replace the thermostat *immediately.*

How? (adverb of manner)

- Add the chemical *cautiously.*

How much? (adverb of degree)

- The *nearly* completed report was deleted from his disk.

Types of Adverbs

An adverb may be a common, a conjunctive, an interrogative, or a numeric **modifier.** Typical *common adverbs* are *almost, seldom, down, also, now, ever,* and *always.*

- I *rarely* work on weekends.

A *conjunctive adverb* is an adverb because it modifies the clause it introduces; it operates as a **conjunction** because it joins two independent clauses. The most common conjunctive adverbs are *however, nevertheless, moreover, therefore, further, then, consequently, besides, accordingly, also,* and *thus.*

- I rarely work on weekends; *however,* this weekend will be an exception.

Two independent clauses joined by a conjunctive adverb require a semicolon before and a comma after the conjunctive adverb because the conjunctive adverb introduces the independent clause that follows it and indicates that clause's relationship with the independent clause preceding the adverb. The conjunctive adverb connects ideas but does not grammatically connect the clause stating the ideas; the semicolon does that. The conjunctive adverb both connects and modifies. As a modifier, it is part of one of the two clauses it connects. The use of the semicolon makes the adverb a part of the clause it modifies.

- The building was not finished on the scheduled date; *nevertheless,* Rogers moved in and began to conduct business from the unfinished offices.

- The new project is almost completed and is already over budget; *however,* we must emphasize the importance of adequate drainage.

Interrogative adverbs ask questions. Common interrogative adverbs are *where, when, why,* and *how.*

- *How* many hours did you work last week?

- *Where* are we going when the new project is finished?

- *Why* did it take so long to complete the job?

Numeric adverbs tell how often. Typical numeric adverbs are *once* and *twice.*

- I have worked overtime *twice* this week.

Comparison of Adverbs

With most one-syllable adverbs, the suffix *-er* is added to show **comparison** with one other item, and the suffix *-est* is added to show comparison with two or more items.

- This copier is *fast*. [positive form]
- This copier is *faster* than the old one. [comparative form]
- This copier is the *fastest* of the three tested. [superlative form]

Most adverbs with two or more syllables end in *-ly*, and most adverbs ending in *-ly* are compared by inserting the comparative *more* or *less* or the superlative *most* or *least* in front of them.

- She moved *more quickly* than the other company's sales representative.
- *Most surprisingly*, the engine failed during the test phase.

A few irregular adverbs require a change in form to indicate **comparison** (*well, better, best; badly, worse, worst; far, farther, farthest*).

- The training program functions *well*.
- Our training program functions *better* than most others in the industry.
- Many consider our training program the *best* known in the industry.

Placement of Adverbs

An adverb usually should be placed in front of the verb it modifies.

- The pilot *meticulously* performed the preflight check.

An adverb may, however, follow the verb (or the verb and its **object**) that it modifies.

- The gauge dipped *suddenly*.
- They repaired the computer *quickly*.

An adverb may be placed between a helping verb and a main verb.

- In this temperature range, the pressure will *quickly* drop.

Adverbs such as *nearly, only, almost, just,* and *hardly* should be placed immediately before the words they limit.

- The color copier with the high-speed document feeder/collator
  ~~only~~ *only* costs $47,000.

Putting the word *only* before the word *costs* is ambiguous because the sentence might be understood to mean that only the copier *with the high-speed document feeder/collator* costs $47,000. Reversing the words *only* and *costs* clearly implies that the $47,000 price is low. (See also **modifiers.**)

advice / advise

Advice is a **noun** that means "counsel" or "suggestion."

* My *advice* is to sign the contract immediately.

Advise is a **verb** that means "give advice."

* I *advise* you to sign the contract immediately.

affect / effect

Affect is a **verb** that means "influence."

* The public utility commission's decisions *affect* all state utilities.

Effect can function either as a verb that means "bring about" or "cause" or as a **noun** that means "result." It is best, however, to avoid using *effect* as a verb. A less formal word, such as *made*, is usually preferable.

* The new manager ~~effected~~ *made* several changes that had a good *effect* on morale.

affectation

Affectation is the use of language that is more formal, technical, or showy than it needs to be to communicate information to the reader. Using unnecessarily formal words (such as *utilization*) and outdated words (such as *herewith*) also are affectation.

- *pursuant to* [instead of *about* or *regarding*]
- *in view of the foregoing* [instead of *therefore*]
- *in view of the fact that* [instead of *because*]

Affected writing forces **readers** to work harder to understand the writer's meaning. It typically contains abstract, highly technical, pseudotechnical, pseudolegal, or foreign words and is often liberally sprinkled with **vogue words. Jargon** can become affectation if it is misused, as can **euphemisms** if their purpose is to hide relevant facts.

Writers are easily lured into affectation through the use of **long variants:** words created by adding **prefixes** and **suffixes** to simpler words (*analyzation* for analysis and *telephonic communication* for telephone call). **Elegant variation**—attempting to avoid repeating a word within a paragraph by substituting a pretentious **synonym**—is also a form of affectation. Another type of affectation is **gobbledygook,** which is wordy, roundabout writing with many pseudolegal and pseudoscientific terms. (See also **conciseness/wordiness, clichés,** and **nominalizations.**)

Understanding the possible reasons for affectation is the first step toward avoiding it. The following are some causes of affectation:

- *Impression.* Some writers use pretentious language in an attempt to impress the reader with fancy words instead of evidence and logic.
- *Insecurity.* Writers who are insecure about their facts, conclusions, or arguments may try to hide behind a smokescreen of pretentious words.
- *Imitation.* Perhaps unconsciously, some writers imitate poor writing they see around them.
- *Intimidation.* A few writers, consciously or unconsciously, try to intimidate or overwhelm their readers with words, often to protect themselves from criticism.
- *Initiation.* Writers who have just completed their training often feel that one way to prove their professional expertise is to use as much technical terminology and jargon as possible.
- *Imprecision.* Writers who are having trouble being precise sometimes find that an easy solution is to use a vague, trendy, or pretentious word.

As those reasons suggest, affectation is a widespread problem in the workplace. Many people feel that affectation lends a degree of authority to their writing, but nothing could be further from the truth. Affectation can alienate customers, clients, and colleagues by placing a barrier between them and the writer. Affectation can also lead to ethical questions because readers may suspect that the writer is attempting to hide information or give a false impression of competence. (See **ethics in writing.**)

The following example and its revision illustrate how eliminating affectation leads to clarity. The first version is taken from a specification soliciting bids from local merchants for the operation of a shop in an Air Force Post Exchange.

AFFECTED   In addition to performing interior housekeeping services, the concessionaire shall perform custodial maintenance on the exterior of the facility and grounds. Where a concessionaire shares a facility with one or more other concessionaires, exterior custodial maintenance responsibilities will be assigned by Post Exchange management on a fair and equitable basis. In those instances where the concessionaire's activity is located in a Post Exchange complex wherein predominant tenancy is by Post Exchange–operated activities, then Post Exchange management shall be responsible for exterior custodial maintenance except for those described in 1, 2, 3, and 4 below. The necessary equipment and labor to perform exterior custodial maintenance, when such a responsibility has been assigned to the concessionaire, shall be furnished by the concessionaire. Exterior custodial maintenance shall include the following tasks:
1. Clean entrance door and exterior of storefront windows daily.
2. Sweep and clean the entrance and customer walks daily.
3. Empty and clean waste and smoking receptacles daily.
4. Check exterior lighting and report failures to the contracting officer's representative daily.

CLARIFIED   The merchant will, with his or her own equipment, perform the following duties daily:
1. Maintain a clean and neat appearance inside the store.

2. Clean the entrance door and the outsides of the store windows.
3. Sweep the entrance and the sidewalk.
4. Empty and clean wastebaskets and ashtrays.
5. Check the exterior lighting and report failures to the representative of the contracting officer.

The merchant will also maintain the grounds surrounding the store. Where two or more merchants share the same building, Post Exchange management will assign responsibility for the grounds. Where the merchant's store is in a building that is occupied predominantly by Post Exchange operations, Post Exchange management will be responsible for the grounds.

affinity

Affinity refers to the attraction of two persons or things to each other.

- The *affinity* between these two elements can be explained in terms of their valence electrons.

Affinity should not be used to mean "ability" or "aptitude."

- She has an ~~affinity~~ *aptitude* for problem solving.

- She has ~~an affinity~~ *a talent* for problem solving.

aforesaid

Aforesaid, like **above,** means "stated previously," but it is also legal **jargon** and should be avoided in writing.

- The ~~aforesaid~~ problems can be solved in six months. *explained in the preceding paragraphs*

agree to/agree with

When you *agree to* something, you "give consent."

- I *agree to* a road test of the new model by August 1.

When you *agree with* something, you are "in accord" with it.

- I *agree with* the recommendations of the advisory board.

agreement

In grammar, *agreement* means the correspondence in form between different elements of a sentence to indicate **number, person, gender,** and **case.**
A subject and its **verb** must agree in number.

- The *design is* acceptable.
  [The singular subject, *design,* requires the singular verb, *is.*]

- The new *products are* going into production soon.
  [The plural subject, *products,* requires the plural verb, *are.*]

A subject and its verb must agree in person.

- *I am* the designer.
  [The first-person singular subject, *I,* requires the first-person singular verb, *am.*]

- *They are* the designers.
  [The third-person plural subject, *they,* requires the third-person plural verb, *are.*]

(See also **agreement of subjects and verbs.**)
A **pronoun** and its antecedent must agree in person, number, gender, and case.

- The *employees* report that *they* are more efficient in the new facility.
  [The third-person plural subject, *employees,* requires the third-person plural pronoun, *they.*]

- *Roberta McGuire* will meet with the staff on Friday, when *she* will assign duties.

[The feminine pronoun *she* is in the subjective case because it agrees with its antecedent, *Roberta McGuire,* which is the subject of the sentence.]

(See also **agreement of pronouns and antecedents.**)

agreement of pronouns and antecedents ㉺

DIRECTORY
Overview   38
Agreement in Gender   39
Agreement in Number   40
Tips for Understanding Agreement of Pronouns and Antecedents ㉺   42

Overview

Every **pronoun** must have an antecedent, that is, a **noun** to which it refers.

* When *employees* are hired, *they* must review the policy manual. [The pronoun *they* refers to the antecedent *employees.*]

Pronoun-reference problems occur for a number of reasons.

* The pronoun *it* in the following sentence has no noun to which it could logically refer.

MISSING ANTECEDENT	Microbiologists must constantly struggle to keep up with the professional literature because *it* is a dynamic science. [The pronoun *it* refers neither to *microbiologists* nor to *professional literature.*]

The solution is to use a noun instead of the pronoun or to provide a noun to which *it* could logically refer.

CLEAR ANTECEDENT	Microbiologists must continue to study *microbiology* because *it* is a dynamic science.

* Using the relative pronoun *which* to refer to an idea instead of a specific noun can be confusing.

AMBIGUOUS	He acted independently on the advice of his consultant, *which* his investors thought unjust.

[Was it the fact that he "acted independently" or was it "the advice" that the investors thought unjust?]

CLEAR   The others thought it unjust for him to act independently.

CLEAR   The others thought it unjust for him to act on the advice of his consultant.

- Using the pronouns *it, they, these, those, that,* and *this* can also lead to vague or uncertain references.

AMBIGUOUS   Studs and thick treads make snow tires effective. *They* are implanted with an air gun.
[Which are implanted with an air gun, studs or thick treads?]

SPECIFIC   Studs and thick treads make snow tires effective. *The studs* are implanted with an air gun.

AMBIGUOUS   The inadequate quality-control procedure has resulted in an equipment failure. *This* is our most serious problem at present.
[Which is our most serious problem at present, the procedure or the failure?]

SPECIFIC   The inadequate quality-control procedure has resulted in an equipment failure. *Quality control* is our most serious problem at present.

SPECIFIC   The inadequate quality-control procedure has resulted in an equipment failure. *The equipment failure* is our most serious problem at present.

Agreement in Gender

A pronoun must agree in **gender** with its antecedent.

- *Mr. Swivet* in the accounting department acknowledges *his* share of the responsibility for the misunderstanding, just as *Ms. Barkley* in the research division must acknowledge *hers.*

Traditionally, a masculine, singular pronoun was used to agree with such indefinite antecedents as *anyone* and *person.*

- *Each* may stay or go as *he* chooses.

It is now understood that such usage ignores or excludes women. When alternatives are available, use them. One solution is to use

the plural. Another is to use both feminine and masculine pronouns, although that combination is clumsy when used too often.

- ~~Every employee~~ *All employees* must sign ~~his~~ *their* time ~~card~~ *cards*.
- Every employee must sign his *or her* time card.

You may be able to avoid the pronoun entirely.

- Every employee must sign ~~his~~ *a* time card.

Do not attempt to avoid expressing gender by resorting to a plural
pronoun when the antecedent is singular.

- ~~A technician~~ *Technicians* can expect to advance on their merit.
- A technician can expect to advance on ~~their~~ *his or her* merit.

You may be able to avoid using a pronoun.

- A technician can expect to advance on merit.

Be careful to avoid gender-related stereotypes in general references,
as in "the nurse . . . *she*" or "the doctor . . . *he.*" What if nurse *is* female or doctor male? (See also **biased language.**)

Agreement in Number

A pronoun must agree with its antecedent in **number.** Many problems of agreement are caused by expressions that are not clear in
number.

- Although the typical *engine* runs well in moderate temperatures,
  ~~they~~ *it* often ~~stall~~ *stalls* in extreme cold.

Use singular pronouns with the antecedents *everybody* and *everyone*
unless to do so would be illogical because the meaning is obviously
plural. (See also **everybody/everyone.**)

- *Everyone* pulled *his or her* share of the load.
- *Everyone* thought my plan should be revised, and I really couldn't
  blame *them.*

Demonstrative adjectives sometimes cause problems with agreement of number. *This* and *that* are used with singular nouns, and *these* and *those* are used with plural nouns. Demonstrative adjectives often cause problems when they modify the nouns *kind, type,* and *sort.* When used with those nouns, demonstrative adjectives should agree with them in number.

- this kind / these kinds
- that type / those types
- this sort / these sorts

Confusion often develops when the **preposition** *of* is added (*"this kind of," "these kinds of"*) and the object of the preposition does not conform in number to the demonstrative **adjective** and its **noun.**

- This kind of retirement ~~plans~~ *plan* is best.
- These kinds of retirement ~~plan~~ *plans* are best.

Avoid that error by remembering to make the demonstrative adjective, the noun, and the object of the preposition—all three—agree in number. The agreement makes the sentence not only correct but also more precise.

Using demonstrative adjectives with words like *kind, type,* and *sort* can easily lead to vagueness. Your writing will be more effective if it is more specific.

- ~~These kinds of~~ *Computer-controlled* hydraulic cranes are best.

A compound antecedent joined by *or* or *nor* is singular if both elements are singular and plural if both elements are plural.

- Neither the *engineer* nor the *technician* could do *his* job until *he* understood the new concept.
- Neither the *executives* nor the *directors* were pleased at the performance of *their* company.

When one of the antecedents connected by *or* or *nor* is singular and the other plural, the pronoun agrees with the closer antecedent.

- Either the *computer* or the *printers* should have *their* serial numbers registered.

TIPS FOR UNDERSTANDING AGREEMENT OF PRONOUNS
AND ANTECEDENTS

If you are a writer of English as a second language, make a special editing pass to check for agreement between pronouns and their antecedents. The following guidelines highlight where English differs from
many languages regarding agreement of pronouns and antecedents.

Each and *every* take singular pronouns, even in compound antecedents.

- *Each* business has *its* own logo.
- *Every* book and article was included on the list with *its* title and date
  of publication.

All and *some* can take singular or plural pronouns, depending on
whether the noun they modify is a count noun or a mass noun.

- *Some* of the chocolate melted; *it* was stored near a furnace.
- *Some* of the prisoners escaped; *they* were left unattended.

A possessive pronoun agrees with the subject, not the object.

- The *father* watches *his* daughter.

The phrase *one of the* . . . takes a singular pronoun even though the
noun that follows must be plural. The pronoun refers to *one*.

- *One* of the accountants lost *her* luggage.

The relative pronoun *who* is used for people; *which* and *that* are used for
objects.

- The *man who* spoke at the shareholders' meeting is the company's
  largest investor.
- Mr. Geoff invests heavily in Exxon *stocks, which* have split twice in recent months.
- The boss wants the *folder that* he left on his desk.

Collective nouns rarely end in -*s,* but they take plural pronouns.

- Those *people* are interested only in seeing *their* investment grow.

Some singular nouns end in -*s* but take singular pronouns.

- The *United States* asked *its* allies to support the effort.

- Either the *printers* or the *computer* should have *its* serial number
  registered.

A compound antecedent with its elements joined by *and* requires a plural pronoun.

- *Seon Ju* and *Juanita* took *their* layout drawings with them.

If both elements refer to the same person, however, use the singular pronoun.

- The noted *scientist* and *author* departed from *her* prepared speech.

Collective nouns may be singular or plural, depending on their meaning.

- The *committee* arrived at the recommended solutions only after *it* had deliberated for days.

- The *committee* quit for the day and went to *their* respective homes.

agreement of subjects and verbs   ESL

A **verb** must agree with its subject. (See also **sentence construction.**) Do not let intervening **phrases** and **clauses** mislead you.

- The program, despite the new instructional materials, still take *takes* several days to install.
  [The verb, *takes,* must agree in **number** with the subject, *program,* rather than *materials.*]

- The use of insecticides and fertilizers, although they offer unquestionable benefits, often result *results* in unfortunate side effects.

  [The verb, *results,* must agree with the subject of the sentence, *use,* rather than with the subject of the preceding clause, *they.*]

Do not make the verb agree with the **noun** immediately in front of it if that noun is not its subject. That problem is especially likely to occur when a plural noun falls between a singular subject and its verb.

- Only *one* of the emergency lights *was* functioning when the accident occurred. [The subject of the verb is *one,* not *lights.*]

- *Each* of the switches *controls* a separate circuit.
  [The subject of the verb is *each*, not *switches*.]

- *Each* of the managers *supervises* a very large region.
  [The subject of the verb is *each*, not *managers*.]

Be careful not to let modifying phrases obscure a simple subject.

- The *advice* of two engineers, one lawyer, and three executives *was* obtained prior to making a commitment.
  [The subject of the verb is *advice*, not *executives*.]

Sentences with inverted word order can cause problems with agreement between subject and verb.

- From this work *have come* several important *improvements*.
  [The subject of the verb is *improvements*, not *work*.]

Such words as *type, part, series,* and *portion* take singular verbs even when they precede a phrase containing a plural noun.

- A *series* of meetings *was* held about the best way to market the new product.

- A large *portion* of most industrial annual reports *is* devoted to promoting the corporate image.

A subject that expresses measurement, weight, mass, or total often takes a singular verb even when the subject word is plural in form. Such subjects are treated as a unit.

- *Four years is* the normal duration of the training program.

- *Twenty dollars is* the wholesale price of each unit.

Indefinite **pronouns** such as *some, none, all, more,* and *most* may be singular or plural, depending on whether they are used with a mass noun (*oil* in the following examples) or with a count noun (*drivers* in the following examples). Mass nouns are singular, and count nouns are plural.

- *None* of the oil *is* to be used.

- *None* of the truck drivers *are* to go.

- *Most* of the oil *has* been used.

- *Most* of the drivers *know* why they are here.

- *Some* of the oil *has* leaked.

- *Some* of the drivers *have* gone.

One and *each* are normally singular.

- *One* of the brake drums *is* still scored.

- *Each* of the original founders *is* scheduled to speak at the dedication ceremony.

A verb following the relative pronoun *who* or *that* agrees in number with the noun to which the pronoun refers (its antecedent).

- Steel is one of those *industries* that *are* most affected by energy costs.
  [*That* refers to *industries*.]

- This is one of those management *problems* that *require* careful analysis.
  [*That* refers to *problems*.]

- She is one of those *employees* who *are* rarely absent.
  [*Who* refers to *employees*.]

The word *number* sometimes causes confusion. When used to mean a specific number, it is singular.

- *The number* of committee members *was* six.

When used to mean an approximate number, it is plural.

- *A number* of people *were* waiting for the announcement.

Relative pronouns (*who, which, that*) take either singular or plural verbs, depending on whether the antecedent is singular or plural. (See also **who/whom.**)

- He is a bookkeeper *who takes* work home at night.

- He is one of those bookkeepers *who take* work home at night.

Some abstract nouns are singular in meaning but plural in form: *mathematics, news, physics,* and *economics.*

- *News* of the merger *is* on page 4 of the *Chronicle*.

- *Textiles is* an industry in need of import quotas.

Some words are always plural, such as *trousers* and *scissors*.

- The *scissors were* ordered last week.

- *A pair* of scissors *is* on order.

A collective subject takes a singular verb when the group is thought of as a unit and a plural verb when the individuals are thought of separately.

A singular subject followed by a phrase or clause containing a plural noun still requires a singular verb.

- One in twenty hard drives we receive from our suppliers a̶r̶e̶ *is* faulty.
  [The subject is *one*, not *hard drives* or *suppliers*.]

The number of a subjective **complement** does not affect the number of the verb — the verb must always agree with the subject.

- The *topic* of his report *was* employee benefits.
  [The subject of the sentence is *topic*, not *benefits*.]

A book with a plural title requires a singular verb.

- *Medical Techniques* is an essential resource.

Compound Subjects

A compound subject is one that is composed of two or more elements joined by a **conjunction** such as *and, or, nor, either . . . or,* or *neither . . . nor.* Usually, when the elements are connected by *and,* the subject is plural and requires a plural verb.

- *Accounting and chemistry are* both prerequisites for this position.

One exception occurs when the elements connected by *and* form a unit or refer to the same thing. In that case, the subject is regarded as singular and takes a singular verb.

- *Bacon and eggs is* a high-cholesterol meal.

- The *red, white, and blue flutters* from the top of the Capitol.

- Our greatest *challenge and business opportunity is* the Internet.

A compound subject with a singular element and a plural element joined by *or* or *nor* requires that the verb agree with the closest element.

- Neither the director nor the *project assistants were* there.

- Neither the project assistants nor the *director was* there.

- Either they or *I am* going to write the report.

- Either I or *they are* going to write the report.

If *each* or *every* modifies the elements of a compound subject, use the singular verb.

- *Each* manager and supervisor *has* a production goal to meet.

- *Every* manager and supervisor *has* a production goal to meet.

all around / all-around / all-round

All-round and *all-around* both mean "comprehensive" or "versatile."

- The company started an *all-round* training program.

Do not confuse the hyphenated words with the two-word phrase *all around,* as in "The security fence was installed *all around* the building."

all right / alright

All right means "all correct," as in "The answers were *all right.*" In formal writing, it should not be used to mean "good" or "acceptable." It is always written as two words, with no hyphen; *alright* is incorrect and should never be used.

- The decision the committee reached was ~~all right~~.
  acceptable

all together / altogether

All together means "all acting together" or "all in one place."

- The necessary instruments were *all together* on the tray.

Altogether means "entirely" or "completely."

- The trip was *altogether* unnecessary.

allude / elude / refer

Allude means to make an indirect reference to something.

- The report simply *alluded* to the problem, rather than stating it clearly.

Elude means to escape notice or detection.

- The discrepancy in the account *eluded* the auditor.

Refer is used to indicate a direct reference to something.

- She *referred* to the chart three times during her speech.

allusion

An allusion is an indirect reference to something from past or current events, literature, or other familiar sources. The use of allusion promotes economical writing because it is a shorthand way of referring to a body of material in a few words or of helping to explain a new and unfamiliar process in terms of one that is familiar. In the following paragraph, the writer sums up a description with an allusion to a well-known biblical story. The allusion, with its implicit reference to "right standing up to might," concisely emphasizes the writer's point.

- As it currently exists, the review process involves the consumer's attorney sitting alone, usually without adequate technical assistance, faced by two or three government attorneys, two or three attorneys from the Compu Systems, and large teams of experts who support the government and the corporation. The entire proceeding is reminiscent of David versus Goliath.

Be sure, of course, that your reader is familiar with the material to which you allude. Allusions should be used with restraint. If overdone, they can lead to **affectation** or can be viewed merely as a **cliché**. (See also **analogy, technical writing style,** and **international correspondence.**)

allusion / illusion

An *allusion* is an indirect reference to something not specifically mentioned.

- The report made an *allusion* to metal fatigue in the support structures.

An *illusion* is a mistaken perception or a false image.

- County officials are under the *illusion* that the landfill will last indefinitely.

almost / most

Do not use *most* as a colloquial substitute for *almost* in your writing.

- New shipments arrive ~~most~~ *almost* every day.

If you can substitute *almost* for *most* in a sentence, *almost* is the word you need.

already / all ready

Already is an **adverb** that means "before this time" or "previously."

- We had *already* shipped the office furniture when the cancellation notice arrived.

All ready is a two-word phrase meaning "completely prepared."

- She was *all ready* to start work on the project when it was suddenly canceled.

also

Also is an **adverb** that means "additionally."

- Two 500,000-gallon tanks have recently been constructed on site. Several 10,000-gallon tanks are *also* available, if needed.

Also should not be used as a connective in the sense of "and."

- He brought the reports, letters, ~~also~~ the section supervisor's recommendations. *and* ^

Avoid opening sentences with *also*. It is a weak transitional word that suggests an afterthought rather than planned writing.

- ~~Also~~ he brought statistical data to support his proposal. *In addition,* ^

- ~~Also he~~ brought statistical data to support his proposal. *He also* ^

ambiguity

A word or passage is ambiguous when it can be interpreted in two or more ways yet provides the reader with no certain basis for choosing among the alternatives.

- Mathematics is more valuable to an engineer than a computer. [Does that mean an engineer is more in need of mathematics than a computer is? Or does it mean that mathematics is more valuable to an engineer than a computer is?]

Ambiguity can take many forms, one of which is ambiguous **pronoun reference.**

AMBIGUOUS
Inadequate quality-control procedures have resulted in more equipment failures. This is our most serious problem at present.
[Ambiguous pronoun reference: Does *this* refer to "quality-control procedures" or "equipment failures"?]

SPECIFIC
Inadequate quality-control procedures have resulted in more equipment failures. *These failures* are our most serious problem at present.

SPECIFIC
Inadequate quality-control procedures have resulted in more equipment failures. *Quality control* is our most serious problem at present.

Incomplete **comparison** and missing or misplaced **modifiers** (including **dangling modifiers**) cause ambiguity.

- Ms. Lee values rigid quality-control standards more than Mr.
  does
  Rosenblum . [Complete the comparison.]
                ^
                      also
- His hobby was cooking. He was especially fond of cocker spaniels.
                           ^
  [Add the missing modifier.]

The placement of some modifiers enables them to be interpreted in either of two ways.

- She volunteered *immediately* to deliver the toxic substance.

By moving the word *immediately,* the meaning can be clarified.

- She *immediately* volunteered to deliver the toxic substance.

- She volunteered to deliver *immediately* the toxic substance.

Imprecise **word choice** (including faulty **idioms**) can cause ambiguity.

- The general manager has denied reports that the plant's recent
                                  rescinded
  fuel allocation cut will be ~~restored~~ . [inappropriate word choice]
                               ^

Ambiguity also can be caused by various forms of **awkwardness.**

---

amount / number (ESL)

Amount is used with things that are thought of in bulk and that cannot be counted (mass **nouns**).

- The *amount* of electricity available for industrial use is limited.

Number is used with things that can be counted as individual items (count nouns).

- The *number* of employees who are qualified for early retirement has increased in recent years.

ampersands

The ampersand (&) is a symbol sometimes used to represent the word *and*, especially in the names of organizations.

- Chicago & Northwestern Railway
- Watkins & Watkins, Inc.

The ampersand is appropriate for footnotes, bibliographies, lists, and references if the ampersand appears in the name being listed. (See also **documenting sources.**) When you are writing the name of an organization in sentences or in an address, however, spell out the word *and* unless the ampersand appears in the official name of the company. (See also **abbreviations.**)

analogy (*see* figures of speech)

and / or

And/or means that either both circumstances are possible or only one of two circumstances is possible. The term is awkward and confusing because it makes the reader stop to puzzle over your distinction.

AWKWARD   Use A *and/or* B.
IMPROVED   Use A or B or both.

ante- / anti-

Ante- means "in front of" or "before."

- *ante*room [outer room or waiting room]
- *ante*date [predate or earlier date]

Anti- means "against" or "opposed to."

- *anti*body, *anti*social

Anti- is hyphenated when joined to proper nouns or to words beginning with the letter *i.*

- *anti-*American, *anti-*intellectual

When in doubt, consult your dictionary. (See also **prefixes.**)

antonyms

An *antonym* is a word with a meaning opposite that of another word.

- good/bad, well/ill, fresh/stale

Many pairs of words that look as if they are antonyms are not. Be careful not to use words like **flammable** and inflammable incorrectly. When in doubt, consult a thesaurus or dictionary.

APA style (*see* documenting sources)

apostrophes   ESL

An apostrophe (') is used to show possession, to indicate the omission of letters, and sometimes to form the plural. Do not confuse the apostrophe used to show the plural with the apostrophe used to show possession.

To Show Possession

An apostrophe is used with an *s* to form the possessive case of some **nouns.**

- The manufacturing plant's output increased this year.

Singular nouns of one syllable that end in *s* form the possessive by adding *'s.*

- The boss's desk was cluttered.

Singular nouns of more than one syllable that end in *s* may form the possessive either with an apostrophe alone or with *'s.* Whichever way you do it, be consistent.

- The hostess' warm welcome
- The hostess's warm welcome

With coordinate nouns, the last noun takes the possessive form to show joint possession.

- *Michelson and Morley's* famous experiment on the velocity of light was conducted in 1887.

To show individual possession with coordinate nouns, each noun should take the possessive form.

- The difference between *Tom's* and *Mary's* test results is statistically insignificant.

Use only an apostrophe with plural nouns ending in *s*.

- The managers' meeting was canceled.

When adding *'s* would result in a noun ending in multiple consecutive *s* sounds, add only an apostrophe.

- The Joneses' test results are available.

Do not use the apostrophe with possessive pronouns.

- yours, its, his, ours, whose, theirs

It's is a contraction of *it is; its* is the possessive form of the **pronoun.** (See **its/it's.**)

- *It's* important that the sales force meet *its* quota.

In names of places and institutions, the apostrophe is usually omitted.

- Harpers Ferry, Writers Book Club

To Form Plurals

The trend for indicating the plural forms of words mentioned as words, of numbers used as nouns, and of abbreviations shown as single or multiple letters is currently to add only *s* rather than using *'s*.

When a word mentioned as a word is italicized, it is current usage to add *s* in roman type.

- There were five *and*s in his first sentence.

Rather than using italics, you may place a word in quotation marks. If you choose this option, use an apostrophe and an *s* ('s).

- There were five "and's" in his first sentence.

To indicate the plural of a number, add *s*.

- 7s
- the late 1990s

For terms that are single letters, whether upper- or lowercase, set the letter in italics and set the *s* in roman type.

- *x*s and *y*s
- *N*s

If the letter and the *s* form a word, you may want to consider using an apostrophe to avoid confusion.

- *A*'s

Use *s* to pluralize an abbreviation that is in all capital letters or that ends with a capital letter.

- IOUs

However, if the abbreviated term contains periods, some writers use an apostrophe to prevent confusion.

- The university awarded seven Ph.D.'s in technical communication last year.
- They will be asked for their I.D.'s.

Whatever practice you follow, be consistent.

To Indicate Omission

An apostrophe is used to mark the omission of letters or **numbers** in a **contraction** or a date.

- can't, I'm, I'll
- the class of '99

appendixes

An appendix, located at the end of a **formal report,** a **proposal,** or another major document, supplements or clarifies the information in the body of the document. (The plural form of the word may be either *appendixes* or *appendices.*) An appendix can provide information that is too detailed or too lengthy to appear in the text without impeding the orderly presentation of ideas for the primary audience. For example, an appendix could contain complex **graphs** and **tables,** profiles of key personnel involved in a proposed project, or documentation of interest only to secondary **readers.** Do not include miscellaneous bits of marginally related information that you were unable to work into the text.

A document may have more than one appendix, with each offering only one type of information. When the document contains more than one appendix, arrange them in the order they are mentioned in the text. Begin each appendix on a new page and identify each with a letter, beginning with the letter *A,* and a unique title, for example, "Appendix A: Sample Questionnaire." (If you have only one appendix, title it simply "Appendix.") List the titles and beginning page numbers of the appendixes in the **table of contents.**

application letters

The letter of application is essentially a sales letter in which you are marketing your skills, abilities, and knowledge. Remember that you may be competing with many other applicants. The immediate **objective** of an application letter and the accompanying **résumé** is to attract the attention of the person who screens and hires job applicants. Your ultimate goal is to obtain a job interview. (See **interviewing for a job** and **job search.**)

The successful application letter accomplishes three tasks: It catches the reader's attention favorably, it convinces the **reader** that you are a qualified candidate for the job, and it requests an interview.

When you are writing a letter of application, do the following:

- Identify the job by title and let the recipient know how you heard about it.

- Summarize your qualifications for the job, specifically your work experience, activities that show your leadership skills, and your education.
- Refer the reader to your enclosed résumé.
- Ask for an interview, stating where you can be reached and when you will be available.
- If you are applying for a specific job, include information pertinent to the position that is not included in your general résumé.

Those who make hiring decisions review many letters of application. To save them time as well as to call attention to your strengths as a candidate, you should state your job objective directly at the beginning of the letter.

- I am seeking a position in your engineering department in which I can use my training in computer science to solve engineering-related problems.

If you have been referred to a company by one if its employees, a career counselor, a professor, or someone else, mention that before stating your job objective.

- During the recent TEKOM convention in Washington, a member of your engineering staff, Karen Jarrett, informed me of a possible opening for a systems analyst in your New Products Division. My extensive background in computer science and my work in the university computer systems division make me highly qualified for the position.

In succeeding paragraphs, expand on the qualifications you mentioned in your opening. Add any appropriate details, highlighting experience listed on your résumé that is especially pertinent to the job you are seeking. Close your letter with a request for an interview. Prepare your letter with utmost care, **proofreading** it carefully.

Three sample letters of application are presented here. Figure A–7 on page 58, by a recent college graduate, is in response to a local newspaper article about the company's plan to build a new computer center. The writer is not applying for a specific job opening, but he describes the position he is seeking. In Figure A–8 on page 59, a letter written by a college senior, the writer does not specify where she learned of the opening because she does not know with certainty whether a position is available. The final example, Figure A–9 on page 60, by a person with many years of

6819 Locustview Drive
Topeka, KS 66614
June 14, 20--

Loudons, Inc.
4619 Drove Lane
Kansas City, KS 63511

Dear Human Resources Director:

I just read an article in the *Kansas Dispatch* about Loudons' new
computer center just north of Kansas City. I would like to apply
for a position as an entry-level programmer at the center.

I understand that Loudons produces both in-house and customer
documentation. My technical-writing skills, as described in the
enclosed résumé, are well suited to your company. I am a recent
graduate of Fairview Community College in Topeka, with an Associate's Degree in Computer Science. In addition to having
taken a broad range of courses, I served as a computer consultant at the college's computer center, where I helped train computer users on new systems.

I will be happy to meet with you at your convenience and discuss how my education and experience will suit your needs. You
can reach me at my home address, at (913) 233-1552, or at
dedwards@cpu.fairview.edu.

Sincerely,

David B. Edwards

David B. Edwards

Enclosure: Résumé

FIGURE A–7. Letter of Application by a Recent College Graduate

2701 Wyoming Street
Atlanta, GA 30307
May 29, 20--

Ms. Laura Goldman
Chief Engineer
Acton, Inc.
80 Roseville Road
St. Louis, MO 63130

Dear Ms. Goldman:

I am seeking a position in your engineering department where I may
use my training in computer sciences to solve engineering problems.
Although I do not know if you have a current opening, I would like
to be part of the department that developed the Internet Selection
System.

I expect to receive a Bachelor of Science degree in Engineering from
Georgia Institute of Technology in June, when I will have completed
the Computer Systems Engineering Program. Since September 19--,
I have been participating, through the university, in the Professional
Training Program at Computer Systems International in Atlanta. In the
program, I was assigned to several staff sections as an apprentice. Most
recently, I have been a programmer trainee in the Engineering Department and have gained a great deal of experience in computer applications. Details of the academic courses I have taken are contained in the
enclosed résumé.

I look forward to hearing from you soon. I can be contacted at my office (404-866-7000, ext. 312), at home (404-256-6320), or at my
email address (vtfrom@aol.com).

Sincerely yours,

Victoria T. Fromme

Victoria T. Fromme

Enclosure: Résumé

FIGURE A–8. Letter of Application by a University Student

522 Beethoven Drive
Roanoke, VA 24017
November 15, 20--

Ms. Cecilia Smathers
Vice President
Hamilton Technologies, Inc.
6194 Main Street
Hampton, VA 23661

Dear Ms. Smathers:

During the recent TEKOM convention in Washington, a member of your
sales staff, Karen Jarrett, informed me of a possible opening for a systems
analyst in your New Products Division. My extensive background in computer science makes me highly qualified for the position.

I was with Technology, Inc., from its formation in 1990 until its closing last
year. During that period, I was involved in all areas of systems analysis
within Technology, Inc. My education and work experience are indicated in
the enclosed résumé.

I would like to discuss my qualifications in an interview at your convenience. Please write to me, telephone me at (804) 449-6743 any weekday,
or email me at gm302.476@sys.com.

Sincerely,

Gregory Mindukakis

Gregory Mindukakis

Enclosure: Résumé

FIGURE A–9. Letter of Application by an Applicant with Years of Experience work experience, opens with an indication of where the writer learned of the job vacancy. (See also **reference letters.**)

appositives

An *appositive* is a **noun** or noun **phrase** that follows and amplifies another noun or noun phrase. It has the same grammatical function as the noun it complements.

* Dennis Gabor, *the famous British scientist,* experimented with coherent light in the 1940s.

* The famous British scientist *Dennis Gabor* experimented with coherent light in the 1940s.

(For detailed information on the use of **commas** with appositives, see **restrictive and nonrestrictive elements.**)

If you are in doubt about the **case** of an appositive, check it by substituting the appositive for the noun it modifies.

* My boss gave the two of us, Jim and *me,* the day off.
[You would not say, "My boss gave *I* the day off."]

articles   ESL

Articles (a, an, the) function as adjectives because they modify the items they designate by either limiting them or making them more precise. There are two kinds of articles, indefinite and definite. (See also **a/an** and **English as a second language.**)

The *indefinite* articles, *a* and *an,* denote an unspecified item.

* *A* program was run on our new computer.
[Not a specific program but an unspecified program. Therefore, the article is indefinite.]

The *definite* article, *the,* denotes a particular item.

* *The* program was run on the computer.
[Not just any program but *the* specific program. Therefore, the article is definite.]

Do not omit all articles from your writing. Including articles costs nothing; eliminating them makes reading more difficult. (See also

Whether to use a definite or an indefinite article is determined by what you can safely assume about your audience's knowledge. Do your readers have enough information — either from their knowledge about the world *or* from the context of your writing — to identify the noun that will be modified by the article? If the answer is yes, use a definite article; if no, use an indefinite article.

In each of these sentences, you can safely assume that the reader can clearly identify the noun.

- *The* sun rises in the east. [There is only one sun.]

- Did you know that yesterday was *the* coldest day of the year so far? [The modified noun refers to *yesterday*.]

- *The* man who left his briefcase in the conference room was a very bright man.
  [The relative phrase *who left his briefcase in the conference room* restricts and, therefore, identifies the meaning of *man*.]

In the following sentence, however, you cannot assume that the reader can clearly identify the noun.

- *An* ice storm is on the way.
  [It is impossible to identify specifically which *ice storm* is meant.]

A more important question for some nonnative speakers of English is when *not* to use articles. These generalizations will help. Do not use articles with:

singular proper nouns
- Queen Elizabeth II, Lake Titicaca, Utah, Main Street, Harvard University, Mount Hood

plural nonspecific countable nouns (when making generalizations)
- Helicopters are the new choice of transportation for the rich and famous.

singular uncountable nouns
- She loves chocolate.

plural countable nouns used as complements
- Those women are physicians.

As you can see, the principles for using definite and indefinite articles appropriately are rather detailed. In fact, most native speakers of English cannot explain the rules for using articles, even though they use the articles correctly. Definite and indefinite articles are small words that are frequently used, but they are extremely difficult to master. (See also **a/an** and **adjectives.**)

telegraphic style.) On the other hand, do not overdo it. An article can be superfluous.

- I'll meet you in *a* half *an* hour.
  [Choose one and eliminate the other.]

- Fill with *a* half *a* pint of fluid.
  [Choose one and eliminate the other.]

Do not capitalize articles in titles except when they are the first word. (See also **adjectives** and **capital letters.**)

- *Time* magazine reviewed *The Old Man and the Sea.*

as

Because *as* can mean so many things (*since, because, for, that, at that time, when,* and *while*) and can be at least four **parts of speech (conjunction, preposition, adverb, pronoun),** it is often overused and misused, especially in speech. (See also **because** and **like/as.**) In writing, *as* is often weak or ambiguous. For example, the following sentence has two possible meanings:

EXAMPLE   *As* we were together, he revealed his plans.

CAN MEAN   *Because* we were together, he revealed his plans.

OR   *While* we were together, he revealed his plans.

The word *as* can also contribute to **awkwardness** by appearing too many times in a sentence.

- *We* ~~As we~~ realized *the moment* ~~as soon as~~ we began the project, ~~the problem~~ *that the problem*

  needed a solution.

as much as / more than

These two phrases are sometimes incorrectly combined, especially when intervening phrases delay the completion of the phrase.

- The engineers had as much ~~, if not more,~~ influence in planning the

  program ~~than~~ *as* the accountants *, if not more* .

as regards (*see* regarding / with regard to)

as such

The phrase *as such* is seldom useful and should be omitted.

- This graphics software is poor~~; as such,~~ ^, and^ it should be eliminated.

as to whether

The phrase *as to whether* (as well as *as to when* and *as to where*) is clumsy and redundant. Either omit it altogether or use only *whether*.

- ~~As to whether~~ *We are still undecided whether* we will redesign the fuel-injection ~~system, we are~~ *system.* ~~still undecided.~~

Be wary of all phrases starting with *as to;* they are often redundant, vague, or indirect.

- ~~As to his policy,~~ I am in full agreement *with his policy*.

as well as

Do not use *as well as* with *both*. The two expressions have similar meanings; use one or the other.

- Both General Motors ~~as well as~~ *and* Ford are developing electric cars.

- ~~Both~~ General Motors, as well as Ford, ~~are~~ *is* developing electric cars.

attribute / contribute

Attribute (with the accent on the second syllable) is a **verb** that means "point to a cause or a source."

- He *attributes* the division's improved safety record to the new training program.

Attribute (with the accent on the first syllable) is a **noun** that means a quality or characteristic belonging to someone or something.

- His mathematical skills are his most valuable *attribute*.

Contribute means "give."

- His mathematical skills will *contribute* much to the project.

audience (*see* readers)

augment / supplement

Augment is a **verb** that means "increase or magnify in size, degree, or effect."

- Many employees *augment* their incomes by freelancing.

Supplement is a **verb** that means "add something to make up for a deficiency."

- The physician told him to *supplement* his diet with vitamins.

average / median

The *average* (or *mean*) is determined by dividing the sum of two or more quantities by the number of items totaled. For example, if one report is 10 pages, another is 30 pages, and a third is 20 pages, their *average* (or *mean*) length is 20 pages. It is incorrect to say that "each report averages 20 pages" because each report is a specific length.

- ~~Each report averages~~ 20 pages.
  The three reports average

The *median* (or *mean*) is the middle number in a sequence of numbers.

- The *median* of the series 1, 3, 4, 7, 8 is 4.

awhile / a while

Awhile is an **adverb** that means "for a short time." It is not preceded by *for* because the meaning of *for* is inherent in the meaning of *awhile*. *A while* is a **noun** phrase that means "a period of time."

- Wait for ~~awhile~~ *a while* before testing the sample.
- Wait ~~for~~ awhile before testing the sample.

awkwardness   (ESL)

Any writing that strikes the reader as awkward—that is, as forced or unnatural—impedes the reader's understanding. Awkwardness has many causes, including overloaded sentences, overlapping **subordination,** grammatical errors, ambiguous statements, and overuse of **expletives.** Other causes are overuse of the passive **voice,** faulty logic, unintentional **repetition, garbled sentences,** and jammed **modifiers.** (See **logic errors.**) To avoid awkwardness, make your writing as direct and as concise as possible.

Writer's Checklist: Awkwardness

The following guidelines will help you smooth out most awkward passages.

- ☑ Use effective **sentence construction** to keep your sentences from becoming unnecessarily complicated.
- ☑ Use the active voice unless you have a reason to use the passive voice.
- ☑ Use **conciseness** and strong **organization** to tighten your writing.
- ☑ During **revision,** strive for **clarity** and **coherence.**

B

bad / badly ESL

Bad is the **adjective** form that follows such linking verbs as *feel* and *look*.

- We don't want our department to look *bad* at the meeting.

Badly is an **adverb.**

- The test model performed *badly* during the trial run.

To say "I feel *badly*" would mean, literally, that your sense of touch was impaired. (See also **good/well.**)

balance / remainder

One meaning of *balance* is "a state of equilibrium"; another meaning is "the amount of money in a bank account after deposits and withdrawals have been credited and debited." *Remainder,* in all applications, is "what is left over." *Remainder* is the more accurate word, therefore, to mean "that which is left over."

- Our research division must maintain a *balance* between innovation and cost-effectiveness.

- The *balance* in the corporate account after the payroll has been met is a matter for concern.

- Round off the fraction to its nearest whole number and drop the *remainder.*

B

be sure and / be sure to

The **phrase** *be sure and* is colloquial and unidiomatic when used for *be sure to*. (See also **idioms.**)

- When you get to the branch office, be sure ~~and~~ phone me.
  to

because

To express cause, *because* is the strongest and most specific connective (others are *for, since, as*). (See also **reason is because.**) *Because* is unequivocal in stating causal relationship.

- We didn't complete the project *because* the raw materials became too costly.
  [The use of *because* emphasizes the cause-and-effect relationship.]

For can express causal relationships, but it is weaker than *because*. It allows the **clause** that follows to be a separate independent clause.

- We didn't complete the project, *for* raw materials became too costly.
  [The use of *for* expresses but does not emphasize the cause-and-effect relationship.]

Since is a weak substitute for *because* as a connective to express cause.

- *Since* the computer failed, paychecks will be delayed.

However, *since* is an appropriate connective when the emphasis is on circumstance, condition, or time rather than on cause and effect.

- *Since* I was in town anyway, I decided to visit the construction site.

- *Since* it went public, the company has earned a profit every year.

As is the least definite connective to indicate cause; its use for that purpose is best avoided.

- I left the office early~~, as~~ I had finished my work.
  because

being as / being that

These **phrases** are nonstandard English and should not be used in writing. Use *because* or *since*.

beside / besides

Besides, meaning "in addition to" or "other than," should be carefully distinguished from *beside*, meaning "next to" or "apart from."

- *Besides* the two of us from the Systems Department, three people from Production were standing *beside* the president when he presented the award.

between / among

Between is normally used to relate two items or persons.

- Preferred stock offers a buyer a middle ground *between* bonds and common stock.

Among is used to relate more than two.

- The subcontracting was distributed *among* the three firms.

between you and me

People sometimes use the incorrect expression *between you and I*. Because the **pronouns** are **objects** of the **preposition** *between*, the objective form of the personal pronoun (*me*) must be used.

- Between you and I̭, John should be taken off the job.

                    me

(See also **case [grammar]**.)

B

bi- / semi-

When used with periods of time, *bi-* means "two" or "every two." *Bimonthly* means "once in two months"; *biweekly* means "once in two weeks." When used with periods of time, *semi-* means "half of" or "occurring twice within a period of time." *Semimonthly* means "twice a month"; *semiweekly* means "twice a week." Both *bi-* and *semi-* normally are joined with the following element without a space or a hyphen.

biannual / biennial

In conventional usage, *biannual* means "twice during the year," and *biennial* means "every other year." (See also **bi-/semi-.**)

biased language   ESL

Biased language refers to words and expressions that offend because they make inappropriate assumptions about gender, ethnicity, physical or mental disability, age, or sexual orientation. To eliminate biased language from your writing, do not make inappropriate assumptions or rely on stereotypes. The easiest way to avoid bias is simply not to mention differences among people unless the differences are relevant to the discussion. Review your writing for bias during **revision,** making a separate pass specifically for that purpose. Carefully review email messages before sending them, too. They are more likely than letters and memos to be written quickly and sent immediately, putting you at risk of inadvertently using biased language. If you are unsure of the appropriateness of an expression or the overall tone of a passage, have several colleagues review the material and give you their honest assessment. Keep abreast of accepted usage by regularly reading newspapers and newsmagazines and general-interest magazines.

Sexist Language

Sexist language can be an outgrowth of sexism, the arbitrary stereotyping of men and women in their roles in life. Sexism, like biased language, can breed and reinforce inequality. To avoid sexism in your writing, treat men and women equally, without making assumptions or falling back on stereotypes about traditional roles. Not all secretaries, nurses, and elementary school teachers are women, nor are all police officers, soldiers, and physicians men. Your language should reflect that reality. Accordingly, use nonsexist occupational descriptions in your writing.

INSTEAD OF:	USE:
cameraman	camera operator, photographer, videographer
chairman	chair, chairperson
fireman	firefighter
foreman	supervisor
mailman	letter carrier
manpower	staff, personnel, workers
policeman/policewoman	police officer
salesman	salesperson
steward/stewardess	flight attendant
telephone lineman	telephone installer
waitress/waiter	server

Use parallel terms to describe men and women.

INSTEAD OF:	USE:
man and wife	husband and wife
Ms. Jones and Bernard Weiss	Ms. Jones and Mr. Weiss; Mary Jones and Bernard Weiss
ladies and men	ladies and gentlemen; women and men

Sexism can creep into your writing by the unthinking use of male pronouns where a reference could apply equally to a man and a woman. One way to avoid such usage is to rewrite the sentence in the plural.

- *All employees* ~~Every employee~~ will have *their supervisors* ~~his supervisor~~ sign *their* ~~his~~ attendance *slips.* ~~slip.~~

Be careful not to change the pronoun to the plural and leave its antecedent in the singular. The pronoun and its antecedent must always agree. (See **agreement of pronouns and antecedents.**)

- *Auditors* ~~An auditor~~ can expect to advance on their merit.

B

Other possible solutions are to use *his or her* instead of *his* alone or to omit the pronoun completely if it is not essential to the meaning of the sentence.

- Everyone must submit his ^*or her* expense report by Monday.

- Everyone must submit ~~his~~ ^*an* expense report by Monday.

He or she can become monotonous when repeated constantly, and a pronoun cannot always be omitted without changing the meaning of a sentence. Another solution is to omit troublesome pronouns by using the imperative **mood** whenever possible.

- ^*Submit all* ~~Everyone must submit his or her~~ expense report^*s* by Monday.

Other Types of Biased Language

It is obvious that you should not use racial, ethnic, or religious stereotyping. In fact, identifying people by such categories is simply not relevant in most workplace writing. Telling readers that an engineer is Native American or that a professor is African-American almost never conveys useful information. It also reinforces stereotypes, implying that it is rare for a person of a certain background to have achieved such a position. It also is inappropriate stereotyping to link a profession or some characteristic to race or ethnicity: a Jewish lawyer, an African-American jazz musician, an Asian-American mathematics prodigy.

Referring to people with disabilities solely by their condition dehumanizes them. If you refer to "a disabled employee," you imply that the part (*disabled*) is as significant as the whole (*employee*). Use "an employee with a disability" instead. Similarly, the preferred usage is "a person who uses a wheelchair" rather than "a wheelchair-bound person"; the latter expression inappropriately equates the condition with the person.

Terms that refer to a person's age are also open to inappropriate stereotyping. Referring to older colleagues as "over the hill" and to younger colleagues as "wet behind the ears" is derogatory, at the least. A person can lack experience, training, or ability for a given task regardless of age, so eliminate those terms and similar ones in your writing. Avoid categorizing older people as "golden agers" or "pretty good for their age." Many people find the former label offensive, and the latter one is not a compliment: It is demeaning and implies that accomplishments by older people are rare and call for special attention.

In matters of sexual orientation, the preferred terms are *gay men, lesbians,* and *heterosexual* or *bisexual men* and *women.* Although *gay* can apply to men and women, the term is more often applied to men. *Lesbian* refers exclusively to women of same-sex orientation.

In most workplace writing, such issues are simply not relevant. Of course, there are contexts in which race, ethnicity, or religion should be identified. For example, if you are writing an Equal Employment Opportunity Commission report about your firm's hiring practices, then the racial composition of the workforce is relevant. In such cases, you need to present the issues in ways that respect and do not demean the individuals or groups to which you refer. (See also **ethics in writing.**)

bibliographies

A bibliography is a list of books, articles, Web sources, and other materials consulted in the preparation of a document. It provides a convenient alphabetical listing of those sources in a standardized form for readers interested in getting further information on the topic or in assessing the scope of the research.

In addition to works actually cited in the text, a bibliography should include works consulted for background information. Entries are listed alphabetically by the author's last name. If an author is unknown, the entry is alphabetized by the first word in the title (following *A, An,* or *The*). Entries also can be arranged in subject categories and then ordered alphabetically within those categories.

An annotated bibliography includes complete bibliographic information about a work (author, title, publisher) followed by a brief description or evaluation of what the work contains. The following annotation concisely summarizes and evaluates a book of historical interest.

Rickard, T. A. *A Guide to Technical Writing.* San Francisco Mining and Scientific Publishers, 1908.

This book is of particular historical interest because it is the first published book on technical writing for the professional. The author comments, "It has been said that in this age the man of science appears to be the only one who has anything to say, and he is the one that least knows how to say it. . . . Write simply and clearly, be accurate and careful; above all, put yourself in the other fellow's place. Remember the reader." Geared to the mining and metallurgical sciences, the 17 short, unnumbered chapters cover matters of language, usage, grammar, and mechanics slanted toward the needs of the technical writer. The book ends with a paper the author read before the American Association for the Advancement of Science, at Denver on August 28, 1901: "A Plea for Greater Simplicity in the Language of Science." "We must remember," the author suggests, "that the language in relation to ideas is a solvent, the purity and clearness of which affect what it bears in solution."

See **formal reports** for guidance on the placement of a bibliography in a report and **documenting sources** for information on the style of bibliographic entries.

blend words

A *blend word* is formed by combining part of one word with part of another.

- motor + hotel = motel
- breakfast + lunch = brunch
- smoke + fog = smog
- electric + execute = electrocute
- chuckle + snort = chortle

Specialists occasionally create blend words (sometimes called "portmanteau words") to meet specific needs, for example, *autopiler* (an automatic compiler in a computer system). Resist creating blend words in your writing unless an obvious need arises; then be sure to define them clearly for your **readers.** (See also **defining terms** and **new words.**)

both . . . and

Statements using the *both . . . and* construction should be balanced grammatically and logically.

- A successful photograph must be *both* clearly focused *and* adequately exposed.

Notice that *both* and *and* are followed logically by ideas of equal weight and of parallel grammatical construction. (See **parallel structure.**)

- For success in engineering, it is necessary both to develop writing
skills and ~~mastering~~ calculus.
  to master

brackets ESL

The primary use of brackets ([ ]) is to enclose a word or words inserted by the writer or an editor into a quotation.

- The text stated, "Hypertext systems can be categorized as either modest [not modifiable] or robust [modifiable], depending on the degree to which users are encouraged to make modifications."

Brackets are used to set off a parenthetical item within parentheses.

- We should be sure to give Emanuel Foose (and his brother Emilio [1812–1882]) credit for his role in founding the institute.

Brackets are also used in academic writing to insert the Latin word *sic,* which indicates that the writer has quoted material exactly as it appears in the original, even though it contains an obvious error.

- Dr. Smith pointed out that "the earth does not revolve around the son [*sic*] at a constant rate."

If you are following MLA style in your writing, use brackets around ellipsis dots to show that some words have been omitted from the original source. (See also **documenting sources** and **ellipses.**)

- "The vast majority of the Internet's inhabitants are males [. . .] between the ages of eighteen and thirty-four" (5).

brainstorming

Brainstorming, a form of free association used to generate ideas, can be done individually or in groups. Brainstorming can stimulate creative thinking about a topic and reveal fresh perspectives and

B

new connections. To brainstorm, jot down as many random ideas as you can think of about the topic. When a group brainstorms, a designated person writes down the words or phrases as the other group members suggest them. Do not stop to analyze ideas or hold back looking for only the "best" ideas; just write down everything that comes to mind. After compiling a list of initial ideas, ask *what, when, who, where, how,* and *why* for each idea, then list additional details that those questions bring to mind. When you run out of ideas, analyze each one you recorded, discarding those that are re-dundant. Then group the items in the most logical order, based on the purpose of the document and the needs of the **readers;** that will result in a tentative **outline** of the document. Although the outline will be sketchy and incomplete, it will show where further brainstorming or research is needed and provide a framework for any new details that additional research yields.

Many writers find a technique called *clustering* (also called *mind mapping*), as shown in Figure B–1, helpful in recording and orga-

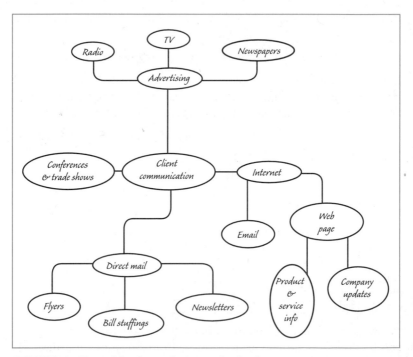

FIGURE B–1. Cluster Map from a Brainstorming Session

nizing ideas created during a brainstorming session. To cluster, begin with a blank sheet of paper or a flip chart. Think of a keyword or phrase that best characterizes your topic and put it in a circle at the center of the paper. Then put the subtopics most closely related to the main topic in circles or boxes around the main topic, connecting each to the center circle, like adding spokes to a wheel hub. Repeat the exercise for each subtopic. Each subtopic should stimulate additional subtopics, which you add in circles connected to the parent subtopic, as you did with the main topic at the center of the page. Continue the process until you exhaust all possible ideas. The resulting "map" will show clusters of terms grouped around the central concept.

bunch

Bunch refers to like things that grow or are fastened together. Do not use the word *bunch* to refer to people.

- A ~~bunch~~ group of trainees toured the site.

C

can / may

In writing, *can* refers to capability, and *may* refers to possibility or permission.

- I *can* have the project finished by the end of the year. [capability]

- I *may* be in Boston on Thursday. [possibility]

- *May* I proceed with the project? [permission]

cannot / can not

Cannot is one word.

- We ~~can not~~ ^{*cannot*} meet the deadline specified in the contract.

cannot help but

Avoid the phrase *cannot help but* in writing.

- We cannot ^{*avoid cutting*} ~~help but cut~~ our staff.

canvas / canvass

Canvas refers to "heavy, coarse, closely woven cotton or hemp fabric." *Canvass* means "to solicit votes or opinions."

- The maintenance crew spread the *canvas* over the equipment.

- The executive committee decided to *canvass* the employees.

capital / capitol

Capital refers either to financial assets or to the city that hosts the government of a state or a nation. *Capitol* refers to the building in which the state or national legislature meets. *Capitol* is often written with a small *c* when it refers to a state building, but it is always capitalized when it refers to the home of the U.S. Congress in Washington, D.C.

capital letters

DIRECTORY
Overview 79
Proper Nouns 79
Common Nouns 80
First Words 80
Specific Groups 80
Specific Places 81
Specific Institutions, Events, and Concepts 81
Titles of Works 82
Personal, Professional, and Job Titles 82
Abbreviations 83
Letters 83
Miscellaneous Capitalizations 83

Overview

The use of capital, or uppercase, letters is determined by custom. Capital letters are used to call attention to certain words, such as proper **nouns** and the first word of a sentence. Care must be exercised in using capital letters because they can affect the meaning of words (march/March, china/China, turkey/Turkey). The proper use of capital letters can help eliminate ambiguity.

Proper Nouns

Proper nouns name specific persons, places, things, concepts, or qualities and are capitalized.

- Physics 101, General Electric, Jennifer Wilde, Argentina

Common Nouns

C

Common nouns name general classes or categories of people, places, things, concepts, or qualities rather than specific ones and are not capitalized.

- a physics class, a company, a person, a country

First Words

The first letter of the first word in a sentence is always capitalized.

- Of all the plans you mentioned, the first one seems the best.

The first word after a **colon** may be capitalized if the statement following is a complete sentence or if it is a formal resolution or question.

- Today's meeting will deal with only one issue: What is the firm's role in environmental protection?

If a subordinate element follows the colon or if the thought is closely related, use a lowercase letter following the colon.

- We had to keep working for one reason: the approaching deadline.

The first word of a complete sentence in **quotation marks** is capitalized.

- Dr. Vesely stated, "It is possible to postulate an imaginary world in which no decisions are made until all the relevant information is assembled."

The first word in the salutation and complimentary close of a letter is capitalized. (See also **correspondence.**)

- Dear Mr. Smith:
- Sincerely yours,

Specific Groups

Capitalize the names of ethnic groups, religions, and nationalities.

- Native American, Italian, Jewish, Romanian

Do not capitalize the names of social and economic groups.

- middle class, working class, unemployed

Specific Places

Capitalize the names of all political divisions.

C

- Ward Six, Chicago, Cook County, Illinois, Ontario, Canada

Capitalize the names of geographical divisions.

- Europe, Asia, North America, the Middle East

Do not capitalize geographic features unless they are part of a proper name.

- The *mountains* in some areas, such as the *Great Smoky Mountains,* make television transmission difficult.

The words *north, south, east,* and *west* are capitalized when they refer to sections of the country. They are not capitalized when they refer to directions.

- I may travel *south* when I relocate.
- We may need a new plant in the *South* next year.

Capitalize the names of stars, constellations, and planets.

- Saturn, Andromeda, Jupiter, Milky Way, Orion

Do not capitalize *earth, sun,* and *moon* except when they are referred to formally as astronomical bodies.

- My workday was so long that I saw the *sun* rise over the lake and the *moon* appear as darkness settled over the *earth.*
- The various effects of the *Sun* on *Earth* and the *Moon* were discussed at the symposium.

Specific Institutions, Events, and Concepts

Capitalize the names of institutions, organizations, and associations.

- The American Society of Mechanical Engineers and the Department of Housing and Urban Development are cooperating on the project.

An organization usually capitalizes the names of its internal divisions and departments.

- Faculty, Board of Directors, Engineering Department

C

Types of organizations are not capitalized unless they are part of an official name.

- Our group decided to form a technical writers' association; we called it the American Association of Technical Writers.

Capitalize historical events.

- Dr. Jellison discussed the Great Depression at the last class.

Capitalize words that designate holidays, specific periods of time, months, or days of the week.

- Labor Day, the Renaissance, the Enlightenment, January, Monday, Ramadan, Easter, Passover

Do not capitalize seasons of the year.

- spring, autumn, winter, summer

Capitalize the scientific names of classes, families, and orders but not the names of species or English derivatives of scientific names.

- Mammalia, Carnivora/mammal, carnivorous

Titles of Works

Capitalize the initial letters of the first and last words of the title of a book, article, play, or film, as well as all major words in the title. Do not capitalize articles (*a, an, the*), coordinating conjunctions (*and* and *but*), or short prepositions (*at, in, on, of*) unless they begin or end the title. Capitalize prepositions that contain more than four letters (*between, because, until, after*). The same rules apply to the subject line of a memo or an email.

- The microbiologist greatly admired the book *The Lives of a Cell*.

Personal, Professional, and Job Titles

Titles preceding proper names are capitalized.

- Ms. March, Professor Galbraith

Appositives following proper names normally are not capitalized. However, the word *President* usually is capitalized when it refers to the chief executive of a national government.

- Chuck Schumer, *senator* from New York [but *Senator* Schumer]
- The *President* called a news conference.

The only exception is an epithet, which actually renames the person.

- Alexander the Great, Solomon the Wise

Job titles used with personal names are capitalized.

- Ho-shik Kim, *Division Manager,* will meet with us on Wednesday.

Job titles used without personal names are not capitalized.

- The *division manager* will meet with us on Wednesday.

Use capital letters to designate family relationships only when they occur before a name or substitute for a name.

- One of my favorite people is *Uncle Fred.*

- Jim and my *uncle* went along.

Abbreviations

Capitalize abbreviations if the words they stand for would be capitalized.

- UCLA (University of California at Los Angeles)

- p. (page)

Letters

Capitalize letters that serve as names or indicate shapes.

- X-ray, vitamin B, T-square, U-turn, I-beam

Miscellaneous Capitalizations

The word *Bible* is capitalized when it refers to the Jewish or Christian Scriptures; otherwise, it is not capitalized.

- He quoted a verse from the *Bible,* then read from Blackstone, the lawyer's *bible.*

All references to deities (Allah, God, Jehovah, Yahweh) are capitalized.

- *God* is the *One* who sustains us.

The first word of a complete sentence enclosed in dashes, brackets, or parentheses is not capitalized when it appears as part of another sentence.

C

- We must make an extra effort in safety this year (accidents last year were up 10 percent).

- We must make an extra effort in safety this year. (Accidents last year were up 10 percent.)

Certain units, such as parts and chapters of books and rooms in buildings, when specifically identified by number, are capitalized.

- Chapter 5, Ch. 5; Room 72, Rm. 72

Minor divisions within such units are not capitalized unless they begin a sentence.

- page 11, verse 14, seat 12

case (grammar)   ESL

DIRECTORY
Overview   84
Subjective Case   85
Objective Case   85
Possessive Case   86
Appositives   86
Determining the Case of Pronouns   87
Who/Whom   87

Overview

Grammatically, *case* indicates the functional relationship of a **noun** or a **pronoun** to the other words in a sentence. Nouns change form only in the possessive case; pronouns may show change for the subjective, the objective, or the possessive case. The case of a noun or a pronoun is always determined by its function in a phrase, clause, or sentence. If it is the subject of a phrase, clause, or sentence, it is in the subjective case; if it is an **object** in a phrase, clause, or sentence, it is in the objective case; if it reflects possession or ownership and modifies a noun, it is in the possessive case. The subjective case can indicate the person or thing acting (*he* sued the vendor), the person or thing acted upon (*he* was sued by the vendor), or the topic of description (*he* is the vendor). The objective case can indicate the thing acted on (the vendor sued *him*) or the person or thing acting but in the objective position (the vendor was sued by *him*). (See also **voice**.) The possessive case indicates the person or thing owning or possessing something (it was *his* company).

The different forms of a noun or a pronoun indicate whether it is functioning as a subject (subjective case), as a **complement** (usually objective case), or as a **modifier** (possessive case).

Subjective Case

A pronoun is in the subjective case (also called the nominative case) when it represents the person or thing acting or is the receiver of the action but in the subject position.

- *I* wrote a letter to that company.

- *I* was praised for my letter writing.

A linking **verb** links a pronoun to its antecedent to show that they identify the same thing. Because they represent the same thing, the pronoun is in the subjective case even when it follows the verb, which makes it a subjective complement.

- *He* is the head of the Quality Control Group. [subject]

- The head of the Quality Control Group is *he*. [subjective complement]

The subjective case is used after the words *than* and *as* because of the understood (although unstated) portion of the clauses in which those words appear.

- George is as good a designer as *I* [am].

- Our subsidiary can do the job better than *we* [can].

Objective Case

A pronoun is in the objective case when it indicates the person or thing receiving the action expressed by a verb in the active voice. (The objective case is also called the accusative case.)

- They informed *me* by letter that they had received my résumé.

Pronouns that follow action verbs (which excludes all forms of the verb *be*) must be in the objective case. Do not be confused by an additional name.

C

- The company promoted *me* in June.
- The company promoted John and *me* in June.

A pronoun is in the objective case when it is the object of a gerund or a **preposition** or the subject of an infinitive.

- Between *you* and *me,* his facts are questionable.
  [objects of a preposition]
- Many of *us* attended the conference. [object of a preposition]
- Training *him* was the best thing I could have done.
  [object of a gerund]
- We asked *them* to recalibrate the instruments.
  [subject of an infinitive]

English does not differentiate between direct objects and indirect objects; both require the objective form of the pronoun. (See also **complements.**)

- The interviewer seemed to like *me.* [direct object]
- They wrote *me* a letter. [indirect object]

Possessive Case

A noun or pronoun is in the possessive case when it represents a person, place, or thing that possesses something. To make a singular noun possessive, add an apostrophe and *s.*

- The *manufacturing plant's* robotic inventory system is now in operation.

With plural nouns that end in *s,* show the possessive by placing an apostrophe after the *s* that forms the plural.

- a manager*s'* meeting

(For other guidelines, see **possessive case.**)

Appositives

An **appositive** is a noun or noun phrase that follows and amplifies another noun or noun phrase. Because it has the same grammatical function as the noun it complements, an appositive should be in the same case as the noun with which it is in apposition.

- Two design engineers, Jim Knight and *I*, were asked to review the drawings. [subjective case]

- The group leader selected two members to represent the department—Mohan Pathak and *me*. [objective case]

Determining the Case of Pronouns

One test to determine the proper case of a pronoun is to try it with some transitive verb such as *resembled* or *hit*. If the pronoun would logically precede the verb, use the subjective case; if it would logically follow the verb, use the objective case.

- *She* (*he, they*) resembled her father. [subjective case]

- Angela resembled *him* (*her, them*). [objective case]

In the following example, try omitting the noun to determine the case of the pronoun.

SENTENCE   (*We/Us*) pilots fly our own airplanes.

CORRECT   *We* fly our own airplanes.
[This correct usage sounds right.]

INCORRECT   *Us* fly our own airplanes.
[This incorrect usage is obviously wrong.]

To determine the case of a pronoun that follows *as* or *than*, mentally add the words that are omitted but understood.

- The other operator is not paid as well as *she* [is paid].
[You would not write, "*Her* is paid."]

- His partner was better informed than *he* [was informed].
[You would not write, "*Him* was informed."]

If pronouns in compound constructions cause problems, try using them singly to determine the proper case.

SENTENCE   (*We/Us*) and the Johnsons are going to the Grand Canyon.

CORRECT   *We* are going to the Grand Canyon.
[You would not write, "*Us* are going to the Grand Canyon."]

Who/Whom

Who and *whom* cause much trouble in determining case. *Who* is the subjective case form, and *whom* is the objective case form. When in

C

doubt about which form to use, try substituting a personal pronoun to see which one fits. If *he, she,* or *they* fits, use *who.*

- *Who* is the representative from the 45th district?
  [*He* is the representative from the 45th district.]

If *him, her,* or *them* fits, use *whom.*

- It depended on *whom?*
  [It depended on *them.*]

It is becoming common to use *who* for the objective case when it begins a clause or sentence, although some readers still object to such a construction, especially in formal contexts.

- *Who* should I call to report a fire?

The best advice is to know your **readers.** (See also **who/whom.**)

case

The word *case* is often merely filler. Be critical of the word and eliminate it if it contributes nothing. (See also **conciseness/wordiness.**)

- An exception was made ~~in the case of~~ those closely connected
  for
  with the project.

cause-and-effect method of development

The cause-and-effect **method of development** is a common strategy to explain why something happened or why you think something will happen. The goal of the cause-and-effect method of development is to make as plausible as possible the relationship between a situation and either its cause or its effect. The conclusions you draw about the relationships should be based on evidence you have gathered. Because not all evidence will be of equal value, keep the following guidelines in mind as you gather evidence.

Evaluating Evidence

The facts and arguments you gather should be relevant to your topic. Be careful not to draw a conclusion that your evidence does not lead to or support. You may have researched some statistics, for example, showing that an increasing number of Americans are licensed to fly small airplanes. You cannot use that information as evidence for a decrease in new car sales in the United States—the evidence does not lead to that conclusion. Statistics on the increase in small-plane licensing may lead to other conclusions, however. You could argue that the upswing has occurred because people have more disposable income and small planes save travel time, provide easy access to remote areas, and, once they are purchased, are economical to operate.

Your evidence should be adequate. Incomplete evidence can lead to false conclusions.

- Driver training classes do not help prevent auto accidents. Two people I know who completed driver training classes were involved in accidents.

Although the evidence cited to support the conclusion may be accurate, there is not enough of it. A thorough investigation of the usefulness of driver training classes in keeping down the accident rate would require many more than two examples. And it would require a comparison of the driving records of those who had completed driver training with drivers who had not.

Your evidence should be representative. If you conduct a survey to obtain your evidence, do not solicit responses only from individuals or groups whose views are identical to yours; be sure you obtain a representative sampling.

Your evidence should also be plausible. Two events that occur close to each other in time or place may or may not be causally related. Thunder and black clouds do not always signal rain, but they do so often enough that if we are outdoors and the sky darkens and we hear thunder, we seek shelter. If you sprain your ankle after walking under a ladder, however, you cannot conclude that a ladder brings bad luck. Merely to say that X caused (or will cause) Y is inadequate. You must demonstrate the relationship with pertinent facts and arguments.

C

Linking Causes to Effects

To show a true relationship between a cause and an effect, you must demonstrate that the existence of the one *requires* the existence of the other. It is often difficult to establish beyond any doubt that one event was the cause of another event. More often, a result will have more than one cause. As you research a subject, your task is to determine which cause or causes are most plausible.

When several probable causes are equally valid, report your findings accordingly, as in the following excerpt from an article on the use of an energy-saving device called a furnace-vent damper. The damper is a metal plate that fits inside the flue or vent pipe of a natural-gas or fuel-oil furnace. When the furnace is on, the damper opens to allow gases to escape up the flue. When the furnace shuts off, the damper closes, thus preventing warm air from escaping up the flue stack. The dampers are potentially dangerous, however. If a damper fails to open at the proper time, poisonous furnace gases could back up into the house and asphyxiate the occupants. Tests run on several dampers showed a number of probable causes for their malfunctioning.

- One damper was sold without proper installation instructions, and another was wired incorrectly. Two of the units had slow-opening dampers (15 seconds) that prevented the [furnace] burner from firing. And one damper jammed when exposed to a simulated fuel temperature of more than 700 degrees.

> —Don DeBat, "Save Energy But Save Your Life, Too," *Family Safety,* Fall 1978, p. 27.

The investigator located more than one cause of damper malfunctions and reported on them. Without such a thorough account, recommendations to prevent malfunctions would be based on incomplete evidence. (See also **logic errors.**)

---

center on / center around

Use the phrase *center on* or *center in,* not *center around.*

- The experiments center ~~around~~ *on* the new discovery.

Often the idea intended by *center on* is better expressed by other words.

dealt with
- The subcommittee hearings on computer security ~~centered on~~ access codes.

chair / chairperson

The terms *chair, chairperson, chairman,* and *chairwoman* are used to refer to a presiding officer. The titles *chair* and *chairperson*, however, avoid any sexual bias that might be implied by the other titles.

- Mary Roberts preceded John Stevens as *chair* of the executive committee.

chronological method of development

The chronological **method of development** arranges the events under discussion in sequential order, beginning with the first event and continuing chronologically to the last. **Trip reports,** laboratory reports, work schedules, some **minutes of meetings,** and certain **trouble reports** are among the types of writing in which information is organized chronologically.

In the memo shown in Figure C–1 on page 92, a retail-store manager describes the steps taken over a one-year period to reduce shoplifting at his store. After providing important background information, the writer presents the steps taken in chronological order.

cite / sight / site

Cite means "acknowledge" or "quote an authority"; *sight* is the ability to see; *site* is a plot of land or the place where something is located.

- The speaker *cited* several famous economists to support his prediction about the stock market.

C

The Rack
Interoffice Memorandum

To: Joanna Sanchez, Vice President for Marketing
From: Larry Brown, Manager, Downtown Branch   *LB*
Date: September 9, 20--
Subject: Reducing Shoplifting at the Downtown Store

Over the past year, my staff and I have taken a number of measures to try to reduce the amount of shoplifting in the downtown store. As you know, we have spent much time, effort, and money on the problem, which we hope will be alleviated during the Christmas shopping season. Let me recap the specific steps we have taken.

Task Force
We formed a task force of salespeople, buyers, managers, and executive staff to recommend ways of curtailing shoplifting and methods of implementing our recommendations. We met four times during January and twice in March to reach our final recommendations.

Mark IV Surveillance System
In April, we installed a Mark IV System, which uses closed-circuit TV cameras at each exit. The cameras, which are linked with our security office, are capable of taping signals from all exits simultaneously. The task force felt the Mark IV System might be useful in detecting a pattern of specific individuals entering and leaving the store. This system, which became operational on April 20, has been very helpful in reducing the number of thefts.

Employee Training
During May and June, we held employee workshops on detecting shoplifters. Security, Inc., a consulting firm, led the workshops and provided not only lectures and tips on spotting shoplifters but also demonstrations of common techniques shoplifters use to divert store personnel. All those who attended thought the workshops were quite helpful.

Other Steps Taken
Because the task force determined that certain items were particularly vulnerable to shoplifters, we decided in July to restructure some of the display areas. Our purpose was to make those areas less isolated from the view of clerks and other store personnel. The remodeling, most of which was relatively minor, was completed over the summer months.

For the fall and holiday sales, we have hired extra uniformed security guards. The guards, also from Security, Inc., should be able to deter first-time shoplifters, although we know that this step will not eliminate the problem altogether.

We believe the steps we have taken will substantially reduce our losses from theft. Of course, after we have reviewed the figures at the end of the year, the task force will meet again in January to assess the success of the methods we have used. If you need more details, please let me know.

FIGURE C–1. Chronological Method of Development

- After the accident, his vision was blurred, and he feared that he might lose his *sight*.

- The *site* for the new factory is three miles from the middle of town.

- Our Web *site* will be developed by the end of the month.

clarity   **ESL**

Clarity is essential in effective communication. Strive to make all your writing direct, orderly, and precise. Many factors contribute to clarity, just as many other elements can defeat it. Logical development, **unity, coherence, emphasis, subordination, pace, transition,** an established **point of view, conciseness,** and **word choice** contribute to clarity. **Ambiguity, awkwardness,** vagueness, poor use of **idiom, clichés,** and inappropriate level of usage detract from clarity.

A logical **method of development** and good **outlining** are essential to clarity. Without a logical method of development and an outline, you communicate only isolated thoughts to your reader, and you cannot effectively achieve your **purpose** by a jumble of isolated thoughts. Use a method of development that puts your thoughts in a logical, meaningful sequence and brings clarity to your writing.

Proper emphasis and subordination are mandatory if you want to achieve clarity. If you do not use those two complementary techniques wisely, all your clauses and sentences will appear to be of equal importance. Your reader will only be able to guess which are most important, which are least important, and which fall somewhere in the middle.

The **pace** at which you present your ideas is important to clarity; if the pace is not carefully adjusted to both the topic and the reader, your writing will appear cluttered and unclear.

Point of view establishes through whose eyes or from what vantage point the reader views the subject. A consistent point of view is essential to clarity; if you switch from the first person to the third person in midsentence, you are certain to confuse your reader.

Clear transition contributes to clarity by providing the smooth flow that enables the reader to connect your thoughts with one another without conscious effort.

C

That conciseness is a requirement of clearly written communication should be evident to anyone who has ever attempted to decipher an insurance policy or a legal contract. Although words are our chief means of communication, too many of them can impede clear communication, just as too many cars on a highway can impede traffic. For the sake of clarity, remove unnecessary words from your writing. **Vague words** can also detract from the conciseness of your writing. Always choose precise words to eliminate ambiguity and awkwardness.

clauses   ⓔⓢⓛ

A clause is a syntactic construction, or group of words, that contains a **subject** and a predicate and that functions as a sentence or as part of a sentence. Every subject-predicate word group in a sentence is a clause, and every sentence must contain at least one independent clause; otherwise, it is a **sentence fragment.** (See also **sentence construction.**)

A clause that could stand alone as a simple sentence is an *independent clause.*

* *The scaffolding fell* when the rope broke.

A clause that could not stand alone if the rest of the sentence were deleted is a *dependent clause.*

* I was at the St. Louis branch *when the decision was made.*

Dependent (or *subordinate*) clauses are useful in making the relationship between thoughts more precise and succinct than if the ideas were presented in a series of simple sentences or compound sentences.

FRAGMENTED	The sewage plant is located between Millville and Darrtown. Both villages use it. [two thoughts of approximately equal importance]
SUBORDINATED	The sewage plant, *which is located between Millville and Darrtown,* is used by both villages. [one thought subordinated to the other]
FRAGMENTED	Title insurance is a policy issued by a title insurance company, and it protects against any title

defects, such as outside claimants.
[two thoughts of approximately equal impor-
tance]

SUBORDINATED   Title insurance, *which protects against any title defects such as outside claimants,* is a policy issued by
a title insurance company.
[one thought subordinated to the other]

Subordinate clauses are especially effective for expressing thoughts
that describe or explain another statement. Too much subordination, however, can be confusing.

- He had selected teachers who taught classes that ~~had a slant that~~ *were*
  ~~was~~ specifically directed ~~toward students who intended to go into~~ *to engineering students.*
  ~~engineering.~~

A clause can be connected with the rest of its sentence by a coordinating **conjunction,** a subordinating conjunction, a relative **pronoun,** or a conjunctive **adverb.**

- It was 500 miles to the facility, *so* we made arrangements to fly.
  [coordinating conjunction]

- Mission control will have to be alert *because* at launch the space
  laboratory will contain a highly flammable fuel.
  [subordinating conjunction]

- It was Robert M. Fano *who* designed and developed the earliest
  multiple-access computer system at MIT. [relative pronoun]

- It was dark when we arrived; *nevertheless,* we began the tour of the
  factory. [conjunctive adverb]

---

clichés

Clichés are expressions that have been used for so long that they are
no longer fresh but come to mind easily because they are so familiar. In addition to being stale, clichés are usually wordy, are often
vague, and can be confusing, especially to nonnative speakers of
English. Each of the following clichés is followed by a better, more
direct word or expression.

INSTEAD OF:	USE:
straight from the shoulder	frank
last but not least	last, finally
run it up the flagpole	see what others think

Writers often use clichés in an attempt to sound impressive (see also **affectation**). Although clichés may come to mind easily while you are **writing a draft,** you should eliminate them during **revision.** (See also **conciseness/wordiness, vague words, vogue words.**)

clipped forms of words

When the beginning or the end of a word is cut off to create a shorter word, the result is a *clipped form* of the word.

- lab, demo, fridge, memo, fax, Net

The word *specification,* for example, is often shortened to *spec.* Although acceptable in conversation, most clipped forms should not appear in writing unless they are commonly accepted as part of the special vocabulary of an occupational group. Some clipped forms, such as *phone, memo,* and *fax,* have nearly replaced their original, longer forms.

Apostrophes normally are not used with clipped forms of words (not *'fridge,* but *fridge*). Because they are not strictly **abbreviations,** clipped forms are not followed by **periods** (not *lab.,* but *lab*).

Do not use clipped forms of spelling (*thru, nite, lite,* and so on) in workplace writing.

coherence

Writing is coherent when the relationships among ideas are clear to the reader. Coherent writing moves logically and consistently from point to point. Each idea should relate clearly to the other ideas, with one idea flowing smoothly to the next. Many elements contribute to smooth and coherent writing, but the major components are (1) a logical sequence of ideas and (2) clear **transitions** between ideas. (See also **methods of development.**)

A logical sequence of presentation is the most important requirement in achieving coherence, and the key to achieving the most logical sequence of presentation is the use of a good outline (see **outlining**). An outline forces you to establish a beginning (**introduction**), a middle (body), and an end (**conclusion**), and that structure alone contributes greatly to coherence. An outline also enables you to lay out the most direct route to your **purpose** without digressing into interesting but only loosely related side issues. Preparing an outline lets you experiment with different sequences and then helps you choose the best one.

Thoughtful transition is also essential; without it, your writing cannot achieve the smooth flow from sentence to sentence and from **paragraph** to paragraph that is required for coherence.

During **revision,** check your draft carefully for coherence. If your writing is not coherent, you are not communicating effectively with your **readers.** (See also **clarity** and **unity.**)

collaborative writing

Collaborative writing occurs when two or more writers work together to produce a single document for which they share responsibility and decision-making authority. The collaborating writers make approximately equal contributions, and they communicate as equals, with no one in a superior or subordinate role.

Collaborative writing teams are formed when (1) the size of a project or the time constraints imposed on it require collaboration, (2) the project involves multiple areas of expertise, or (3) the project requires the melding of divergent views into a single perspective that is acceptable to the whole team or another group.

The members of the collaborative writing team need to recognize and use each member's expertise to their collective advantage. Team members must respect one another's professional capabilities and strive to achieve a compatible working relationship while acknowledging that some conflict is a natural part of any group interaction.

The team should designate one person as its "leader." The leader does not have decision-making authority—he or she merely assumes the extra responsibility of coordinating the team members' activities and organizing the final project. Team leadership can be determined by mutual agreement of the team members or assigned on a rotating basis if the team works together to produce multiple documents on a regular basis.

Tasks of the Collaborative Writing Team

The collaborative writing team normally performs four tasks: planning the document, researching and **writing a draft,** reviewing the drafts of other team members, and revising drafts on the basis of those reviews. (See also **research** and **revision.**)

Planning.   The team collectively identifies the **reader,** the **purpose,** and the **scope** of the project, as well as its goals and the most effective organization for the whole document. The team analyzes the overall project, conceptualizes the document to be produced, creates a broad outline of the document, divides the document into segments, and assigns each segment to individual team members, often on the basis of expertise. (See **outlining.**)

In the planning stage, the team projects a schedule and sets any writing style standards that team members are expected to follow. The schedule includes due dates for drafts, for reviews of the drafts by other team members, for revisions, and for the final document. It is important that deadlines be met, even if a draft is not as polished as the individual writer would like, because one team member's missed deadline can hold up the work of the entire team.

Research and Writing.   Planning is followed by research and writing, a period of intense independent activity by the individual members of the team. Each member researches his or her assigned segment of the document, fleshes out the broad outline in greater detail, and produces a draft from the detailed outline. Then, by the deadline established for drafts, the individual writers submit copies of the drafts to their teammates for review.

Reviewing.   During the review stage, team members assume the role of the reading audience to address any problems that might arise for the readers. Each team member critically yet diplomatically reviews the work of the other team members. The reviewers evaluate their colleagues' drafts, from the **organization** of each segment to the **clarity** of each paragraph and sentence. They offer advice to help the writer improve his or her segment of the document. Team members can easily solicit feedback from their colleagues by sharing files on a network system, emailing documents back and forth, or exchanging disks. Redlining, a software editing capability, allows a reviewer to show suggested changes without deleting the original text. The author can then easily accept or reject the proposed changes.

Revising.   In this stage, the individual writers evaluate their colleagues' reviews and accept or reject their suggestions. Once each member revises his or her draft, all drafts can be consolidated into a final master copy maintained by the team leader.

Team members must evaluate their colleagues' suggestions objectively—on the basis of merit—without reacting emotionally. This is often a touchy part of the collaborative writing process; writers must be careful not to let their egos impede their good judgment. The ability to accept criticism and use it productively is one of the critical differences between an effective team member and an ineffective one.

Conflict

Team members may not agree on every subject, and differing perspectives can easily lead to conflict. Conflicts ranging from relatively mild differences over minor points to major showdowns may occur. A team that tolerates some disharmony, considers all viewpoints, and works through conflicting opinions to reach consensus or a compromise usually produces a better document than a team with no conflict. Although mutual respect among team members is necessary, too much deference can inhibit challenges and thus reduce creativity. Writers must be willing to challenge one another but must do so tactfully and diplomatically.

Writer's Checklist: Collaborative Writing

☑   Respect each team member's professional capabilities.

☑   Accept conflict as a natural part of any interaction.

☑   Tolerate some disharmony but temper it with mutual respect.

☑   Designate one person as the team leader.

☑   Collectively identify the audience, purpose, and scope of the project.

Writer's Checklist: Collaborative Writing (continued)

C

- ☑ Create a working outline of the document.
- ☑ Establish a schedule: due dates for drafts, revisions, and final documents.
- ☑ Assign a segment or task to each team member.
- ☑ Research and write drafts of document segments.
- ☑ Exchange segments for team reviews.
- ☑ Challenge team segments diplomatically.

collective nouns (*see* nouns)

colloquialisms (*see* English, varieties of)

colons   ESL

The colon (:) is a mark of anticipation and introduction that alerts readers to the close connection between the first statement and what follows.

A colon is used to connect a list or series to a word, clause, or phrase with which it is in apposition.

- Three topics will be discussed: the new accounting system, the new bookkeeping procedures, and the new payroll software.

Do not, however, place a colon between a **verb** and its **objects.**

- Three fluids that clean pipettes are ̷ water, alcohol, and acetone.

One common exception is made when a verb is followed by a stacked **list.**

- Corporations that manufacture computers include:

Apple	Compaq	Micron
IBM	Dell	Gateway

Do not use a colon between a **preposition** and its object.

- I would like to be transferred to ̷ Tucson, Boston, or Miami.

A colon is used to link one statement to another statement that develops, explains, amplifies, or illustrates the first.

- Any organization is confronted with two separate, though related, information problems: It must maintain an effective internal communication system, and it must see that an effective external communication system is maintained.

A colon is used to link an **appositive** phrase to its related statement if more emphasis is needed and if the phrase comes at the end of the sentence.

- There is only one thing that will satisfy Mr. Sturgess: our finished report.

Colons are used to link numbers that signify different nouns.

- Matthew 14:1 [chapter 14, verse 1]

- 9:30 a.m. [9 hours, 30 minutes]

In proportions, colons indicate the ratio of amounts to each other.

- The cement is mixed with the water and sand at 7:5:14.

Colons are often used in mathematical ratios.

- $7:3 = 14:x$

In **documenting sources,** colons link the place of publication with the publisher and may perform other specialized functions.

- Watson, R. L. *Statistics for Accountants and Electrical Engineers.* Englewood: EEE, 2001.

A colon follows the salutation in **correspondence,** even when the salutation refers to a person by first name.

- Dear Ms. Jeffers:

- Dear George:

The initial **capital letter** of a **quotation** is retained following a colon if the quoted material originally began with a capital letter.

- The senator issued the following statement: "We are not concerned about the present. We are worried about the future."

A colon always goes outside **quotation marks.**

C

- This was the real meaning of his "suggestion": the division must show a profit by the end of the year.

When quoting material that ends in a colon, drop the colon and replace it with an **ellipsis.**

- "Any large corporation is confronted with two separate, though related, information problems . . . ."

The first word after a colon may be capitalized if the statement following the colon is a complete sentence or a formal resolution or question.

- The conference passed a single resolution: Voting will be open to associate members next year.

If the element following the colon is subordinate, use a lowercase letter to begin that element.

- There is only one way to stay within our present budget: to reduce expenditures for research and development.

comma splice   ESL

A comma splice is a grammatical error in which two independent **clauses** are joined by only a comma.

INCORRECT   It was 500 miles to the facility, we arranged to fly.

A comma splice can be corrected in several ways.

Substitute a **semicolon** or a semicolon and a conjunctive **adverb.**

- It was 500 miles to the facility ; we arranged to fly.

- It was 500 miles to the facility ; therefore, we arranged to fly.

  (See also **commas.**)

Add a coordinating **conjunction** following the comma.

- It was 500 miles to the facility, *so* we arranged to fly.

Create two sentences. (Keep in mind that putting a **period** between two closely related and brief statements may result in two weak sentences.)

- It was 500 miles to the facility. We arranged to fly.

Subordinate one clause to the other.

- *Because* it was 500 miles to the facility, we arranged to fly.

commas   ESL

DIRECTORY
Overview   103
Linking Independent Clauses   104
Enclosing Elements   104
Introducing Elements   105
Separating Items in a Series   106
Clarifying and Contrasting   107
Showing Omissions   107
Using with Other Punctuation   108
Using with Numbers and Names   108
Avoiding Unnecessary Commas   110

Overview

Like all **punctuation,** the comma (,) helps readers understand the writer's meaning and prevents **ambiguity.** Notice how the comma helps make the meaning clear in the following example:

AMBIGUOUS	To be successful managers with MBAs must continue to learn. [At first glance, this sentence seems to be about "successful managers with MBAs."]
CLEAR	To be successful, managers with MBAs must continue to learn. [The comma makes clear where the main part of the sentence begins.]

Do not follow the old myth that you should insert a comma wherever you would pause if you were speaking. Although you would pause wherever you encounter a comma, you should not insert a comma wherever you might pause. As the previous example illustrates, effective use of commas depends on an understanding of **sentence construction.**

Linking Independent Clauses

Use a comma before a coordinating **conjunction** (*and, but, or, nor,* and sometimes *so, yet,* and *for*) that links independent **clauses.**

- Human beings have always prided themselves on their unique capacity to create and manipulate symbols, *but* today computers manipulate symbols.

However, if two independent clauses are short and closely related—and there is no danger of confusing the reader—the comma may be omitted. Both of the following examples are correct.

- The cable snapped and the power failed.

- The cable snapped, and the power failed.

Enclosing Elements

Commas are used to enclose nonrestrictive clauses and phrases and parenthetical elements. (For other means of punctuating parenthetical elements, see **dashes** and **parentheses.** See also **restrictive and nonrestrictive elements.**)

- Our new factory, *which began operations last month,* should add 25 percent to total output. [nonrestrictive clause]

- The accountant, *working quickly and efficiently,* finished early. [nonrestrictive phrase]

- We can, *of course,* expect their lawyer to call us. [parenthetical element]

Yes and *no* are set off by commas in such uses as the following:

- I agree with you, *yes.*

- *No,* I do not think we can finish as soon as we would like.

A **direct address** should be enclosed in commas.

- You will note, *Mark,* that the surface of the brake shoe complies with the specifications.

A phrase in apposition (which identifies another expression) is enclosed in commas.

C

- Our company, *Blaylok Precision Company,* did well this year.

Interrupting parenthetical and transitional words or **phrases** are
usually set off with commas.

- The report, it turns out, was incorrect.

- We must wait for the written authorization to arrive, *however,* before we can begin work on the project.

Commas are omitted when the word or phrase does not interrupt
the continuity of thought.

- I *therefore* suggest that we begin construction.

Introducing Elements

Clauses and Phrases.   It is generally a good rule of thumb to put
a comma after an introductory clause or phrase. Identifying where
the introductory element ends helps indicate where the main part of
the sentence begins.

Always place a comma after a long introductory clause.

- *Because many rare fossils seem never to occur free from their matrix,* it
is wise to scan every slab with a hand lens.

A long modifying phrase that precedes the main clause should always be followed by a comma.

- *During the first series of field-performance tests last year at our Colorado
proving ground,* the new engine failed to meet our expectations.

When an introductory phrase is short and closely related to the
main clause, the comma may be omitted.

- *In two seconds* a temperature of 20 degrees Fahrenheit is created in
the test tube.

A comma should always follow an introductory absolute phrase.

- *The tests completed,* we organized the data for the final report.

Words and Quotations.   Certain types of introductory words are
followed by a comma. One such is a proper noun used in direct
address.

C

- *Nancy,* enclosed is the article you asked me to review.

An introductory **interjection** (such as *oh, well, why, indeed, yes,* and *no*) is followed by a comma.

- *Yes,* I will make sure your request is approved.

- *Indeed,* I will be glad to send you further information.

A transitional word or phrase like *moreover* or *furthermore* is usually followed by a comma to connect the following thought with the preceding clause or sentence.

- *Moreover,* steel can withstand a humidity of 99 percent, provided that there is no chloride or sulfur dioxide in the atmosphere.

- *In addition,* we can expect a better world market as a result of this move.

- *However,* we should expect business with Latin America to decline due to the global economic climate.

When **adverbs** closely modify the **verb** or the entire sentence, they should not be followed by a comma.

- *Perhaps* we can still solve the environmental problem.

- *Certainly* we should try.

Use a comma to separate a direct quotation from its introduction.

- Morton and Lucia White said, "People live in cities but dream of the countryside."

Do not use a comma when giving an indirect quotation. (See also **quotation marks.**)

- Morton and Lucia White said that people dream of the countryside, even though they live in cities.

Separating Items in a Series

Although the comma before the last item in a series is sometimes omitted, it is generally clearer to include it. The **ambiguity** that may result from omitting the comma is illustrated in the following sentence.

CONFUSING   Random House, Bantam, Doubleday and Dell were individual publishing companies.

[Does "Doubleday and Dell" refer to one company or two?]

CLEAR   Random House, Bantam, Doubleday, and Dell were individual publishing companies.

Phrases and clauses in coordinate series, like words, are punctuated with commas.

- Plants absorb noxious gases, act as receptors of dirt particles, and cleanse the air of other impurities.

When **adjectives** modifying the same noun can be reversed and make sense, or when they can be separated by *and* or *or*, they should be separated by commas.

- The drawing was of a *modern, sleek, swept-wing* airplane.

When an adjective modifies a phrase, no comma is required.

- She was investigating his *damaged radar beacon system.*
  [The adjective *damaged* modifies the phrase *radar beacon system.*]

Never separate a final adjective from its noun.

- He is a conscientious, honest, reliable, worker.

Clarifying and Contrasting

If you find you need a comma to prevent misreading when a word is repeated, rewrite the sentence.

AWKWARD   The assets we had, had surprised us.

IMPROVED   We had been surprised at our assets.

Use a comma to separate two contrasting thoughts or ideas.

- The project was finished on time, but not within the budget.

Use a comma after an independent clause that is only loosely related to the dependent clause that follows it.

- I should be able to finish the plan by July, even though I lost time because of illness.

Showing Omissions

A comma sometimes replaces a verb in certain elliptical constructions.

- Some were punctual; *others, late.* [The comma replaces *were.*]

It is better, however, to avoid such constructions in workplace writing.

Using with Other Punctuation

Conjunctive adverbs (*however, nevertheless, consequently, for example, on the other hand*) that join independent clauses are preceded by a **semicolon** and followed by a comma. Such adverbs function both as modifiers and as connectives.

- Your idea is good; *however,* your format is poor.

Use a semicolon to separate phrases or clauses in a series when one or more of the phrases or clauses contain commas.

- Our new products include amitriptyline, which has sold very well; dipyridamole, which has not sold well; and cholestyramine, which was just released.

When an introductory phrase or clause ends with a **parenthesis,** the comma separating the introductory phrase or clause from the rest of the sentence always appears outside the parenthesis.

- Although we left late (at 7:30 p.m.), we arrived in time for the keynote address.

Commas always go inside **quotation marks.**

- The operator placed the discharge bypass switch at "normal," which triggered a second discharge.

Except with **abbreviations,** a comma should not be used with a **period, question mark, exclamation mark,** or **dash.**

- "Have you finished the project?/" I asked.

Using with Numbers and Names

Commas are conventionally used to separate distinct items. Use commas between the elements of an address written on the same line (but not between the state and the zip code).

- Kristen James, 4119 Mill Road, Dayton, Ohio 45401

A date can be written with or without a comma following the year if the date is in the month-day-year format.

- October 26, 2000, was the date the project began.

- October 26, 2000 was the date the project began.

If the date is in the day-month-year format, as is typical in **international correspondence,** do not set off the date with commas.

- The date was 26 October 2000 that the project began.

Use commas to separate the elements of Arabic numbers.

- 1,528,200 feet

However, because many countries use the comma as the decimal marker, use spaces or periods rather than commas in international documents.

- 1 528 200 meters

- 1.528.200 meters

A comma may be substituted for the colon in the salutation of a personal letter. Do not, however, use a comma in a business letter, even if you use the person's first name.

- Dear Marie, [personal letter]

- Dear Marie: [business letter]

Use commas to separate the elements of geographical names.

- Toronto, Ontario, Canada

Use a comma to separate names that are reversed or that are followed by an abbreviation.

- Smith, Alvin

- Jane Rogers, Ph.D.

- LMB, Inc.

Use commas to separate certain elements of bibliography, footnote, and reference entries. (See also **bibliographies** and **documenting sources.**)

- Hall, Walter P., ed. *Handbook of Communication Methods.* New York: Stoddard Press, 1999. [bibliography entry]

- [1]Walter P. Hall, ed., *Handbook of Communication Methods* (New York: Stoddard Press, 1999) 30. [footnote]

Avoiding Unnecessary Commas

A number of common writing errors involve placing commas where they do not belong. As stated earlier, such errors often occur because writers assume that a pause in a sentence should be indicated by a comma.

Be careful not to place a comma between a subject and verb or between a verb and its **object.**

* The cold conditions at the test site in the Arctic/ made accurate readings difficult.

* She has often said/ that one company's failure is another's opportunity.

Do not use a comma between the elements of a compound subject or a compound predicate consisting of only two elements.

* The director of the engineering department/ and the supervisor of the quality-control section were opposed to the new schedules.

* The engineering director listed five major objections/ and asked that the new schedule be reconsidered.

Placing a comma after a coordinating conjunction such as *and* or *but* is a common error.

* The chairperson formally adjourned the meeting, *but/* the members of the committee continued to argue.

Do not place a comma before the first item or after the last item of a series.

* The new products we are considering include/ calculators, scanners, and cameras.

* It was a fast, simple, inexpensive/ process.

Do not use a comma to separate a prepositional phrase from the rest of the sentence unnecessarily.

* We discussed the final report/ on the new project.

committee

Committee is a collective **noun** that takes a singular **verb.**

- The *committee* is to meet at 3:30 p.m.

If you want to emphasize the individuals on the committee, use the *members of the committee* with the plural verb form.

- The *members of the committee* were in agreement.

common nouns (*see* **nouns**)

comparative degree (*see* **adjectives** *and* **adverbs**)

compare / contrast

When you *compare* things, you point out similarities or both similarities and differences. When you *contrast* things, you point out only the differences. In either case, you compare or contrast only things that are part of a common category.

- He *compared* all the features of the two brands before making his choice.
- Their styles of selling *contrasted* sharply.

When *compare* is used to establish a general similarity, it is followed by *to*.

- *Compared to* the computer, the abacus is a primitive device.

When *compare* is used to indicate a close examination of similarities or differences, it is followed by *with*.

- We *compared* the features of the new capacitor very carefully *with* those of the old one.

Contrast is normally followed by *with*.

C

- The new policy *contrasts* sharply *with* the earlier one in requiring that sealed bids be submitted.

When the **noun** form of *contrast* is used, one speaks of the *contrast between* two things or of one thing being in *contrast to* the other.

- There is a sharp *contrast between* the old and new policies.
- The new policy is in sharp *contrast to* the earlier one.

comparison   ESL

When you are making a comparison, be sure that both or all the elements being compared are clearly evident to your **reader.**

- The Nicom 3 software is better. *than the Nicom 2 software.*

The things being compared must be of the same kind.

- Imitation alligator hide is almost as tough as a real alligator. *hide*

Be sure to point out the parallels or differences between the things being compared. Do not assume your reader will know what you mean.

- Washington is farther from Boston than Philadelphia. *it is from*

A double comparison in the same sentence requires that the first comparison be completed before the second one is stated.

- The discovery of electricity was one of the great ~~if not the greatest~~ scientific discoveries in history. *, if not the greatest*

Do not attempt to compare things that are not comparable.

- Farmers say that ~~storage space is reduced by 40 percent compared with~~ baled hay. *requires 40 percent less storage space than loose hay requires*

  [*Storage space* is not comparable to *baled hay.*]

comparison method of development

C

As a **method of development,** comparison points out similarities and differences between the elements of your subject. The comparison method of development is especially effective because it can explain a difficult or unfamiliar subject by relating it to a simpler or more familiar one.

You must first determine the basis for the **comparison.** For example, if you were comparing bids from contractors for a remodeling project at your company, you most likely would compare such factors as price, previous experience, personnel qualifications, availability at a time convenient for you, and completion date.

Once you have determined the basis (or bases) for comparison, you can determine the most effective way to structure your comparison: whole by whole or part by part. In the *whole-by-whole method,* all the relevant characteristics of one item are discussed before those of the next item are considered. In the *part-by-part method,* the relevant features of each item are compared one by one. The following discussion of typical woodworking glues, organized according to the whole-by-whole method, describes each type of glue and its characteristics before going on to the next type.

- *White glue* is the most useful all-purpose adhesive for light construction, but it cannot be used on projects that will be exposed to moisture, high temperature, or great stress. Wood that is being joined with white glue must remain in a clamp until the glue dries, which takes about 30 minutes.

  Aliphatic resin glue has a stronger and more moisture-resistant bond than white glue. It must be used at temperatures above 50°F. The wood should be clamped for about 30 minutes. . . .

  Plastic resin glue is the strongest of the common wood adhesives. It is highly moisture resistant, though not completely waterproof. Sold in powdered form, this glue must be mixed with water and used at temperatures above 70°F. It is slow setting, and the joint should be clamped for four to six hours. . . .

  Contact cement is a very strong adhesive that bonds so quickly it must be used with great care. It is ideal for mounting sheets of plastic laminate on wood. It is also useful for attaching strips of veneer to the edges of plywood. Because this adhesive bonds immediately when two pieces are pressed together, clamping is not necessary, but the parts to be joined must be carefully aligned before being placed together. Most brands are flammable, and the fumes

C

can be harmful if inhaled. To meet current safety standards, this type of glue must be used in a well-ventilated area, away from flames or heat.

The purpose of this whole-by-whole method of comparison is to weigh the advantages and disadvantages of each glue for certain kinds of woodworking. The comparison could be expanded, of course, by comparing additional types of glue.

If your purpose were to consider, one at a time, the various characteristics of all the glues, the information might be arranged according to the part-by-part method, as in the following example.

- Woodworking adhesives are rated primarily according to their bonding strength, moisture resistance, and setting times.

  Bonding strengths are categorized as very strong, moderately strong, or adequate for use with little stress. Contact cement and plastic resin glue bond very strongly, while aliphatic resin glue bonds moderately strongly. White glue provides a bond least resistant to stress.

  The *moisture resistance* of woodworking glues is rated as high, moderate, and low. Plastic resin glues are highly moisture resistant, aliphatic resin glues are moderately moisture resistant, and white glue is least moisture resistant.

  Setting times for the glues vary from an immediate bond to a four-to-six-hour bond. Contact cement bonds immediately and requires no clamping. Because the bond is immediate, surfaces being joined must be carefully aligned before being placed together. White glue and aliphatic resin glue set in 30 minutes; both require clamping to secure the bond. Plastic resin, the strongest wood glue, sets in four to six hours and also requires clamping.

The part-by-part method could accommodate further comparison. Comparisons might be made according to temperature ranges, special warnings, common uses, and so on.

complaint letters

Companies sometimes make mistakes when they provide goods and services, and customers write complaint letters asking that such situations be corrected. The **tone** of a complaint letter is important; the most effective ones do not sound complaining. If you write a letter that reflects only your annoyance and anger, you may not be taken seriously—you may simply seem petty and irrational. Re-

BAKER MEMORIAL HOSPITAL

Diagnostic Services Department
501 Main Street
Springfield, OH 45321
(513) 683-8100
Fax (513) 683-8000

September 23, 20--

Manager, Customer Relations
Computer Solutions, Inc.
521 West 23rd Street
New York, NY 10011

Subject: HV3 Monitors

On July 9, I ordered nine HV3 monitors for your model MX-15 scanner. The monitors were ordered from your Web site.

On August 2, I received from your Newark, New Jersey, parts warehouse seven HL monitors. I immediately returned these monitors with a note indicating the mistake that had been made. However, not only have I failed to receive the HV3 monitors I ordered, but I have also been billed repeatedly.

Would you please either send me the monitors I ordered or cancel my order. I have enclosed a copy of my original order letter and the most recent bill.

Sincerely,

Paul Denlinger

Paul Denlinger
Manager

Enclosures

FIGURE C–2. Complaint Letter

C

member that the person who receives your letter may not be the one who was directly responsible for the situation about which you are complaining. An effective complaint letter should assume that the recipient will be conscientious in correcting the problem.

Although the circumstances and the severity of the problem may vary, effective complaint letters generally follow this pattern.

1. They identify the faulty item or items and include relevant information such as invoice numbers, part names, and dates. It is a good idea to include a copy of the bill or contract.
2. They explain logically, clearly, and specifically what went wrong, especially for a problem with a service. (Avoid expressing an opinion of why you *think* some problem occurred if you have no way of knowing.)
3. They state what you expect the reader to do to solve the problem to your satisfaction.

Large organizations often have special departments called Customer Service, Consumer Affairs, or Adjustments to handle complaints. If you address your letter to one of those departments, it should reach someone in the company who can help you. In smaller organizations, you might write to a vice president in charge of sales or service. For very small businesses, write directly to the owner. As a last resort, you may find that sending copies of a complaint letter to more than one person in the company will get faster results. Each employee who receives the letter will know (because of the notation at the bottom of the page) that others, possibly higher in the organization, have received the letter and will take note of whether the problem is solved. Figure C–2 on page 115 shows a typical complaint letter. (See also **adjustment letters, correspondence,** and **refusal letters.**)

complement / compliment

Complement means "anything that completes a whole" (see also **complements**). It is used as either a **noun** or a **verb.**

- A *complement* of four employees would bring our staff up to its normal strength. [noun]

- The two programs *complement* one another perfectly. [verb]

Compliment means "praise." It too is used as either a noun or a verb.

- The manager *complimented* the staff on its efficient job. [verb]
- The manager's *compliment* boosted staff morale. [noun]

complements   ESL

A complement is a word, **phrase,** or **clause** used in the predicate of a sentence to complete the meaning of the sentence. (See also **sentence construction.**)

- Pilots fly *airplanes*. [word]
- To live is *to risk death*. [phrase]
- John knew *that he would be late*. [clause]

Four kinds of complements are generally recognized: *direct object, indirect object, objective complement,* and *subjective complement.*

A *direct object* is a **noun** or noun equivalent that receives the action of a transitive **verb;** it answers the question "what?" or "whom?" after the verb.

- Grace built *an antenna*. [noun]
- I like *to work*. [verbal]
- I like *it*. [pronoun]
- I like *what I saw*. [noun clause]

An *indirect object* is a noun or noun equivalent that occurs with a direct **object** after certain kinds of transitive verbs such as *give, wish, cause,* and *tell.* It answers the question "to whom or what?" or "for whom or what?"

- We should buy *the Milwaukee office* a *color copier*.
  [*Color copier* is the direct object and *Milwaukee office* is the indirect object.]

An *objective complement* completes the meaning of a sentence by revealing something about the object of the transitive verb. An objective complement may be either a noun or an **adjective.**

- They call him *a genius*. [noun]
- We painted the building *white*. [adjective]

C

A *subjective complement,* which follows a linking verb rather than a transitive verb, describes the subject. A subjective complement may be either a noun or an adjective.

* His sister is *an engineer.* [noun]
* His brother is *ill.* [adjective]

complex sentences (*see* sentence construction)

compose / comprise

Compose means "create" or "make up the whole." The parts *compose* the whole.

* The 13 moving parts *compose* the mechanism.
* The mechanism is *composed* of 13 moving parts.

Comprise means "include," "contain," or "consist of." The whole *comprises* the parts.

* The mechanism *comprises* 13 moving parts.

compound sentences (*see* sentence construction)

compound words   ESL

A compound word is made from two or more words that are hyphenated, written as one word, or written as separate words that are so closely related as to constitute a single concept. (If you are not certain whether a compound word should be hyphenated, check a dictionary. See also **hyphens.**)

* nevertheless, editor-in-chief, courthouse, run-of-the-mill, low-level, high-energy, post office, Web site, online

Be careful to distinguish between compound words and words that frequently appear together but do not constitute compound words, such as *green house* and *greenhouse.*

Plurals of compound words are usually formed by adding an *s* to the last letter.

C

- bedrooms, masterminds, overcoats, bookcases, post offices

However, when the first word of the compound is more important to its meaning than the last, the first word takes the *s*. (When in doubt, check your dictionary.)

- editors-in-chief, fathers-in-law

Possessives are formed by adding *'s* to the end of the compound word.

- the *vice president's* speech, his *sister-in-law's* car, the *pipeline's* diameter, the *antibody's* action

compound-complex sentences (*see* **sentence construction**)

computer graphics (*see* **graphs**)

concept / conception

A *concept* is a thought or an idea. A *conception* is the sum of a person's ideas, or concepts, on a subject.

- This final *conception* of the whole process evolved from many smaller *concepts*.

- From the *concept* of combustion evolved the *conception* of the internal combustion engine.

conciseness / wordiness

Conciseness means that extraneous words, phrases, clauses, and sentences have been removed without sacrificing **clarity** or appropriate detail. Conciseness is not a synonym for brevity; a report may be concise, while its **abstract** is brief as well as concise. Conciseness is always desirable, but brevity may or may not be desirable in a given passage, depending on the writer's objective. Although concise sentences are not guaranteed to be effective, wordy sentences always lose some of their readability and coherence because of the extra load they carry.

C

Causes of Wordiness

Modifiers that repeat an idea implicit or present in the word being modified contribute to wordiness by being redundant.

- *active* consideration
- *final* outcome
- *present* status
- *completely* finished

- *basic* essentials
- worthy *of merit*
- *advance* planning
- the reason *is because*

Coordinated synonyms that merely repeat each other contribute to wordiness.

- each and every
- basic and fundamental

- finally and for good
- first and foremost

Excess qualification also contributes to wordiness, as the following examples demonstrate:

- *perfectly* clear
- *completely* accurate
- *radically* new

Expletives, relative pronouns, and relative adjectives, although they have legitimate purposes, often result in wordiness.

> WORDY   *There are* [expletive] many supervisors in the area *who* [relative pronoun] are planning to attend the workshop, *which* [relative adjective] is scheduled for Friday.
>
> CONCISE   Many supervisors in the area plan to attend the workshop scheduled for Friday.

Circumlocution (a long, indirect way of expressing things) is a leading cause of wordiness.

> WORDY   The payment to which a subcontractor is entitled should be made promptly so that in the event of a subsequent contractual dispute we, as general contractors,

may not be held in default of our contract by virtue of nonpayment.

CONCISE   Pay subcontractors promptly. Then if a contractual dispute occurs, we cannot be held in default of our contract because of nonpayment.

When conciseness is overdone, writing can become choppy and ambiguous. (See also **telegraphic style.**) Be aware that too much conciseness can produce a style that is not only too brief but also blunt. For example, if you do not understand a written request and respond by writing, "Your request was unclear" or "I don't understand your question," you risk offending your **reader.** Instead of attacking the writer's ability to phrase a request, ask for more information.

• I will need more information before I can answer your request. Specifically, can you give me the title and the date of the report you are looking for?

That version is a little longer than the other two responses, but it is both clear and more polite.

Writer's Checklist: Achieving Conciseness

Wordiness is natural when you are **writing a draft,** but it should not survive **revision.** When you revise, you can achieve conciseness not only by eliminating the wordy phrases illustrated earlier, but also through a number of other means.

☑ Use **subordination** to achieve conciseness.

WORDY   The chemist's report was carefully illustrated, *and it covered five pages.*

CONCISE   The chemist's *five-page* report was carefully illustrated.

☑ Use simpler words and phrases.

WORDY   It is the policy of the company to provide the proper equipment to enable employees to conduct the telephone communication necessary to discharge their responsibilities; such should not be utilized for personal communications.

CONCISE   The telephones are for company business; do not use them for personal calls.

Writer's Checklist: Achieving Conciseness (continued)

C

☑ Eliminate redundancy.

WORDY Post-installation testing, which is offered to all our customers at no further cost to them whatsoever, is available
with each Line Scan System One purchased from this
company.

CONCISE Free post-installation testing is offered with each Line
Scan System One.

☑ Change the passive **voice** to the active and the indicative **mood** to
the imperative. The following example does both.

WORDY Bar codes normally are used when an order is intended to
be displayed on a computer, and inventory numbers normally are used when an order is to be placed with the
manufacturer.

CONCISE Use bar codes to display the order on a computer, and
use inventory numbers to place the order with the manufacturer.

☑ Eliminate or replace wordy introductory phrases or pretentious
words and phrases. (See **affectation.**)

* It may be said that
* In the case of
* It appears that
* Needless to say

REPLACE in order to, so as to, so as to be able to, with a view to
WITH *to*

REPLACE due to the fact that, for the reason that, owing to the
fact that, the reason for
WITH *because*

REPLACE by means of, by using, utilizing, through the use of
WITH *by* or *with*

REPLACE at this time, at this point in time, at present, at the present
WITH *now*

☑ Eliminate the overuse of intensifiers (such as *very, more, most, best,
quite*) and the excessive use of adjectives and adverbs (such as
great, really, especially). Instead, provide useful and specific details.

WORDY It was a *very* good meeting that was *quite* productive.
SPECIFIC The meeting was productive. We decided to . . .

WORDY Our Web site is *really* popular because of its *great* graphics.
SPECIFIC Our Web site attracted 5,000 visitors this week. Several
clients praised the animated graphics.

conclusions

C

The conclusion of a document ties all the main ideas together and can do so emphatically by making a final significant point. The final point may be to recommend a course of action, make a prediction or a judgment, or merely summarize the document's main points.

The way you conclude depends on both the **purpose** of your writing and your **reader's** needs. For example, a committee **report** about possible locations for a new production facility might end with a recommendation. A particularly lengthy document often concludes with a summary of its main points. Study the following examples.

RECOMMENDATION	These results indicate that you need to alter your testing procedure to eliminate the impurities we found in specimens A through E.
JUDGMENT	Although our estimate calls for a substantially higher budget than in the three previous years, we believe it is reasonable given our planned expansion.
PREDICTION	Although I have exceeded my original estimate for equipment ($60,000) by $6,900, I have reduced my original labor estimate ($180,000) by $10,500; therefore, I will easily stay within the limits of my original bid. In addition, I see no difficulty in having the arena finished for the December 23 holiday program.
SUMMARY	As this letter has indicated, we would attract more recent graduates if we did the following:

1. Establish a Web site where students can register and submit online résumés
2. Increase our advertising in local student newspapers
3. Expand our local co-op program
4. Send a representative to career fairs at local colleges
5. Invite local college instructors to teach in-house courses here at the facility

The concluding statement may merely present ideas for consideration, call for action, or deliberately provoke thought.

IDEAS FOR CONSIDERATION	The new prices become effective the first of the year. Price adjustments are routine for

C

Finding

Finding

Finding

Implication
of finding

Implication
of finding

CONCLUSION

The results from this study show that information on the toxicity of sites does influence housing prices. It appears that information from sources including the Environmental Protection Agency (EPA) and local community groups does impact sales prices; thus at least in this case the EPA announcements seem to provide some additional information to the residents. Official announcements that sites will be cleaned up are not necessarily believed by residents, or the agency may not be perceived as able to eliminate the externality effectively. It is possible that after the cleanup is completed and residents have an opportunity to assess the success of the process, prices would respond. Regardless, studies that have looked at prices for brief periods before and after official announcements have potentially missed the source of the information as well as the timing of the movement in house prices.

The results from this study could be used to provide dollar estimates of the costs of a toxic waste site to nearby homeowners from the damage to house values. It appears that the EPA does contribute to the market adjustment process during the announcement phase, and that perhaps there is a role for federal agencies in disseminating information in the siting of facilities such as incinerators. However, it does not appear that the EPA affects the market process in the cleaning phases. If prices do respond to cleanup efforts in the future, benefits would give government officials at all levels a more complete framework within which they can allocate scarce financial resources for cleaning up toxic sites in a more efficient and equitable way. It would also allow for the design of a more complete compensation program to help those who have been harmed by the existence of these sites.

From K. A. Kiel, "Measuring the Impact of the Discovery and Cleaning of Identified Hazardous Waste Sites on House Values," *Land Economics* 71 (1995): 428–35.

FIGURE C–3. Sample Conclusion

the company, but some of your customers will not consider them so. Please bear in mind the needs of both your customers and the company as you implement these new prices.

CALL FOR ACTION   Send us a check for $250 now if you wish to keep your account active. If you have not responded to our previous letters because of some special hardship, I will be glad to work out a solution with you personally.

THOUGHT-	Can we continue to accept the losses incurred by
PROVOKING	inefficiency? Must we accept inefficiency as in-
STATEMENT	evitable? Or should we consider steps to control
	it now?

Be careful not to introduce a new topic when you conclude. A conclusion should always relate to and reinforce the ideas presented earlier in your writing. In a large document, the conclusion section both pulls together the results or findings and interprets them in the light of the study's purpose and the methods by which the study was conducted. The evidence for the findings is discussed in the body of the document, and the conclusions must grow out of the information discussed there. Moreover, the conclusions must be consistent with what the **introduction** promised the report would examine (its purpose) and how it would do so (its method). If the introduction stated that the report's objective was to assess the market for a new product, the conclusion should not discuss the reorganization needed to produce the product. The conclusion in Figure C–3 on page 124 is from an article assessing the impact of hazardous waste sites on housing values.

For guidance about the location of the conclusions section in a report, see **formal reports.** For letter and other short closings, see **correspondence** and entries on specific types of documents.

conjunctions

A conjunction connects words, **phrases,** or **clauses** and can also indicate the relationship between the elements it connects.

A *coordinating conjunction* joins two sentence elements that have identical functions. The coordinating conjunctions are *and, but, or, for, nor, yet,* and *so.*

- Nature *and* technology are only two conditions that affect petroleum operations around the world. [joins two **nouns**]

- To hear *and* to listen are two different things. [joins two phrases]

- I would like to include the test results, *but* that would make the report too long. [joins two clauses]

Correlative conjunctions are used in pairs. The correlative conjunctions are *either . . . or, neither . . . nor, not only . . . but also, both . . . and,* and *whether . . . or.*

C

- The inspector will arrive *either* on Wednesday *or* on Thursday.

A *subordinating conjunction* connects sentence elements of different weights, normally independent and dependent clauses. The most frequently used subordinating conjunctions are *so, although, after, because, if, where, than, since, as, unless, before, that, though, when,* and *whereas.*

- I left the office *after* I had finished writing the report.

A *conjunctive adverb* is an **adverb** that has the force of a conjunction because it joins two independent clauses. The most common conjunctive adverbs are *however, moreover, therefore, further, then, consequently, besides, accordingly, also,* and *thus.*

- The engine performed well in the laboratory; *however,* it failed under road conditions.

Coordinating conjunctions in the titles of books, articles, plays, and movies should not be capitalized unless they are the first or last word in the title. (See also **capital letters.**)

- The play *Romeo and Juliet* was written by William Shakespeare.

Occasionally, a conjunction may begin a sentence; in fact, conjunctions can be strong transitional words and at times can provide **emphasis.** (See also **transition.**)

- I realize that the project is more difficult than expected and that you also have encountered personnel problems. *But* we must meet our deadline.

---

conjunctive adverbs (*see* adverbs)

---

connected with / in connection with

Connected with and *in connection with* are wordy and often vague phrases that usually can be replaced or shortened.

- He is ~~connected with~~ the TFT Corporation.
  employed by

- The fringe benefits ~~in connection~~ with the job are quite good.

connotation / denotation

The *denotation* of a word is its dictionary definition. The *connotations* of a word are its emotional associations. For example, *cheap* and *frugal* both refer to a reluctance to spend money, but they have different connotations, as the following sentences show.

- Her company was *cheap*. [negative connotation]
- Her company was *frugal*. [positive connotation]

Clear writing requires words with both the most accurate denotations and the most appropriate connotations. (See also **defining terms.**)

consensus

Because *consensus* normally means "harmony of opinion," the phrases *consensus of opinion* and *general consensus* are redundant. The word *consensus* can be used to refer only to a group, never to one or two people.

- The consensus ~~of opinion~~ of the members was that the committee should change its name.
- The ~~general~~ consensus of the members was that the committee should change its name.

continual / continuous

Continual means "happening over and over" or "frequently repeated."

- Writing well requires *continual* practice.

Continuous means "occurring without interruption" or "unbroken."

- The *continuous* roar of the engines was deafening.

contractions (ESL)

A contraction is a shortened spelling of a word or phrase with an **apostrophe** substituting for the missing letter or letters.

- cannot   →   can't
- will not   →   won't
- have not   →   haven't
- it is   →   it's

Contractions are often used in speech but should rarely be used in reports, formal letters, and most on-the-job writing. (See also **technical writing style.**)

coordinating conjunctions (*see* conjunctions)

copyright (ESL)

Copyright is a grant of limited exclusive control by an author of his, her, or (in the case of a corporation or organization) its work. The author, whether individual or corporate, retains the right to benefit from the work created. That grant gives authors (or others to whom they transfer ownership of copyright) control over the reproduction and dissemination of their works. A copyrighted work or a portion of the work may not be indiscriminately reproduced unless the copyright owner gives permission, usually in exchange for royalties or other compensation, or unless the fair use provision of copyright law applies. For works created after January 1978, copyright lasts for the life of the work's author plus 70 years. Where the "author" is a corporation or a publisher, copyright lasts 95 years from the year of its first publication or 120 years from the year of its creation, whichever is shorter. After the copyright expires, the work goes into the public domain, that is, it is no longer copyrighted.

The copyright law provides a "fair use" provision that allows teachers, librarians, reviewers, and others to reproduce a portion of copyrighted materials for educational and certain other purposes without notifying or compensating the copyright owners. The law refers to such purposes as "criticism, comment, news reporting, teaching (including multiple copies for classroom use), scholarship, or research." The law includes guidelines for the fair use of copyrighted materials in the classroom. Those guidelines set three standards: brevity, spontaneity, and cumulative effect. How a copying situation adheres to those standards determines whether the instructor needs permission to use multiple copies. Although the law does not exactly define fair use, it sets out four criteria for determining whether a given use is fair:

- the purpose and nature of the use
- the amount of material used in relation to the copyrighted work as a whole
- the nature of the copyrighted work
- the effect of the use on the potential market for or value of the copyrighted work

Publications that explain the copyright law in detail are available from the Copyright Office, Library of Congress, Washington, DC 20559, or at their Web site, (http://lcweb.loc.gov/copyright/). The Library of Congress Copyright Web site provides links to other informative sites, including one that gives guidelines for locating U.S. copyright holders: (www.lib.utexas.edu/Libs/HRC/WATCH/locating.html).

Writer's Checklist: Copyright

If you use copyrighted materials in your written work or electronic documents posted on the **Internet,** be aware of the following guidelines.

☑ In general, give credit to any source from which material is taken, unless it is "boilerplate" or "common knowledge," as described in the entry **plagiarism.** (See also **documenting sources.**)

☑ A small amount of material from a copyrighted source may be used without permission or payment as long as the use satisfies the fair use criteria. The amount varies in proportion to the length of the original work. Several paragraphs may be acceptable from a book, whereas only a few lines or a stanza may be acceptable from a song or poem.

☑ Any work first published after March 1, 1989, receives copyright protection whether or not it bears a notice of copyright, although

Writer's Checklist: Copyright (continued)

all published materials generally contain such a notice, usually on the back of the title page. (For an example of a copyright notice, turn to the back of the title page of this book.) Works published before March 1, 1989, without a notice of copyright are in the public domain.

☑ All publications created by U.S. government agencies are in the public domain and thus may be used without requesting permission.

☑ Copyright law applies to electronic works just as it does to their print counterparts. Assume that material found on the Internet is copyrighted unless you are certain that it is not.

☑ If you plan to reproduce or further distribute copyrighted works posted on the Internet, you must obtain permission from the copyright holder, unless the fair use provision of copyright law applies to your intended use. To obtain permission to use copyrighted material, email your request directly to the site containing the information you wish to use. As with printed works, document the source of material (text, graphics, tables) obtained on the Internet.

correlative conjunctions (*see* conjunctions)

correspondence

The process of writing letters involves many of the same steps that go into most other documents, as described in "Five Steps to Successful Writing" that follows the preface. One important consideration in correspondence is the impression conveyed to **readers.** To convey the right impression—of yourself as well as of your company or organization—take particular care with both the **tone** and the **style** of your writing. (See also **technical writing style.**)

Audience: Tone and Style

To focus the relevance of any correspondence for the reader, identify your subject in the **opening.** Remember that your reader may not immediately recognize or see the importance of your topic — indeed, he or she may be preoccupied with some other business.

- Yesterday, I received your letter and the defective pager, number AJ 50172, that you described. I sent the pager to our quality-control department for tests.

    Carol Moore, our lead technician, reports that preliminary tests indicate . . .

Your closing should either let the reader know what he or she should do next or establish goodwill — or often both.

- Thanks again for the report, and let me know if you want me to send you a copy of the tests.

Because a closing is in a position of emphasis, be especially careful to avoid **clichés.** Of course, some commonly used closings are so precise, even though they are clichés, that they are hard to replace.

- Thank you for your help.

- If you have further questions, please let me know.

Do not use such a closing just because it is easy. Make your closing work for you. It may be helpful to provide prompts to which the reader can respond.

- If you would like further information, such as a copy of the questionnaire we used, you can email me at delgado@prn.aol.com.

Written directly to another person, correspondence is always more personal than **reports** or other forms of business and technical writing. To achieve a conversational and natural style, imagine your reader sitting across the desk from you as you write; write to the reader as if you were talking to him or her in person.

Take into account your reader's needs. Ask yourself, "How might I feel if I were the recipient of such a letter?" to gain some insight into your reader's needs and feelings, then tailor your message to fit those needs and feelings. Furthermore, each letter you write offers a chance to build goodwill for your business. Many companies spend large sums to promote a favorable public image. An impersonal and unfriendly letter to a customer can tarnish that image, but a thoughtful and sincere letter can greatly enhance it. Suppose, for example, you received a refund request from a customer who forgot to enclose the receipt with the request. In a letter to that customer, you might write the following:

C

- The sales receipt must be enclosed with the merchandise before we can process a refund.

If you consider how to keep the customer's goodwill, you might word the request this way:

- Please enclose the sales receipt with the merchandise so we can send your refund promptly.

You can put the reader's needs and interests at the center of the letter by writing from the reader's perspective. Often, although not always, doing so means using the words *you* and *your* rather than *we, our, I,* and *mine*—a technique called the **"you" viewpoint.** Consider the following revision, which is written with the "you" viewpoint.

- So you can receive your refund promptly, please enclose the sales receipt with the merchandise.

That version stresses the reader's benefit and interest. By emphasizing the reader's needs, the writer is more likely to accomplish the objective (see **purpose**): to get the reader to act.

Writer's Checklist: Building Goodwill

The following guidelines will help you achieve a tone that builds goodwill with the recipients of your correspondence.

☑ Be respectful, not demanding.

| DEMANDING | Submit your answer in one week. |
| RESPECTFUL | I would appreciate your answer within one week. |

☑ Be modest, not arrogant.

| ARROGANT | My report is thorough, and I'm sure you won't be able to continue without it. |
| MODEST | I have tried to be as thorough as possible in my report, and I hope you find it useful. |

☑ Be polite, not sarcastic.

| SARCASTIC | I just received the shipment we ordered six months ago. I'm sending it back—we can't use it now. Thanks! |
| POLITE | I am returning the shipment we ordered on March 12, 20--. Unfortunately, it arrived too late for us to be able to use it. |

Writer's Checklist: Building Goodwill (continued)

☑ Be positive and tactful, not negative and condescending.

NEGATIVE   Your complaint about our prices is way off target. Our prices are definitely not any higher than those of our competitors.

TACTFUL   Thank you for your suggestion concerning our prices. We believe, however, that our prices are competitive with, and in some cases are below, those of our competitors.

Good News and Bad News Patterns

Although the relative directness of correspondence may vary, it is generally more effective to present good news directly and bad news indirectly, especially if the stakes are high.* This principle is based on the fact that (1) a reader forms impressions and attitudes early in a letter and (2) as a writer you may want to subordinate the bad news to reasons that make the bad news understandable. Further, when you are writing **international correspondence,** you need to be aware that far more cultures in the world are indirect in business messages than direct.

Consider the thoughtlessness and direct rejection in Figure C–4 on page 134. Although the letter is concise and uses the pronouns *you* and *your,* the writer apparently did not consider how the recipient would feel as she read the letter. The letter is, in short, rude: Its pattern is (1) bad news, (2) explanation, (3) close. (See also **refusal letters.**)

A better general pattern for a "bad news" letter is (1) an opening that provides context, (2) an explanation, (3) the bad news, and (4) goodwill. The opening introduces the subject and establishes a professional tone. The body provides an explanation by reviewing the facts that make the bad news understandable. Although bad news is never pleasant, information that either puts the bad news in perspective or makes the bad news seem reasonable maintains goodwill between the writer and the reader. The closing should establish or reestablish a positive relationship through goodwill or helpful information. Consider, for example, the rejection letter

---

*Gerald J. Alred, "'We Regret to Inform You': Toward a New Theory of Negative Messages," *Studies in Technical Communication*, ed. Brenda R. Sims (Denton: U of North Texas and NCTE, 1993), pp. 17–36.

Southtown Dental Center
3221 Ryan Road   San Diego, CA 92217
Phone: (714) 321-1579   Fax: (714) 321-1222
sdc@exec3.com

November 11, 20--

Ms. Barbara L. Mauer
157 Beach Drive
San Diego, CA 92113

Dear Ms. Mauer:

Your application for the position of records administrator at Southtown Dental
Center has been rejected. We have found someone more qualified than you.

Sincerely,

Mary Hernandez

Mary Hernandez
Office Manager

FIGURE C–4. A Poor "Bad News" Letter

shown in Figure C–5 on page 135. The revised letter carries the
same disappointing news as the first one, but the writer is careful to
thank the reader for her time and effort, explain why she was not
accepted for the job, and offer her encouragement in finding a position in another office.

Presenting good news is, of course, easier. Present good news
early—at the beginning of the letter, if at all possible. The pattern
for "good news" letters should be (1) good news, (2) explanation of
facts, (3) goodwill. By presenting the good news first, you increase
the likelihood that the reader will pay careful attention to details,

C

> **Southtown Dental Center**
> 3221 Ryan Road   San Diego, CA 92217
> Phone: (714) 321-1579   Fax: (714) 321-1222
> sdc@exec3.com

November 11, 20--

Ms. Barbara L. Mauer
157 Beach Drive
San Diego, CA 92113

Dear Ms. Mauer:

Context

Thank you for your time and effort in applying for the position of records administrator at Southtown Dental Center.

Explanation leading to bad news

Because we need someone who can assume the duties here with a minimum of training, we have selected an applicant with over ten years of experience.

Goodwill

I am sure that with your excellent college record you will find a position in another office.

Sincerely,

Mary Hernandez

Mary Hernandez
Office Manager

FIGURE C–5. A Courteous "Bad News" Letter

and you achieve goodwill from the start. Figure C–6 on page 136 is a good example of a "good news" letter.

Writing Style and Accuracy

Letter-writing style varies from informal, as in a letter to a close associate, to formal (or restrained), as in a letter to someone you do not know.

Southtown Dental Center
3221 Ryan Road   San Diego, CA 92217
Phone: (714) 321-1579   Fax: (714) 321-1222
sdc@exec3.com

November 11, 20--

Ms. Barbara L. Mauer
157 Beach Drive
San Diego, CA 92113

Dear Ms. Mauer:

Good news — Please accept our offer of the position of records administrator at Southtown Dental Center.

Explanation — If the terms we discussed in the interview are acceptable to you, please come in at 9:30 a.m. on November 15. At that time, we will ask you to complete our personnel form, in addition to . . .

Goodwill — I, as well as the others in the office, look forward to working with you. Everyone was favorably impressed with you during your interview.

Sincerely,

Mary Hernandez

Mary Hernandez
Office Manager

FIGURE C–6. A "Good News" Letter

INFORMAL It worked! The new process is better than we had dreamed.

RESTRAINED You will be pleased to know that the new process is more effective than we had expected.

You probably will find yourself using the restrained style more frequently than the informal one. Remember that an overdone attempt to sound casual or friendly can sound insincere. On the other hand, do not adopt so formal a style that your letters read like legal contracts; that type of writing appears wordy, pompous, and affected. (See **affectation.**)

AFFECTED   In response to your query, I wish to state that we no longer have an original copy of the brochure requested. Be advised that a photographic copy is enclosed herewith. Address further correspondence to this office for assistance as required.

IMPROVED   Because we are out of original copies of our brochure, I am sending you a copy. If I can be of further help, please let me know.

The improved version is more concise, and **conciseness** is always important. However, do not be so concise that you become blunt. Responding to a written request that is vague with "Your request was unclear" or "I don't understand" could easily offend your reader. What you need to do is ask for more information and establish goodwill to encourage your reader to provide the information.

- I will need more information before I can answer your request. Specifically, can you give me the title and the date of the report you are looking for?

Although this version is a bit longer, it promotes goodwill and will elicit a faster response. (See also **telegraphic style.**)

A letter is a written record, so it must be accurate. Facts, figures, and dates that are incorrect or misleading can cost time, money, and goodwill. Check for appropriate punctuation. Remember, when you sign a letter, you are accepting responsibility for it. Therefore, allow yourself time to review correspondence carefully before you mail or email it.

Formats and Parts

Although **memo** and **email** formats are often established by your organization or the software you are using, letter format may vary. Word-processing templates provide basic forms, but they may not give you the specific dimensions and spacing you need. The following guidelines should help you include the parts of a letter that will be useful to you as well as achieve a professional appearance.

C

Center the letter on the page vertically and horizontally. Although one-inch margins are the default standard in many word-processing programs, it is more important to establish a "picture frame" of blank space surrounding the page of text. When you use organizational letterhead stationery, consider the bottom of the letterhead as the top edge of the paper. The left and right margins should be approximately equal; to give a fuller appearance to very short letters, increase both margins to about an inch and a half. Use your computer's "full-page" or "print preview" feature to help you achieve proportion.

The two most common formats of business letters are the full-block style, shown in Figure C–7 on page 139, and the modified-block style, shown in Figure C–8 on page 140. All other letter styles are variations of those two styles. In the full-block style, which is suitable only with organizational letterhead, the entire letter is aligned at the left margin. In the modified-block style, the return address, the date, and the complimentary closing begin at the center of the page, and the other elements are aligned at the left margin.

If your employer recommends or requires a particular format, use it. Otherwise, follow the guidelines provided here and in Figures C–7 and C–8.

Heading.   Place your full address and the date in the heading. Because your name appears at the end of the letter, it need not be included in the heading. Spell out words like *street, avenue, first,* and *west* rather than abbreviating them. You may either spell out the name of the state in full or use the standard Postal Service abbreviation. The date usually goes directly beneath the last line of the return address. Do not abbreviate the name of the month. If you are using company letterhead that gives the address, enter only the date three lines below the last line of printed copy, or two inches from the top of the page.

Inside Address.   Include the recipient's full name, title, and address in the inside address, two to six lines below the date, depending on the length of the letter. The inside address should be aligned with the left margin, and the left margin should be at least one inch wide.

Salutation.   Place the salutation, or greeting, two lines below the inside address and aligned with the left margin. In most business letters, the salutation contains the recipient's personal title (*Mr., Ms., Dr.,* and so on) and last name, followed by a colon. If you are

C

Letterhead	**EVANS & ASSOCIATES** 520 Niagara Street Braintree, MA 02184 Phone: (781) 787-1175 Fax: (781) 787-1213 Email: 92000.121@CompuServe.com
Date	May 15, 20--
Inside address	Mr. George W. Nagel Director of Operations Boston Transit Authority 57 West City Avenue Boston, MA 02210
Salutation	Dear Mr. Nagel:
	Enclosed is our final report evaluating the safety measures for the Boston Intercity Transit System.
Body	We believe that the report covers the issues you raised and that it is self-explanatory. However, if you have any further questions, we would be happy to meet with you at your convenience.
	We would also like to express our appreciation to Mr. L. K. Sullivan of your committee for his generous help during our trips to Boston.
Complimen tary close	Sincerely,
Signature	*Carolyn Brown*
Typed name Title	Carolyn Brown, Ph.D. Director of Research
Additional information	CB/ls bt515.doc Enclosure: Final Safety Report cc: ITS Safety Committee Members

FIGURE C–7. Full-Block Style (with Letterhead)

C

center

Heading
from center to
right

3814 Oak Lane
Dedham, MA 02180
December 8, 20--

Inside
address

Dr. Carolyn Brown
Director of Research
Evans & Associates
520 Niagara Street
Braintree, MA 02184

Salutation

Dear Dr. Brown:

Body

Thank you very much for allowing me to tour your testing facilities. The information I gained from the tour will be of
great help to me in preparing the report for my class at Marshall Institute. The tour has also given me some insight into
the work I may eventually do as a laboratory technician.

I especially appreciated the time and effort Vikram Singh
spent in showing me your facilities. His comments and advice were most helpful.

Again, thank you.

Complimentary
close aligned
with heading

Sincerely,

Leslie Warden

Signature

Typed name

Leslie Warden

center

FIGURE C–8. Modified-Block Style (without Letterhead)

on a first-name basis with the recipient, use only the first name in the salutation.

Address women as *Ms.*, unless they have expressed a preference for *Miss* or *Mrs.* However, other titles (such as *Dr., Senator, Major*) may take precedence over *Ms.* If you do not know whether the recipient is a man or a woman, use a title appropriate to the context of the letter. The following are examples of the kinds of titles you may find suitable:

- Dear Customer: [letter from a store]
- Dear Homeowner: [letter from insurance agent seeking business]
- Dear Parts Manager: [letter to an auto-parts dealer]

In the past, correspondence to large companies or organizations was customarily addressed to "Gentlemen." Today, writers who do not know the name or title of the recipient often address the letter to an appropriate department or identify the subject in a subject line and use no salutation.

- National Business Systems
  501 West National Avenue
  Minneapolis, MN 55407

  Attention: Customer Relations Department

  I am returning three pagers that failed to operate. . . .

- National Business Systems
  501 West National Avenue
  Minneapolis, MN 55407

  Subject: Defective Parts for SL-100 Pagers

  I am returning three pagers that failed to operate. . . .

When a person's first name could be feminine or masculine, one solution is to use both the first and last names in the salutation.

- Dear Pat Smith:

Avoid "To Whom It May Concern," which is impersonal and dated.

Body. The body of the letter should begin two lines below the salutation (or any element that precedes the body, such as a subject or attention line). Single-space within paragraphs and

C

double-space between paragraphs. To provide a fuller appearance to a very short letter, you can double-space throughout and indicate paragraphs by indenting the first line of each paragraph five spaces from the left, or you can adjust the margins.

Complimentary Closing.   Type the complimentary closing two spaces below the body. Use a standard expression like *Yours truly, Sincerely,* or *Sincerely yours.* (If the recipient is a friend as well as an associate, you can use a less formal complimentary closing, such as *Best wishes, Best regards,* or simply *Best.*) Capitalize only the initial letter of the first word and follow the expression with a comma. Place your full name four lines below, aligned with the closing. On the next line, include your professional title, if it is appropriate to do so. Sign the letter in the space between the complimentary closing and your name. If you are writing to someone with whom you are on a first-name basis, it is acceptable to sign only your given name; otherwise, sign your full name.

Second Page.   If a letter requires a second page, always carry at least two lines of the body text over to that page. Use plain (non-letterhead) paper of quality equivalent to that of the letterhead stationery for the second page. It should have a header with the recipient's name, the page number, and the date. The header can go in the upper left corner or across the page, as shown in Figure C–9 on page 143.

Additional Information.   Business letters sometimes require additional information: initials of the person keying the letter (other than the writer), an enclosure notation, or a notation that a copy of the letter is being sent to one or more other persons. Place such information at the left margin, two spaces below the typed name and title of the writer in a long letter, four spaces below in a short letter.

If the person keying the letter is not the writer, show the letter writer's initials in capital letters and the initials of the person keying the letter in lowercase letters, as shown in Figure C–7 on page 139. When the writer is also the person keying the letter, no initials are needed.

File name notation is sometimes included to indicate where the document is stored in the computer memory (for example, *bt515.doc,* as shown in Figure C–7).

Enclosure notations indicate that the writer is sending material along with the letter (an invoice, an article, and so on). Enclosure

C

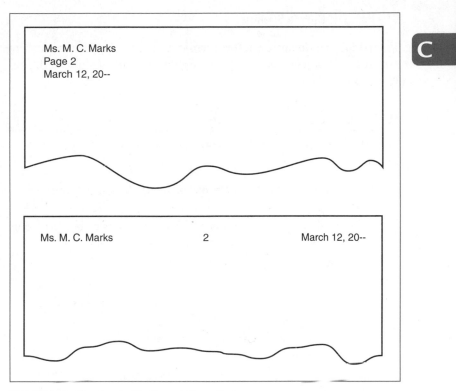

Ms. M. C. Marks
Page 2
March 12, 20--

Ms. M. C. Marks   2   March 12, 20--

FIGURE C-9. Headers for the Second Page of a Letter

notations can take several forms; choose the form that seems most helpful to your readers.

- Enclosure: Final Safety Report

- Enclosures (2)

- Enc. or Encs.

Enclosure notations are included in long, formal letters or in any letters where the enclosed items would not be obvious to the reader. Remember, though, that an enclosure notation cannot stand alone. You must mention the enclosed material in t'he body of the letter.

A *copy notation (cc:)* tells the reader that a copy of the letter is being sent to the named recipients (see Figure C–7 on page 139). Use a *blind copy notation (bcc:)* when you do not want the addressee to know that a copy is being sent to someone else. A blind copy notation appears only on the copy, not on the original.

C

- bcc: Dr. Brenda Shelton

For additional details on letter format and design, you may wish to consult a guide such as *The Gregg Reference Manual* (ninth edition) by William A. Sabin.

Writer's Checklist: Correspondence

The following guidelines will help you to write clear, well-organized correspondence.

☑ Establish your purpose, analyze your audience, and establish your **scope.**

☑ Prepare an outline, even if it is only a list of points to be covered in the order you want to cover them (see **outlining**).

☑ Write the first draft (see **writing a draft**).

☑ Use the appropriate or prescribed format.

☑ Allow for a "cooling" period, especially before you write correspondence that addresses a problem.

☑ Revise the draft, checking for key problems in **clarity** and **coherence.** (See **revision.**)

☑ Check for accuracy: A letter is a written record, and facts, figures, and dates that are incorrect or misleading can cost time, money, and goodwill.

☑ Check for appropriate **punctuation** and use good **proofreading** techniques.

☑ Allow yourself time to review the letter before you mail or email it.

☑ Remember that when you sign a letter, initial a memo, or send an email, you are accepting responsibility for it.

For information on specific types of correspondence, see the Topical Key.

cover letters

A *cover letter* identifies the item being sent, the person to whom it is being sent, and the reason for its being sent. A cover letter provides a permanent record of the transmittal for both the writer and the **reader.**

C

ECOLOGY SYSTEMS AND SERVICES

39 Beacon Street
Boston, Massachusetts 02106
(617) 351-1212
Fax: (617) 351-2121

May 23, 20--

Mario Espinoza, Chief Engineer
Louisiana Chemical Products
3452 River View Road
Baton Rouge, LA 70893

Dear Mr. Espinoza:

Enclosed is the final report on our installation of pollution control
equipment at Eastern Chemical Company, which we send with
Eastern's permission. Please call me collect (ext. 1206) or send
me an email message at the address below if I can answer any
questions.

Sincorcly,

Susan Wong

Susan Wong, Ph.D.
Technical Services Manager
swong.ecology@omni.com

SW/ls
Enclosure: Report

FIGURE C–10. Short Cover Letter

C

WATERFORD PAPER PRODUCTS

P.O. Box 413
WATERFORD, WI 53474

Phone: (414) 738-2191
Fax: (414) 738-9122

January 16, 20--

Mr. Roger Hammersmith
Ecology Systems, Inc.
1015 Clarke Street
Chicago, IL 60615

Dear Mr. Hammersmith:

Enclosed is the report estimating our power consumption for the year as requested by John Brenan, Vice President, on September 4.

The report is a result of several meetings with the Manager of Plant Operations and her staff and an extensive survey of all our employees. The survey was delayed by the transfer of key staff in Building "A." We believe, however, that the report will provide the information you need to furnish us with a cost estimate for the installation of your Mark II Energy Saving System.

We would like to thank Diana Biel of ESI for her assistance in preparing the survey. If you need any more information, please let me know.

Sincerely,

James G. Evans

James G. Evans
New Projects Office
jge.waterford@omni.com

Enclosure

FIGURE C–11. Long Cover Letter

Keep your remarks brief in a cover letter. Your **opening** should explain what is being sent and why. In an optional second paragraph, you might include a summary of the information you are sending. A letter accompanying a **proposal,** for example, might point out any sections in the proposal of particular interest to the reader. The letter could then go on to present a key point or two explaining why the writer's firm is the best one to do the job. This paragraph could also mention the conditions under which the material was prepared, such as limitations of time or budget. The closing paragraph should contain acknowledgments, offer additional assistance, or express the hope that the material will fulfill its purpose.

The example in Figure C–10 on page 145 is brief and to the point. Figure C–11 on page 146 is a bit more detailed because it touches on the manner in which the information was gathered. (See also **correspondence.**)

credible / creditable

Something is *credible* if it is believable.

* The statistics in this report are *credible*.

Something is *creditable* if it is worthy of praise or credit.

* The lead engineer did a *creditable* job.

criterion / criteria

Criterion means "an established standard for judging or testing." *Criteria* and *criterions* are both acceptable plural forms of *criterion,* but *criteria* is generally preferred.

* In evaluating this job, we must use three *criteria*. The most important *criterion* is quality of workmanship.

critique

A *critique* is a written or oral evaluation of something. Avoid using *critique* as a **verb** meaning "criticize."

* Please critique his job description.
  (prepare a / of)

D

dangling modifiers   (ESL)

A phrase that does not clearly and logically refer to the correct **noun** or **pronoun** is called a *dangling modifier*. A dangling modifier usually appears at the beginning of a sentence as an introductory phrase.

> **DANGLING** *While eating lunch in the cafeteria,* the computer malfunctioned.
> [The sentence neglects to mention *who* was eating lunch in the cafeteria.]
>
> **CORRECT** While the *programmer* was eating lunch in the cafeteria, the computer malfunctioned.

Dangling modifiers can appear at the end of the sentence as well.

> **DANGLING** The program gains in efficiency *by eliminating the superfluous instructions.*
> [An action is stated, but no one is identified who could perform the stated action.]
>
> **CORRECT** The program gains in efficiency *when you* eliminate the superfluous instructions.

To test whether a **phrase** is a dangling modifier, turn it into a **clause** with a subject and a **verb.** If the expanded phrase and the independent clause do not have the same subject, the phrase is dangling.

> **DANGLING** After finishing the research, the paper was easy to write.
> [The implied subject of the phrase is *I,* and the subject of the independent clause is *paper;* therefore, the sentence contains a dangling modifier.]
>
> **CORRECT** After finishing the research, *I found that* the paper was easy to write.
> [Changing the subject of the independent clause to

agree with the implied subject of the introductory phrase eliminates the dangling modifier.]

CORRECT   After I finished the research, the paper was easy to write.
[This version changes the phrase to a dependent clause with an explicit subject.]

D

dashes ⓔˢˡ

The dash (—) can perform all the duties of punctuation: linking, separating, and enclosing. It is an emphatic mark that is easily overused. Use the dash cautiously to indicate more informality, **emphasis,** or abruptness than the other punctuation marks would show.

A dash can emphasize a sharp turn in thought.

- The project will end January 15 — unless the company provides additional funds.

A dash can indicate an emphatic pause.

- The job will be done — after we are under contract.

Sometimes, to emphasize contrast, a dash is used with *but*.

- We may have produced work more quickly — but the result was not as good.

A dash can be used before a final summarizing statement or before repetition that has the effect of an afterthought.

- It was hot near the ovens — steaming hot.

Such a statement may also complete the meaning of the clause preceding the dash.

- We try to speak as we write — or so we believe.

A dash can be used to set off an explanatory or appositive series.

- Three of the applicants — John Evans, Rosalita Fontiana, and Kyong-Shik Choi — seem well qualified for the job.

Dashes set off parenthetical elements more sharply and emphatically than **commas.** Unlike dashes, **parentheses** tend to reduce the importance of what they enclose. Compare the following sentences:

D

- Only one person — the president — can authorize such activity.
- Only one person, the president, can authorize such activity.
- Only one person (the president) can authorize such activity.

The first word after a dash is never capitalized unless it is a proper **noun.**

data / datum

In much informal writing, *data* is considered a collective singular noun. In formal scientific and scholarly writing, however, *data* is generally used as a plural, with *datum* as the singular form. Base your decision on whether your reader should consider the data as a single collection or as a group of individual facts. Whatever you decide, be sure that your pronouns and verbs agree in **number** with the selected **usage.**

- The *data are* voluminous. *They indicate* a link between smoking and lung cancer. [formal]

- The *data is* now ready for evaluation. *It is* in the mail. [less formal]

(See also **English, varieties of.**)

dates   ESL

In the United States, dates are generally indicated by the month, day, and year, with a comma separating the figures.

- October 26, 2013

The day-month-year system used in other parts of the world and by the U.S. military does not require commas.

- 26 October 2013

A date can be written with or without a comma following the year if the date is in the month-day-year format.

- October 26, 2013, is the payoff date.
- October 26, 2013 is the payoff date.

Use the strictly numerical form for dates (10/26/99) sparingly and never in business letters or formal documents, because it is not always immediately clear. In fact, the numerical form may be confusing in **international correspondence** because many countries write the day before the month and year. In England, for example, 1/11/99 means 1 November 1999; in the United States, it means January 11, 1999. Writing out the name of the month makes the entire date immediately clear to all readers.

Centuries

Confusion often occurs because the spelled-out numbers of centuries do not correspond to the numeral designations.

- The twentieth century is the 1900s (1900–1999).

When the century is written as a **noun,** do not use a hyphen.

- The sixteenth century produced great literature.

When the century is written as an **adjective,** however, use a hyphen.

- Twentieth-century technology relies on clear communications.

decided / decisive

Something that is *decided* is "clear-cut," "without doubt," "unmistakable." Something that is *decisive* is "conclusive" or "has the power to settle a dispute."

- The executive committee's *decisive* action gave our firm a *decided* advantage over the competition.

decreasing-order-of-importance method of development

Decreasing order of importance is a **method of development** in which the writer's major points are arranged in descending order of importance. It begins with the most important fact or point, then goes to the next most important, and so on, ending with the least

D

important. It is especially appropriate for a report addressed to a busy decision maker, who may be able to reach a decision after considering only the most important points, or for a report addressed to various readers, some of whom may be interested in only the major points and some who may need all the information.

The advantages of this method of development are that (1) it gets the reader's attention immediately by presenting the most important point first, (2) it makes a strong initial impression, and (3) it ensures that even the most hurried reader will not miss the most important point.

The example shown in Figure D–1 uses decreasing order of importance.

Memorandum

To: Tawana Shaw, Director, Human Resources Department
From: Frank W. Nemitz, Manager, Product Marketing *FWN*
Date: November 13, 20--
Subject: Selection of Manager of the Technical Writing Department

The most qualified candidate for Manager of the Technical Writing Department is Mildred Bryant, who is currently acting manager of the department. In her 12 years in the department, Ms. Bryant has gained wide experience in all facets of its operations. She has maintained a consistently high production record and has demonstrated the skills and knowledge required for the supervisory duties she is now handling in an acting capacity. Another consideration is that she has continually been rated "outstanding" in all categories in her job-performance appraisals. However, her supervisory experience is limited to her present three-month tenure as acting manager of the department, and she lacks the college degree required by the job description.

Michael Bastick, Graphics Coordinator, my second choice, also has strong potential for the position. An able administrator, he has been with the company for seven years. Further, he is enrolled in a management-training course at Metro State University's downtown campus. He is ranked second because he lacks supervisory experience and because his most recent work as a technical writer has been limited.

Jane Fine, my third-ranking candidate, has shown herself to be an exceptionally skilled writer in her three years with the Public Relations Department. Despite her obvious potential, she does not yet have the breadth of experience in technical writing that is required to manage the Technical Writing Department. Jane Fine also lacks on-the-job supervisory experience.

FIGURE D–1. Decreasing Order of Importance

de facto / de jure

De facto means that something exists or is a fact and therefore is accepted for practical purposes. *De jure* means that something legally exists.

- The law states that no signs should be erected along Highway 127. Store owners have disregarded that law, and many signs exist along Highway 127. The presence of the signs along Highway 127 is *de facto* but not *de jure*—their presence is "a fact," but it is not "lawful."

Limit the use of Latin and legal terms since they can easily become an **affectation.**

defective / deficient

If something is *defective,* it is faulty.

- The wiring was *defective.*

If something is *deficient,* it is lacking or is incomplete in an essential component.

- The compound was found to be *deficient* in calcium.

defining terms

Good on-the-job writing ensures that readers understand key terms and concepts in what they are reading. Writers must be careful to define terms that readers may not understand.

Terms can be defined either formally or informally, depending on your purpose and your readers. A *formal definition* is a form of classification. A term is defined by placing it in a category and then identifying the features that distinguish it from other members of the same category.

TERM	CATEGORY	DISTINGUISHING FEATURES
A *spoon* is	an eating utensil	that consists of a small, shallow bowl on the end of a handle.

D

TERM	CATEGORY	DISTINGUISHING FEATURES
An *auction* is	a public sale	in which property passes to the highest bidder through successively increased offers.
An *annual* is	a plant	that completes its life cycle, from seed to natural death, in one growing season.

An *informal definition* explains a term by giving a more familiar word or phrase as a **synonym.**

- An *invoice* is a *bill.*

- Plants have a *symbiotic,* or *mutually beneficial,* relationship with certain kinds of bacteria.

State definitions positively; focus on what the term *is* rather than on what it is not.

NEGATIVE   In a legal transaction, *real property* is not personal property.
[This definition does not tell the reader exactly what *real property* is.]

POSITIVE   *Real property* is legal terminology for the right or interest a person has in land and the permanent structures on that land.

For a discussion of when negative definitions are appropriate, see **definition method of development.**

Avoid circular definitions, which merely restate the term to be defined and therefore fail to clarify it.

CIRCULAR   *Spontaneous combustion* is fire that begins spontaneously.

REVISED   *Spontaneous combustion* is the self-ignition of a flammable material through a chemical reaction.

In addition, avoid "is when" and "is where" definitions. Such definitions fail to include the category and are too indirect.

- A contract is ~~when two or more people agree to something.~~ *a binding agreement between two or more people.*

- A day-care center ~~is where working parents can leave their~~ *provides care during the workday for children of working parents.* ~~preschool children during the day.~~

Even informal writing occasionally requires terms used in a sense that your readers may not understand; such terms should always be defined.

- In these specifications, the term *safety can* refers to a five-gallon container having a spring-closing spout cover designed to relieve internal pressure.

D

definite / definitive

Definite and *definitive* both apply to what is precisely defined, but definitive more often refers to what is complete and authoritative.

- When the committee took a *definite* stand, the president made it the *definitive* company policy.

definition method of development

DIRECTORY
Overview   155
Extended Definition   156
Definition by Analogy   156
Definition by Cause   156
Definition by Components   157
Definition by Exploration of Origin   157
Negative Definition   158

Overview

To define something is to identify precisely its fundamental qualities. In technical writing, clear and accurate definitions are critical. Sometimes simple, dictionarylike definitions suffice (see **defining terms**), but at other times definitions need to be expanded with additional details, examples, comparisons, or other explanatory devices. The most common explanatory devices are (1) extended definition, (2) definition by analogy, (3) definition by cause, (4) definition by components, (5) definition by exploration of origin, and (6) negative definition.

Extended Definition

When more than a phrase or a sentence or two is needed to explain an idea, use an extended definition, which explores a number of qualities of the item being defined. How an extended definition is developed depends on your readers' needs and on the complexity of the subject. Readers familiar with a topic might be able to handle a long, fairly complex definition, whereas less familiar readers might require simpler language and more basic information.

The easiest way to give an extended definition is with specific examples. Listing examples gives readers easy-to-picture details that help them see and thus understand the term being defined.

- Form, which is the shape of landscape features, can best be represented by both small-scale features, such as *trees* and *shrubs,* and by large-scale elements, such as *mountains* and *mountain ranges.*

Definition by Analogy

Another useful way to define a difficult concept, especially when you are writing for nonspecialists, is to use an **analogy** to link the unfamiliar concept with a simpler or more familiar one. (An analogy can help the reader understand an unfamiliar term by showing its similarities with a more familiar term.) In the following description of radio waves in terms of their length (long) and frequency (low), notice how the writer develops an analogy to show why a low frequency is advantageous.

- The low frequency makes it relatively easy to produce a wave having virtually all its power concentrated at one frequency. Think, for example, of a group of people lost in a forest. If they hear sounds of a search party in the distance, they all will begin to shout for help in different directions. Not a very efficient process, is it? But suppose all the energy that went into the production of this noise could be concentrated into a single shout or whistle. Clearly the chances that the group will be found would be much greater.

Definition by Cause

Some terms are best defined by an explanation of their causes. In the following example from a professional journal, a nurse describes an apparatus used to monitor blood pressure in severely ill patients. Called an indwelling catheter, the device displays blood-pressure readings on an oscilloscope and on a numbered scale. Users of the device, the writer explains, must understand what a *dampened wave form* is.

D

- The dampened wave form, the smoothing out or flattening of the pressure wave form on the oscilloscope, is usually caused by an obstruction that prevents blood pressure from being freely transmitted to the monitor. The obstruction may be a small clot or bit of fibrin at the catheter tip. More likely, the catheter tip has become positioned against the artery wall and is preventing the blood from flowing freely.

Definition by Components

Sometimes a formal definition of a concept can be made simpler by breaking the concept into its component parts. In the following example, the formal definition of *fire* is given in the first paragraph, and the component parts are given in the second.

FORMAL DEFINITION   Fire is the visible heat energy released from the rapid oxidation of a fuel. A substance is "on fire" when the release of heat energy from the oxidation process reaches visible light levels.

COMPONENT PARTS   The classic fire triangle illustrates the elements necessary to create fire: *oxygen, heat,* and *burnable material (fuel)*. Air provides sufficient oxygen for combustion; the intensity of the heat needed to start a fire depends on the characteristics of the burnable material. A burnable substance is one that will sustain combustion after an initial application of heat to start the combustion.

Definition by Exploration of Origin

Under certain circumstances, the meaning of a term can be clarified and made easier to remember by an exploration of its origin. Medical terms, because of their sometimes unfamiliar Greek and Latin roots, benefit especially from an explanation of this type. Tracing the derivation of a word also can be useful when you want to explain why a word has favorable or unfavorable associations, particularly if your goal is to influence your reader's attitude toward an idea or an activity (see also **persuasion**).

D

- Efforts to influence legislation generally fall under the head of *lobbying,* a term that once referred to people who prowl the lobbies of houses of government, buttonholing lawmakers and trying to get them to take certain positions. Lobbying today is all of this, and much more, too. It is a respected—and necessary—activity. It tells the legislator which way the winds of public opinion are blowing, and it helps inform [legislators] of the implications of certain bills, debates, and resolutions [that they must face].
  —Bill Vogt, *How to Build a Better Outdoors*

Negative Definition

In some cases, it is useful to point out what something is not to clarify what it is. A negative definition is effective only when the reader is familiar with the item with which the defined item is contrasted. If you say *x* is not *y,* your readers must understand the meaning of *y* for the explanation to make sense. In a crane operator's manual, for instance, a negative definition is used to show that, for safety reasons, a hydraulic crane cannot be operated in the same manner as a lattice boom crane.

- A hydraulic crane is *not* like a lattice boom crane in one very important way. In most cases, the safe lifting capacity of a lattice boom crane is based on the *weight needed to tip the machine.* Therefore, operators of friction machines sometimes depend on signs that the machine might tip to warn them of impending danger. This practice is very dangerous with a hydraulic crane. . . .
  —*Operator's Manual* (Model W-180), Harnishfeger Corporation

demonstrative adjectives (*see* adjectives)

demonstrative pronouns (*see* pronouns)

dependent clauses (*see* clauses)

description

Effective description uses words to transfer a mental image from the writer's mind to the reader's. The key to effective description is the accurate presentation of details, even with relatively simple descriptions. The purchase order that follows is a typical example of simple descriptive writing. Notice that the order for infectious-waste disposal bags used in hospitals includes five specific descriptive details in addition to the part number.

D

Purchase Order

PART NO.	DESCRIPTION	QUANTITY
IW 8421	Infectious-waste bags, 12″ × 14″, heavy-gauge polyethylene, red double closures with self-sealing adhesive strips	5 boxes containing 200 bags per box

---

Complex descriptions, of course, involve more details. In describing a mechanical device, for example, it is best to describe the whole device before giving a detailed description of how each part works. The description should conclude with an explanation of how each part contributes to the functioning of the whole.

In descriptions intended for readers who are unfamiliar with the topic, details are crucial. A description for such readers benefits from showing or demonstrating (as opposed to telling), primarily through the use of images and details. You can also use analogy to explain unfamiliar concepts in terms of familiar ones. (See **figures of speech.**)

Illustrations can be powerful aids in descriptive writing, especially when they show details too intricate to explain completely in words. (For a discussion of how to incorporate visual material into text, see **illustrations** and **layout and design.**) The example in Figure D–2 on page 160 describes a mechanical assembly mechanism. Intended for the assembler/repairperson, the description includes an illustration of the mechanism.

Note that the description in Figure D–2 concentrates on the number of pieces—their sizes, shapes, and dimensions—and on their relationship to one another to perform their function. It also specifies the materials of which the hardware is made. Because the description is illustrated (with identifying labels) and is intended for technicians who have been trained on the equipment, it does not require the use of "bridging devices" to explain the unfamiliar in relation to the familiar. Even so, an important term is defined (*chad*), and crucial alignment dimensions are specified (± .003″).

D

Number of features and relative size differences noted

The Die Block Assembly consists of two machined block sections, eight Code Pins, and a Feed Pin. The larger section, called the Die Block, is fashioned of a hard, noncorrosive beryllium-copper alloy. It houses the eight Code Pins and the smaller Feed Pin in nine finely machined guide holes.

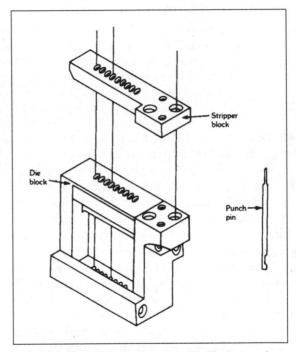

The guide holes in the upper part of the Die Block are made smaller to conform to the thinner tips of the Feed Pins. Extending over the top of the Die Block and secured to it at one end is a smaller, armlike block called the Stripper Block. The Stripper Block is made from hardened tool steel, and it also has been drilled through with nine finely machined guide holes. It is carefully fitted to the Die Block at the factory so that its holes will be precisely above those in the Die Block and so that the space left between the blocks will measure .015" ($\pm$ .003"). The residue from the operation, called chad, is pushed out through the top of the Stripper Block and guided out of the assembly by means of a plastic Residue Collector and Residue Collector Extender.

Use of analogy

Definition

FIGURE D–2. A Complex Description

design and layout (*see* layout and design)

desktop publishing (*see* layout and design)

despite / in spite of

Although there is no literal difference between *despite* and *in spite of*, *despite* suggests an effort to avoid blame.

- *Despite* our best efforts, the plan failed.
  [We are not to blame for the failure.]

- *In spite of* our best efforts, the plan failed.
  [We did everything possible, but failure overcame us.]

Despite and *in spite of* (both meaning "notwithstanding") should not be blended into *despite of*.

- ~~Despite of~~ our best efforts, the plan failed.
  In spite of

diacritical marks   (ESL)

Diacritical marks are **symbols** added to letters to indicate their specific sounds or pronunciation values. Diacritical marks include the phonetic symbols used in dictionaries as well as the marks used with foreign words. When you write **international correspondence,** be sure to use appropriate diacritical marks, especially for names.

- Sabine Müller, Marina Méndez-Santalla

Some dictionaries place a reference list of common marks and sounds at the bottom of each page. Figure D–3 on page 162 lists common diacritical marks with their equivalent sound values. (See also **foreign words in English.**)

D

NAME	SYMBOL	EXAMPLE	MEANING
macron	‾	cāke	indicates the standard pronunciation ("long" sound) of the letter
breve	˘	brăcket	indicates the "short" sound of the letter, in contrast to the standard pronunciation
dieresis	¨	coöperate	indicates that the second of two consecutive vowels is to be pronounced separately; this mark is also used as an "umlaut" to indicate blended vowel sounds, especially in German (universität = universitaet)
acute accent	´	cliché	indicates a primary vocal stress on the indicated letter
grave accent	`	crèche	indicates a deep sound articulated toward the back of the mouth
circumflex	^	crêpe	indicates a very soft sound
tilde	~	cañon	indicates the palatal nasal *ny* sound; a Spanish diacritical mark
cedilla	¸	garçon	placed beneath the letter *c* in French, Portuguese, and Spanish; indicates an *s* sound

FIGURE D–3. Diacritical Marks

diagnosis / prognosis

Because they sound somewhat alike, these words are often confused with each other. *Diagnosis* means "an analysis of the nature of something" or "the conclusions reached by such analysis."

- The CEO *diagnosed* the problem as the allocation of too few dollars for research.

Prognosis means "a forecast or prediction."

- He offered his *prognosis* that the problem would be solved next year.

diction

The term *diction* is often misunderstood because it means both "the choice of words used in writing and speech" and "the degree of distinctness of enunciation of speech." In discussions of writing, diction applies to choice of words. (See also **word choice**.)

dictionaries

A dictionary is a reference, in book or electronic form, that lists words arranged in alphabetical order. It gives information about the words' meanings, etymologies (origin and history), forms, pronunciations, **spellings**, uses as **idioms**, and functions as parts of speech. (See Figure D–4.) For certain words, a dictionary lists

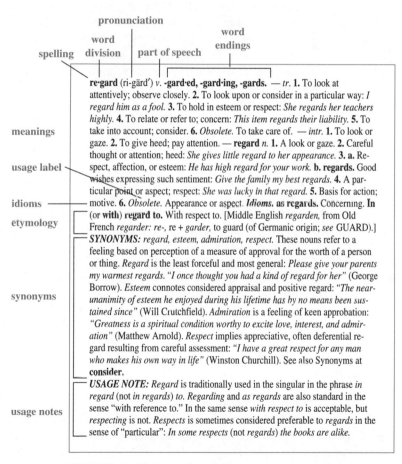

pronunciation

word division

part of speech

word endings

spelling

meanings

usage label

idioms

etymology

synonyms

usage notes

re·gard (ri-gärd') *v.* **-gard·ed, -gard·ing, -gards.** — *tr.* **1.** To look at attentively; observe closely. **2.** To look upon or consider in a particular way: *I regard him as a fool.* **3.** To hold in esteem or respect: *She regards her teachers highly.* **4.** To relate or refer to; concern: *This item regards their liability.* **5.** To take into account; consider. **6.** *Obsolete.* To take care of. — *intr.* **1.** To look or gaze. **2.** To give heed; pay attention. — **regard** *n.* **1.** A look or gaze. **2.** Careful thought or attention; heed: *She gives little regard to her appearance.* **3. a.** Respect, affection, or esteem: *He has high regard for your work.* **b. regards.** Good wishes expressing such sentiment: *Give the family my best regards.* **4.** A particular point or aspect; respect: *She was lucky in that regard.* **5.** Basis for action; motive. **6.** *Obsolete.* Appearance or aspect. *Idioms.* **as regards.** Concerning. **In** (or **with**) **regard to.** With respect to. [Middle English *regarden,* from Old French *regarder: re-,* re + *garder,* to guard (of Germanic origin; *see* GUARD).]
SYNONYMS: regard, esteem, admiration, respect. These nouns refer to a feeling based on perception of a measure of approval for the worth of a person or thing. *Regard* is the least forceful and most general: *Please give your parents my warmest regards.* "*I once thought you had a kind of regard for her*" (George Borrow). *Esteem* connotes considered appraisal and positive regard: "*The near-unanimity of esteem he enjoyed during his lifetime has by no means been sustained since*" (Will Crutchfield). *Admiration* is a feeling of keen approbation: "*Greatness is a spiritual condition worthy to excite love, interest, and admiration*" (Matthew Arnold). *Respect* implies appreciative, often deferential regard resulting from careful assessment: "*I have a great respect for any man who makes his own way in life*" (Winston Churchill). See also Synonyms at **consider.**
USAGE NOTE: Regard is traditionally used in the singular in the phrase *in regard* (not *in regards*) *to. Regarding* and *as regards* are also standard in the sense "with reference to." In the same sense *with respect to* is acceptable, but *respecting* is not. *Respects* is sometimes considered preferable to *regards* in the sense of "particular": *In some respects* (not *regards*) *the books are alike.*

FIGURE D–4. Dictionary Sample Entry

synonyms and shows how their meanings are similar but different. It may also provide illustrations, if appropriate, such as tables, maps, photographs, and drawings. There are two types of dictionaries: desk dictionaries and unabridged dictionaries.

Elements of a Dictionary Entry

Meaning and Etymology.   For most words, the enumeration of meanings dominates the length of the dictionary entry. Each meaning is listed in order by number, but the significance of the ordering varies among dictionaries. Some dictionaries give the most widely accepted current meaning first; others list the meanings in historical order, with the oldest meaning first and the current meaning last. A dictionary's preface indicates whether current or historical meanings are listed first. In some dictionaries, a meaning specific to a certain field of knowledge is so labeled (function, mathematics; parallel, music). Information about the origin and history of a word (its etymology) is given in brackets.

Spelling, Word Division, Pronunciation.   The entry gives the preferred spelling of the word followed by variant spellings (*catalog/catalogue*), if any. The entry also shows how the word is divided into syllables, by dots inserted between syllables (*ri·gard*). The pronunciation of the word is given in parentheses following the boldfaced entry word. The phonetic symbols (called **diacritical marks**) are explained in a key in the front pages of the dictionary, and an abbreviated symbol key, showing how the symbols should be pronounced, is found at the bottom of each page. Words sometimes have two pronunciations; although both are accurate, the one given first is the preferred pronunciation.

Part of Speech and Word Endings.   The entry gives the word's part of speech (*n,* noun; *v,* verb; *adj,* adjective; and so on). Any irregular forms, such as plurals and verb tense changes, are shown in boldface.

Synonyms.   Many entries include a listing of synonyms, words similar in meaning to the entry word. The listing usually includes a brief discussion of how the meaning of each synonym is distinct from the others.

Usage.   Usage labels indicate the appropriate (or, in some cases, inappropriate) use of a particular meaning of a word. The most common labels are *nonstandard, colloquial, informal, slang, dialect, archaic, vulgar, British,* and *obsolete.* Some dictionaries also include usage notes, which discuss interesting or problematic aspects of the correct use of a word. Usage notes are based on the advice of a panel of experts assembled by the dictionary's editors and on actual usage in books, newspapers, and periodicals.

D

Illustrations and Graphics.   Some dictionaries include illustrations (usually photographs and line drawings, sometimes charts) to help clarify the meaning of certain words or to put them in context.

Types of Dictionaries

Desk and Multimedia Dictionaries.   Desk and multimedia dictionaries are often abridged versions of larger dictionaries. There is no single "best" dictionary, but there are several guidelines for selecting a good one. Choose the most recent edition. The older the dictionary, the less likely it is to have the up-to-date information you need. Select a dictionary with upward of 125,000 entries. Some dictionaries available on CD-ROM include a human-voice pronunciation of each word. Many general and specialized dictionaries arc available at a variety of Internet sites. Excellent sites are also available that translate words from one language to another; one such site is "Research It!" (http://www.itools.com/research-it/).
The following are considered good desk dictionaries.

The American Heritage College Dictionary, book and CD
     editions, 1997
Encarta World English Dictionary, book and CD editions, 1999
Merriam Webster's Collegiate Dictionary, 10th indexed ed.,
     1998
Merriam Webster's Collegiate Dictionary and Thesaurus,
     electronic ed., 1998
Random House Webster's College Dictionary, 2nd ed., 1997

Unabridged Dictionaries.   Unabridged dictionaries provide complete and authoritative linguistic information. They are impractical for desk use because of their size and expense, but they are available in libraries and are important reference sources.

- *The Oxford English Dictionary,* 2nd ed., is the standard historical dictionary of the English language. Its 20 volumes contain over 600,000 words and give the chronological developments of over 240,000 words, providing numerous examples of uses and sources. It is also available on CD-ROM for Windows.
- *The Random House Dictionary of the English Language,* 2nd ed., contains about 315,000 entries and uses copious examples. It gives a word's most current meaning first and includes biographical and geographical names.
- *Webster's Third New International Dictionary of the English Language, Unabridged,* contains over 450,000 entries. Word meanings are listed in historical order, with the current meaning given last. This dictionary does not list biographical and geographical names, nor does it include usage information.

ESL Dictionaries.   English-as-a-second-language (ESL) dictionaries are more helpful to the nonnative speaker than are regular English dictionaries or bilingual dictionaries. The pronunciation symbols in ESL dictionaries are based on the international phonetic alphabet rather than on English phonetic systems, and useful grammatical information is included in both the entries and special grammar sections. In addition, the definitions usually are easier to understand than those in regular English dictionaries; for example, a regular English dictionary defines *opaque* as "impervious to the passage of light," while an ESL dictionary defines the word as "not allowing light to pass through." The definitions in ESL dictionaries also are usually more thorough than those in bilingual dictionaries. For example, a bilingual dictionary might indicate that *obstacle* and *blockade* are synonymous but not indicate that although both words can refer to physical objects ("The soldiers went around the *blockade* [or *obstacle*] in the road"), only *obstacle* can be used for abstract meanings ("Lack of money can be an *obstacle* [not a *blockade*] to a college education").

The following dictionaries and reference books provide helpful information for nonnative speakers of English.

Longman Dictionary of American English: A Dictionary for Learners of English

Longman Dictionary of Contemporary English, Paul
   Proctor (editor)
Longman Dictionary of English Idioms, Laurence Urdang
Scott, Foresman Dictionary, E. L. Thorndike and Clarence
   E. Barnhart

D

ESL TIPS FOR USING A BILINGUAL DICTIONARY

As a nonnative speaker of English, you may periodically want to consult
a bilingual dictionary. Be careful; using a bilingual dictionary is not as
simple as looking up the word in your native language and copying
down the first entry—you could easily end up with the wrong equivalent
or an inappropriate form.

Here are some tips for using a bilingual dictionary:

1. Look up the word in your first language.
2. Carefully review the dictionary entry. What part of speech are you interested in? A noun? A verb? A participle? Try to match the part of
   speech you are looking for with the parts of speech listed in the dictionary entry. If you need help with the abbreviations, you can usually find the legend at the bottom of the page or in the introduction
   to the dictionary.
3. If your dictionary lists fields of knowledge, scan the entry to see if
   any field seems particularly suited to your situation. Is the way you
   want to use the word specifically related to one of the fields?
4. Once you determine that you have located the correct English word,
   turn to the English section of the dictionary and look up that word.
   Read the entry carefully. Does the meaning of the definition in your
   first language match your intended meaning in English? If not, look
   the word up in your native language section again and select another
   word. Repeat step 4 until you find the word that expresses the
   meaning you intended.
5. As you write, make sure you have used the appropriate form of the
   new word.

   Professor Robert Beard of Bucknell University has compiled one of
the best and most extensive collections of online bilingual dictionaries
and grammars.

DICTIONARIES	http://www.facstaff.bucknell.edu/rbeard/ diction.html
GRAMMARS	http://www.facstaff.bucknell.edu/rbeard/ grammars.html

D

differ from / differ with

Differ from suggests that two things are not alike.

- The vice president's background *differs from* the general manager's background.

Differ with indicates disagreement between persons.

- Our attorney *differed with* theirs over the selection of the last jury member.

different from / different than

In formal writing, the **preposition** *from* is used with *different*.

- The Quantum PC is *different from* the Macintosh computer.

Different than is acceptable when it is followed by a **clause.**

- The job cost was *different than* we had estimated it.

direct address

Direct address refers to a sentence or phrase in which the person being spoken or written to is explicitly named. It is often used in speech and in **email** messages.

- *John,* call me as soon as you arrive at the airport.
- Call me, *John,* as soon as you arrive at the airport.

Notice that the person's name in a direct address is set off by commas.

discreet / discrete

Discreet means "having or showing prudent or careful behavior."

- Because the matter was personal, he asked Bob to be *discreet* about it at the office.

Discrete means something is "separate, distinct, or individual."

- The advertising department was a *discrete* unit of the marketing division.

disinterested / uninterested

Disinterested means "impartial, objective, unbiased."

- Like good judges, scientists should be passionately interested in the problems they tackle but completely *disinterested* when they seek to solve those problems.

Uninterested means simply "not interested."

- Despite Asha's enthusiasm, her manager remained *uninterested* in the project.

division-and-classification method of development

An effective way to organize information about a complex subject is to divide it into manageable parts and then discuss each part separately. You might use this approach, called *division*, to describe a physical object, such as the parts of a fax machine; to examine an organization, such as a company; or to explain the components that make up the Internet global network. The emphasis in division as a **method of development** is on breaking down a complex whole into a number of like units—it is easier to consider smaller units and to examine the relationship of each to the other than to attempt to discuss the whole.

If you were a financial planner describing the types of mutual funds available to your investors, you could divide the variety available into three broad categories: money-market funds, bond funds, and stock funds. Such division would be accurate, but it would be only a first-level grouping of a complex whole. The three broad categories could, in turn, be subdivided into additional groups based on investment strategy, as follows:

D

Money-market funds
- taxable money-market funds
- tax-exempt money-market funds

Bond funds
- taxable bond funds
- tax-exempt bond funds
- balanced funds (mix of stocks and bonds)

Stock funds
- balanced funds (mix of stocks and bonds)
- equity-income funds
- growth and income funds
- domestic growth funds
- international growth funds
- aggressive growth funds
- small-capitalization funds
- specialized funds

Specialized funds could be further subdivided as follows:

Specialized funds
- communications
- energy
- financial services
- technology
- environmental services
- gold
- worldwide capital goods
- health services
- utilities

The process of *classification* is similar to the process of division. While division is the separation of a whole into its component parts, classification is the grouping of a number of units (such as people, objects, or ideas) into related categories. Consider the following list:

triangular file	steel tape ruler	needle-nose pliers
vise	pipe wrench	keyhole saw
mallet	tin snips	C-clamps
rasp	hacksaw	plane
glass cutter	ball-peen hammer	steel square
spring clamp	claw hammer	utility knife
crescent wrench	folding extension ruler	slip-joint pliers
crosscut saw	tack hammer	utility scissors

To group the items in the list, you would first determine what they have in common. The most obvious characteristic they share is that they all belong in a carpenter's tool chest. With that observation as a starting point, you can begin to group the tools into related categories. Pipe wrenches belong with slip-joint pliers because both tools grip objects. The rasp and the plane belong with the triangular file because all three tools smooth rough surfaces. By applying this kind of thinking to all the items in the list, you can group the tools according to function (Figure D–5).

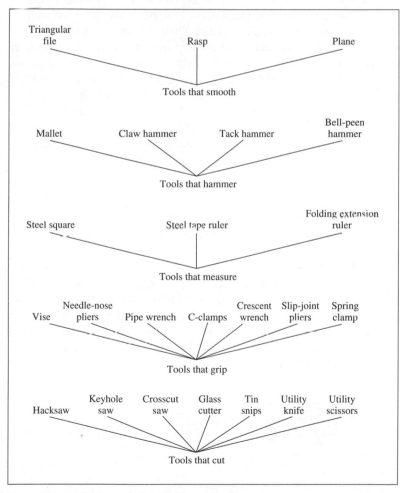

FIGURE D–5. Division and Classification of Tools

To divide or classify a subject, you must first divide the subject into its largest number of comparable units. The basis for division depends, of course, on your subject and your purpose. For explaining the functions of tools, the classifications in Figure D–5 (smoothing, hammering, measuring, gripping, and cutting) are excellent. For recommending which tools a new homeowner should buy first, however, those classifications are not helpful — each group contains tools that a new homeowner might want to purchase right away. To give homeowners advice on purchasing tools, you probably would clarify the types of repairs they most likely will have to do. That classification could serve as a guide to tool purchase.

Once you have established the basis for the division, apply it consistently, putting each item in only one category. For example, it might seem logical to classify needle-nose pliers as both a tool that cuts and a tool that grips because most needle-nose pliers have a small section for cutting wires. However, the primary function of needle-nose pliers is to grip. So listing them only under "tools that grip" would be most consistent with the basis used for listing the other tools. (See also **instructions** and **process explanation.**)

documentation

The word *documentation* in academic work refers to the giving of formal credit to sources used or quoted in research papers, **trade journal articles,** and **reports** — that is, the forms used in **bibliographies,** footnotes, and references. (See **copyright, documenting sources,** and **plagiarism.**) In the workplace, *documentation* also may refer to information that is recorded, or "documented," on paper or in electronic files. The technical manuals that manufacturers provide to customers are examples of such documentation. Even flowcharts and engineering drawings are considered documentation.

documenting sources

DIRECTORY
Overview   173
MLA-Style Parenthetical Documentation   173
   Citation Format for Works Cited   175
   Sample Entries (MLA Style)   176

APA-Style Parenthetical Documentation   179
   Citation Format for Reference List   181
   Sample Entries (APA Style)   183
Style Manuals   185
   Specific Areas   185
   General   186

D

Overview

Documenting sources achieves three important purposes. First, full, consistent, and accurate documentation allows readers to locate and consult the information given and find further information on the subject. Second, when used judiciously, the documenting of sources enables writers to support their assertions about key points of evidence in such documents as **proposals, investigative reports,** and **trade journal articles.** Finally, by identifying where they obtained facts, ideas, **quotations,** and paraphrases and giving proper credit to others, writers avoid **plagiarism.** (See also **bibliographies** and **copyright.**)

This entry describes two principal systems for documenting sources, one developed by the Modern Language Association (MLA) and the other by the American Psychological Association (APA). In MLA parenthetical documentation, a brief citation of the author and page number appears in parentheses in the text and full information is provided in a list of the works cited. In APA parenthetical documentation, a brief citation of the author, year of publication, and page number (for direct quotations only) appears in parentheses in the text, and full information is provided in a reference list. Many professional organizations and journals publish style manuals that describe their own formats for documenting sources. If you are writing for publication in a professional field, consult the manual for that particular field or the style sheet for the journal to which you are submitting your article. A list of some of those manuals appears at the end of this entry.

Use the directory to this entry to locate information quickly. Samples of formats appear at the end of each discussion.

MLA-Style Parenthetical Documentation

The parenthetical method recommended by the MLA is detailed in the *MLA Handbook for Writers of Research Papers* (fifth edition) and the *MLA Style Manual and Guide to Scholarly Publishing* (second edition). This method gives an abbreviated reference to a source

D

parenthetically in the text and lists full information about the source in a separate alphabetical section titled "Works Cited." Information on citing Internet sources can be found at the MLA home page (http://www.mla.org/).

When you cite a source in text, give only the author's last name and the page or page numbers in parentheses (38–39, 144–48, 2436–37, etc.). If the author's name is mentioned in the text, give only the page number of the source in parentheses. If two or more authors have the same last name, include their initials or first names to avoid confusion. The parenthetical citation should include no more information than is necessary to enable readers to find the corresponding entry in the list of works cited. When you are referring to an entire work, rather than to a particular page in a work, mention the author's name in the text and omit the parenthetical citation. The following passages contain sample parenthetical citations in MLA style.

- The earth returns to space as much energy as it absorbs from the sun (Sagan 101).

- Sagan explains why the earth's average temperature is directly related to the condition of the atmosphere (102).

Place a parenthetical citation in text between the closing quotation mark or the last word of the sentence (or clause) and the end punctuation mark (usually a period). Use the spacing shown in the examples. If the parenthetical citation refers to an indented quotation, however, place it outside the last sentence of the quotation, as shown.

- . . . a close collaboration with the physics and technology staff is essential. (Minsky 42)

If you are citing a page or pages of a multivolume work, give the volume number, followed by a colon, a space, and the page number(s): (Jones 2: 53–56). If the entire volume is being cited, identify the author and the volume, as follows: (Smith, vol. 3).

If your list of works cited includes more than one work by the same author, give the title of the work (or a shortened version if the title is long) in the parenthetical citation, unless you mention it in the text. If, for example, your list of works cited includes more than one work by Carl Sagan, a proper parenthetical citation for his book *Billions and Billions: Thoughts on Life and Death at the Brink of the*

Millennium would appear as (Sagan, *Billions* 101). Use only one space between the title and the page number.

Citation Format for Works Cited. The list of works cited should begin on the first new page following the end of the text. Each new entry should begin at the left margin, with the second and subsequent lines within an entry indented five spaces. Double-space within and between entries.

AUTHOR.   List entries in alphabetical order by the author's last name (by the last name of the first listed author if the work has more than one author). For multiple works by the same author, alphabetize the entries by the first major word of the title (following *a*, *an*, or *the*) and put three hyphens and a period in place of the author's name for the second and subsequent entries.

> Sagan, Carl. Billions and Billions: Thoughts on Life and Death at the Brink
> of the Millennium. New York: Random, 1997.

> ---, Contact. New York: Pocket, 1997.

If the author is a corporation, alphabetize the entry by the name of the corporation. If the author is a government agency, alphabetize the entry by the government entity, followed by the agency (for example, "United States. Dept. of Health and Human Services."). Some sources require more than one agency name (for example, "United States. Dept. of Labor. Bureau of Labor Statistics."). If no author is given, begin the entry with the title, alphabetized by the first significant word in the title. An editor's name is followed by the abbreviation *ed.* (or *eds.* for more than one editor).

> Terzian, Yervant, and Elizabeth Bilson, eds. Carl Sagan's Universe. Cam-
> bridge: Cambridge UP, 1997.

TITLE.   The second element of the entry is the title of the work. Capitalize the first word of the title and each significant word thereafter. Underline the title of a book or pamphlet. Place quotation marks around the title of an article in a periodical, an essay in a collection, or a paper in proceedings. The title should be followed by a period.

PERIODICALS.   For an article in a journal, list the volume number, the date, and the page numbers immediately after the title of the periodical. For an article in a magazine or newspaper, omit the volume number.

D

SERIES OR MULTIVOLUME WORKS.   For works in a series, give the name of the series and the series/volume number of the work after the title. If the edition used is not the first, specify the edition.

PUBLICATION INFORMATION.   The final elements of the entry for a book, a pamphlet, or conference proceedings are the place of publication, publisher, and date of publication. Use a shortened form of the publisher's name (for example, *St. Martin's* for St. Martin's Press, Inc.; *Random* for Random House; *Oxford UP* for Oxford University Press). If publication information cannot be found in the work, use the abbreviations *n.p.* (no publication place), *n.p.* (no publisher), and *n.d.* (no date). For familiar reference works, list only the edition and year of publication.

ONLINE SOURCES.   Citations for online sources are similar to citations for printed sources. To cite an online source, begin with the author, title, and, if the information is included in a printed version, publication information. Indicate the date the information was retrieved and include in angle brackets the Web address (the Uniform Resource Locator, or URL) or enough address information to allow the reader to retrieve the source. Personal communications, such as email messages and bulletin board postings, follow the format for letters and interviews, with URL information provided for archived material.

Standards continue to evolve for citations of online and electronic sources. When citing online information, keep in mind that the two primary goals are to give credit to the author and, whenever possible, to enable readers to retrieve the source.

Sample Entries (MLA Style)

BOOK, ONE AUTHOR

Davies, P. C. W. About Time: Einstein's Unfinished Revolution. New York: Simon, 1995.

BOOK, TWO OR MORE AUTHORS

Hey, Tony, and Patrick Walters. Einstein's Mirror. Cambridge: Cambridge UP, 1997.

BOOK, CORPORATE AUTHOR

Microsoft Corporation. Microsoft Sourcebook for the Help Desk: Techniques and Tools for Support Organization Design and Management. 3rd ed. Redmond: Microsoft, 1999.

D

WORK IN AN EDITED COLLECTION

Thorne, Kip S. "Do the Laws of Physics Permit Wormholes for Interstellar Travel and Machines for Time Travel?" Carl Sagan's Universe. Ed. Yervant Terzian and Elizabeth Bilson. Cambridge: Cambridge UP, 1997. 215-39.

BOOK EDITION, IF NOT THE FIRST

Kirkpatrick, James M. The AutoCAD Book: Drawing, Modeling, and Applications, Including Release 14. 4th ed. Englewood Cliffs: Prentice, 1998.

TRANSLATED WORK

Daudel, Raymond. The Realm of Molecules. Trans. Nicholas Hartmann. New York: McGraw, 1999.

MULTIVOLUME WORK

Schmittroth, Linda, Mary Reilly McCall, and Bridget Travers, eds. Eureka! 6 vols. New York: UXL, 1995.

WORK IN A SERIES

Knapp, Brian. Silicon. Grolier Educ. Elements Ser. 6. Danbury: Grolier, 1996.

REPORT

Bertot, John Carlo, Charles R. McClure, and Douglas L. Zweizig. The 1996 National Survey of Public Libraries and the Internet: Progress and Issues: Final Report. Washington: GPO, 1996.

THESIS OR DISSERTATION

Van Alkemade, K. I. "The Problem of Knowledge in Computers and Composition." Diss. U of Wisconsin-Milwaukee, 1997.

ENCYCLOPEDIA ARTICLE

Gibbard, Bruce G. "Particle Detector." World Book Encyclopedia. 1997 ed.

PROCEEDINGS

Johnson, Peggy, and Bonnie MacEwan, eds. Collection Management and Development: Issues in an Electronic Era. Proc. of the Advanced

Collection Management and Dev. Inst., 26–28 Mar. 1999, Chicago. Chicago: Amer. Lib. Assn., 1999.

PAPER IN PROCEEDINGS

Cline, Nancy M. "Staffing: The Art of Managing Change." Collection Management and Development: Issues in an Electronic Era. Proc. of the Advanced Collection Management and Dev. Inst., 26–28 Mar. 1999, Chicago. Ed. Peggy Johnson and Bonnie MacEwan. Chicago: Amer. Lib. Assn., 1999. 13–28.

PAPER PRESENTED AT A CONFERENCE

Kim, R. "Management of Technology." Intl. Conf. on Systems Engineering, U of Lima. 2 May 1999.

JOURNAL ARTICLE

Zeng, Ganwen, and Ahmed Hemami. "An Overview of Robot Force Control." Robotica: International Journal of Information, Education and Research in Robotics and Artificial Intelligence 15 (1997): 473–82.

MAGAZINE ARTICLE

Frank, Adam. "Quantum Honeybees." Discover Nov. 1997: 80–86.

ANONYMOUS ARTICLE (IN WEEKLY PERIODICAL)

"Plane Soars to New Heights on Sunbeams." New Scientist 21 June 1997: 13.

NEWSPAPER ARTICLE

Broad, William J. "New Work Proposed for Shuttles: Salvage in Space." The New York Times 16 Sept. 1997, natl. ed.: B9+.

LETTER TO AUTHOR

Paluch, Dustin. Letter to the author. 20 Sept. 1997.

LETTER TO OTHER PERSON(S)

Viets, Hermann. Letter to all MSOE students, fac., and staff. 1 Sept. 1998.

PERSONAL INTERVIEW

Sariolgholam, Mahmood. Personal interview. 29 Nov. 2000.

EMAIL MESSAGE

Kahl, Jonathan D. "Re: Web page." E-mail to the author. 2 Oct. 2001.

PUBLICATION ON CD-ROM

Microsoft Visual C++ 5.0, Learning ed. CD-ROM. Redmond: Microsoft, 1997.

CONFERENCE PROCEEDINGS AT A WEB SITE (ELECTRONIC VERSION)

Mehringer, D. M., R. L. Plante, and P. A. Roberts, eds. Astronomical Data
Analysis Software and Systems VIII. Proc. of the Eighth Annual Conf.
on Astronomical Data Analysis Software and Systems, 1–4 Nov.
1998, Urbana. Astronomical Soc. of the Pacific Conf. Ser. 172.
1999. Astronomical Soc. of the Pacific. 2 Sept. 2000 <http://monet.
astro.uicuc.edu/adass98/>.

CONFERENCE PROCEEDINGS AT A WEB SITE (PRINTED VERSION)

Mehringer, D. M., R. L. Plante, and P. A. Roberts, eds. Astronomical Data
Analysis Software and Systems VIII. Proc. of the Eighth Annual Conf.
on Astronomical Data Analysis Software and Systems, 1–4 Nov. 1998,
Urbana. Astronomical Soc. of the Pacific Conf. Ser. 172. San Francisco: Astronomical Soc. of the Pacific, 1999. 2 Oct. 2000 <http://
monet.astro.uicuc.edu/adass98/>.

ARTICLE IN AN ELECTRONIC NEWSLETTER

Shevlin, Diana. "Treasury Proposes EFT Regulation: New EFT 99 Web Site
Launches Rule's Release." Financial Connection 6.2 (1997).
5 Oct. 1997 <http://www.fms.treas.gov/finconn/fcsep97.html>.

DOCUMENT RETRIEVED BY GOPHER

United States. Dept. of Labor. Bureau of Labor Statistics. Fifty Years in
Print: 1996–97 Edition of The Occupational Outlook Handbook
Published. USDL 96–78. 15 Mar. 1996. 9 Oct. 1997 <gopher://
hopi2.bls.gov:70/00hopiftp.dev/news.release/ecopro1.txt>.

USENET NEWSGROUP POSTING

Armitage, Phillip. "AutoCAD Memory Requirements." Online posting. 25
June 1997. 2 Oct. 1997 <news:comp.cad.autocad>.

APA-Style Parenthetical Documentation

The parenthetical style recommended by the APA is described in
the *Publication Manual of the American Psychological Association*
(fourth edition). This method gives abbreviated references to
sources parenthetically in the text and lists full information about
the sources in a separate reference list.

When you are documenting direct quotations in text, give the
author's last name, the year of publication, and the page number in
parentheses. The page number is optional for paraphrased information and ideas. If the author's name is mentioned in the text, give
only the year of publication and the page number in parentheses.

The parenthetical citation should include no more information than is necessary to enable readers to find the corresponding entry in the reference list. If the author's name and the year of publication are mentioned in the text, give only the page number for direct quotations and omit parenthetical information for paraphrased material. The following passages contain sample APA parenthetical citations.

- World War II was the occasion of radar's first application in warfare, and Great Britain led the way in radar research (Butrica, 1996).

- According to Butrica (1996), the use of radar as an offensive and defensive warfare agent made World War II "the first electronic war" (p. 2).

- The "first electronic war" (Butrica, 1996, p. 2) was fought as much in the research laboratory as on the battlefield.

When APA parenthetical citations are added in midsentence, place them after the closing quotation marks and continue with the rest of the sentence. If the APA parenthetical citation follows a block quotation, place the citation after the final punctuation mark. Use the spacing shown in the examples. Within the citation itself, separate the name, date, and page number with commas. Allow one space after each comma. Use the abbreviation *p.* or *pp.* before page numbers.

If your reference list includes more than one work by the same author published in the same year, add the lowercase letters a, b, c, and so forth to the year in both the reference list entries and the text citations: (Ostro, 1993b, p. 347). When a work has two authors, cite both names joined by an ampersand: (Hey & Walters, 1997). For the first citation of a work with three, four, or five authors, include all names. For subsequent citations, include only the name of the first author followed by *et al.* When two or more works by different authors are cited in the same parentheses, list the citations alphabetically and use semicolons to separate the citations: (Hey & Walters, 1997; Knapp, 1996; Ostro, 1993a). The works cited in the examples would be listed alphabetically, as follows, in a reference list.

Butrica, A. J. (1996). To see the unseen: A history of planetary radar astronomy. The NASA History Series. Washington, DC: U.S. Government Printing Office.

Hey, T., & Walters, P. (1997). Einstein's mirror. Cambridge, England: Cambridge University Press.

Knapp, B. (1996). Grolier educational elements series: Vol. 6. Silicon. Danbury, CT: Grolier Educational.

Ostro, S. J. (1993a). Planetary radar astronomy. Reviews of Modern Physics, 65, 1235–1279.

Ostro, S. J. (1993b). Radar astronomy. In S. P. Parker & J. M. Pasachoff (Eds.), McGraw-Hill encyclopedia of astronomy (pp. 347–348). New York: McGraw-Hill.

D

Citation Format for Reference List.   The reference list should begin on the first new page following the end of the text. Each new entry should begin at the left margin, with the second and subsequent lines indenting five spaces from the left margin. (Your instructor may require you to use a paragraph indent instead. If so, begin at a paragraph indent, with subsequent lines continuing at the left margin.)* Double-space within and between entries.

Do not include forms of personal communication, such as letters, email, messages from electronic bulletin boards, and telephone conversations, in the reference list. Those information sources should be cited only in the text. Include the communicator's initials and surname and the date: (M. Sariolgholam, personal communication, November 28, 2000).

AUTHOR.   Entries appear in alphabetical order by the author's last name (by the last name of the first author if the work has more than one author). Give the author's surname and initials only. Works by the same author should be listed chronologically according to the year of publication, from earlier to later dates. If the author is a corporation, alphabetize the entry by the name of the corporation; if the author is a government agency, alphabetize the entry by the government entity that published the work (for example, "U.S. Department of Health and Human Services"). Some sources require more than one agency name (for example, "U.S. Department of Health and Human Services, Indian Health Service"). However, if an agency that is part of a higher department is well-known, it is not necessary to include the higher department (for example, "Food and Drug Administration"). If no author is given, begin the entry with the title, alphabetized by the first

---

*At the printing of this book, the APA accepts the use of either a hanging indent (as shown here) or a regular paragraph indent for references, as long as the chosen format is used consistently throughout.

significant word. For edited works, place the editor's name at the beginning of the entry, followed by *Ed.* (or *Eds.*) in parentheses.

PUBLICATION DATE.   The second element of the entry is the work's date of publication. Place the copyright date or the date of production in parentheses. For periodicals other than journals, include the year, a comma, and the month, week, or day. Place a period after the parentheses. For journals and books, give only the year.

TITLE.   Give the title of the work after the date of publication. For titles of articles, chapters, or books, capitalize only the first word of the title and the subtitle and any proper names. Underline titles of books but do not underline or use quotation marks for titles of articles or chapters. Follow the title with a period.

PERIODICALS.   For an article in a periodical (journal, magazine, or newspaper), give the title of the periodical in upper- and lower-case letters. The volume number and the page numbers should immediately follow the periodical title. Underline the periodical title and volume number. Separate the elements by commas and end with a period.

SERIES OR MULTIVOLUME WORKS.   For works in a series, the series number of the work follows the title. For multivolume works, the number of volumes follows the title. If the edition is not the first, specify the edition.

PUBLISHING INFORMATION.   The final elements of the entry for a book, a pamphlet, or conference proceedings are the place of publication and the publisher. Use a shortened form of the publisher's name when possible and abbreviate names of associations, corporations, and university presses. Omit terms such as *Publisher, Co.,* and *Inc.* but do not omit or abbreviate the words *Books* and *Press.*

ONLINE SOURCES.   When citing online sources, begin with the author, the date of publication, and the title, just as for printed sources. At the end of the reference, indicate that the work was "retrieved [date of retrieval] from [source]." (The source may be the World Wide Web or a database such as EBSCO.) Include enough of the URL to allow interested readers to retrieve the source. Cite personal communications, such as email messages and bulletin board postings, in the text only. Include the communicator's initials and surname and the date: (J. D. Kahl, personal communication, October 2, 2000).

Standards continue to evolve for citations of online and electronic sources. When citing online information, keep in mind that the two primary goals are to give credit to the author and, whenever possible, to allow the reader to find the source. For updates on APA style, visit APA's Web site at (http://www.apa.org/journals).

D

Sample Entries (APA Style)

BOOK, ONE AUTHOR

Davies, P. C. W. (1995). About time: Einstein's unfinished revolution. New York: Simon and Schuster.

BOOK, TWO OR MORE AUTHORS

Hey, T., & Walters, P. (1997). Einstein's mirror. Cambridge, England: Cambridge University Press.

BOOK, CORPORATE AUTHOR

Microsoft Corporation (1999). Microsoft sourcebook for the help desk: Techniques and tools for support organization design and management (3rd ed.). Redmond, WA: Microsoft Press.

WORK IN AN EDITED COLLECTION

Thorne, K. S. (1997). Do the laws of physics permit wormholes for interstellar travel and machines for time travel? In Y. Terzian & E. Bilson (Eds.), Carl Sagan's universe (pp. 121–134). Cambridge, England: Cambridge University Press.

BOOK EDITION, IF NOT THE FIRST

Kirkpatrick, J. M. (1998). The AutoCAD book: Drawing, modeling, and applications, including release 14 (4th ed.). Englewood Cliffs, NJ: Prentice Hall.

TRANSLATED WORK

Daudel, R. (1999). The realm of molecules (N. Hartmann, Trans.). New York: McGraw-Hill.

MULTIVOLUME WORK

Schmittroth, L., McCall, M. R., & Travers, B. (Eds.). (1995). Eureka! (Vols. 1–6). New York: UXL.

WORK IN A SERIES

Knapp, B. (1996). Grolier educational elements series: Vol. 6. Silicon. Danbury, CT: Grolier Educational.

D

REPORT

Bertot, J. C., McClure, C. R., & Zweizig, D. L. (1996). The 1996 national survey of public libraries and the Internet: Progress and issues: Final report. Washington, DC: U.S. Government Printing Office.

THESIS OR DISSERTATION

Michelman, T. S. (1993). Contingent valuation methodology and the bounded rationality perspective. Unpublished master's thesis, University of Rhode Island, Kingston.

ENCYCLOPEDIA ARTICLE

Gibbard, B. G. (1997). Particle detector. In World Book encyclopedia (Vol. 15, pp. 186–187). Chicago: World Book.

PROCEEDINGS

Johnson, P., & MacEwan, B. (Eds.). (1999). Collection management and development: Issues in an electronic era. Proceedings of the Advanced Collection Management and Development Institute, Chicago, IL, March 26–28, 1999. Chicago: American Library Association.

PAPER IN PROCEEDINGS

Cline, N. M. (1999). Staffing: The art of managing change. In P. Johnson & B. MacEwan (Eds.), Collection management and development: Issues in an electronic era. Proceedings of the Advanced Collection Management and Development Institute, Chicago, IL, March 26–28, 1999 (pp. 13–28). Chicago: American Library Association.

PAPER PRESENTED AT A CONFERENCE

Kim, R. (1995, May). Management of technology. Paper presented at the International Conference on Systems Engineering, University of Lima, Peru.

JOURNAL ARTICLE

Zeng, G., & Hemami, A. (1997). An overview of robot force control. Robotica: International Journal of Information, Education and Research in Robotics and Artificial Intelligence, 15, 473–482.

MAGAZINE ARTICLE

Frank, A. (1997, November). Quantum honeybees. Discover, 18, 80–86.

ANONYMOUS ARTICLE (IN WEEKLY PERIODICAL)

Plane soars to new heights on sunbeams. (1997, June 30). New Scientist, 130, 13.

NEWSPAPER ARTICLE

Broad, W. J. (1997, September 16). New work proposed for shuttles: Salvage in space. The New York Times, pp. B9, B14.

PUBLICATION ON CD-ROM

Microsoft Visual C++ 5.0, learning edition [CD-ROM]. (1997). Redmond, WA: Microsoft.

CONFERENCE PROCEEDINGS AT A WEB SITE

Mehringer, D. M., Plante, R. L., & Roberts, P. A. (Eds.). (1999). Astronomical data analysis software and systems VIII. Proceedings of the Eighth Annual Conference on Astronomical Data Analysis Software and Systems, November 1–4, 1998. Urbana, IL (Astronomical Society of the Pacific Conference Ser., Vol. 172). Retrieved December 1, 1999 from the World Wide Web: http://monet.astro.uicuc.edu/adass98/

ARTICLE IN AN ELECTRONIC NEWSLETTER

Shevlin, D. (1997, September). Treasury proposes EFT regulation: New EFT 99 Web site launches rule's release. Financial Connection 6(2). Retrieved September 1, 1999 from the World Wide Web: http://www.fms.treas.gov/finconn/fcsep97.html

DOCUMENT RETRIEVED BY GOPHER

Bureau of Labor Statistics. (1996, March 15). Fifty years in print: 1996–97 edition of the Occupational Outlook Handbook published (USDL 96-78). Retrieved June 1, 1997 from Gopher: gopher://hopi2.bls.gov:70/00hopiftp.dev/news.release/ecopro1.txt

Style Manuals

Many professional societies, publishing companies, and other organizations publish manuals that prescribe bibliographic reference formats for their publications or publications in their fields. In addition, several general style manuals are well-known and widely used.

Specific Areas

American Chemical Society. *ACS Style Guide: A Manual for Authors and Editors.* 2nd ed. Washington, D.C.: Amer. Chemical Soc., 1998.

American Institute of Physics, Publications Board. *AIP Style Manual.* 4th ed. New York: Amer. Institute of Physics, 1990.

D

American Mathematical Society. *A Manual for Authors of Mathematical Papers*. Providence: Amer. Mathematical Soc., 1990.

American Medical Association. *American Medical Association Manual of Style*. 9th ed. Baltimore: Williams, 1998.

American Psychological Association. *Publication Manual of the American Psychological Association*. 4th ed. Washington, D.C.: Amer. Psychological Assn., 1994.

Council of Biology Editors. *Scientific Style and Format: The CBE Manual for Authors, Editors, and Publishers*. 6th ed. New York: Cambridge UP, 1994.

U.S. Geological Survey. *Suggestions to Authors of the Reports of the United States Geological Survey*. 7th ed. Washington, D.C.: GPO, 1991.

General

The Chicago Manual of Style. 14th ed. Chicago: Univ. of Chicago Press, 1993.

Gibaldi, Joseph. *MLA Handbook for Writers of Research Papers*. 5th ed. New York: Mod. Lang. Assn. of Amer., 1999.

Gibaldi, Joseph. *MLA Style Manual and Guide to Scholarly Publishing*. 2nd ed. New York: Mod. Lang. Assn. of Amer., 1998.

National Information Standards Organization. *Scientific and Technical Reports—Elements, Organization, and Design*. Bethesda, Md.: Natl. Information Standards Organization, 1995. ANSI Z39.18-1995.

Skillin, M. E., and R. M. Gay. *Words into Type*. 3rd ed. Englewood Cliffs, N.J.: Prentice, 1974.

Turabian, K. L. *A Manual for Writers of Term Papers, Theses, and Dissertations*. 6th ed. Chicago: Univ. of Chicago Press, 1996.

double negatives   ESL

A *double negative* is the use of an additional negative word to reinforce an expression that is already negative. It is an attempt to emphasize the negative, but it should not be used in workplace writing.

- I ~~haven't~~ got none. *have*

Barely, hardly, and *scarcely* cause problems because writers some-times do not recognize that those words are already negative.

* I ~~don't~~ hardly ever have time to read these days.

Not unfriendly, not without, and similar constructions are not double negatives because in such constructions two negatives are meant to suggest the gray area between negative and positive meanings. Be careful how you use such constructions; they can be confusing to the reader and should be used only if they serve a purpose.

* He is *not unfriendly.* [He is neither hostile nor friendly.]

* It is *not without* regret that I offer my resignation.
  [I have mixed feelings rather than only regret.]

The correlative **conjunctions** *neither* and *nor* may appear together in a clause without creating a double negative, so long as the writer does not attempt to use the word *not* in the same clause.

* It was ~~not~~ , as a matter of fact, ~~neither~~ *neither* his duty nor his desire to fire the employee.

* It was not, as a matter of fact, ~~neither~~ *either* his duty ~~nor~~ *or* his desire to fire the employee.

* She ~~did not~~ neither ~~care~~ *cared* about nor ~~notice~~ *noticed* the error.

Negative forms are full of traps that often entice inexperienced writers into **logic errors,** as illustrated in the following example:

> ILLOGICAL There is *nothing* in the book that has *not* already been published in some form, but some of it is, I believe, very little known.

In this sentence, "some of it" logically can refer only to "*nothing* in the book that has *not* already been published." The sentence can be corrected in one of two ways. The pronoun *it* can be replaced by a specific noun.

> LOGICAL There is nothing in the book that has not already been published in some form, but some of the *information* is, I believe, very little known.

Or the idea can be stated positively. (See also **positive writing.**)

LOGICAL   Everything in the book has been published in some form, but some of it is, I believe, very little known.

D drawings

A drawing represents objects or forms through hand-drawn or computer-generated lines and can focus on details or relationships that a **photograph** cannot capture. A drawing can emphasize the significant part or function of a mechanism and omit what is not significant. However, if the actual appearance of an object is necessary, include a photograph.

The three types of drawings discussed in this entry are conventional line drawings, cutaway drawings, and exploded-view drawings. Each type of drawing has unique advantages—the type of drawing you use should be determined by its purpose.

A conventional drawing like Figure D–6 on page 189 is appropriate if your readers need a representation of an object's general appearance or an overview of a series of steps. Use a cutaway drawing, like the one in Figure D–7 on page 190, to show the internal parts of a piece of equipment and illustrate their relationship to the whole. To show the proper sequence in which parts fit together or to show the details of individual parts, use an exploded-view drawing, like that shown in Figure D–8 on page 191.

Drawings that require a high degree of accuracy and precision generally are prepared by graphics specialists. If you need only general-interest images to illustrate newsletters and brochures or create presentation overheads, look in clip art libraries or programs like PowerPoint. Clip art libraries contain thousands of noncopyrighted symbols, shapes, and images of people, equipment, furniture, buildings, and so on. Figure D–9 on page 191 shows examples of clip art images.

Think about your need for drawings during the **research** and **organization** stages of writing. Treat drawings as an integral part of your **outline,** indicating approximately where each should appear throughout the outline with "drawing of . . ." enclosed in brackets. As with any part of an outline, the notations can be moved or deleted as required. Planning your drawings from the beginning stages of your outline ensures that they are integrated into successive versions of the draft to the finished piece of writing. For additional help with drawings, see the Writer's Checklist that follows on page 190.

If You Do Fight the Fire, Remember the Word PASS

PULL

the pin: Some extinguishers require releasing a lock latch, pressing a puncture lever, or taking another first step.

AIM

low: Point the extinguisher nozzle (or its horn or hose) at the base of the fire.

SQUEEZE

the handle: This releases the extinguishing agent.

SWEEP

from side to side: Keep the extinguisher aimed at the base of the fire and sweep back and forth until it appears to be out. Watch the fire area. If fire breaks out again, repeat the process.

FIGURE D–6. Conventional Drawing That Illustrates Instructions

D

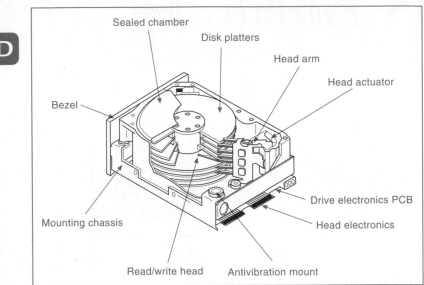

FIGURE D-7. Cutaway Drawing of a Hard Disk Drive

Writer's Checklist: Creating and Using Drawings

Many organizations have their own format specifications for creating drawings. In the absence of such specifications, follow the guidelines in **illustrations.** In addition, follow these guidelines specific to drawings.

☑ Show equipment and other objects from the point of view of the person who will use them.

☑ When illustrating a subsystem, show its relationship to the larger system of which it is a part.

☑ Draw the different parts of an object in proportion to one another or indicate that certain parts are enlarged.

☑ When a sequence of drawings is used to illustrate a process, arrange them from left to right or from top to bottom on the page.

☑ Label parts in the drawing so that text references to them are clear and consistent.

☑ Depending on the complexity of what is shown, label the parts themselves or use a letter/number key (see Figure D-8 on page 191).

D

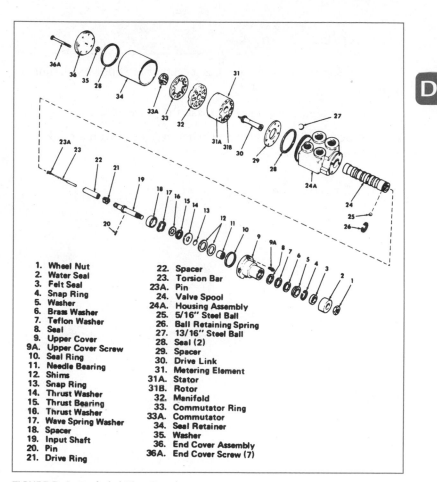

1. Wheel Nut
2. Water Seal
3. Felt Seal
4. Snap Ring
5. Washer
6. Brass Washer
7. Teflon Washer
8. Seal
9. Upper Cover
9A. Upper Cover Screw
10. Seal Ring
11. Needle Bearing
12. Shims
13. Snap Ring
14. Thrust Washer
15. Thrust Bearing
16. Thrust Washer
17. Wave Spring Washer
18. Spacer
19. Input Shaft
20. Pin
21. Drive Ring

22. Spacer
23. Torsion Bar
23A. Pin
24. Valve Spool
24A. Housing Assembly
25. 5/16" Steel Ball
26. Ball Retaining Spring
27. 13/16" Steel Ball
28. Seal (2)
29. Spacer
30. Drive Link
31. Metering Element
31A. Stator
31B. Rotor
32. Manifold
33. Commutator Ring
33A. Commutator
34. Seal Retainer
35. Washer
36. End Cover Assembly
36A. End Cover Screw (7)

FIGURE D–8. Exploded-View Drawing

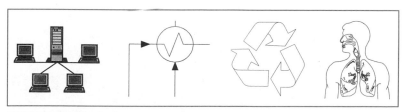

FIGURE D–9. Clip Art Images

D

due to / because of

Due to (meaning "caused by") is acceptable following a linking **verb.**

- His short temper was *due to* stress at the office.

Due to is not acceptable, however, when it is used with a nonlinking verb to replace *because of.*

- He left work ~~due to~~ illness.
                because of
                ^

E

each

When *each* is used as a subject, it takes a singular **verb** or **pronoun.**

- *Each* of the reports *is* to be submitted ten weeks after *it* is assigned.

- *Each* worked as fast as *his or her* ability would permit.

When *each* occurs after a plural subject with which it is in apposition, it takes a plural verb or pronoun. (See also **agreement.**)

- The reports *each have* white embossed titles on *their* covers.

each and every

Although the phrase *each and every* is commonly used in speech in an attempt to emphasize a point, it is redundant and should be eliminated from your writing. Replace it with *each* or *every.*

- Each ~~and every~~ part was correctly labeled.

- *Every*
  ~~Each and every~~ part was correctly labeled.
  ^

economic / economical

Economic refers to the production, development, and management of material wealth. *Economical* simply means "not wasteful or extravagant."

- The currency devaluation had an *economic* impact on several countries.

- With the company's recent financial losses, management has requested that employees be as *economical* as possible in their equipment purchases.

editing (*see* revision)

e.g./i.e.

The **abbreviation** *e.g.* stands for the Latin *exempli gratia*, meaning "for example"; *i.e.* stands for the Latin *id est*, meaning "that is." Since the English expressions (*for example* and *that is*) serve the same purpose, there is no need to use the Latin expressions or abbreviations except to save space in notes and **illustrations.**

- Some terms of the contract (e.g., *for example* duration and job classification) were settled in the first two bargaining sessions.

- We were a fairly heterogeneous group; i.e., *that is* there were managers, supervisors, and vice presidents at the meeting.

If you must use *i.e.* or *e.g.*, punctuate it as follows. If *i.e.* or *e.g.* connects two independent clauses, a semicolon should precede it and a comma should follow. If *i.e.* or *e.g.* connects a **noun** and an **appositive,** a comma should precede and follow it.

- We were a fairly heterogeneous group; *i.e.,* we were managers, supervisors, and vice presidents.

- We were a fairly heterogeneous group, *i.e.,* managers, supervisors, and vice presidents.

elegant variation

The attempt to avoid repeating a noun in a paragraph by substituting other words (often pretentious **synonyms**) is called *elegant variation*. Avoid the practice—repeat a word if it says what you mean, or use a suitable pronoun. (See also **affectation** and **long variants.**)

- The use of robotics in the assembly process has increased produc-
  The use of robotics
  tion. ~~Robotic utilization~~ has also cut costs.
       ^

- The use of robotics in the assembly process has increased produc-
       , and it
  tion~~.~~ ~~Robotic utilization~~ has also cut costs.
       ^

(See also **thesaurus.**)

E

ellipses

When you omit words in quoted material, use a series of three spaced periods, called *ellipsis dots,* to indicate the omission. (See also **quotations.**) Do not use ellipsis dots for any purpose other than to indicate omission.

WITHOUT OMISSION   "Material distributed for promotional use is sometimes charged for, particularly in high-volume distribution to schools, although prices for these publications are not uniformly based on the cost of developing them."

WITH OMISSION   "Material distributed for promotional use is sometimes charged for . . . although prices for these publications are not uniformly based on the cost of developing them."

If you are following MLA style in your writing, use **brackets** around the ellipsis dots to show that words have been omitted from the original source.

- "The vast majority of the Internet's inhabitants are males [. . .] between the ages of eighteen and thirty-four" (5).

When the beginning of the sentence is omitted, begin the quotation with a lowercase letter and without ellipsis dots.

WITHOUT OMISSION   "When the programmer has determined a system of runs, he or she must create a system flowchart to provide a picture of the data flow through the system."

WITH OMISSION   The letter states that the programmer "must create a system flowchart to provide a picture of the data flow through the system."

When the omission comes at the end of a sentence and the quotation continues following the omission, use four dots. The first dot is the period that ends the sentence, and the other dots indicate the omission. (See also **sentence construction.**)

WITHOUT OMISSION   "In all publications departments except ours, publications funds — once they are initially allocated by higher management — are controlled by publications personnel. Our company is the only one to have nonpublications people control funding within a budget period. In addition, all publications departments control printing funds as well."

WITH OMISSION   "In all publications departments except ours, publications funds — once they are initially allocated by higher management — are controlled by publications personnel. . . . In addition, all publications departments control printing funds as well."

Use a full line of dots across the page to indicate the omission of one or more paragraphs.

WITHOUT OMISSION   "A computer system operates with two types of programs: the software programs that manufacturers supply and the programs users create. The manufacturer's software consists of a number of programs that enable the system to perform complicated manipulations of data from the relatively simple instructions specified in the user's program.

Programmers must systematically approach solving any data-processing problem. They must first clearly identify the problem and then define a system of runs that will solve it. For example, a given problem may require a validation and sort run to handle account numbers or employee numbers, a computation and update run to manipulate the data for the output reports, and a print run to produce the output reports.

When a system of runs has been determined, the programmer must create a system flowchart to provide a picture of the
data flow through the system."

WITH OMISSION   "A computer system operates with two types
of programs: the software programs that
manufacturers supply and the programs
users create. The manufacturer's software
consists of a number of programs that enable
the system to perform complicated manipulations of data from the relatively simple instructions specified in the user's program.
. . . .
When a system of runs has been deter-
mined, the programmer must create a system flowchart to provide a picture of the
data flow through the system."

email

DIRECTORY
Overview   197
Writing Style   198
Design Considerations   198
Signature Blocks   199
Netiquette   200
Review and Confidentiality Implications   202

Overview

Email, the transmission of messages through a computer network,
is especially useful for employees in large organizations who frequently write and send **correspondence** (especially **memos**),
reports, meeting notices, and even **questionnaires** to colleagues
throughout the organization. Email can also be used to send attached documents and data to remote locations, for example, to
colleagues at different branches of the company or to clients at
different companies. The **Internet** permits email links worldwide.
Because the information is digital, email enables members of a
collaborative writing team to forward drafts easily to other team

members for feedback and to incorporate feedback into the final document. (See also **collaborative writing.**)

Writing Style

Although email messages are similar to other workplace writing in some respects, they tend to be less formal and more conversational — somewhere between a telephone conversation and a memo in **tone.** In fact, the speed at which messages are transmitted and the back-and-forth interchange that commonly occurs between two or more people make email messages similar to conversations. Nevertheless, you still need to think carefully about your **readers** and the accuracy and appropriate level of detail of the information you transmit. When you send email, carefully consider the **purpose** of the message, how much information to include, and the audience's familiarity with the topic.

Design Considerations

Many email systems do not offer the array of typographical cues that most word-processing programs provide. For that reason, avoid using boldface, italics, and a variety of fonts, because your recipient's email system may not be able to read them. Instead, use a variety of alternative highlighting devices but be consistent. For example, capital letters or asterisks, used sparingly, can substitute for boldface, italics, and underlines as **emphasis:**

- Dr. Wilhoit's suggestions benefit doctors AND patients.

- Although the proposal is sound in *theory,* it will never work in *practice.*

Intermittent underlining can replace solid underlining or italics in references to published works.

- My report follows the format outlined in _Handbook of Technical Writing_.

Keep in mind the following special considerations when you are sending email.

- Tables and bulleted lists, like many other formatting features, do not transmit well. If sending a document with such features is necessary, do so in an attachment to your message, after making sure that the recipient has compatible software to view and save the attachment.

- Always fill in the subject line with a concise phrase that describes the topic of your message. The recipients can then decide at a glance when they need to read it. Subject lines also help recipients organize their incoming messages by topic.

- Put your response to someone else's email message at the beginning (or top) of the email window. Do not make the recipient scroll to the end of the original message to find your response.

- In quoting the message you are replying to, include only those parts relevant to your reply. To clearly indicate the difference between your response and the text quoted, mark the beginning of the quoted text with a greater-than symbol (>).

Signature Blocks

Because writers often use email messages to communicate with people both within and outside their organization, many companies and individual writers include *signature blocks* (called *signatures* for short) at the bottom of their messages. Signatures, which writers can usually preprogram to appear on every email they send, supply information that company letterhead usually provides in other correspondence. If your organization recommends a certain format or restricts the content of signatures, adhere to that standard. Otherwise, choose a signature that shows your readers your full name, official title, department or division, and the organization for which you work. Other items often included in a signature are telephone and fax numbers, mailing address, and Internet address.

When you use a signature block at the bottom of an email message, use typographical cues to separate the signature from the message. Two or three line spaces between the message and the signature can provide the necessary white, or blank, space to keep the parts distinct, but many other cues can highlight the signature. A line of hyphens (-), underlines (_), equals signs (=), tildes (~), or asterisks (*) can effectively separate a signature from the email message. Usually, typographical highlights begin at the left margin and continue until the end of the signature's longest line, as in this example:

E

```
======================
```
Daniel J. Vasquez, Publications Manager
Medical Information Systems
TechCom Corporation
P.O. Box 5413   Salinas CA 93962
Office Phone 888.229.4511 (x 341)
General Office Fax 888.229.1132
http://www.tcc.org/
```
======================
```

As with all **layout and design** features, the cues you decide to use should be appropriate for both the **tone** of your usual correspondence and the image you want to present of yourself and your organization.

Netiquette

Netiquette (*Internet + etiquette*) is especially important, because email often replaces other forms of workplace communication. Keep in mind the following conventions governing netiquette.

- Do not overwhelm your reader with dense, lengthy passages. If your message runs longer than 200 to 250 words—about half a page of printed text—send it as an attached file, along with a brief transmittal memo in the email message itself. Be sure to break the text into brief paragraphs. No one wants to read long, dense blocks of text on a computer screen. (See also **conciseness/wordiness.**)
- Keep workplace email professional—do not send off-color jokes, use **biased language,** or discuss office gossip. Likewise, do not send *flames,* emails that contain abusive, obscene, or derogatory language to attack someone. If you are tempted to dash off an angry message, wait until you are in a better frame of mind so your message focuses on the issues rather than a personal attack.
- Be scrupulous about typing email addresses and otherwise ensuring that the intended recipient gets the message. Many email addresses are made up of the recipient's name or initials. It is easy to transpose letters when you type an address and have the message returned or sent to the wrong person, thereby wasting time and jeopardizing confidentiality. The workplace is filled with stories of unflattering or critical messages inadvertently sent to or intercepted by the person being criticized be-

cause the sender used the wrong address or clicked "Reply" rather than "Forward."

- Do not write in all uppercase letters. Such text is difficult to read and is the electronic equivalent of shouting. Use upper- and lowercase letters, as you would for any other written document.
- Use *emoticons* sparingly. Emoticons are sideways faces made with punctuation marks and letters used to represent the writer's mood. Some commonly used emoticons appear in Fig- ure E–1. Use them only for casual messages between friends or in chat rooms. They are not appropriate for most business and professional email messages, especially those going to one's su- periors or to clients, customers, or suppliers outside the organi- zation. Avoid using emoticons in messages to nonnative speak- ers of English or international readers. (See also **international correspondence.**)

E

:-)	a smile	:-D	a laugh
:-(	a frown	;-)	a wink
>:-<	a mad face	:-<	a sad face
:-@	a scream	%-)	a confused look
:-O	a yawn	:-7	tongue in cheek

FIGURE E–1. Selected Email Emoticons

- Use email abbreviations sparingly. Figure E–2 lists a sampling of abbreviations commonly used to express certain routine phrases. As with all types of abbreviations, use them only when you are sure your readers will understand them. Like emoti- cons, email abbreviations will likely be incomprehensible to nonnative English speakers.
- Check your incoming email regularly and respond as quickly as you can. If you receive an assignment by email that will take a few days or longer to complete, send an interim response saying so.

FYI	for your information	IOW	in other words
BTW	by the way	FWIW	for what it's worth
TIA	thanks in advance	WRT	with respect to
IMHO	in my humble opinion	PMFJI	pardon me for jumping in

FIGURE E–2. Email Abbreviations

Review and Confidentiality Implications

Email is a quick and easy way to communicate, but avoid the temptation to dash off a first draft and send it as is. As with other workplace correspondence, maintain a high level of professionalism when you send email: The information should not contain grammatical or factual errors, ambiguities, or unintended implications. Nor should it inadvertently omit crucial details. When you send an informal message to a colleague, you can correct any misunderstandings relatively easily but at the expense of wasted time. Take even more care when sending messages to superiors in your organization or to people outside the organization. Time spent reviewing your email can save a great deal of time and embarrassment sorting out misunderstandings resulting from a careless message. (See also **revision.**)

Confidentiality is another issue to keep in mind when you send email. All messages sent by email, no matter how personal, sensitive, or proprietary, can be intercepted by someone other than the intended recipient. Keep in mind that email messages are never truly deleted, even when you think you have removed them from your computer. Not only can the information be printed, circulated, and forwarded, but most companies back up and save all company email on computer tape. Be aware that employers can legally monitor email communications to protect their assets. Some companies make such a policy known, but others do not. Companies can also be compelled legally to provide email messages to a third party, like a court. Consider the content of all your messages in light of those possibilities. The potential for the unintended release of inappropriate information makes a careful review of your email text before you click "Send" all the more important.

eminent / imminent

Someone or something that is *eminent* is outstanding or distinguished.

- She is an *eminent* scientist.

If something is *imminent,* it is about to happen.

- He paced nervously, knowing that the committee's decision was *imminent.*

emphasis

Emphasis is the principle of stressing the most important ideas in writing. There are several means of achieving emphasis:

- position within a sentence, paragraph, or document
- variation in the length of sentences (see **sentence construction**)
- **repetition**
- sentence type
- climactic order within a sentence (**sentence variety**)
- punctuation (the **dash**)
- **intensifiers**
- mechanical devices, such as italics and capital letters
- direct statements (terms like *most important* and *foremost*)

Emphasis can be achieved by *position* because the first and last words of a sentence, paragraph, or document stand out in the reader's mind.

- *Moon*
  ~~Because they reflect geological history, moon~~ craters are impor-
  because they reflect geological history
  tant to understanding the earth's history.

Notice that the revised version of the sentence emphasizes *moon craters* simply because the term appears at the beginning of the sentence and *geological history* because it is at the end of the sentence.

Another way to achieve emphasis is to *vary sentence length*. A very short sentence that follows a very long sentence or a series of long sentences stands out in the reader's mind.

- We have already reviewed the problem the accounting department has experienced during the past year. We could continue to examine the causes of our problems and point an accusing finger at all the culprits beyond our control, but in the end it all leads to one simple conclusion. *We must cut costs.*

Emphasis can be achieved by the *repetition* of key words and phrases.

- Similarly, atoms *come and go* in a molecule, but the molecule *remains;* molecules *come and go* in a cell, but the cell *remains;* cells

come and go in a body, but the body *remains;* persons *come and go* in an organization, but the organization *remains.*
—Kenneth Boulding, *Beyond Economics*

Different emphasis can be achieved by the *selection of sentence type:* a compound sentence, a complex sentence, or a simple sentence.

- The report turned in by the police detective was carefully illustrated, and it covered five pages of single-spaced copy.
[This compound sentence carries no special emphasis because it contains two coordinate independent clauses.]

- The police detective's report, which was carefully illustrated, covered five pages of single-spaced copy.
[This complex sentence emphasizes the size of the report.]

- The carefully illustrated report turned in by the police detective covered five pages of single-spaced copy.
[This simple sentence emphasizes that the report was carefully illustrated.]

Emphasis can be achieved by a *climactic order* of ideas or facts within a sentence, listing them in sequence from least to most important.

- Over subsequent weeks the Human Resources Department worked diligently, management showed tact and patience, and the employees demonstrated remarkable support for the policy changes.

Emphasis can be achieved by setting an item apart with a *long dash* (also called an *em dash*).

- The job will be done — after we are under contract.

Emphasis can be achieved by the use of intensifiers (*most, very, really*), but this technique is so easily abused that it should be used with caution.

- The final proposal is *much* more persuasive than the first.

Emphasis also can be achieved by *mechanical devices,* such as italics, bold type, underlining, and capital letters, but those techniques are easily abused; for example, **email** messages that overuse capital letters are referred to as "shouting."

Finally, emphasis can be achieved by *direct statement,* such as *most important* and *foremost,* as well as **direct address.**

- *Most important* of all, keep in mind that everything you do affects the company's bottom line.

- *Andreas,* I believe we should rethink our plans.

(See also **subordination.**)

E

English as a second language (ESL)   ⓔⓢⓛ

DIRECTORY
Overview   205
Count and Mass Nouns   205
Tips for Understanding Real-Life Sources of English ⓔⓢⓛ   206
Articles   206
Gerunds and Infinitives   207
Adjective Clauses   208
Present Perfect Verb Tense   208
Present Progressive Verb Tense   209
ESL Entries   210

Overview

Learning to write well in a second language takes a great deal of effort and practice. The most effective way to improve your command of written English is to read widely beyond the reports and professional articles your job requires. Read magazine and newspaper articles, novels, biographies, short stories, and any other writing that interests you. In addition, listening carefully to native speakers on television, on radio, and in person can carry over to your written work.

Count and Mass Nouns

Count nouns refer to things that can be counted: *tables, pencils, projects, specialists.* Mass nouns identify things that cannot be counted: *electricity, water, oil, air, wood, love, loyalty, pride, harmony.*

The distinction between whether something can or cannot be counted determines the form of the noun to use (singular or plural), the kind of article that precedes it (*a, an, the,* or no article), and the kind of limiting **adjective** it requires (*fewer* or *less, much* or *many,* and so on). The distinction can be confusing with words like

206 English as a second language (ESL)

---

electricity and *oil*. Although we can count watt hours of electricity and bottles of water, counting becomes inappropriate when we use the words *electricity* and *water* in a general sense, as in "Water is an essential resource." (See also **fewer/less.**)

Articles

This discussion of articles applies only to common nouns (not to proper nouns, such as the names of people) because count and mass nouns are always common nouns.

The general rule is that every count noun must be preceded by an article (*a, an,* or *the*), a demonstrative adjective (*this, that, these, those*), a possessive adjective (*my, your, her, his, its, their*), or some expression of quantity (such as *one, two, several, many, a few, a lot of, some,* or *no*). The article, adjective, or expression of quantity appears either directly in front of the noun or in front of the whole noun phrase.

- Beth read *a* book last week. [article]

- *Those* books Beth read were long and boring.
  [demonstrative adjective]

- *Their* book was long and boring. [possessive adjective]

- *Some* books Beth read were long and boring. [indefinite adjective]

The articles *a* and *an* are used with count nouns that refer to one item of the whole class of like items.

- Matthias has *a* pen. [Matthias could have *any* pen.]

The article *the* is used with nouns that refer to a specific item that both the reader and the writer can identify.

- Matthias has *the* pen.
  [Matthias has a *specific* pen, and both the reader and the writer know which specific pen it is.]

- Matthias needs *the* space for his work.
  [Matthias needs a *specific* space, and both the reader and the writer know what space Matthias needs.]

When making generalizations with count nouns, writers can either use *a* or *an* with a singular count noun or use no article with a plural count noun. Consider the following generalization using an article.

- An egg is a good source of protein.
  [*any egg, all eggs, eggs in general*]

However, the following generalization uses a plural count noun with no article.

- Eggs are good sources of protein. [*any egg, all eggs, eggs in general*]

When you are making a generalization with a mass noun, do not use an article in front of the mass noun.

- Sugar is bad for your teeth.

(See also the discussion of articles in the entry **adjectives.**)

Gerunds and Infinitives

Nonnative writers are often puzzled by which form of a **verbal** (a verb used as another part of speech) to use when it functions as the direct **object** of a verb. No consistent rule exists for distinguishing between the use of an infinitive and a gerund as the object of a verb. Sometimes a verb takes an infinitive as its object, sometimes it takes a gerund, and sometimes it takes either an infinitive or a gerund. At times, even the base form of the verb is used.

Gerund as a complement

- He enjoys *working*.

Infinitive as a complement

* She promised *to fulfill* her part of the contract.

Gerund or infinitive as a complement

* It began *raining* soon after we arrived. [gerund]
* It began *to rain* soon after we arrived. [infinitive]

Basic verb form as a complement

* The president had the manager *assign* her staff to another project.

To make such distinctions accurately, rely on what you hear native speakers use or what you read. You might also consult a reference book for ESL students.

Adjective Clauses

Because of the variety of ways adjective clauses are constructed in different languages, they can be particularly troublesome for nonnative English writers. The following guidelines will help you form adjective clauses correctly.

Place an adjective clause directly after the noun it modifies.

* The tall woman ^*who is standing across the room*^ is a vice president of the company ~~who is standing across the room.~~

The adjective clause *who is standing across the room* modifies *woman,* not *company,* and thus comes directly after *woman.*

Avoid using a relative pronoun with another pronoun in an adjective clause.

* The man who ~~he~~ sits at that desk is my boss.

* The woman whom we met ~~her~~ at the meeting is on the board of directors.

Present Perfect Verb Tense

As a general rule, use the present perfect **tense** to refer to events completed in the past that have some implication for the present. As a rule, when a specific time is mentioned, use the simple past. Notice the difference in the following two sentences.

PRESENT PERFECT	I *have written* the letter and I am waiting for an answer. [No specific time is mentioned, but the action, *have written,* affects the present.]
SIMPLE PAST	I *wrote* the letter yesterday. [The time when the action took place is specified, and the action, *wrote,* has no relation to the present.]

E

Use the present perfect tense to describe actions that occurred in the past and have some bearing on the present.

- She *has revised* that report three times. [She might revise it again.]

- The president and his chief adviser *have met* many times over the past few months. [They may meet again.]

Use the present perfect with a *since* or *for* phrase to describe actions that began in the past and continue in the present.

- This company *has been* in business *for* ten years.

- This company *has been* in business *since* 1983.

Present Progressive Verb Tense

The present progressive tense is especially troubling for those whose native language does not use this tense. The present progressive tense is used to describe some action or condition that is ongoing (or in progress) in the present and may continue into the future.

- I *am searching* for an error in the document. [The search is occurring now and may continue.]

In contrast, the simple present tense more often relates to habitual actions.

- I *search* for errors in my documents. [I regularly search for errors in my documents, but I am not necessarily searching now.]

ESL Entries

Throughout this book, entries of particular relevance to speakers of English as a second language are marked with an ESL symbol (ESL). The following is a complete list of those entries.

a/an
abbreviations
abstract words/concrete words
adjectives
adverbs
agreement
agreement of pronouns and antecedents
agreement of subjects and verbs
amount/number
apostrophes
awkwardness
bad/badly
biased language
brackets
capital letters
case (grammar)
clarity
clauses
clipped forms of words
coherence
colons
comma splice
commas
comparison
complements
compound words
conciseness/wordiness
conjunctions
contractions
copyright
dangling modifiers
dashes
dates
diacritical marks
dictionaries
documenting sources
double negatives

English as a second language (ESL)
English, varieties of
everybody/everyone
exclamation marks
expletives
fewer/less
foreign words in English
former/latter
gender
good/well
hyphens
idioms
intensifiers
interjections
it
its/it's
lay/lie
like/as
lowercase and uppercase letters
may be/maybe
Miss/Mrs./Ms.
mixed constructions
modifiers
mood
none
nor/or
nouns
number
numbers
objects
OK/okay
one
paragraphs
parallel structure
parts of speech
periods
person
phrases

plagiarism
possessive case
prefixes
prepositions
pronoun reference
pronouns
punctuation
question marks
quotation marks
quotations
restrictive and nonrestrictive
  elements
run-on sentences
semicolons
sentence construction
sentence faults
sentence fragments
sentence variety
sexism in language
spelling

style
subordination
substantives
suffixes
synonyms
syntax
telegraphic style
tense
that
that/which/who
their/there/they're
to/too/two
tone
transition
usage
verbals
verbs
voice
word choice

E

English, varieties of

Written English includes two broad categories: standard and non-standard. Standard English is used in business, industry, government, education, and all professions. It has rigorous and precise criteria for capitalization, **diction, punctuation, spelling,** and **usage.**

Nonstandard English does not conform to such criteria; it is often regional in origin or reflects the special usages of a particular ethnic or social group. As a result, although it may be vigorous and colorful, the usefulness of nonstandard English as a means of communication is limited to certain contexts and to people already familiar and comfortable with it in those contexts. It rarely appears in printed material except for special effect. Nonstandard English is characterized by inexact or inconsistent punctuation, capitalization, spelling, diction, and usage choices; thus, you should avoid using it in workplace writing.

Colloquial English

Colloquial English is spoken English or writing that uses words and expressions common to casual conversation ("We need to get him up to speed"). Colloquial English is appropriate to some kinds of writing (personal letters, notes, some **email**) but not to most workplace writing.

Dialectal English

Dialectal English is a social or regional variety of the language that is comprehensible to people of that social group or region but that may be incomprehensible to outsiders. Dialect, which is usually nonstandard English, involves distinct **word choices,** grammatical forms, and pronunciations. For example, residents of southern Louisiana who descended from French colonists speak a dialect often referred to as Cajun.

Localisms

A localism is a regional wording or phrasing. For example, a large sandwich on a long split roll is variously known throughout the United States as a *hero, hoagie, grinder, poor boy, submarine,* and *torpedo.* The word *poke,* meaning a *bag* or *sack,* is peculiar to the southern region of the United States. Such words normally should be avoided in workplace writing because not all readers will be familiar with the local meanings.

Slang

Slang is an informal vocabulary composed of **figures of speech** and colorful words used in humorous or extravagant ways. There is no objective test for slang, and many standard words are given slang applications. For instance, slang may be a familiar word used in a new way ("She told him to chill out," meaning she told him to relax), or it may be a new word ("He's a wonk," meaning he works or studies excessively).

Most slang is short-lived and has meaning only for a narrow audience. Sometimes, however, slang becomes standard because the word fills a legitimate need. *Skyscraper* and *date* (as in "go on a date"), for example, were once considered slang expressions. Nevertheless, although slang may be valid in informal and personal writing or fiction, it generally should be avoided in workplace writing. (See also **jargon** and **technical writing style.**)

equal / unique / perfect

Logically, *equal* (meaning "having the same quantity or value as another"), *unique* (meaning "one of a kind"), and *perfect* (meaning "a state of highest excellence") are **absolute words** and therefore should not be compared. However, colloquial usage of *more* and *most* as **modifiers** of *equal, unique,* and *perfect* is so common that an absolute prohibition against such use is impossible.

- Your coin is *more unique* (or *perfect*) than mine.

Some writers try to overcome the problem by using *more nearly equal (unique, perfect).* When clarity and preciseness are critical, the use of comparative degrees with *equal, unique,* and *perfect* can be misleading. The best rule of thumb is to avoid using comparative degrees with absolute terms.

MISLEADING   Ours is a *more equal* percentage split than theirs.
PRECISE   Our percentage split is 51–49; theirs is 54–46.

-ese

The suffix *-ese* is used to designate types of **jargon** or certain languages or literary styles (computer*ese*, Chin*ese*, official*ese*, journal*ese*). Using *-ese* to coin words can become an **affectation.**

etc.

Etc. is an abbreviation for the Latin *et cetera,* meaning "and others" or "and so forth"; therefore, *etc.* should not be used with *and.*

- He brought pencils, pads, erasers, a calculator, ~~and~~ etc.

Do not use *etc.* at the end of a list or series introduced by the phrase *such as* or *for example* because those phrases already indicate that there are other things of the same category that are not named.

- She brought ~~camping items, such as~~ backpacks, sleeping bags, tents, etc., even though he didn't need them.

- She brought camping items, such as backpacks, sleeping bags,
  and
  ‸ tents, ~~etc.,~~ even though he didn't need them.

In technical writing, *etc.* should be used only when there is logical progression (1, 2, 3, etc.) and when at least two items are named. It is often better to avoid *etc.* altogether, because the reader may not be able to infer what other items the list might include.

- He brought backpacks, sleeping bags, tents, and other camping items.

ethics in writing   ESL

Ethics refers to the choices we make in what we say and do that affect others for good or for ill. Ethical issues are inherent in writing and speaking because what we write and say can influence others by the information we communicate and the language we use to communicate that information. Further, how you express ideas affects your readers' perceptions of you and your company's ethical stance. Obviously, no book can tell you how to act ethically in every situation, but here are some typical ethical lapses to watch for and address during **revision.***

- *Avoid language that attempts to evade responsibility.* Some writers use the passive **voice** because they hope to avoid responsibility or obscure an issue.

  - Several mistakes were made. [Who made them?]

  - It has been decided. [Who has decided?]

  - The product will be inspected. [Who will inspect it?]

- *Avoid language that could possibly mislead readers.* Be as precise as you can in your choice of language. Do not use words with more than one meaning, especially as a means to circumvent the truth. Consider the company document that stated, "A

---

*Adapted from Brenda R. Sims, "Linking Ethics and Language in the Technical Communication Classroom," *Technical Communication Quarterly* 2.3 (Summer 1993): 28.5–99.

nominal charge will be assessed for using our facilities." When clients objected that the charge was actually very high, the writer pointed out that the word *nominal* means "the named amount" as well as "very small." In that situation, clients had a strong case in charging that the company was attempting to be deceptive. Various **abstract words,** technical and legal **jargon,** and **euphemisms** are unethical when they are used to mislead readers or to hide a serious or dangerous situation, even though technical or legal experts could interpret them as accurate. (See also **word choice.**)

E

- *Do not deemphasize or suppress important information.* Not including information that a reader would want to have, such as safety hazards or costs for which a customer might be responsible, is unethical. For example, failing to mention the disadvantages of a product could easily mislead readers. Not only is that kind of half-truth unethical, it may be illegal as well. (See also **copyright** and **plagiarism.**) Use **layout and design** to highlight, not hide, information that is important to readers. Deemphasizing a negative feature of a product or service or obscuring a customer fee by using a very small typeface or a footnote to bury the information could be perceived as suppressing important information. Even highlighting advantages in a bulleted list and then burying a disadvantage in the middle of a paragraph could unfairly mislead readers.

- *Do not emphasize misleading or incorrect information.* Similarly, avoid the temptation to highlight a feature or service that readers would find attractive but that is available only with some product models or at extra cost. (See also **logic errors** and **positive writing.**) Readers could justifiably object that you have given them a false impression in order to sell a product or service, especially if you also deemphasized the extra cost or other special conditions. This technique is as unsavory as the bait-and-switch tactic, in which an advertised product is "sold out" when customers ask for it, and customers are then directed to a similar but more expensive product.

- *Treat others fairly and respectfully.* Adhere to the golden rule in your writing by treating others — people, companies, projects — fairly and with respect. Avoid language that is biased, racist, or sexist or that perpetuates stereotypes. Men and women work together, as do people of different ethnic

E

backgrounds, cultures, religious practices, and native languages. Respect those differences rather than writing in ways that emphasize them inappropriately and that could be offensive to others. (See also **biased language.**) Likewise, be fair and respectful to colleagues and professional competitors alike. Do not make damaging, malicious statements about others without regard for the truth. You would wish no less from others (and you and your company could be held liable for such statements).

In technical writing, guard against false, fabricated, or plagiarized research and test results. As an author, technical reviewer, or editor, your ethical obligation is to correct or point out any misrepresentations of fact before publication, whether the publication is a technical journal article, laboratory **test report,** or product handbook. The stakes of such ethical oversights are high because of the potential for technology to affect the health and safety of others. Those at risk can include unwary consumers and workers injured because of faulty products or unprotected exposure to toxic materials.

Writer's Checklist: Ethical Writing

To avoid ethical problems, ask the following questions as you revise your writing.

- ☑ *Is the document honest and truthful?* Scrutinize findings and conclusions carefully. Make sure that the data support them.

- ☑ *Am I acting in my employer's best interest? My client's or the public's best interest? My own best long-term interest?* Your writing reflects on you and your employer. Review it from the perspective of its intended effect and have someone outside your company review and comment on what you have said.

- ☑ *What if everybody acted or communicated in this way?* Apply the golden rule: If you were the intended audience, would the message be acceptable and respectful?

- ☑ *Am I willing to take responsibility, publicly and privately, for what the document says?* Will you stand behind what you have written? To your employer? To your family and friends?

Writer's Checklist: Ethical Writing (continued)

☑ *Does the document violate anyone's rights?* Have people from different backgrounds review your writing and consider what it says from their perspective.

☑ *Am I ethically consistent in my writing?* Only by the consistent application of the principles sketched here and those you have assimilated throughout your life can you meet this standard.

euphemisms

A *euphemism* is a word that is an inoffensive substitute for one that could be distasteful, offensive, or too blunt.

• *remains* for *corpse*

• *passed away* for *died*

• *previously owned* or *preowned* for *used*

Used judiciously, a euphemism can help you avoid embarrassing or offending someone. Overused, however, euphemisms can hide the facts of a situation (such as *incident* or *event* for *accident*) or be a form of **affectation.** (See also **ethics in writing** and **word choice.**)

everybody / everyone

Both *everybody* and *everyone* are usually considered singular and take singular **verbs** and **pronouns.**

• *Everybody is* happy with the new contract.

• *Everyone* here *leaves* at 4:30 p.m.

• *Everybody* at the meeting made *his or her* proposal separately.

• *Everyone* went *his or her* separate way after the meeting.

But the meaning can be obviously plural.

• *Everyone* thought the report should be revised, and I really couldn't blame *them.*

When the use of singular verbs and pronouns would be offensive by implying sexual bias, it is better to use plural verbs and pronouns or to use the expression "his or her."

- *They all* ~~Everyone~~ went ~~his~~ *their* separate ~~way~~ *ways* after the meeting.
- Everyone went his *or her* separate way after the meeting.

Although normally written as one word, *everyone* is written as two words to emphasize each individual in a group.

- *Every one* of the team members contributed to this discovery.

exclamation marks   ESL

The exclamation mark (!) indicates strong feeling. The most common use of an exclamation mark is after a word, phrase, clause, or sentence to indicate urgency, elation, or surprise. (See also **interjections.**)

- Hurry!
- Great!
- Wow!

In technical writing, the exclamation mark is often used in cautions and warnings.

- Notice!
- Stop!
- Danger!

An exclamation mark can be used after a whole sentence or an element of a sentence.

- The subject of this meeting—please note it well!—is our budget deficit.

Keep in mind that an exclamation mark cannot make an argument more convincing, lend force to a weak statement, or call attention to an intended irony.

An exclamation mark can be used after a title that is an exclamatory word, phrase, or sentence.

- "Our International Perspective Must Change!" is an article by Richard Moody.

When used with quotation marks, the exclamation mark goes outside, unless what is quoted is an exclamation.

- The manager yelled, "Get in here!" Then Ben, according to Ray, "jumped like a kangaroo"!

executive summaries

An executive summary consolidates the principal points of a **report** or **proposal.** An executive summary must accurately and concisely reflect the original document so the reader can glean the document's significance without having to read it in full. Because they are comprehensive, executive summaries tend to be proportional in length to the works they summarize. The typical summary is about 10 percent the length of the original and follows the same sequence.

Write the executive summary so it can be read independently of the report or proposal. Do not refer by number to figures, tables, or references contained elsewhere in the original. Executive summaries may occasionally contain a figure, table, or footnote if that information is integral to the summary. Because executive summaries frequently are read in place of the full document, spell out all uncommon **symbols, abbreviations,** and **acronyms.**

Writer's Checklist: Executive Summaries

- ☑ Write the executive summary after you have completed the original document.
- ☑ Avoid using terminology that may not be familiar to your readers.
- ☑ Make the summary concise but not brusque. Be especially careful not to omit transitional words and phrases (such as *however, moreover, therefore, for example,* and *in summary*).
- ☑ Finally, include only information discussed in the original document.

The sample executive summary in Figure E–3 on pages 220–22 is adapted from a ten-page report that describes how electric utilities in a select group of countries provide engineering expertise to control room operators at nuclear power plants.

For guidance about the location of an executive summary in a report or other document, see **formal reports.** (See also **abstracts.**)

E

EXECUTIVE SUMMARY

Introduction

Purpose and scope

This report describes the experiences and practices of utilities in a selected group of countries with providing engineering expertise on shift in nuclear power plants.[1] The report also discusses the extent to which engineering expertise is made available and the alternative models of providing such expertise. The implications of the foreign experience for plants in the United States is described, particularly with reference to the shift technical adviser position and to a proposed shift engineer position.

Methods

The relevant information for this study came from the open literature, interviews with utility staff, and utility reports.

The major conclusions that emerge from this study include the following:

- The basis for the initial decision of whether to include a graduate engineer in the operations shift complement is unclear for most countries. However, the reasons for changes in a given system have been identified. Generally, changes were introduced in response to specific problems (such as high turnover rates among crew members, or an accident seen as related to available shift engineering expertise).

Major findings

- Two primary models are being used to provide engineering expertise on shift:

   1. The graduate engineer as line manager of shift operations (usually in a shift supervisor position)
   2. The graduate engineer in a nonsupervisory position on shift

- A comparison of these two models did not indicate that one inherently functions more effectively than the other. However, each alternative appears to affect the following specific areas differently:

---

[1]For the purpose of this discussion, "engineering expertise on shift" refers to the use of university-degreed engineers during round-the-clock shift work.

FIGURE E–3. Sample Executive Summary

1. Crew relationships and performance
2. Labor supply, recruitment, and retention
3. System implementation problems

- Data were not available for analyzing whether alternative approaches to providing engineering expertise on shift could be directly linked to plant safety in terms of accident frequency and severity. Indirect relationships to operational safety appear to be primarily through impacts on operational functioning (for example, turnover and crew relationships) that ultimately are likely to affect safe operations.

Findings

- The determination of which alternative model would be most appropriate for a given country needs to be made from a systems perspective.

- Decisions about engineering expertise on shift should not be made independently of staffing patterns and organizational design regarding the following issues:

  1. Effects on incumbent reactor operations staff of changed career opportunities
  2. Availability of degreed engineers and reactor operators
  3. Career paths for graduate engineers at nuclear power plants
  4. Organizational relationships between the engineering and operations sections of nuclear power plants

Engineering Expertise Alternatives

In the countries surveyed, essentially two approaches are being employed to make engineering expertise available on shift: (1) a graduate engineer occupies a line management position, and (2) a specific engineering position is created to provide expertise to the operations staff. Both approaches are relevant to issues under consideration in the United States. In the United States, proposals have been made to require that all licensed reactor operators have engineering degrees. This proposed requirement would apply to reactor operators, senior reactor operators, and shift supervisors. Proposals have also been made to create a position for a degreed shift engineer in addition to the current requirement for a shift technical adviser.

Currently, Spain, Italy, and Japan do not, as a rule, use graduate engineers on shift. On the other hand, Canada, Germany, and the United Kingdom do have an engineering graduate on shift in operations line management positions or have made the decision to establish this policy. The use of a separate engineer position on shift is found in Switzerland, France, and Sweden.

Conclusions and Recommendations

The foreign experiences described in this report suggest some key issues relevant to the requirement of a shift technical adviser and

FIGURE E-3. Sample Executive Summary (*continued*)

E

requiring baccalaureate engineering degrees for shift supervisors. Both models—operations line management and a separate engineering position—involve some trade-offs with regard to system advantages.

One important issue concerns whether the separation of engineers from routine operational functions enhances or detracts from their ability to diagnose problems and make appropriate decisions in situations of operational failure. Insufficient evidence exists to answer this question. Both options have adherents based on positive experience with a given system.

Using a degreed engineer as shift supervisor appears to have advantages for crew integration and the combination of technical expertise and operations experience in the major position of authority on shift. However, significant problems have been experienced with the retention of graduate engineers as shift supervisors. Additionally, if this requirement were imposed in the United States, it would limit the career prospects for incumbent reactor operators, which could have adverse consequence on crew performance.

Recom-
mendations

Creating a separate nonsupervisory engineering position could have advantages in providing job functions and a career path that more fully utilize engineering expertise than the shift supervisor position does. In addition, such a position does not require a change in the established recruitment and career path open to existing reactor operators.

FIGURE E–3. Sample Executive Summary (*continued*)

explaining a process (*see* process explanation)

expletives   ESL

An *expletive* is a word that fills the position of another word, phrase, or clause. *It* and *there* are common expletives.

- *It* is certain that he will go.

In the example, the expletive *it* occupies the position of subject in place of the real subject, *that he will go.* Expletives are sometimes necessary to avoid **awkwardness,** but they are commonly overused, and most sentences can be better stated without them.

I did it for
- ~~There are~~ several reasons ~~that I did it~~ .
  ^

Many   *were*
- ~~There were many~~ orders lost because of a computer failure.
  ^   ^

- *We lost*
  ~~There were~~ many orders ~~lost~~ because of a computer failure.
  ⌃

In addition to its usage as a grammatical term, the word *expletive* means an exclamation or oath, especially one that is profane.

explicit/implicit

An *explicit* statement is one expressed directly, with precision and clarity. An *implicit* meaning is one that is not directly expressed.

- His directions to the Wausau facility were *explicit*, and we found it with no trouble.

- Although the CEO did not mention the company's financial condition, the danger of overconfidence was *implicit* in her speech.

exposition

Exposition, or expository writing, informs readers by presenting facts and ideas in direct and concise language; it usually relies less on colorful or figurative language than does writing meant to be expressive or persuasive. Expository writing attempts to explain to readers what the subject is, how it works, and how it relates to something else. Exposition is aimed at the readers' understanding rather than at their imagination or emotions; it is a sharing of the writer's knowledge. Exposition aims to provide accurate, complete information and to analyze it for the readers.

Because it is the most effective **form of discourse** for explaining difficult subjects, exposition is widely used in **reports, memos,** and other types of technical writing. To write exposition, you must have a thorough knowledge of your subject. As with all writing, how much of that knowledge you pass on to your readers depends on the **readers'** needs and your **purpose.**

F

fact

Expressions containing the word *fact* ("due to the *fact* that," "except for the *fact* that," "as a matter of *fact*," or "because of the *fact* that") are often wordy substitutes for more accurate terms.

- ~~Due to the fact that~~ *Because* the technical staff has a high turnover rate, our training program has suffered.

Do not use the word *fact* to refer to matters of judgment or opinion.

- ~~It is a fact that~~ *In my opinion,* our research has improved because we now have a capable technical staff.

The word *fact* is, of course, valid when facts are what is meant.

- Our tests have uncovered numerous *facts* to support your conclusion.

(See also **conciseness/wordiness** and **logic errors.**)

fax (facsimile)

Fax (for facsimile transmission) uses telephone connections to transmit images of whole pages — text and graphics — from one fax machine to another. A fax page takes about 20 seconds to reach its destination, be it local, coast to coast, or international, at the cost of a phone call. Faxes are used primarily to transmit information to locations without **Internet** access or to send documents with nondigital elements, such as handwritten corrections and notes, that must be viewed as originally created. When you have to send a nondigital drawing or diagram or a document like a contract that contains one or more signatures, a fax is the preferred medium because it ensures authenticity. Keep in mind, however, that faxes are almost never as clear as the original document.

When you transmit fax messages, also be aware that they can be intercepted by persons other than those for whom they were intended because fax machines are often in central locations. If you have to fax confidential or sensitive messages, call the intended recipient first so he or she can be waiting at the machine as you transmit the fax. Many people today prefer to send information as **email** attachments because they not only provide greater confidentiality but also offer the benefit of digital transmission and storage.

F

feasibility reports

When managers plan to undertake a new project (develop a new product or service, expand a customer base, purchase new equipment, or consider a move), they first try to determine the project's chances for success. A feasibility report documents the study a manager conducts to help make that determination and presents evidence about the practicality of the proposed project: What are the costs involved? How soon can the costs be recovered? Is sufficient staff available? Are there any legal or other special requirements? Based on the evidence, the feasibility report writer recommends whether the project should be carried out.

Before beginning to write a feasibility report, state clearly and concisely the **purpose** of the study.

- The purpose of this study is to determine the best approaches to expanding our international operations.

Report Sections

Every feasibility report should contain the following sections: (1) an **introduction,** (2) the body, (3) a **conclusion,** and (4) a recommendation. (See also **proposals** and **formal reports.**)

Introduction.   The introduction states the purpose of the report, describes the circumstances that led to the report, and includes any pertinent background information. It may also discuss the scope or extent of the report and any procedures or methods used in the analyses of alternatives. Note any limitations on the study here as well.

Body.   The body of the report presents a detailed evaluation of all the alternatives under consideration. Evaluate each alternative according to specific criteria. Ordinarily, separate subsections cover the individual evaluations.

Conclusion.   The conclusion summarizes the evaluation of alternatives, usually in the order in which they are discussed in the body of the report. This summary of relative strengths and weaknesses usually points to one alternative as the best or most feasible.

Recommendation.   This section presents the writer's opinion on which alternative best meets the criteria as summarized in the conclusion.

Typical Feasibility Report

Consider a scientific and engineering consulting firm that needs to upgrade its primary computer system and Internet capability. The staff of this firm might conduct a feasibility study to determine which mainframe with peripherals and enterprise software would best suit their requirements. The organization's requirements establish the criteria by which each alternative is evaluated. The following outline shows how the preliminary topics for the feasibility study might be organized, and sections from the resulting feasibility report are presented in Figure F–1 on pages 227–30.

  I. Introduction
    A. Purpose (to determine which of two proposed options would be best for our technical, Internet, and operational needs)
    B. Scope (detailed discussion of the costs and benefits of two options)
 II. Criteria for comparison
    A. Costs of systems
    B. System reliability and readiness
    C. Capacity of systems
III. Alternatives under consideration
    A. Second ARC 98 processor
    B. Purchase HRS 60/EP package

IV. Conclusion
   A. Relative cost of options
   B. Analysis of options
V. Recommendation (HRS 60/EP)

F

Introduction

The purpose of this report is to determine which of two proposed options would best enable ACM Technology Consulting to upgrade its mainframe computer system and its Internet capacity to meet its increasing data and communication requirements.

Background

In October 1998, the Information Development and Technical Support Group at ACM put the MISSION System into operation. Since then, the volume of processing transactions has increased fivefold (from 1,000 to 5,000 updates per day). This increase has severely impaired system response time; in fact, average response time has increased from less than 10 seconds to 120 seconds. Further, our new Web-based client services system has increased exponentially the demand for processing speed and access capacity.

Diminished performance is also apparent in the backlog of batch-processing transactions. During a recent check, 70 real-time and approximately 2,000 secondary transactions were backlogged. In addition, the ARC 98 processor that runs MISSION is three years old and requires frequent maintenance. Downtime caused by those repairs must be made up in overtime. In a recent 10-day period in January, processor downtime averaged 25 percent during work hours (7:30 a.m. to 6:00 p.m.). In February, the system was down often enough to endanger the entire schedule.

Scope

Two alternative solutions to provide increased processing capacity have been investigated: (1) purchase of a new ARC 98 processor to supplement the first, and (2) purchase of an HRS 60/EP with PRS enterprise software and expandable Internet peripherals to replace the current ARC 98. The two alternatives are evaluated here primarily according to cost and, to a lesser extent, according to expanded capacity for future operations.

Purchasing a Second ARC 98 Processor

This alternative would require additional annual maintenance costs, salary for an additional computer specialist, increased energy costs, and a one-time construction cost for necessary remodeling as well as installing Internet and other connections.

FIGURE F–1. Typical Feasibility Report

Annual maintenance costs	$35,000	
Annual costs for computer specialist	75,000	
Annual increased energy costs	7,500	
Annual operating costs		$117,500
Construction cost (one-time)		50,000
Total first-year cost		$167,500

The costs for the installation and operation of another ARC 98 processor are expected to produce savings in system reliability and readiness.

System Reliability

A second ARC 98 would reduce current downtime periods from four to two per week. Downtime recovery averages 30 minutes and affects 40 users. Assuming that 50 percent of users require the system at a given time, we determined that the following reliability savings would result:

$$2 \text{ downtimes} \times 0.5 \text{ hours} \times 40 \text{ users} \times 50\%$$
$$\times \$12.00/\text{hour overtime} \times 52 \text{ weeks} = \$12,480 \text{ (annual savings)}$$

System Readiness

Currently, an average of one day of batch processing per week cannot be completed. That gap prevents online system readiness when users report to work and affects all users at least one hour per week. Improved productivity would yield these savings:

$$40 \text{ users} \times 1 \text{ hour/week} \times \$9.00/\text{hour average wage rate}$$
$$\times 52 \text{ weeks} = \$18,720 \text{ (annual savings)}$$

Summary of Savings

System reliability	$12,480
System readiness	18,720
Total annual savings	$31,200

. . . .

ARC 98 Capacity

By adding a second ARC 98 processor, current capacity would be doubled. Each processor could process 2,500 transactions per day while cutting response time from 120 seconds to 60 seconds. However, if new systems essential to increased plant productivity are added to the MISSION System, efficiency could be degraded to its current level in the next three to five years. That estimate is based on the assumption that the new systems would add between 250 and 500 transactions per day immediately. Those figures could increase tenfold in the next several years if current rates of expansion continue.

Purchasing an HRS 60/EP Package

Purchasing an HRS 60/EP with PRS enterprise software and expandable Internet peripherals requires additional annual maintenance costs, increased energy costs, and a one-time facility adaptation cost.

Annual maintenance costs	$75,000	
Annual increased energy costs	9,000	
Annual operating costs		$84,000
Cost of adapting existing facility		24,500
Total first-year cost		$108,500

The installation costs are expected to produce savings in system reliability, system readiness, and staffing for the Information Systems and Support Group.

System Reliability
Annual savings will be the same as those for the ARC 98 processor: $12,480.

System Readiness
Annual savings will be the same as those for the ARC 98 processor: $18,720.

Wages for the Information Systems and Support Group
New system efficiencies would permit the following wage reductions in the department:

One computer specialist (wages and fringe benefits)	$75,000
One-shift overtime premium (at $200/week × 52 weeks)	10,400
Total annual wage savings	$85,400

. . . .

HRS 60/EP Capacity
The HRS 60/EP processor can process 5,000 transactions per day with an average response time of 10 seconds per transaction. Should the volume of future transactions double, the HRS 60/EP could process 10,000 transactions per day without exceeding 20 seconds per transaction on average. The increase in capacity over the present system would permit implementation of plans to add four new systems to MISSION.

Conclusion

A comparison of costs for both systems indicates that the HRS 60/EP would cost $2,200 more in first-year costs.

ARC 98 Costs

Net additional operating	$56,300
One-time (construction)	50,000
First-year total	$106,300

HRS 60/EP Costs

Net additional operating	$84,000
One-time (facility)	24,500
First-year total	$108,500

Installation of a second ARC 98 processor would permit the present information-processing systems to operate relatively smoothly and efficiently.

FIGURE F–1. Typical Feasibility Report (*continued*)

It would not, however, provide the expanded processing capacity that the HRS 60/EP processor would for implementing new subsystems required to increase processing speed and Internet access.

Recommendation

The HRS 60/EP processor should be purchased because of the long-term savings and because its additional capacity and flexibility will allow for greater expansion in the future.

FIGURE F–1. Typical Feasibility Report (*continued*)

female

The term *female* is usually restricted to scientific, legal, and medical contexts (a female patient, a female suspect). Keep in mind that the term sounds cold and impersonal. *Girl, woman,* and *lady* are acceptable substitutes in other contexts, but keep in mind that they have connotations involving age, dignity, and social position. (See also **biased language** and **male.**)

few / a few

In certain contexts, *few* carries more negative overtones than the phrase *a few* does.

POSITIVE   There are *a few* good points in your report.
NEGATIVE   There are *few* good points in your report.

fewer / less   ESL

Fewer refers to items that can be counted (count **nouns**).

- *Fewer* employees took the offer than we expected.

Less refers to mass quantities or amounts (mass nouns).

- The crop yield decreased this year because we had *less* rain than necessary for an optimum yield.

figuratively / literally

These two words are often confused. *Literally* means "really" and should not be used in place of *figuratively*, which means "metaphorically." To say that someone "literally turned green with envy" would mean that the person actually changed color.

- In the winner's circle the jockey was, *figuratively* speaking, ten feet tall.

- When he said, "Let's kill our competitors," he did not mean it *literally*.

Avoid the use of *literally* to reinforce the importance of something.

- She was ~~literally~~ the best of the group.

figures of speech

A figure of speech is an imaginative comparison, either stated or implied, between two things that are basically not alike but have at least one thing in common. For example, if a device is cone-shaped and has an opening at the top, you might say that it looks like a volcano.

Technical writers sometimes use figures of speech to clarify the unfamiliar by relating a new and difficult concept to one with which their readers are familiar. In that respect, figures of speech help establish a common ground of understanding between the specialist and the nonspecialist. Technical writers also use figures of speech to help translate the abstract into the concrete; in the process of doing so, figures of speech also make writing more colorful and graphic. (See also **abstract words/concrete words.**) A figure of speech must be appropriate, however, to achieve the desired effect.

INAPPROPRIATE	Without the fuel of tax incentives, our economic engine would operate less efficiently. [It would not operate at all without fuel.]
APPROPRIATE	Without the fuel of tax incentives, our economic engine would sputter and die. [This is not only apt, it also states a rather dry fact in a colorful manner.]

A figure of speech also must be consistent to be effective.

INCONSISTENT   We must get our research program *back on track,* and we are counting on you to *carry the ball.*

CONSISTENT   We must get our research program back on track, and we are counting on you to do it.

A figure of speech should not overshadow the point the writer is trying to make by attracting more attention to itself.

- The whine of the engine sounded like ten thousand cats having their tails pulled by ten thousand mischievous children.

Trite figures of speech, which are called *clichés,* defeat the purpose of a figure of speech—to be fresh, original, and vivid. A surprise that comes "like a bolt out of the blue" seems stale and not much of a surprise. It is better to use no figure of speech at all than to use a trite one. (See **clichés** and **trite language.**)

Types of Figures of Speech

Analogy is a comparison between two objects or concepts that shows the ways in which they are similar. The following example explains a computer search technique for finding information in a data file by comparing it to the method used by most people to find a word in a dictionary.

- The search technique used in *indexed sequential processing* is similar to the search technique used to look up a word in a dictionary. To locate a specific word, you scan the keywords located at the top of each dictionary page that identify the first and last words on each page until you find the keywords that encompass the word you seek. Assume that all the keywords that refer to the last word on each dictionary page were placed in a file, along with their corresponding page numbers, and that all the words in the dictionary were placed in another file. To locate any dictionary word, you simply scan the first (keyword) file until you find a keyword that alphabetically follows the word you seek and go to the place in the second (dictionary) file indicated by the keyword.

Antithesis is a statement in which two contrasting ideas are set off against each other in a balanced syntactic structure.

- Art is long, but life is short.

Hyperbole is gross exaggeration used to achieve an effect or emphasis.

- She *murdered* me on the tennis court.

Litotes are understatements, for emphasis or effect, achieved by denying the opposite of the point you are making.

- Ninety dollars is not a small price for a book.

Metaphor is a figure of speech that points out similarities between two things by treating them as though they were the same thing.

- He is the sales department's *utility infielder.*

Metonymy is a figure of speech that uses one aspect of a thing to represent it, such as *the blue* for the sky and *wheels* for an automobile.

- *The hard-hat* sector of the labor force was especially hurt by the terms of the contract.

Personification is a figure of speech that attributes human characteristics to nonhuman things or abstract ideas. We characteristically speak of the *birth* of a planet and the *stubbornness* of an engine that will not start.

- She said that the proposal was *dead* on arrival.

Simile is a direct comparison of two essentially unlike things, linking them with the word *like* or *as.*

- His feelings about his business rival are so bitter that in recent conversations with his staff he returned to the subject compulsively, *like someone scratching an itch.*

fine

When used in expressions such as "I feel *fine*" or "a *fine* day," *fine* is colloquial. The colloquial use of *fine*, like that of *nice*, is too vague for technical writing. Use the word *fine* to mean "refined," "delicate," or "pure."

- A *fine* film of oil covered the surface of the water.

- She made a *fine* distinction between the two possible meanings of the disputed passage.

- *Fine* crystal is made in Austria.

F

first / firstly

Firstly is an unnecessary attempt to add the *-ly* **suffix** to an **adjective** to form an **adverb.** *First* is an adverb in its own right and sounds much less stiff than *firstly*. The same is true of other ordinal **numbers.**

- *First*
  ~~Firstly~~, we should ask for an estimate.
  ^

flammable / inflammable / nonflammable

Both *flammable* and *inflammable* mean "capable of being set on fire." Because the *in-* **prefix** usually causes the base word to take its opposite meaning (*incapable, incompetent*), use *flammable* instead of *inflammable* to avoid possible misunderstanding.

- The cargo of gasoline is *flammable*.

Nonflammable is the opposite, meaning "not capable of being set on fire."

- The asbestos suit was *nonflammable*.

flowcharts

A flowchart is a diagram of a process that involves stages, shown in sequence from beginning to end. A flowchart provides an overview of a process and allows the reader to identify its essential steps quickly and easily. The process could range from the steps involved in assembling a bicycle to the stages in programming instructions for converting a software application from one computer system to another.

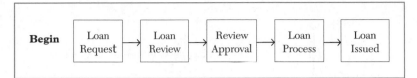

FIGURE F–2. Flowchart Using Labeled Blocks

Flowcharts can take several forms: The steps might be represented by labeled blocks, as shown in Figure F–2; pictorial symbols,
as shown in Figure F–3; or International Organization for Stan-
dardization (ISO) symbols (see "Writer's Checklist: Flowcharts" at
the end of this entry), as shown in Figure F–4 on page 236.

For advice on integrating flowcharts into your text, see also **illustrations, global graphics,** and **graphs.**

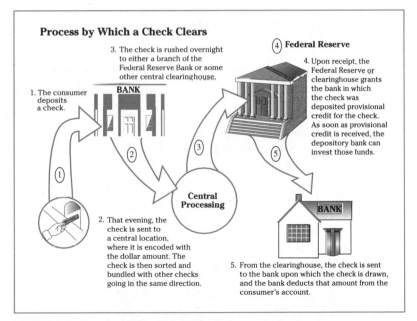

FIGURE F–3. Flowchart Using Pictorial Symbols

F

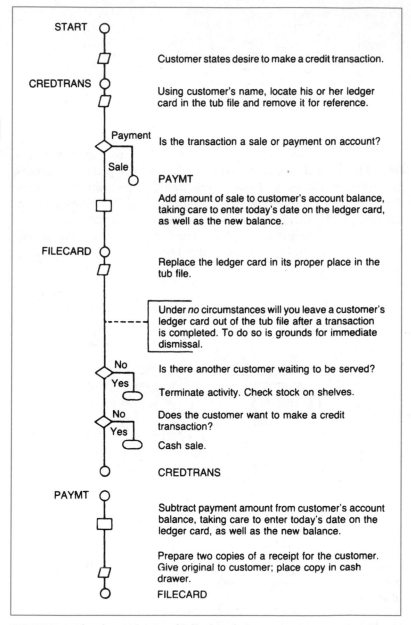

START

Customer states desire to make a credit transaction.

CREDTRANS

Using customer's name, locate his or her ledger card in the tub file and remove it for reference.

Payment

Is the transaction a sale or payment on account?

Sale

PAYMT

Add amount of sale to customer's account balance, taking care to enter today's date on the ledger card, as well as the new balance.

FILECARD

Replace the ledger card in its proper place in the tub file.

Under *no* circumstances will you leave a customer's ledger card out of the tub file after a transaction is completed. To do so is grounds for immediate dismissal.

No

Is there another customer waiting to be served?

Yes

Terminate activity. Check stock on shelves.

No

Does the customer want to make a credit transaction?

Yes

Cash sale.

CREDTRANS

PAYMT

Subtract payment amount from customer's account balance, taking care to enter today's date on the ledger card, as well as the new balance.

Prepare two copies of a receipt for the customer. Give original to customer; place copy in cash drawer.

FILECARD

FIGURE F–4. Flowchart Using Standardized Symbols

Writer's Checklist: Flowcharts

When creating a flowchart, keep the following guidelines in mind.

- ☑ Label each step in the process or identify it with labeled blocks, pictorial representations, or standardized symbols.
- ☑ Follow the standard flow directions: left to right and top to bottom. When the flow is otherwise or occurs in both directions, be sure to indicate that with arrows.
- ☑ Include a key if the flowchart contains symbols your readers may not understand.
- ☑ For flowcharts that document computer programs and other information-processing procedures, use standardized symbols set forth in *Information Processing — Documentation Symbols and Conventions for Data, Program and System Flowcharts, Program Network Charts, and System Resources Charts,* ISO publication 5807-1985 (E).

F

footnotes (*see* documenting sources)

forceful/forcible

Although *forceful* and *forcible* are both **adjectives** meaning "characterized by or full of force," *forceful* is usually limited to persuasive ability and *forcible* to physical force.

- John made a *forceful* presentation at the committee meeting.

- A witness said that the thief made a *forcible* entry into the apartment.

foreign words in English   ESL

The English language has a long history of borrowing words from other languages. Most borrowings occurred so long ago that we seldom recognize the borrowed terms (also called *loan words*) as being of foreign origin.

GERMAN	kindergarten
LATIN	animal
GREEK	church

Foreign expressions should be used only if they serve a real need. The overuse of foreign words in an attempt to impress your reader or achieve elegance is **affectation.** Effective communication can be accomplished only if your readers understand what you write; therefore, choose foreign expressions only when they make an idea clearer.

Words not fully assimilated into the English language are set in **italics** (or underlined when italic type is not available).

- *sine qua non, coup de grâce, in res, in camera*

Words and abbreviations that have been fully assimilated need not be italicized. When in doubt, consult a current dictionary.

- cliché, etiquette, vis-à-vis, de facto, résumé, habeas corpus, per diem

As foreign words are absorbed into English, their plural forms give way to English plurals.

- *formulae* becomes *formulas*

- *agenda* becomes *agendas*

Additionally, accent marks tend to be dropped from words (especially French words) the longer they are used in English. However, not all foreign words shed their accent marks with the passage of time, so consult a dictionary whenever you are in doubt. (See also **e.g./i.e.** and **etc.**)

---

foreword / forward

Although the pronunciation is the same, the spellings and meanings of these two words are quite different. The word *foreword* is a **noun** meaning "introductory statement at the beginning of a book or other work."

NOUN   The department chair was asked to write a *foreword* for the professor's book.

The word *forward* is an **adjective** or **adverb** meaning "at or toward the front."

| ADJECTIVE | Move the lever to the *forward* position on the panel. |
| ADVERB | Turn the dial until the needle begins to move *forward*. |

foreword / preface (*see* **formal reports**)

formal reports

Formal reports are written accounts of major projects. Most formal reports are divided into three primary parts—front matter, body, and back matter—each of which contains a number of elements. The number and arrangement of the elements may vary, depending on the subject, the length of the **report,** and the kinds of material covered.

Many organizations have a preferred style for formal reports and furnish guidelines for report writers to follow. If your employer does not, use the format recommended in this entry. The following list includes most of the elements a formal report might contain, in the order of their appearance in the report. Often, a **cover letter** or **memo** precedes the front matter.

Front matter
  Title page
  Abstract
  Table of contents
  List of figures
  List of tables
  Foreword
  Preface
  List of abbreviations and symbols
Body
  Executive summary
  Introduction
  Text (including headings)
  Conclusions
  Recommendations
  References

Back matter
  Appendixes
  Bibliography
  Glossary
  Index

Front Matter

F

The front matter serves several purposes: It gives the **reader** a general idea of the author's **purpose;** it gives an overview of the type of information in the report; and it lists where specific information is covered in the report. Not all formal reports include every element of front matter described here. A title page and a table of contents are usually mandatory, but the **scope** of the report and its intended audience determine whether the other elements are included.

Title Page.   Although the formats of title pages for formal reports vary, a title page should include the following information:

* *The full title of the report.* The title should reflect the topic, scope, and purpose of the report. Follow these guidelines when creating the title:
  * Avoid titles that begin with *Notes on, Studies on, A Report on,* or *Observations on.* Those phrases are redundant and state the obvious. However, phrases such as *Annual Report* and *Feasibility Study* should be used in a title or subtitle because they help define the purpose and scope of the report.
  * Do not use abbreviations in the title. Use **acronyms** only when the report's audience is familiar with the topic.
  * Do not include the time period covered by the report in the title; include that information in a subtitle.

EFFECTS OF PROPOSED HIGHWAY
CONSTRUCTION ON PROPERTY VALUES
Annual Report, 20--

* *The name of the writer, principal investigator, or compiler.* Sometimes contributors identify themselves by their job title in the organization (Lydia R. Zermeno, Cost Analyst; John T. Yeung, Head, Research and Development), and sometimes they identify themselves by their tasks in contributing to the report (Gina Hobbs, Principal Investigator; Matthew Hutz, Compiler).

- *The date or dates of the report.* For one-time reports, list the date when the report is to be distributed. For periodic reports, such as those issued monthly or quarterly, list in a subtitle the period that the report covers; elsewhere on the title page, list the date when the report is to be distributed.
- *The name of the organization for which the writer works.*
- *The name of the organization to which the report is being submitted.* This information is included if the report is written for a customer or client, as shown in Figure F–5.

The title page, although unnumbered, is considered page *i* (small roman numeral *i*). The back of the title page, which is blank and unnumbered, is page *ii,* and the abstract falls on page *iii.* The body of the report begins with Arabic number 1, and new chapters or large sections typically begin on a new right-hand (odd-numbered) page. Reports with printing on only one side of each sheet can be numbered consecutively regardless of where new sections begin. Throughout the report, center page numbers at the bottom of the page.

Internet Development and Implementation:
Web Site Design, Telecommunication Infrastructure, and Personnel Training

Enviropaper Corporation

Gina L. Hobbs, Principal Investigator
Internet Solutions and Consulting
April 11, 20--

FIGURE F–5. Typical Title Page

Abstract. An **abstract,** which normally follows the title page, highlights the major points of the report, enabling readers to decide whether to read the entire report.

Table of Contents. A **table of contents** lists all the major sections or **headings** of the report in their order of appearance, along with their page numbers.

List of Figures. When a report contains more than five figures, list them, along with their page numbers, in a separate section beginning on a new page and immediately following the table of contents. Number figures consecutively with Arabic numbers. Figures include all **illustrations—drawings, photographs, maps,** charts, and **graphs**—contained in the report.

List of Tables. When a report contains more than five **tables,** list them, along with their titles and page numbers, in a separate section immediately following the list of figures (if there is one). Number tables consecutively with Arabic numbers.

Foreword. A foreword is an optional introductory statement written by someone other than the author. The foreword author is usually an authority in the field or an executive of the company. The foreword author's name and affiliation and the date the foreword was written appear at the end of it. The foreword generally provides background information about the publication's significance and places it in context of other works in the field.

Preface. The preface is an optional introductory statement written by the author of the report. It announces the purpose, background, and scope of the report. Typically, it highlights the relationship of the report to a given project or program and discusses any special circumstances leading to the work. A preface may also specify the audience for whom the report is intended, and it may acknowledge those who helped during the course of the project or in the preparation of the report. Finally, a preface may cite permission obtained for the use of copyrighted works. (See also **copyright.**)

List of Abbreviations and Symbols. When the report uses numerous abbreviations and symbols and there is a chance that readers will not be able to interpret them, the front matter should include a list of all symbols and abbreviations (including acronyms) and what they stand for.

Body

The body is the section of the report in which the author describes in detail the methods and procedures used to generate the report, demonstrates how results were obtained, describes the results, and draws conclusions and makes recommendations.

Executive Summary. The body of the report begins with the **executive summary,** which provides a more complete overview of the report than an abstract does.

Introduction. The purpose of the **introduction** is to give readers any general information necessary to understand the detailed information in the rest of the report.

Text. The text of the body presents the details of how the topic was investigated, how the problem was solved, what alternatives were explored, and how the best choice among them was selected. The information is often clarified and further developed by illustrations and tables and may be supported by references to other publications.

Conclusions. The **conclusions** section pulls together the results of the **research** and offers conclusions based on the analysis.

Recommendations. Recommendations, which are sometimes combined with the conclusions, state what course of action should be taken based on the results of the study. The recommendations section may state, for example, "I think we should pursue new markets in . . ." or "I recommend we expand our marketing efforts to the Internet."

References. If in your report you refer to material in or quote directly from published works or other research sources, including online sources, you must provide a list of references, or "works cited," in a separate section. If your employer has a preferred reference style, follow it; otherwise, use the guidelines provided in the entry **documenting sources.** For a relatively short report, the references should go at the end of the body of the report. For a report with a number of sections or chapters, a reference section should fall at the end of each major section or chapter. In either case, the reference section should be labeled as such and should start on a new page. If a particular reference appears in more than one section or chapter, it should be repeated in full in each appropriate section.

Using references achieves three important purposes. First, full, consistent, and accurate documentation allows readers to locate and consult the primary sources as well as find further information on the subject. Second, used judiciously, documenting sources enables you to support your assertions about key points of evidence. Finally, by identifying where you obtained the facts, ideas, **quotations,** and paraphrases and giving proper credit to others, you avoid **plagiarism.**

Back Matter

The back matter of a formal report contains supplementary material, such as where to find additional information about the topic (bibliography), and expands on certain subjects (appendixes). Other back matter elements clarify the terms used (glossary) and how to easily locate information in the report (index).

Appendixes.   An **appendix** contains information that clarifies or supplements the text. An appendix provides information that is too detailed or lengthy to appear in the text without impeding the orderly presentation of ideas for the primary audience. Be careful: Do not use appendixes for miscellaneous bits of information that you were unable to work into the text.

Bibliography.   A bibliography is an alphabetical list of all sources that were consulted (not just those cited) in researching the report. A bibliography is not necessary if the reference listing contains a complete list of sources. (See **bibliographies.**)

Glossary.   A **glossary** is an alphabetical list of selected terms used in the report and their definitions.

Index.   An index is an alphabetical list of all the major topics and their subcategories discussed in the report. It cites the page numbers where discussion of each topic can be found and allows readers to find information on topics quickly and easily. The index is always the final section of a report. (See also **indexing.**)

format

Format refers to both the organization of information in a document and the physical arrangement of information on the page.

In one sense, format refers to the fact that some types of job-related writing, such as **formal reports** and **correspondence,** are characterized by conventions that govern the scope and placement of information. For example, in formal reports, the **table of contents** precedes the preface but follows the title page and the **abstract.** Likewise, although variations exist, parts of letters — such as inside address, salutation, and complimentary close — typically are arranged in standard patterns.

Format also refers to the general physical appearance (as discussed in **layout and design**) of a finished document, whether printed or electronic. (See also **email** and **Web page design.**)

former / latter   ESL

Former and *latter* should be used to refer to only two items in a sentence or paragraph.

- The president and his aide emerged from the conference, the *former* looking nervous and the *latter* looking glum.

Because these terms make the reader look to previous material to identify the reference, they complicate reading and are best avoided.

forms of discourse

There are four *forms of discourse:* exposition, description, persuasion, and narration. **Exposition** is the straightforward presentation of facts and ideas with the purpose of informing the readers;

description is an attempt to re-create an object or situation with words so the reader can mentally visualize it; **persuasion** attempts to convince the reader that the writer's point of view is the correct or desirable one; and **narration** is the presentation of a series of events in chronological order. These types of writing rarely exist alone in pure form; they usually appear in combination.

F

formula / formulae / formulas

The plural form of *formula* is either *formulae* (a Latin derivative) or *formulas*. *Formulas* is more common because it is the more natural English plural. (See also **foreign words in English.**)

- You may present the underlying theory in your introduction but save proofs or *formulas* for the body of your report.

fortuitous / fortunate

When an event is *fortuitous,* it happens by chance or accident and without plan. Such an event may be lucky, unlucky, or neutral.

- My encounter with the general manager in Denver was entirely *fortuitous;* I had no idea he was there.

When an event is *fortunate,* it happens by good fortune or happens favorably.

- Our chance meeting had a *fortunate* outcome.

fragment (*see* **sentence fragments**)

functional shift

Many words shift easily from one **part of speech** to another, depending on how they are used. When they do, the process is called a *functional shift,* or a shift in function.

- It takes ten minutes to *walk* from the sales office to the accounting department. However, the long *walk* from the sales office to the accounting department reduces efficiency.
  [*Walk* shifts from **verb** to **noun.**]

- I talk to the Chicago office on the *phone* every day. He was concerned about the office *phone* expenses. He will *phone* the home office from London.
  [*Phone* shifts from **noun** to **adjective** to **verb.**]

- *After* we discuss the project, we will begin work. *After* lengthy discussions, we began work. The partners worked well together forever *after*.
  [*After* shifts from **conjunction** to **preposition** to **adverb.**]

Jargon is often the result of functional shifts. In medicine, for example, an *attending physician* is referred to as the "attending" (a shift from an adjective to a noun). Likewise, in nuclear plant construction, a *reactor containment building* is called a "containment" (a shift from an adjective to a noun). Do not shift the function of a word indiscriminately merely to shorten a phrase or expression. (See also **affectation** and **conciseness/wordiness.**)

F

G

garbled sentences

A garbled sentence is one that is so tangled with structural and grammatical problems that it cannot be repaired.

- My job objectives are accomplished by my having a diversified background which enables me to operate effectively and efficiently, consisting of a degree in mechanical engineering, along with twelve years of experience, including three years in Staff Engineering-Packaging sets a foundation for a strong background in areas of analyzing problems and assessing economical and reasonable solutions.

A garbled sentence often results from an attempt to squeeze too many ideas into one sentence. Do not try to patch such a sentence; rather, analyze the ideas it contains, list them in a logical sequence, and then construct one or more entirely new sentences.

An analysis of the preceding example yields the following five ideas:

- My job requires that I analyze problems to find economical and workable solutions.
- My diversified background helps me accomplish my job.
- I have a mechanical engineering degree.
- I have twelve years of job experience.
- Three of those years have been in Staff Engineering-Packaging.

Using those five ideas—together with **parallel structure, sentence variety, subordination,** and **transition**—the writer might have described the job as follows:

- My job requires that I analyze problems to find economical and workable solutions. Both my training and my experience help me achieve this goal. Specifically, I have a mechanical engineering degree and twelve years of job experience, three of which have been in the Staff Engineering-Packaging Department.

(See also **clarity** and **sentence construction.**)

gender ⓔⓢⓛ

In English grammar, gender refers to the classification of nouns and pronouns as masculine, feminine, and neuter. The gender of most words can be identified only by the choice of the appropriate **pronoun** (*he, she, it*). Only these pronouns and a select few nouns (*heir/heiress, duke/duchess*) reflect gender, although many have been replaced by single terms for both sexes. (See also **biased language** and **he/she.**)

Gender is important to writers because they must be sure that nouns and pronouns within a grammatical construction agree in gender. A pronoun, for example, must agree with its noun antecedent in gender. We must refer to a woman as *she* or *her*, not as *it*; to a man as *he* or *him*, not as *it*; to a barn as *it*, not as *he* or *she*. (See also **agreement.**)

G

ⓔⓢⓛ TIPS FOR ASSIGNING GENDER

The English language system has an almost complete lack of gender distinctions. That can be confusing for a nonnative speaker of English whose native language may be marked for gender. In the few cases in which English does make a gender distinction, there is a close connection between the assigning of gender and the sex of the subject. The few instances in which gender distinctions are made in English are summarized as follows:

Subject pronouns	he/she
Object pronouns	him/her
Possessive adjectives	his/her(s)
Some nouns	king/queen, boy/girl, cow/bull, etc.

When a noun, such as *doctor*, can refer to a person of either sex, you need to know the sex of the person to which the noun refers to determine the gender-appropriate pronoun.

- The doctor gave *her* patients lots of attention. [Doctor is female.]
- The doctor gave *his* patients lots of attention. [Doctor is male.]

When the sex of the noun antecedent is unknown, be sure to follow the guidelines for nonsexist writing in the entry **biased language.** (*Note:* Some English speakers refer to vehicles and countries as *she*, but contemporary usage tends to use *it*.)

general-to-specific method of development

In a general-to-specific method of development, you begin with a general statement and then provide facts or examples to develop and support that statement. For example, if you begin a report with the general statement "Companies that diversify are more successful than those that do not," the remainder of the report would offer examples and statistics that prove to the reader that companies that diversify are, in fact, more successful than companies that do not.

A memo or short report organized entirely in a general-to-specific sequence discusses only one point. All other information in the document supports the general statement, as in the following example from a memo about locating additional computer chip suppliers.

Subject: Expanding Our Supplier Base for
Computer Chips

GENERAL STATEMENT On the basis of information presented at the supply meeting on April 14, we recommend that the company initiate relationships with computer chip manufacturers. Several events make such an action necessary.

SUPPORTING INFORMATION Our current supplier, Datacom, is experiencing growing pains and is having difficulty shipping the product on time. Specifically, we can expect a reduction of between 800 and 1,000 units per month for the remainder of this fiscal year. The number of units should stabilize at 15,000 units per month thereafter.

Domestic demand for our computers continues to grow. Demand during the current fiscal year is up 500,000 units over the last fiscal year. Our sales projections for the next five years show that demand should peak next year at about 830,000 units given the consumer demand, which will increase exponentially.

Finally, our expansion into European and Asian markets will require additional shipments of at least 750,000 units per quarter for the remainder of this fiscal year. Sales Department projections put global computer sales at double that rate, or 1,500,000 units per fiscal year, for the next five years.

gerunds (*see* **verbals**)

global communication

The prevalence of multinational corporations, the international subsidiaries of many companies, multinational trade agreements, the emergence of Europe as a giant single market, the increasing diversity of the U.S. workforce, and increases in immigration mean that the ability to reach audiences from varied cultural backgrounds is essential. The audiences for such communications—whether in **international correspondence** or **presentations**—include clients, business partners, colleagues, and current and potential employees.

For global audiences, gestures and body language have different meanings, depending on the cultural interpretation. For example, North Americans often use hand gestures such as the victory sign (✌) and the OK sign (👌) as positive motivators. In Australia, however, the victory sign conveys the same rude meaning as holding up the middle finger in North America. Similarly, the gesture that means "OK" in North America can mean "worthless" in France or "money" in Japan and is a sexual insult in many other parts of the world. (See Figure G–1.)

Body Part	Gesture	Country	Interpretation
Head	Nodding up and down	Bulgaria	"No"
Left hand	Showing palm	Muslim countries	"Dirty," "unclean"
Index finger	Pointing to others	Venezuela, Sri Lanka	Rude
Index finger and thumb	Circular "OK"	Germany, Netherlands	Rude
Ankle	Crossing over knee	Indonesia and Syria	Rude
Eye	Touching finger below eye	Honduras	"Caution"

FIGURE G–1. Cultural Differences in Interpreting Body Language

Writer's Checklist: Global Communication

These guidelines will help build your awareness of global and cross-cultural communication. However, this list cannot begin to cover the many issues (from gift giving to appropriate dress) important to cross-cultural communication.

- ☑ Learn about gesture usage in other cultures. Always research gesture meanings in a particular culture before you meet with a person from that culture.

- ☑ Consult with someone from your intended audience's culture. Many gestures and visual elements are so subtle that only someone who is very familiar with the culture can explain the effect they will have on others from that culture.

- ☑ Acknowledge diversity within your organization. Discussing the differing cultures within your company or region will reinforce the idea that not everyone interprets body language and gestures in the same way.

- ☑ Invite global and intercultural communication experts to speak to your employees. Companies in your area may have employees who could be resources for cultural discussions.

- ☑ Consult publications about cross-cultural technical communication, such as those available through Intercultural Press (http://www.bookmasters.com/interclt/).

- ☑ Understand that the key to effective communication with global audiences is recognizing that cultural differences, despite the challenges they present, offer growth for both you and your organization.

global graphics

In the global business and technological environment, **graphics** require the same careful attention given to other aspects of **global communication.** Although they can help communicate important safety warnings and even replace difficult-to-translate terms, graphics used internationally present unique challenges.

Symbols, images, and even colors are not free from cultural associations: They depend on context, and context is culturally determined. For instance, symbols that carry simultaneous religious and nonreligious meanings have long been used in North America, where it is common to use the cross as a symbol for "first aid" or "hospital." In Muslim countries, however, a cross (red or otherwise) represents Christianity, while a crescent (usually green) signifies first aid. A manual for export to Honduras could indicate "caution" by using a picture of a person touching a finger below the eye. In France, however, that gesture would have a totally different meaning: "You can't fool me."

Careful attention to the connotations that visual elements may have for a diverse audience makes translations easier, prevents embarrassment, and earns respect for the company and its products and services. (See also **illustrations** and **presentations.**)

Writer's Checklist: Global Graphics

The complex cultural connotations of visuals challenge writers to think beyond their own experience. These guidelines will help you communicate effectively through the use of graphics to global and multicultural audiences.

☑ Consult with someone from your intended audience's country. Many visual elements are so subtle that only someone very familiar with the culture can recognize and explain the effect those elements will have on readers.

☑ Organize visual information for the intended audience. For example, North American readers tend to read visuals from left to right in clockwise rotation. Middle Eastern cultures read visuals from left to right in counterclockwise rotation.

☑ Be sure that the graphics you use have no religious implications.

☑ Carefully consider how you depict people in visuals. Nudity in advertising, for example, is generally acceptable in Europe but much less so in North America and Asia. In some cultures, such as Islamic and fundamentalist religious groups, showing even isolated bare body parts can alienate audiences.

☑ Use outlines or neutral abstractions to represent human beings. For example, use stick figures for bodies with a circle for the head. Avoid representing men and women.

☑ Examine how you display body positions in signs and visuals. Body positioning can carry unintended cultural meanings very different from your own. For example, some Middle Eastern cultures regard the display of the soles of one's shoes to be disrespectful and

Writer's Checklist: Global Graphics (continued)

offensive. Therefore, a visual that attempts to demonstrate the ease of performing a task by showing a figure in a reclining chair with feet up would be inappropriate in parts of the Middle East.

☑ Try to use neutral colors in your graphics. Generally, black-and-white and gray-and-white illustrations work well, whereas colors can be problematic. For example, red commonly indicates warning or danger in North America, Europe, and Japan. In China, however, red symbolizes joy. In Europe and North America, blue generally has a positive connotation; in Japan, blue represents villainy.

☑ Check your use of punctuation marks, which are as language specific as symbols. For example, in North America the question mark generally represents the need for information or the help function in a computer manual or program. In many countries, that symbol is not understood at all.

☑ Create simple visuals and use consistent labels for all visual items. In most cultures, simple shapes with fewer elements are easier to read.

☑ Explain the meaning of icons or symbols. Include a **glossary** to explain technical symbols that cannot be changed, such as company logos.

☑ Test icons and symbols in context with members of your target audience. Usability testing with cultural experts is critical.

glossaries

A glossary is an alphabetical list of definitions of terms used in a **formal report, technical manual,** or other long document.

If you are writing a **report** that will go to readers who are not familiar with many of the terms you use, you may want to include a glossary. If you do, keep the entries concise and be sure they are clear enough that all readers can understand the definitions.

- *Amplitude modulation:* Varying the amplitude of a carrier current with an audio-frequency signal.

- *Carbon microphone:* A microphone that uses carbon granules as a means of varying resistance with sound waves.

Arrange the terms alphabetically, with each entry beginning on a new line. The definition follows the term, dictionary style. In a **formal report,** the glossary, labeled as such, appears directly after the appendix(es) and begins on a new page.

Inclusion of a glossary does not relieve you of the responsibility of defining in the text any terms your reader will not know when those terms are first mentioned. (See also **defining terms.**)

gobbledygook

G

Gobbledygook is writing that suffers from an overdose of traits guaranteed to make it stuffy, pretentious, and wordy. Such traits include the overuse of big and mostly **abstract words, affectation,** inappropriate **jargon, long variants, clichés, euphemisms,** stacked **modifiers, vogue words,** and deadwood. Gobbledygook is writing that attempts to sound official (officialese), legal (legalese), or scientific; it tries to make a "natural elevation of the geosphere's outer crust" out of a molehill. Consider the following statement from an auto repair release form.

LEGALESE   I hereby authorize the above repair work to be done along with the necessary material and hereby grant you and/or your employees permission to operate the car or truck herein described on streets, highways, or elsewhere for the purpose of testing and/or inspection. An express mechanic's lien is hereby acknowledged on above car or truck to secure the amount of repairs thereto.

DIRECT   You have my permission to do the repair work listed on this work order and to use the necessary material. You may drive my vehicle to test its performance. I understand that you will keep my vehicle until I have paid for all repairs.

(See also **affectation, conciseness/wordiness,** and **word choice.**)

good / well   ESL

The confusion about the use of *good* and *well* can be cleared up by remembering that *good* is an **adjective** and *well* is usually an **adverb.**

| ADJECTIVE | Janet presented a *good* plan. |
| ADVERB | The plan was presented *well*. |

Well also can be used as an adjective to describe someone's health.

| ADJECTIVE | He is not a *well* man. |

(See also **bad/badly.**)

G

government proposals (*see* proposals)

grammar

Grammar is the systematic description of the way words work together to form a coherent language. In that sense, it is an explanation of the structure of a language. However, grammar is popularly taken to mean the set of rules that governs how a language ought to be spoken and written. In that sense, it refers to the **usage** conventions of a language.

Those two meanings of grammar—how the language functions and how it ought to function—are easily confused. To clarify the distinction, consider the expression *ain't*. Unless used intentionally to add colloquial flavor, *ain't* is unacceptable because its use is considered nonstandard. Yet taken strictly as a **part of speech,** the term functions perfectly well as a verb. Whether it appears in a declarative sentence ("I ain't going.") or an interrogative sentence ("Ain't I going?"), it conforms to the normal pattern for all verbs in the English language. Although readers may not approve of its use, they cannot argue that it is ungrammatical in such sentences.

To achieve clarity, you need to know both grammar (as a description of the way words work together) and the conventions of usage. Knowing the conventions of usage helps you select the appropriate over the inappropriate word or expression. (See also **word choice.**) A knowledge of grammar helps you diagnose and correct problems arising from how words and phrases function in relation to one another. For example, knowing that certain words and phrases function to modify other words and phrases gives you a basis for correcting those **modifiers** that are not doing their job. Understanding **dangling modifiers** helps you avoid or correct a construction that obscures the intended meaning. In short, an understanding of grammar and its special terminology is valuable chiefly because it enables you to recognize and correct problems so you can communicate clearly and precisely.

graphs

DIRECTORY
Overview   257
Line Graphs   257
Bar Graphs   260
Pie Graphs   262
Picture Graphs   264

G

Overview

A graph presents numerical data in visual form. Presenting data visually has several advantages over presenting data within the text or in **tables.** Trends, movements, distributions, comparisons, and cycles are more readily apparent in graphs than they are in tables. By providing a more visual way to compare data, a graph often highlights differences and similarities otherwise buried in text or a table. However, although graphs present data in a more interesting and comprehensible form than tables do, they are less precise. For that reason, graphs are often accompanied by tables that give exact data. The most common types of graphs are *line graphs, bar graphs, pie graphs,* and *picture graphs.* (For additional advice, see **illustrations;** for information about using presentation graphics, see **presentations.**)

Line Graphs

A line graph shows the relationship between two variables or sets of numbers by plotting points in relation to two axes drawn at right angles. Once plotted, the points are connected to form a continuous line. In that way, the abstract mathematical data set is placed in context and further interpreted through the graphic representation of the relationship between the two variables.

The vertical axis usually represents amounts, and the horizontal axis usually represents increments of time, as shown in Figure G–2.

Line graphs that portray more than one set of variables are common because they allow for comparisons between two sets of statistics for the same period of time. In creating such graphs, be certain to identify each line with a label or a legend, as shown in Figure G–3 on page 259. You can emphasize the difference between the two lines by shading the space between them.

Be especially careful to proportion the vertical and horizontal scales so they give a precise presentation of the data that is free of visual distortion. To do otherwise is not only inaccurate but potentially unethical. (See also **ethics in writing.**) In Figure G–4 on page 259, because the scale is compressed and some of the data is omitted, the graph on the left gives the appearance of a dramatic decrease; the graph on the right, which includes more data, is accurate.

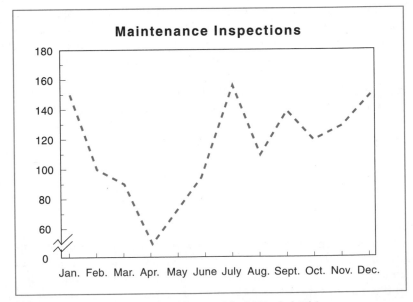

FIGURE G–2. Single-Line Graph (with a Break in the Vertical Axis)

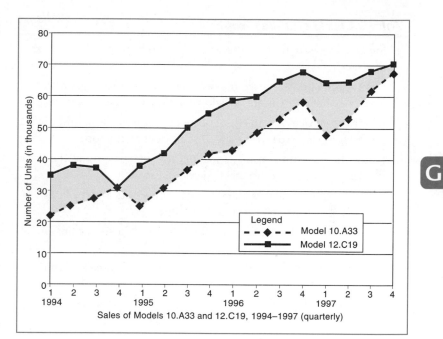

FIGURE G–3. Double-Line Graph with Legend

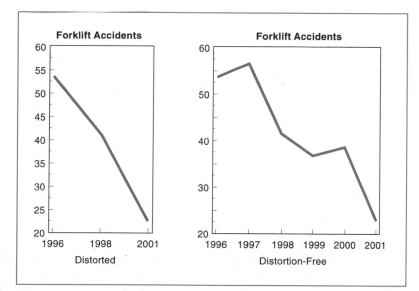

FIGURE G–4. Distorted Expression of Data (*left*) and Distortion-Free
Expression of Data (*right*)

Writer's Checklist: Line Graphs

☑ Indicate the zero point of the graph (the point where the two axes intersect).

☑ If the range of data shown makes it inconvenient to begin at zero, insert a break in the scale, as shown in Figure G-2 on page 258.

☑ Divide the vertical axis in equal portions, from the least amount at the bottom (or zero) to the greatest amount at the top.

☑ Divide the horizontal axis in equal units from left to right. If a label is necessary, center it directly beneath the scale.

☑ When necessary, include a key or legend that lists and explains symbols, as shown in Figure G-3 on page 259.

☑ If the data comes from another source, include a source line under the graph at the lower left.

☑ Place explanatory **footnotes** directly below the figure caption or label.

☑ Make all lettering read horizontally if possible, although the caption or label for the vertical axis is usually positioned vertically.

Bar Graphs

Bar graphs consist of horizontal or vertical bars of equal width, scaled in length to represent some quantity. They are commonly used to show (1) quantities of the same item at different times, (2) quantities of different items at the same time, and (3) quantities of the different parts of an item that make up a whole.

Figure G–5 on page 261 is an example of a bar graph that shows varying quantities of the same item over the years. Each bar represents a different quantity of the same item (airbags).

Some bar graphs, like Figure G–6 on page 261, show the quantities of different items for the same period of time. A bar graph can be horizontal, as in Figure G–6, or vertical, as in Figure G–5.

Bar graphs can also show the different portions of an item that make up the whole, as shown in Figure G–7 on page 262. Such a bar graph is divided according to the appropriate proportions of the subcomponents of the item. This type of graph can be constructed vertically or horizontally and can indicate multiple items, each with subcomponents, when comparisons are necessary. Where such items

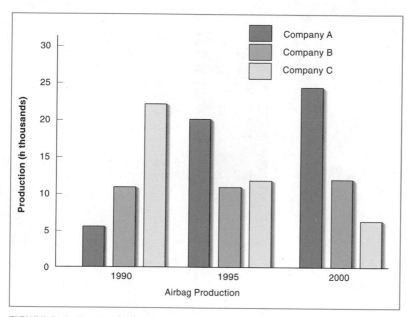

FIGURE G-5. Bar Graph Showing Quantities of Same Item over Time

G

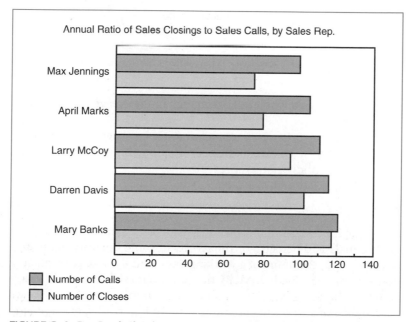

FIGURE G-6. Bar Graph Showing Quantities of Different Items during Fixed Period

G

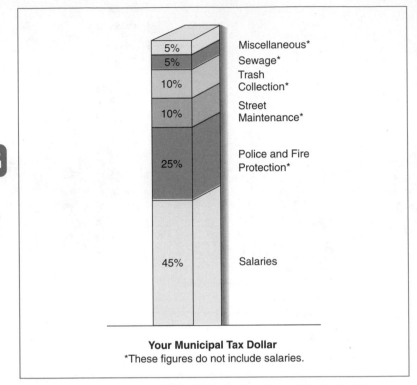

Your Municipal Tax Dollar
*These figures do not include salaries.

FIGURE G–7. Bar Graph Showing the Different Parts That Make Up the Whole

represent parts of a whole, the segments in the bar graph must total 100 percent.

In addition to labels, each subdivision of a bar graph must be marked clearly by color, shading, or crosshatching, with a key that identifies the subdivisions represented.

Pie Graphs

A pie graph presents data as wedge-shaped sections of a circle. The circle equals 100 percent, or the whole, of some quantity (a tax dollar, a bus fare, the hours of a working day), and the wedges represent how the whole is divided. Many times, the data shown in a bar graph could also be depicted in a pie graph. For example, Figure G–7 shows percentages of a whole in bar graph form. In Figure G–8 on page 263, the same data are converted into a pie graph, dividing "Your Municipal Tax Dollar" into wedge-shaped sections that represent percent-

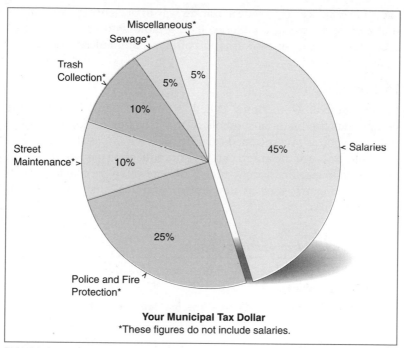

FIGURE G-8. Pie Graph Showing the Same Data as Figure G-7

Writer's Checklist: Pie Graphs

☑ Make sure that the complete circle is equivalent to 100 percent.

☑ When possible, begin at the 12 o'clock position and sequence the wedges clockwise, from largest to smallest.

☑ Keep in mind that if too many items are presented in a pie graph, the graph can look cluttered or the slices can be too thin to be clear. If too few items are presented, the graph will not be useful.

☑ Give each wedge a distinctive color, pattern, shade, or texture.

☑ Label each wedge with its percentage value and keep all callouts (labels that identify the wedges) horizontal.

☑ If you want to draw attention to a particular segment of the pie graph, detach that slice.

ages. Pie graphs also provide a quicker way of presenting information that can be shown in a table; in fact, a table with a more detailed breakdown of the same information often accompanies a pie graph.

Picture Graphs

Picture graphs (sometimes called *pictograms*) are modified bar graphs that use pictorial symbols of the item portrayed. Each symbol corresponds to a specified quantity of the item, as shown in Figure G–9. Because the graph can present only approximate figures, it also gives precise figures. Picture graphs are often used to add interest to documents, such as **newsletter articles,** that are aimed at wide audiences.

Writer's Checklist: Picture Graphs

☑ Make sure the symbol you choose is easily recognizable (see also **global graphics**).

☑ Have each symbol represent a specific number of units.

☑ Because it is difficult to judge relative sizes accurately, show larger quantities by increasing the number of symbols rather than by creating a larger symbol.

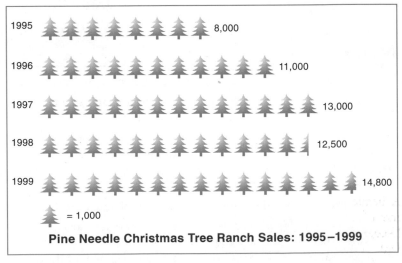

FIGURE G–9. Picture Graph

H

he / she

The use of either *he* or *she* to refer to both sexes excludes half of the population. To avoid this problem, you could use the phrases *he or she* and *his or her*.

- Whoever is appointed will find *his or her* task difficult.

However, *he or she* and *his or her* are clumsy when used repeatedly, and the *s/he* construction is awkward. One solution is to reword the sentence to use a plural pronoun; if you do, be sure to change the noun to which the pronoun refers to its plural form.

- ~~The administrator~~ *Administrators* cannot do ~~his or her job~~ *their jobs* until ~~he or she~~ *they* ~~understands~~ *understand* the concept.

Other times, you may be able to avoid using a **pronoun** altogether.

- Whoever is appointed will find ~~his or her~~ *the* task difficult.

(See also **biased language.**)

headers and footers

A header in a **report, technical manual,** letter, or other document is identifying information placed at the top of each page. The header normally contains the topic (or topic and subtopic) dealt with in that section of the document.

A footer is identifying information placed at the bottom of each page. The footer generally contains the date of the document, the page number, and sometimes the document's name and section title.

Although the information included in headers and footers varies greatly from one organization to the next, the header and footer shown in Figure H–1 are fairly typical.

H

Header

FIGURE H–1. Header and Footer

headings

Headings (also called *heads*) are titles or subtitles within the body of a document that divide the material into manageable segments, highlight the main topics, and signal topic changes. A **formal report** or **proposal** may need several levels of headings to indicate major divisions, subdivisions, and even smaller units. Although some documents may need as many as five levels of headings, that many are rarely necessary (and often confusing). In fact, it is better not to use more than three levels of headings. (See also **layout and design.**)

Headings typically represent the major topics of an **outline.** In a short document, you can use the major divisions of your outline as headings; in a longer document, you may need to use both major and minor divisions.

General Heading Style

There is no one correct format for headings. Often a company settles on a standard format, which everyone in the company follows. Sometimes a customer for whom a report or proposal is being prepared requires a particular format. In the absence of specific guidelines, the following system, illustrated in Figure H–2 on page 268, should serve you well.

- The first-level heading is in bold uppercase letters, centered on a line by itself, and separated by blank lines from the material it follows and the material it introduces.
- The second-level heading is in bold uppercase letters, placed flush to the left margin on a line by itself, and separated by blank lines from the material it follows and the material it introduces.
- The third-level heading is in bold upper- and lowercase letters, placed flush to the left margin on a line by itself, and separated by blank lines from the material it follows and the material it introduces.
- The fourth-level heading is in italic upper- and lowercase letters, indented from the left margin, and followed by a period. The heading is "run in" on the same line with the first sentence of the text it introduces. A blank line may or may not precede a fourth-level heading.

H

First-level
head

DISTRIBUTION CENTER LOCATION REPORT

The committee initially considered 30 possible locations for the proposed new distribution center. Of these, 20 were eliminated almost immediately for one reason or another (unfavorable zoning regulations, inadequate transportation infrastructure, etc.). Of the remaining ten locations, the committee selected for intensive study the three that seemed most promising: Chicago, Minneapolis, and Salt Lake City. We have visited these three cities, and our observations and recommendations follow.

Second-
level head

CHICAGO

Of the three cities, Chicago seems to the committee to offer the greatest advantages, although we wish to examine it more carefully before making a final recommendation.

Third-level
head

Selected Location

Though not at the geographical center of the United States, Chicago is the demographic center to more than three-quarters of the U.S. population. It is within easy reach of our corporate headquarters in New York. And it is close to several of our most important suppliers of components and raw materials, such as those in Columbus, Detroit, and St. Louis. Several considerations were considered essential to the location, although some may not have had as great an impact on the selection. . . .

Fourth-level
heads

Air Transportation.   Chicago has two major airports (O'Hare and Midway) and is contemplating building a third. Both domestic and international air cargo service are available. . . .

Sea Transportation.   Except during the winter months when the Great Lakes are frozen, Chicago is an international seaport. . . .

Rail Transportation.   Chicago is served by the following major railroads. . . .

FIGURE H–2. Headings Used in a Document

Decimal Numbering System

The decimal numbering system, which is common in some technical publications as well as in policies and procedures, uses a combination of numbers and decimal points to differentiate among levels of headings. The following outline shows the correspondence between different levels of headings and the decimal numbers used:

   1 FIRST-LEVEL HEADING
      1.1 Second-Level Heading
      1.2 Second-Level Heading
         1.2.1 Third-Level Heading
         1.2.2 Third-Level Heading
            1.2.2.1 Fourth-Level Heading
            1.2.2.2 Fourth-Level Heading
      1.3 Second-Level Heading
         1.3.1 Third-Level Heading
         1.3.2 Third-Level Heading
   2 FIRST-LEVEL HEADING

Although the second-, third-, and fourth-level headings are indented in an outline or **table of contents,** they are flush with the left margins when they function as headings in the body of a report. Every heading starts on a new line, with an extra line of space above and below the heading.

H

Writer's Checklist: Using Headings

Headings are important because they help **readers** find information, reveal the structure of a document, break the text into digestible segments, and highlight key points. Keep these important points in mind when using headings:

- ☑ A heading should signal a new topic or, if it is a lower-level heading, a new subtopic within the larger topic.

- ☑ Avoid too many headings or levels of headings, because they clutter a document; on the other hand, too few headings fail to provide necessary structure.

- ☑ Within the unit of text they subdivide, headings at the same level should be of relatively equal importance.

- ☑ When higher-level headings are subdivided, use at least two lower-level headings when possible.

Writer's Checklist: Using Headings (continued)

☑ Subdivide sections only as needed; not every section requires lower-level headings.

☑ Make all headings at any one level in **parallel structure.**

☑ Do not allow a heading to "float" on the final line of a page. If two lines of text cannot fit below a heading, start the section at the top of the next page.

☑ A heading does not substitute for discussion; the text should read as if the heading were not there.

helping verbs (*see* verbs)

herein / herewith (*see* **affectation**)

hyperbole (*see* **figures of speech**)

hypertext

Hypertext is the system that allows users of the **World Wide Web,** CD-ROMs, and other electronic media to move from one document to another or to navigate and find online information. Web pages, for example, contain hypertext indicators — words, graphics, and icons — that link to other pages. (See also **Web page design.**)

The hypertext system includes three main components: *text, nodes,* and *links.* Text, the informational content of the system (words, graphics, sound, and video), is collected into self-contained units called nodes. The nodes are connected in a variety of ways by electronic links. Hypertext is similar to the structure of this handbook. Each entry is a discrete unit (node) that contains text and perhaps graphical information. Many *Handbook* entries include a cross-reference (link) in the form of a boldfaced word or a "see also" statement that points readers to related information. That is the basic structure of a hypertext document; electronic hypertext also allows for the inclusion of animation segments, audio clips, video clips, and real-time conferencing.

Hypertext systems are categorized as either *modest* (not modifiable) or *robust* (modifiable), depending on the degree to which users are encouraged to make modifications. In modest versions, users navigate information by following existing paths created by hypertext designers. That read-only approach enables users to retrieve specific information by navigating predetermined paths as well as to retrace their steps. Applications common to modest approaches include tutorials, guided tours, and help systems. In robust versions, users can add to, delete, and edit text within nodes, as well as construct, eliminate, and reconstruct the links between them. Using modifiable hypertext systems, a user is both writer and reader. Robust hypertext applications include real-time conferencing sessions and **collaborative writing** activities.

Many applications combine modest and robust hypertext systems by allowing users to modify some portions of the hypertext while restricting their ability to modify other portions. The online help systems used by software programs that allow users to create and link their own text to existing help screens are examples of such a hybrid approach.

hyphens ESL

The hyphen (-) serves both to link and to separate words. The hyphen's most common linking function is to join **compound words.**

- able-bodied
- self-contained
- self-esteem

A hyphen is used to form compound numbers from twenty-one through ninety-nine and fractions when they are written out.

- forty-one
- three-quarters

Hyphens Used with Modifiers

Two- and three-word **modifiers** that express a single thought are hyphenated when they precede a **noun.**

- It was a *well-written* report

- We need a *clear-cut* decision

However, a modifying phrase is not hyphenated when it follows the noun it modifies.

- The report was *well written*.

- The decision was *clear cut*.

If each of the words can modify the noun without the aid of the other modifying word or words, do not use a hyphen.

H

- a *new laser* printer

If the first word is an **adverb** ending in -*ly*, do not use a hyphen.

- a *newly* minted coin

- a *badly* needed scanner

A hyphen is always used as part of a letter or number modifier.

- 5-cent

- 9-inch

- A-frame

- H-shaped

In a series of unit modifiers that all have the same term following the hyphen, the term following the hyphen need not be repeated throughout the series; for greater smoothness and brevity, use the term only at the end of the series.

- The third-, fourth-, and fifth-floor rooms were recently painted.

Hyphens Used with Prefixes and Suffixes

A hyphen is used with a **prefix** when the root word is a proper noun.

- pre-Columbian

- anti-American

- post-Newtonian

A hyphen may be used when the prefix ends and the root word begins with the same vowel.

- re-elect
- re-enter
- anti-inflammatory

A hyphen is used when *ex-* means "former."

- ex-president
- ex-spouse

A hyphen may be used to emphasize a prefix.

- She was anti-everything.

The **suffix** *-elect* is hyphenated.

- president-elect
- commissioner-elect

Hyphens and Clarity

The presence or absence of a hyphen can alter the meaning of a sentence.

> AMBIGUOUS   We need a biological waste management system.

That sentence could mean one of two things: (1) We need a system to manage "biological waste," or (2) We need a "biological" system to manage waste.

> CLEAR   We need a biological-waste management system.
> CLEAR   We need a biological waste-management system.

To avoid confusion, some words and modifiers should always be hyphenated. *Re-cover* does not mean the same thing as *recover*, for example; the same is true of *re-sent* and *resent*, *re-form* and *reform*, *re-sign* and *resign*.

Other Uses of the Hyphen

Hyphens should be used between letters showing how a word is spelled.

- In his letter, he misspelled *believed* b-e-l-e-i-v-e-d.

A hyphen can stand for *to* or *through* between letters and numbers.

- pp. 44-46

- the Detroit-Toledo Expressway

- A-L and M-Z

Hyphens also are used to divide words at the end of a line. Most word-processing programs give you the option of automatically hyphenating words at the end of a line according to your default settings. To avoid improper end-of-line hyphenation, follow these general guidelines.

H

- Do not divide one-syllable words.
- Divide words at syllable breaks, which you can determine with a dictionary.
- Do not divide a word if only one letter would remain at the end of a line or if fewer than three letters would start a new line.
- Do not divide a word at the end of a page.
- If a word already has a hyphen in its spelling, try to divide the word at the existing hyphen.
- When dividing Web addresses at the end of a line, try to break the address after a slash. Inserting a hyphen into the address may confuse readers.

I

An idiom is a group of words that has a special meaning apart from its literal meaning. Someone who "runs for office" in the United States, for example, need not be a track star. The same candidate would "stand for office" in the United Kingdom. In both nations, the person is seeking public office; however, the idioms of the two countries differ. Because such expressions are specific to a culture, nonnative speakers must memorize them. The native writer has little trouble understanding idioms and need not attempt to avoid them in writing, provided the readers are equally comfortable with them.

Idioms often provide helpful shortcuts. In fact, they can make writing more natural and vigorous. However, if there is any chance that your writing might be translated into another language or be read in other English-speaking countries, eliminate obvious idiomatic expressions that might puzzle readers.

Idiom also refers to the practice of using certain **prepositions** following some verbs, nouns, and adjectives. Because there is no sure system to explain such usages, the best advice is to check a dictionary. The following common pairings give writers trouble. (See also **international correspondence** and **English as a second language.**)

absolve from [responsibility]	acquaint with
absolve of [crimes]	
	adapt for [a purpose]
accordance with	adapt from [change]
according to	adapt to [a situation]
accountable for [actions]	adhere to
accountable to [a person]	
	adverse to
accused by [a person]	
accused of [a deed]	affinity between, with

275

agree on [terms]
agree to [a plan]
agree with [a person]

angry at, about [a thing]
angry with [a person]

approve of

apply for [a position]
apply to [contact]

argue for, against [a policy]
argue with [a person]

arrive at [a specific location,
    a conclusion]
arrive in [a city, a country]

based upon

blame for [an action]
blame on [a person]

compare to [things that are
    similar but not the same
    kind]
compare with [things of the
    same kind to determine
    similarities and differ-
    ences]

comply with

concur in [consensus]
concur with [a person]

conform to

consist of

convenient for [a purpose]
convenient to [a place]

correspond to, with [a thing]
correspond with [a person]

deal with

depend upon

deprive of

devoid of

differ about, over [an issue]
differ from [a thing]
differ on [amounts, terms]
differ with [a person]

different from

disagree on [an issue, a plan]
disagree with [a person]

disappointed in

disapprove of

disdain for

divide between, among
divide into [parts]

engage in

exclude from

expect from [things]
expect of [people]

expert in

identical with, to

impatient for [something]
impatient with [someone]

imply that

impose on

improve on

inconsistent with

independent of

infer from

inferior to

necessary for [an action]
necessary to [a state of being]

occupied by [things, people]
occupied with [actions]

opposite of [qualities]
opposite to [positions]

part from [a person]
part with [a thing]

proceed to [begin]
proceed with [a project]

proficient in

profit by [things]
profit from [actions]

prohibit from

qualify as [a person]
qualify by [experience, actions]
qualify for [a position, an award]

rely on

responsibility for, of

reward for [an action]
reward with [a gift]

similar to

surrounded by [people]
surrounded with [things]

talk to [a group]
talk with [a person]

wait at [a place]
wait for [a person, an event]

illegal / illicit

If something is *illegal,* it is prohibited by law. If something is *illicit,* it is prohibited by either law or custom. *Illicit* behavior may or may not be *illegal,* but it does violate social convention or moral codes and therefore usually has a clandestine or immoral **connotation.**

- Their *illicit* behavior caused a scandal, and the district attorney charged that *illegal* acts were committed.

- Explicit sexual material may be *illicit* without being *illegal.*

illustrations

The objective of using illustrations or visuals is to help readers understand the information being presented. Each type of illustration has unique strengths and weaknesses. The most common types of illustrations are discussed in the following entries: **flowcharts, graphs, maps, organizational charts, photographs,** and **tables.** (See also **presentations** and **Web page design.**)

If you plan to use illustrations, consider your **purpose** and your **readers** carefully. For example, you would need different illustrations for an automobile owner's manual than you would for a mechanic's diagnostic guide. Take illustration requirements into account even before you begin writing—when you are planning the **scope** and **organization** of your document. Make illustrations an integral part of your document outline, noting approximately where each should appear. At each place in the outline, include a simple sketch or a brief description of the illustration.

Many of the qualities of good writing—simplicity, **clarity, conciseness,** directness—are equally important in the creation and use of visuals. Presented with clarity and consistency, illustrations can help the reader focus on key portions of your document. Be aware, though, that even the best visual only enhances or supports the text. Your writing must provide context for the visual and point out its significance.

Writer's Checklist: Creating and Integrating Illustrations

The following guidelines apply to most visual materials you might use to supplement or clarify the information in your text. These tips will help you design and integrate illustrations effectively.

- ☑ Clarify why the illustration is included in the text. The amount of description varies, depending on the illustration's complexity and the reader's background. Nonexperts usually require lengthier explanations than experts do.

- ☑ Use consistent terminology. Do not refer to something as a "proportion" in the text and as a "percentage" in the illustration. Define all acronyms in the text, figure, or table. If any symbols are not self-explanatory, as in graphs, include a key that defines them.

- ☑ Place an illustration as close as possible to the text where it is discussed, especially if the illustration is central to the discussion. No illustration should precede its first text mention. If the illustration is lengthy and detailed, place it in an **appendix** and refer to it in the text.

- ☑ Give each illustration a concise title that clearly describes its content.

- ☑ Assign figure and table numbers, particularly if your document contains more than one illustration and table. The figure or table number precedes the title:

    - Figure 1. Projected sales for 2002–2010

Writer's Checklist: Creating and Integrating Illustrations (continued)

☑ Refer to illustrations in the text of your document by their figure or table numbers. (Note that in **reports** and many other documents, graphical illustrations — photographs, drawings, maps — are generically labeled "figures," while tables are labeled "tables.")

☑ In documents with more than five illustrations or tables, include a section titled "List of Figures" or "List of Tables" that identifies each by number, title, and page number. The list should follow the **table of contents.** (See also **formal reports.**)

☑ Keep the illustration direct and simple by including only information necessary to the discussion in the text and by eliminating unnecessary labels, arrows, boxes, and lines.

☑ Specify the units of measurement used or include a scale of relative distances, when appropriate. Make sure relative sizes are clear or indicate distance with a scale, as on a map.

☑ Position the lettering of any explanatory text or labels horizontally for ease of reading, if possible.

☑ Allow adequate white space around and within the illustration. (See also **layout and design.**)

☑ If you want to use an illustration from a copyrighted publication, first obtain written permission from the **copyright** holder. Acknowledge borrowed material in a source or credit line below the caption for a figure and below any footnotes at the bottom of a table.

☑ You can use information that is public, such as demographic or economic data, which are frequently illustrated in federal government publications, without obtaining written permission to reproduce it. Still, acknowledge the source in a credit line.

☑ When writing a **trade journal article,** consult the editorial guidelines of the particular journal or style manual recommended.

imply / infer

If you *imply* something, you hint or suggest it. If you *infer* something, you reach a conclusion on the basis of evidence.

• Her email *implied* that the project would be delayed.

- The manager *inferred* from the email that the project would be delayed.

in / into

In means "inside of"; *into* implies movement from the outside to the inside.

- The equipment was *in* the test chamber, so she sent her assistant *into* the chamber to get it.

in order to

The phrase *in order to* is sometimes essential to the meaning of a sentence.

- If the vertical scale of a graph line would not normally show the zero point, use a horizontal break in the graph *in order to* include the zero point.

In order to also helps control the pace of a sentence, even when it is not essential to the meaning of the sentence.

- The committee must know the estimated costs *in order to* evaluate the feasibility of the project.

Most often, however, *in order to* is just a meaningless filler phrase that is dropped into a sentence without thought. (See **conciseness/wordiness.**)

- ~~In order to~~ start the engine, open the choke and throttle and then
  To
  press the starter.

in terms of

When used to indicate a shift from one kind of language or terminology to another, the phrase *in terms of* can be useful.

- *In terms of* gross sales, the year has been relatively successful; however, *in terms of* net income, it has been discouraging.

When simply dropped into a sentence because it easily comes to mind, *in terms of* is meaningless **affectation**. (See also **conciseness/wordiness**.)

- She was thinking ~~in terms of~~ about subcontracting much of the work.

- She was thinking ~~in terms~~ of subcontracting much of the work.

inasmuch as / insofar as

Inasmuch as, meaning "because," is a weak connective for stating causal relationships.

- ~~Inasmuch as~~ Because the heavy spring rains delayed construction, the office building will not be completed on schedule.

Insofar as, meaning "to the extent that," should be reserved for that explicit use rather than to mean *since*.

- *Insofar as* the report deals with causes of highway accidents, it will be useful in preparing future plans for highway construction.

- ~~Insofar as~~ Since you are here, we will review the material.

(See also **affectation**.)

increasing-order-of-importance method of development

When you want the most important of several ideas to be freshest in readers' minds after they finish reading your writing, organize your information by increasing order of importance. This method of development is appropriate when you want to save your strongest points until the end. The sequence begins with the least important point or fact, then moves to the next least important, and builds finally to the most important point.

Writing organized by increasing order of importance has the disadvantage of beginning weakly, with the least important information. The reader may become impatient or distracted before

---

INTER-OFFICE MEMORANDUM

To: Sun-Hee Kim, Vice President, Operations
From: Harry Matthews, Human Resources Department *HM*
Date: May 19, 20--
Subject: Recruiting Qualified Electronics Technicians

As our company continues to expand, and with the planned opening of the Lakeland Facility late next year, we need to increase and refocus our recruiting program to keep our company staffed with qualified electronics technicians. Over the past three years, we have relied on our in-house internship program and on local and regional technical school graduates to fill these positions.

Although our in-house internship program provides a qualified pool of employees, technical school enrollments in the area have in the past provided candidates who are already trained. Each year, however, fewer technical school graduates are being produced, and even the most vigorous Career Day recruiting has yielded disappointing results.

In the past, we relied heavily on the recruitment of skilled veterans from all branches of the military. This source of qualified applicants all but disappeared when the military offered attractive re-enlistment bonuses for skilled technicians in uniform. As a result, we need to become aggressive in our attempts to reach this group through advertising. I would like to meet with you soon to discuss the details of a more dynamic recruiting program for skilled technicians leaving the military.

I am certain that with the right recruitment campaign, we can find the skilled employees essential to our expanding role in electronics products and consulting.

FIGURE I–1. Increasing-Order-of-Importance Method of Development

reaching your main point. However, for writing in which the ideas lead point by point to an important **conclusion,** increasing order of importance is effective. Many oral **presentations** benefit from this **method of development.**

In the example given in Figure I–1, the writer begins with the least productive source of applicants and builds up to the most productive source.

---

indefinite adjectives (*see* **adjectives**)

indefinite pronouns (*see* **pronouns**)

indentation

Text that is indented is set in from the margin. The most common use of indentation is at the beginning of a paragraph, where the first line is usually indented five spaces, unless the full-block style is used, as in **correspondence.** Another use of indentation is in outlining, in which each subordinate entry is indented under its major entry.

A long quotation may be indented in a manuscript instead of being enclosed in quotation marks. Indent each line ten spaces from the left margin if following MLA style, and five to seven spaces from the left margin if following APA style. Since editorial guidelines vary, always check the style manual prescribed. If you are not following a specific style manual, you may block indent ten spaces from both the right and left margins for reports and other documents. (See also **quotations.**)

independent clauses (*see* **clauses**)

indexing

An index is an alphabetical list of all the major topics and sometimes subtopics discussed in a written work. It cites the pages where each topic can be found and allows readers to find information on particular topics quickly and easily. The index always comes at the very end of the work.

The key to compiling a useful index is selectivity. Instead of listing every possible reference to a topic, select references to passages where the topic is discussed fully or where a significant point is made about it. For actual index entries choose those words or phrases that best represent a topic. Key terms are those that a reader would most likely look for in an index. For example, the key terms in a reference to the development of legislation about environmental impact statements would probably be *legislation* and *environmental impact statement*. In selecting terms for index entries, use chapter or section titles only if they include such key words. For index entries on **tables** and **illustrations,** use the key words in their titles.

Compiling an Index

Do not attempt to compile an index until the final manuscript is completed because terminology and page numbers will not be accurate before then. The best way to compile a list of topics is to read through your written work from the beginning; each time a key term appears in a significant context, list the term and its page number on a 3-by-5 index card. An index entry can consist solely of a main entry and its page number.

- Aquatic monitoring programs, 42

An index entry can also include a main entry, subentries, and even sub-subentries, as shown in Figure I–2. A subentry indicates pages where a specific subcategory or subdivision of the main topic can be found.

When you have completed this process for the entire work, sort the main entries alphabetically, then sort all subentries and sub-subentries alphabetically beneath their main entries.

Wording Index Entries

The first word of an index entry should be the principal word, because the reader will look for topics alphabetically by their main

Monitoring programs, 27–44 —————————— Main entry
    aquatic, 42
    ecological, 40 ————————— Subentries
    meteorological, 37
       operational, 39
       preoperational, 37 ————— Sub-subentries
    radiological, 30
    terrestrial, 41, 43–44 ————— Subentries
    thermal, 27

FIGURE I–2. Index Entry with Main Entry, Subentries, and Sub-subentries

words. Selecting the right word to list first is easier for some topics than for others. For instance, *tips on repairing electrical wire* would not be a suitable index entry because a reader looking for information on electrical wire would not look under the word *tips*. Ordinarily, an entry with two key words, like *electrical wire,* should be indexed under each word (*electrical wire* and *wire, electrical*). An index entry should be written as a **noun** or a noun **phrase** rather than as an **adjective** alone.

- Electrical *wire, 20-21*
    grounding, 21
    insulation, 20
    repairing, 22
    size, 21

Cross-Referencing

Cross-references in an index help readers find other related topics in the text. A reader looking up *technical writing,* for example, might find cross-references to *report* or *email*. Cross-references do not include page numbers; they merely direct readers to another index entry, where they can find page numbers. There are two kinds of cross-references: *see* references and *see also* references.

See references are most commonly used with topics that can be identified by several different terms. Listing the topic page numbers by only one of the terms, the indexer then lists the other terms throughout the index as *see* references.

- Economic costs. *See* Benefit-cost analyses

See references also direct readers to index entries where a topic is listed as a subentry.

- L-shaped fittings. *See* Elbows, L-shaped fittings

See also references indicate other entries that include additional information on a topic.

- Ecological programs, 40–49
    See also Monitoring programs

Writer's Checklist: Indexing

The following tips will help you style your index.

- ☑ Capitalize the first word of a main entry and all other terms normally capitalized in the text. Do not capitalize the first words in subentries and sub-subentries unless they are proper nouns.
- ☑ The cross-reference terms *see* and *see also* should appear in italics.
- ☑ Place each subentry in the index on a separate line, indented from its main entry. Indent sub-subentries from the preceding subentry. Indentations allow readers to scan a column quickly for pertinent subentries or sub-subentries.
- ☑ Separate entries from page numbers with commas.
- ☑ Format the index with double columns, as is done in the index to this book.

indiscreet / indiscrete

Indiscreet means "lacking in prudence or sound judgment."

- His public discussion of the proposed merger was *indiscreet*.

Indiscrete means "not divided or divisible into parts."

- The separate departments, once combined, become *indiscrete*.

(See also **discreet/discrete.**)

individual

Using *individual* as a noun is an **affectation.** Use *people* or another appropriate term.

- Several ~~individuals~~ *people* on the committee did not vote.
- Several ~~individuals on~~ *members of* the committee did not vote.

Individual is most appropriate when used as an **adjective** to distinguish a single person from a group.

- The *individual* employee's obligations to the firm are detailed in the booklet that describes company policies.

(See also **persons/people.**)

infinitive phrases (*see* **phrases**)

infinitives (*see* **verbals**)

ingenious / ingenuous

Ingenious means "marked by cleverness and originality"; *ingenuous* means "straightforward" or "characterized by innocence and simplicity."

- Seon Ju's *ingenious* plan, which streamlined production in Department L, was the beginning of her rise in the company.

- I believe that the *ingenuous* co-op students bring freshness to the company.

inquiries and responses

An inquiry letter or **email** can be as simple as a request for a free brochure or as complex as asking a consultant to define the specific requirements for establishing a **usability testing** lab.

There are two broad categories of inquiries. One kind provides a benefit (or potential benefit) to the reader: You may ask, for instance, for information about a product that a company has recently advertised. The second kind of inquiry primarily benefits the writer; an example is a request to a public utility for information on an energy-related project you are developing. The second kind of letter requires the use of **persuasion** and special consideration of your readers' needs.

Writing Inquiries

Your purpose in writing inquiries will probably be to obtain answers to specific questions. You will be more likely to receive a

University of Dayton
P.O. Box 113
Dayton, OH 45409
March 11, 20--

Jane E. Metcalf
Engineering Services
Miami Valley Power Company
P.O. Box 1444
Miamitown, OH 45733

Dear Ms. Metcalf:

Could you please send me some information on heating systems for a computerized, energy-efficient house that a team of engineering students at the University of Dayton is designing?

The house, which contains 2,000 square feet of living space (17,600 cubic feet), meets all the requirements stipulated in your brochure "Insulating for Efficiency." We need the following information:

1. The proper-size heat pump to use in this climate for such a home.
2. The wattage of the supplemental electrical heating units that would be required for this climate.
3. The estimated power consumption and current rates of those units for one year.

We will be happy to send you a copy of our preliminary design report. If you have questions or suggestions, please contact me at kjp@fly.ud.edu or call 513-229-4598.

Thank you for your help.

Sincerely,

Kathryn J. Parsons

Kathryn J. Parsons

FIGURE I-3. Inquiry Letter

prompt, helpful reply if you make it easy for the reader to respond. To do that, follow these guidelines:

- Keep your questions specific and clear but concise.
- Phrase your questions so the reader will immediately know the type of information you are seeking, why you need it, and how you will use it.

```
Date: Mon, 19 Apr 2000 11:42:25 -0500 (EDT)
From: Jane E. Metcalf (metcalf@mvpc.org)
To: Kathryn J. Parsons (kjp@fly.ud.edu)
Cc: mwang@mvpc.org
Subject: Design Questions


Dear Kathryn Parsons:

Thank you for inquiring about the heating system we
recommend for use in homes designed according to the
specifications outlined in our brochure "Insulating for
Efficiency."

Because I cannot answer your specific questions, I have
forwarded your inquiry to Michael Wang, Engineering
Assistant in our Development Group. He should be able
to answer the questions you have raised. You should be
hearing from him shortly.

Best wishes,

Jane Metcalf

=================================================
Jane E. Metcalf, Director of Public Information
Miami Valley Power Company
P.O. Box 1444 ~ Miamitown, OH 45733
Office 513.264.4880 ~ Fax 513.264.4889
Web http://www.enersaving.com
=================================================
```

FIGURE I-4. Response Indicating That an Inquiry Has Been Forwarded

- If possible, present your questions in a numbered list to make it easy for your reader to deal with them.
- Keep the number of questions to a minimum.
- Offer some inducement for the reader to respond, such as promising to share the results of what you are doing.
- Promise to keep responses confidential, if appropriate.

At the end of the letter, thank the reader for taking the time to respond. In addition, make it convenient for the recipient to respond by providing contact information, such as a phone number or an email address, as shown in Figure I-3 on page 288.

MIAMI VALLEY POWER COMPANY
P.O. BOX 1444
MIAMITOWN, OH 45733
(513) 264-4800

March 24, 20--

Ms. Kathryn J. Parsons
University of Dayton
P.O. Box 113
Dayton, OH 45409

Dear Ms. Parsons:

Jane Metcalf forwarded to me your letter of March 11 about the house that your engineering team is designing. I can estimate the insulation requirements of a typical home of 17,600 cubic feet as follows:

1. For such a home, we would generally recommend a heat pump capable of delivering 40,000 Btus. Our model AL-42 (17 kilowatts) meets that requirement.
2. With the AL-42 efficiency, you don't need supplemental heating units.
3. Depending on usage, the AL-42 unit averages between 1,000 and 1,500 kilowatt-hours from December through March. To determine the current rate for such usage, check with the Dayton Power and Light Company.

I can give you an answer that would apply specifically to your house only with information about its particular design (such as number of stories, windows, and entrances). If you send me more details, I will be happy to provide more precise figures. Your project sounds interesting.

Sincerely,

Michael Wang

Michael Wang
Engineering Assistant
mwang@mvpc.org
http://www.enersaving.com

FIGURE I–5. Response to an Inquiry

Responding to Inquiries

When you receive an inquiry, determine whether you have both the information and the authority to respond. If you are the right person in your organization to respond, answer as promptly as you can and be sure to answer every question asked. How long and how detailed your response should be depend on the nature of the questions and the information provided in the letter about the writer. Even if the writer has asked a question that seems silly or one to which you feel the answer is obvious, answer it courteously and as completely as you can. You may tactfully point out that the reader has omitted or misunderstood something.

If you have received a letter you feel you cannot answer, find out who can and forward the letter to that person. Notify the letter writer that you have forwarded the letter, as shown in Figure I–4 on page 289. The person to whom an inquiry has been forwarded should state in the first paragraph of his or her response why someone else is answering the original inquiry, as shown in Figure I–5 on page 290. (See also **correspondence.**)

inside / inside of

In the phrase *inside of,* the word *of* is redundant and should be omitted.

- The switch is just inside ~~of~~ the door.

Using *inside of* to mean "in less time than" is colloquial and should be avoided in writing.

- They were finished ~~inside of~~ *in less than* an hour.

insoluble / insolvable

The words *insoluble* and *insolvable* are sometimes used interchangeably to mean "incapable of being solved." *Insoluble* also means "incapable of being dissolved." *Insolvable* means only "incapable of being solved."

- Rubber gloves are *insoluble* in most household solvents.

- Until yesterday, the production problem seemed *insolvable.*

instructions

When you explain how to perform a specific task, you are giving instructions. To write accurate and easy-to-follow instructions, you must thoroughly understand the task you are describing. If you are unfamiliar with the task, watch someone who is familiar with it go through each step before you begin to write.

Writing Instructions

When you are writing instructions, keep in mind your **readers'** level of knowledge. If you know that all your readers have good backgrounds in the topic, you can use fairly specialized words. If that is not the case, it is more appropriate to use simple, everyday language.

The clearest, simplest instructions are written as commands in the imperative **mood.**

Raise
- ~~The operator should raise~~ the access lid.
  ^

Phrase the instructions concisely. You can make sentences shorter by leaving out articles (*a, an, the*), some pronouns (*you, this, these*), and some verbs, but such sentences are **telegraphic** and often hard to understand. For example, the first version of the following instruction for placing a document in a scanner tray is confusing and difficult to understand.

TELEGRAPHIC   Place document in tray printed side facing opposite.

CLEAR   Place the document to be scanned in the document tray with the printed side facing away from you.

One good way to make instructions easy to follow is to divide them into short, simple steps. Be sure to order the steps in the proper sequence. Steps can be organized with numbers or with words. You can number each step in the sequence:

1. Connect each black cable wire to a brass terminal.
2. Attach one 4-inch green jumper wire to the back.
3. Connect the jumper wire to the bare cable wire.

Or you can use words that indicate time or sequence:

- *First,* determine the problem the customer is having with the computer. *Next,* observe the system in operation. *At that time,* question the operator until you are sure that the problem has been explained completely. *Then* analyze the problem and make any necessary adjustments.

Plan ahead for your reader. If the instructions in step 2 will affect a process in step 9, say so in step 2. Sometimes your instructions have to make clear that two operations must be performed simultaneously. Either state that fact in an **introduction** to the specific instructions or include both operations in one step.

CONFUSING	1. Hold down the CONTROL key.
	2. Press the RETURN key before releasing the CONTROL key.
CLEAR	1. While holding down the CONTROL key, press the RETURN key.

If your instructions involve a great many steps, break them into stages, each with a separate heading so that each stage begins again with step 1. Using **headings** as dividers is especially important if your reader is likely to be performing the operation as he or she reads the instructions.

Using Illustrations

Clear and well-planned **illustrations** can make complex instructions easy to understand. Illustrations often simplify instructions by reducing the number of words necessary to explain a process or procedure. **Drawings** and diagrams enable your reader to identify parts and the relationships between parts more easily than long explanations.

The instructions in Figure I–6 on page 294 guide the reader through the steps of "streaking" a saucer-sized disk of material (called *agar*) used to grow bacteria colonies. The object is to thin out the original specimen (the inoculum) so the bacteria will grow in small, isolated colonies. The streaking process makes certain that part of the saucer is inoculated heavily and the remaining portions are inoculated progressively more lightly. The streaking is done by hand with a thin wire, looped at one end for holding a small sample of the inoculum.

STREAKING AN AGAR PLATE

Distribute the inoculum over the surface of the agar in the following manner:

1. Beginning at one edge of the saucer, thin the inoculum by streaking back and forth over the same area several times, sweeping across the agar surface until approximately one-quarter of the surface has been covered. *Sterilize the loop in an open flame.*

2. Streak at right angles to the originally inoculated area, carrying the inoculum out from the streaked areas onto the sterile surface with only the first stroke of the wire. Cover half of the remaining sterile agar surface. *Sterilize the loop.*

3. Repeat as described in Step 2, covering the remaining sterile agar surface.

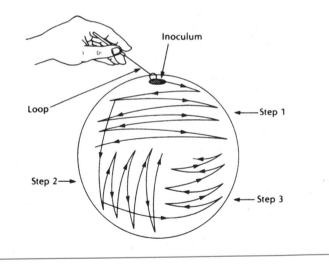

FIGURE I–6. Illustrated Instructions

Warning Readers

Alert your readers to any potentially hazardous steps or materials. If the instructions call for materials that are flammable or give off noxious fumes, let the readers know before they reach the step for which the material is needed. Readers who will be handling hazardous materials must be cautioned about requirements for special clothing, tools, equipment, or other measures they must take to safely complete their task. Visually highlight warnings, cautions, and precautions to make them stand out from the surrounding text. For instance, warning notices can be presented in a box. In Figure I–7, instructions written for maintenance and repair crews of heavy industrial equipment, note the instructions, separated by rules (solid black lines) from other steps, that direct crews to shut off power sources and "bleed" any residual energy from the equipment before they begin work. Other attention-getting devices include all capital letters, large and distinctive fonts, and color.

REMOVING RESIDUAL ENERGY

Follow these steps for a typical lockup before all maintenance and repair work:

Step 1: Alert the operator and floor supervisor that you are ready.

Step 2: Identify all sources of residual energy on the machinery.

WARNING

Before beginning work, perform these procedures in the following order:

Step 3: Place padlocks on the switch, lever, or valve to lock the equipment in the OFF position.

Step 4: "Bleed" all hydraulic or pneumatic pressure and all electrical current (capacitance) so that machine components will not accidentally move.

Step 5: Test operator controls.

Step 6: After work is finished, replace all machine safeguards that were removed. Secure and check them to make sure they fit properly.

Step 7: Finally, remove padlocks and clear the machine for operation.

FIGURE I–7. Set of Instructions with Precautions Made Clear

Experiment with font style, size, and color to determine which devices are most effective. (See also **layout and design.**)

Testing Instructions

Finally, to test the accuracy and **clarity** of your instructions, ask someone who is not familiar with the operation to follow your written directions. A first-time user can spot missing steps or point out passages that should be worded more clearly. As you observe your tester, note any steps that seem especially puzzling or confusing and revise the instructions accordingly. (See also **process explanation** and **usability testing.**)

insure / ensure / assure

Insure, ensure, and *assure* all mean "make secure or certain." *Assure* refers to people, and it alone has the connotation of setting a person's mind at rest. *Ensure* and *insure* mean "make secure from harm." Only *insure* is widely used in the sense of guaranteeing the value of life or property.

- I *assure* you that the equipment will be available.

- We need all the data to *ensure* the success of the project.

- We should *insure* the contents of the building.

intensifiers   ESL

Intensifiers are adverbs that emphasize degree, such as *very, quite, rather, such,* and *too.* Although they serve a legitimate and necessary function, they are often overused. Too many intensifiers weaken your writing. When revising your draft, either eliminate intensifiers that do not make a definite contribution or replace them with specific details.

- The team was ~~quite~~ happy to ~~receive the very good news~~ *learn* that it
  had been awarded a ~~rather substantial monetary~~ *$5,000* prize for its
  design.

The difference is not that the original sentence is wrong and the revised one right, but that the intensifiers "quite," "very," and "rather" add nothing to the sentence and impair **conciseness.**

Some words (such as *unique, perfect, impossible, final, permanent, infinite,* and *complete*) do not logically permit intensification because, by definition, they do not permit degrees of comparison. Although **usage** often ignores that logical restriction, be aware that ignoring it is contrary to the basic meanings of those words.

- It was ~~quite~~ impossible for the part to fit into its designated position.

(See also **absolute words** and **equal/unique/perfect.**)

interface

An *interface* is a surface that provides a common boundary between two bodies or areas. The bodies or areas may be physical (the *interface* of a piston and a cylinder) or conceptual (the *interface* of mathematics and statistics).

Do not use *interface* as a substitute for *cooperate, interact,* or even *work.*

- The Water Resources Department will ~~interface~~ *work* with the Department of Marine Biology on the proposed project.

(See also **affectation.**)

interjections   ESL

An interjection is a word or phrase standing alone or inserted into a sentence to exclaim or command attention. Grammatically, it has no connection to the sentence. An interjection can be strong (*Hey! Ouch! Wow!*) or mild (*oh, well, indeed*). A strong interjection is followed by an **exclamation mark.**

- *Wow!* Those test results are incredible!

A weak interjection is followed by a comma.

- *Well,* we need to rethink the proposal.

An interjection inserted into a sentence may need either a comma before it and after it or no commas at all.

- We must, *indeed,* rethink the proposal.

- What *in the world* should we do next?

Because they get their main expressive force from sound, interjections are more common in speech than in writing. They are rarely appropriate to technical writing.

internal proposals (*see* proposals)

international correspondence

With organizations participating in the increasingly global marketplace, you may need to write letters, **memos,** or **email** messages to readers whose native language is not English. Such readers may be outside your organization, as in the case of customers, suppliers, and distributors. Or they may be inside your organization, for example, at branch offices outside the United States.

Because English is widely taught and used in international business, you will be able to send most international correspondence in English. If you must use a translator, however, be sure the translator understands the **purpose** of your **correspondence.** It is also prudent to let your reader know (in the letter itself or in a postscript) that a translator helped write the letter. For first-time contacts, consider sending both the English version and a translation in the reader's native language.

Culture and Business Writing Style

Just as American business writing style has changed over time, ideas about appropriate business writing style vary from culture to culture. You must be alert to the needs and expectations of readers from different cultural and linguistic backgrounds. For example, in the United States, direct, concise writing may demonstrate courtesy by not wasting another person's time; in other cultures (in countries such as Spain and India), directness and brevity may suggest that the writer dislikes the reader so much that he or she wants to make the communication as short as possible. Although business writing should always be courteous and efficient, experienced writers understand that the forms of courtesy and ideas about efficiency vary from country to country.

In American business correspondence, traditional salutations such as *Dear* and complimentary closings such as *Yours truly* have, through custom and long use, acquired meanings quite distinct from their dictionary definitions. Understanding the unspoken meanings of these forms and using them naturally is routine in American business culture. But such customary expressions vary from culture to culture. Therefore, when you read correspondence from businesspeople in other cultures or countries, be alert to the differences and consider how you can use expressions from those other cultures in your own writing. Japanese business writers, for example, often use traditional openings that reflect on the season, compliment the reader's success, and offer hopes for the reader's continued prosperity. Such traditional openings may strike some American readers as being overly elaborate, literary, or even insincere. Whereas an American business writer might consider one brief letter sufficient to communicate a request, a writer in another culture may expect an exchange of three or four longer letters to pave the way for action.

The first step to avoiding misunderstandings is to be aware that differences exist and to learn how they affect communication. Japanese business writers, for example, express negative messages and **refusal letters** indirectly to avoid embarrassing the recipient. Be sensitive to the expectations of readers who judge effective and appropriate communication from the perspective of a different culture. (To learn more about this subject, see **global communication** and **global graphics.** Also, use the term *intercultural communication* to search library and **Internet** sources.)

Language and Usage in International Correspondence

Take special care in international correspondence to avoid American **idioms** ("it's a slam dunk," "give a heads up," and the like), unusual **figures of speech,** and **allusions** to events or attitudes particular to American life. Such expressions could easily confuse your reader. Avoid humor, irony, and sarcasm—they are easily misunderstood outside their cultural context.

Pretentious or overly ornate writing, as in **affectation,** will also impede the reader's understanding. If you plan to use **jargon** or technical terminology, ask yourself whether the words you choose can be found in abbreviated English-language dictionaries. (See also **word choice.**)

Generally, write clear and complete sentences (see **sentence construction**). Unusual word order or rambling sentences will frustrate and confuse a nonnative reader of English. Read your writing aloud to identify overly long sentences and to eliminate any misplaced **modifiers** or **awkwardness.** Long sentences that contain more information than the reader can comfortably absorb should be divided into two or more sentences. However, avoid using an overly simplified "storybook" style. A reader who has studied English as a second language would be insulted by a condescending tone and childish language.

Finally, **proofreading** is crucial; a misspelled word, such as *there* for *their* or *discreet* for *discrete,* will be particularly troublesome for someone reading in a second language, especially if that reader turns to a **dictionary** for help and cannot find the word because it is misspelled. (See also **English as a second language.**)

Dates, Time, and Measurement

Countries differ in their use of formats to represent **dates,** time, and other kinds of measurement. To represent dates, most countries typically write the day before the month and year. For example, 1/11/99 means 1 November 1999 in most parts of the world; in the United States, it means January 11, 1999. Write out the name of the month to make the entire date immediately clear to all international readers. Time poses similar problems, so you may need to specify time zones or refer to international standards, such as Greenwich Mean Time (GMT) or Universal Time Coordinated (UTC), for clarity.

Use other international standards, such as commas for decimal points and the metric system (used in all countries except the United States). For up-to-date information about accepted conventions for numbers and symbols in chemical, electrical, data-processing, pharmaceutical, and other fields, consult guides and manuals specific to the subject matter.

Cross-Cultural Examples

Figure I–8 and Figure I–9 on pages 301–04 are two versions of a letter written to a Japanese businessman, Mr. Ichiro Katsumi,

Sun West Corporation, Inc.

2565 North Armadillo
Tucson, AZ 85719
Phone: (602) 555-6677
Fax: (602) 555-6678   Email: sunwest@aol.com

February 27, 20--

Ichiro Katsumi, Investment Director, *?*
Toshiba Investment Comp. *?*
1-29-10 Ichiban-cho
Tokyo 105, Japan

Dear Ichiro:

sp How are you? I've just heard through the grapevine *[Slang]* that you'l *[sp]*
be coming to visiit us in Tucson next month. That's great,
sp we've been looking forward to seeing you for some time now,
especially since we heard you're intereested in investing in
our company because as you know, cash flow is very important to any company, especially a small one like ours. *[Slang]*

Jargon I've been asked by the head honchos here to confirm your
flight reservations. A temp took the original information, but *[Slang]*
sp you know how hard it is to get good help nowadays, so I
need to confirm it agian. You'll be coming in on 3/20/02 on *[U.S. date format]*
Delta, flight no. 435 at 2:00 p.m. And we'll send someone to
pick you up at the airport. I'm sure you'll be tired, you'll probably have some computer equipment with you and lots of *[sp/Slang]*
luggage, so be sure to tell the skycab to help you and we'll
Slang reimburse you for that and anything else you spend money
on. When you get off the plane, just go to the baggage
sp cliam, then get your stuff, then go outside to the limo area
and our driver will be there wiht a sign with your name on it. *[sp]*

FIGURE I-8. Inappropriate International Correspondence Marked for Revision

Ichiro Katsumi   2   February 27, 2002

Slang —

Now, to the important (stuff.) In all honesty, we are very ex- (*sp*)
cited that you are coming to invest in our company. I (thikn)
this will provide us with a much-needed infusion of funds
with which to not only stabilize but spur growth of our little
Slang — company. Our products are unique and we could never expand (stuff) without your help since Tucson's a growing place
with (lots of people) moving in here. And, of course, you'd end   *Idiom*
up being the recipients of the (fruits of your labor,) too. So if all
works out well, we should realize immense profits in two
years or so. And despite what other people around the world
say about Americans, we really are hard workers, especially (*sp*)
my boss who heads up the company — (nose to the grind-
Informal — stone) every day and (burning the midnight oil) everynight!
Idioms

Anyway, (I)'ve enclosed a guidebook and map of Tucson and
material on our company. If you see anything you'd like to do
in town, let me know. And if you have any questions about
the company before we see you, just drop us a quick email
or fax (I don't think (snail mail) will get back to us in time).

Have a safe trip,) *?*
                                                        — *Jargon*

Ty Smith

Ty Smith

FIGURE I–8. **Inappropriate International Correspondence Marked for Revision**
(*continued*)

Sun West Corporation, Inc.

2565 North Armadillo
Tucson, AZ 85719
Phone: (602) 555-6677
Fax: (602) 555-6678   Email: sunwest@aol.com

March 1, 20--

Ichiro Katsumi
Investment Director
Toshiba Investment Company
1-29-10 Ichiban-cho
Tokyo 105, Japan

Dear Mr. Katsumi:

I hope that you and your family are well and prospering in the new year. We at Sun West Corporation are very pleased that you will be coming to visit us in Tucson this month. It will be a pleasure to meet you, and we are very gratified and honored that you are interested in investing in our company.

So that we can ensure that your stay will be pleasurable, we have taken care of all of your travel arrangements. You will

- Leave Narita-New Tokyo International Airport on Delta Airlines flight #75 at 5:00 p.m. on March 20, 2002
- Arrive at Los Angeles International Airport at 10:50 a.m. local time and depart for Tucson on Delta flight #186 at 12:05 p.m.
- Arrive at Tucson International Airport at 1:30 p.m. local time on March 20
- Depart Tucson International Airport on Delta flight #123 at 6:45 a.m. on March 27
- Arrive in Salt Lake City, Utah, at 10:40 a.m. and depart at 11:15 a.m. on Delta flight #34 and arrive in Portland, Oregon, at 12:10 p.m. local time
- Depart Portland, Oregon, on Delta flight #254 at 1:05 p.m. and arrive in Japan at 3:05 p.m. local time on March 28

If this information is not accurate or if you need additional information about your travel plans or information on Sun West Corporation, please call, fax, or email me directly. That way, we will receive your message in time to make the appropriate changes or additions.

FIGURE I–9. Appropriate International Correspondence

After you arrive in Tucson, a chauffeur from Skyline Limousines will be waiting for you at Gate 12. He or she will be carrying a card with your name, will help you collect your luggage from the baggage claim area, and will then drive you to the Loews Ventana Canyon Resort. This is one of the most prestigious resorts in Tucson, with spectacular desert views, high-quality amenities, and one of the best golf courses in the city. The next day, the chauffeur will be back at the Loews at 9:00 a.m. to drive you to Sun West Corporation.

We at Sun West Corporation are very excited to meet you and introduce you to all the members of our hard-working and growing company family. After you meet everyone, you will enjoy a catered breakfast in our conference room. At that time, you will receive a schedule of events planned for the remainder of your trip. Events include presentations from the president of the company and from departmental directors on

- The history of Sun West Corporation
- The uniqueness of our products and current success in the marketplace
- Demographic information and benefits of being located in Tucson
- The potential for considerable profits for both our companies with your company's investment

We encourage you to read through the enclosed guidebook and map of Tucson. In addition to events planned at Sun West Corporation, you will find many natural wonders and historical sites to see in Tucson and in Arizona in general. If you see any particular event or place that you would like to visit, please let us know. We will be happy to show you our city and all it has to offer.

Again, we are very honored that you will be visiting us, and we look forward to a successful business relationship between our two companies.

Sincerely,

Susan Roberts

Susan Roberts
Vice President

Encl. as stated

FIGURE I–9. Appropriate International Correspondence (*continued*)

Investment Director of Toshiba Investment Company. Mr. Katsumi is interested in investing in Sun West Corporation of Tucson, Arizona, and plans to visit Tucson for a week to meet with company officials, tour the company, and examine the company's products and financial records. Mr. Ty Smith of Sun West Corporation has been asked to write a letter to Mr. Katsumi confirming travel arrangements and mentioning the benefits that an investment in Sun West Corporation would bring.

Ty Smith asked a colleague who is a specialist in language, usage, and international communication to review his letter and note any problems in the margin. His colleague indicated that the letter in Figure I–8 would be confusing to Mr. Katsumi or to any global or nonnative recipient. Aside from the spelling errors, the letter is filled with slang (*grapevine, head honchos*), clichés (*fruits of your labor, nose to the grindstone, burning the midnight oil*), jargon (*snail mail*), and the inappropriate use of *stuff*, an American expression that is not specific enough to communicate anything useful.

Compare that letter to the one in Figure I–9, which was rewritten by a writer more sensitive to the needs of a multicultural audience. The second version of the letter is written in language that is literal and specific. The slang, clichés, and jargon are gone. For easier comprehension and translation, the sentences are shorter, bulleted lists are used to break up the **paragraphs, contractions** are eliminated, and the expression of dates is clearer.

Writing correspondence for audiences in other cultures requires careful thought because we are all, to greater or lesser degrees, embedded in the language habits of our cultures. When you are writing for international readers, rethink the ingrained habits that define how you express yourself. Doing so will help you achieve clarity and mutual understanding with nonnative English speakers.

Internet

The Internet is a system of public and private networks that allows people around the world to communicate, find and share information, and offer commercial services online. Internet resources include electronic mail (email), discussion groups, chat environments, the **World Wide Web** (WWW or the Web), gopher, file transfer protocol (FTP), and telnet. (For guidelines on locating and evaluating information on the Internet, see **Internet research**.)

Communicating with Others

Email.   Email is one of the Internet's most successful and effective services. (For guidance on the effective use of this technology, see **email.**)

Discussion Groups.   Internet discussion groups provide forums for exchanging information with other users about topics of mutual interest. Some discussion groups are conducted through an exchange of email, while others function as postings on a public bulletin board that anyone can read. Internet Relay Chat (IRC) permits users to participate in discussions in real time. Unless you are using Internet discussion groups for professional purposes, such as gathering information for a report, they are generally inappropriate for the workplace.

EMAIL DISCUSSION GROUPS.   Topic-oriented email discussion groups are relatively easy to locate. The simplest way is to conduct a search using the term "mailing list" in a search service such as AltaVista or Yahoo! You can also use Liszt (http://www.liszt.com/), a mailing list directory that is organized into topical areas, such as books, computers, education, humanities, music, news, and recreation (Figure I–10 on page 307). Each broad topic heading indicates the number of lists found there and shows sample subtopics. The heading Computers (250 lists) includes the subtopics Hardware, Database, and Programming, each of which leads to increasingly specific topics in those categories.

To join an email discussion group (often called "listserv"), send a request to the list administrator asking that your address be added to the list. Thereafter, you will receive all email sent to the group by other members, and your questions and comments will be sent to all other list members. Discussion groups also are often used to distribute information to other members: monthly newsletters, the status of legislation, new developments in a field, upcoming meetings and conferences, and the like.

NEWSGROUPS.   Newsgroups are another type of online discussion group in which people exchange ideas, advice, and information. They are part of the Internet distribution network called "Usenet." You can visit a newsgroup site to read and post messages to other people with a shared interest in a specific topic. Unlike email discussion groups, in which messages are sent directly to your

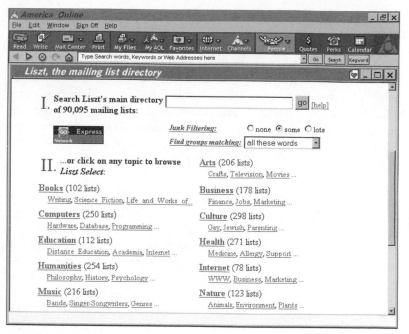

FIGURE I–10. Liszt Home Page

email in-box, with Usenet newsgroups, you must actively seek the online bulletin board, where messages are posted. Once you have reached a site, you can bookmark it for future access.

What topics are available? They number in the thousands and for ease of access are organized into broad topic categories: biz (business), comp (computers), sci (science), rec (recreation), soc (social issues), news (Usenet-related topics), and more. Each broad category is subdivided into increasingly specific topics.

The Liszt Usenet Newsgroup Directory (http://www.liszt.com/news/) provides links to thousands of newsgroups and gives detailed guidance on how to access and participate in newsgroup discussions.

INTERNET RELAY CHAT.   IRC, or "chat rooms," permits online "conversations" over topic-related channels that take place in real time. That is, the messages are read as they are typed rather than being stored and read afterward, as with listserv and Usenet messages. One advantage of using IRC is that it can function like a telephone conference call by allowing you to carry on simultaneous conversations or interviews with several people in remote locations. The Liszt IRC Chat Directory (http://www.liszt.com/chat/) provides access to over 38,000 IRC topics and gives instructions for participating in a conversation and conducting IRC subject searches.

Other more elaborate chat environments include Multi-User Dungeons (MUDs) and Multi-User Dungeons Object Oriented (MOOs), which are text-based thematic virtual environments. In those environments, users assume online personas, chat with other people, navigate "rooms" and other settings, and, in more advanced cases, design their own virtual worlds. These environments tend to be used more for recreational chat than for business-oriented activities.

WEB FORUMS.   Web forums allow participants to share information directly over the Web. They function as bulletin or message boards similar to Usenet sites in that they permit users to post and respond to messages. Unlike Usenet, Web forums are Web based and do not rely on a separate Internet system. That difference allows Web forums more flexibility to customize their use and to limit solicitations for get-rich-quick schemes and other unwelcome messages.

Most forums require you to register before you can post messages, although they waive that requirement for read-only access. Web forums are not real-time transmissions, like IRC, so you need not be online at the same time as other members to communicate with them.

You can locate Web forum topics at Forum One (http://www.forumone.com/) by clicking on one of the broad topics listed (Current Events, Society and Culture, Business and Finance, Computers, Health, Science, Education, among others) or by entering a keyword in the search window to retrieve a list of active discussions on that topic. The discussions range from forums for health care, computer, and engineering professionals to forums devoted to hobbies and entertainment.

Connecting with Other Systems

Telnet.   Telnet is a system that allows you to connect to remote computers via the Internet and run them and their software applications as if you were physically there. For example, if you are on a business trip and want to check your email messages back at the office, you can use telnet to log on to the computer supporting your account to access your messages. You can also use telnet to connect to public-access computers and databases, such as those in libraries, to search their bibliographic databases.

File Transfer Protocol.   FTP is a resource for moving files around the Internet quickly. Using an FTP client program, you can acquire software applications and data files for your personal computer and download them to your hard drive. Although some FTP sites require passwords, most allow anonymous connections. FTP lets you log on to a host computer, find the files you want, and transfer them to your own computer. FTP does not permit users to view the information before transferring it unless the host computer makes software available online for doing so.

Gopher Search System.   Gopher sites allow you to search for and retrieve information on the Internet. When you use gopher, you need not know the address of the computer containing the information you seek. Instead, you navigate up and down a menu and select an item, and gopher then locates it for you or takes you through a series of menus that narrow the search. Unlike FTP, gopher allows you to view the information before you transfer it to your computer.

The gopher system is being superseded by the World Wide Web, and many institutions no longer maintain their gopher sites.

World Wide Web.   The Web is a system that simplifies navigation around the Internet and combines the capabilities of all the other Internet access systems. The Web supports applications for email, discussion groups, and chat environments, as well as for searching, retrieving, and providing information. The Web is a multimedia environment: It brings together vast amounts of information in text, graphic, audio, and video form on a global scale. (See also **World Wide Web.**)

Internet research

The Internet and the **World Wide Web** provide a staggering amount of information: books, journal and newspaper articles, technical reports, speeches, photographs, maps, government reports, laws and proposed legislation, court decisions, meeting transcripts, business and technical material, and public-access terminals for many libraries' collections. You can also conduct research by participating in Internet and email discussion groups and newsgroups. Thousands of such groups exist; to access them, see the entry **Internet**. **Email** lets you target information requests to specific persons or groups faster and at lower cost than sending queries and **questionnaires** by conventional mail to the same recipients.

The lack of overall organization on the Internet and the fact that many Internet databases change can present challenges to locating the information you need. Unlike a library, the Internet has no single indexing system or catalog that brings the information together for browsing or easy access. Fortunately, a wide variety of search services can help you locate the resources you need. The type of information you seek largely determines which search tool is most efficient. To locate specific subjects, you can use two types of search tools: search engines and indexes.

Search Engines

A search engine is a computer program that helps you locate information in and across electronic databases. In their simplest form, search engines compare user requests with indexed database information and list the files that match. (For more on search strategies and specific search engines on the Web, see **search engines**.)

Indexes (Subject Directories)

An index, also known as a subject directory, organizes information by broad subject categories (for example, business, entertainment, health, sports) and related subtopics (for example, the business subtopics marketing, finance, investing). An index search eventually produces a list of specific sites that contain information about the topics you request. Once you locate a site of interest that you want to revisit, you can bookmark it.

For example, say you are searching for information about "plagiarism and legal issues for the Web." You might go to the main search and index page for Yahoo! (http://www.yahoo.com/). There you will find a number of broad topic links, as shown in Figure I–11 on page 311. A click on one of the topics, such as Computers & Internet, will retrieve a search screen that lists a broad range of related topics, as shown in Figure I–12 on page 311. A click on Ethics will

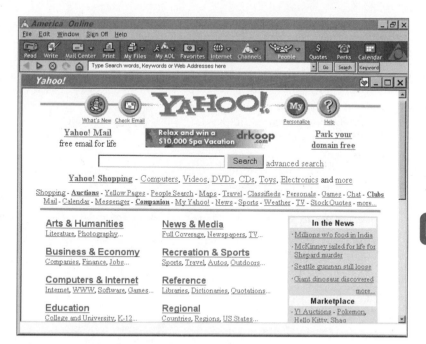

FIGURE I–11. Yahoo! Home Page

FIGURE I–12. Yahoo! Computers & Internet Topics

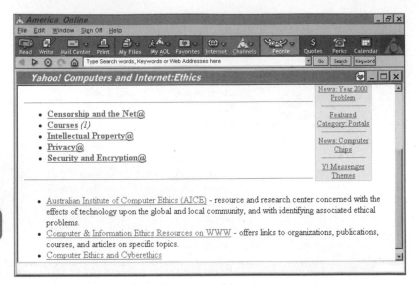

FIGURE I-13. Yahoo! Ethics Subtopics and Links

retrieve a list of five subtopics and links to a number of Web sites, as shown in Figure I-13.

Technical Information on the Web

In addition to general topic indexes and search engines, the Web also includes numerous sites devoted to specific subject areas. To get started with research on a technical topic, you could begin at one of the following sites.

- *National Science Foundation* (http://www.nsf.gov/). Offers links to all the sciences and government-related agencies.
- *WWW Virtual Library: Engineering* (http://arioch.gsfc.nasa.gov/wwwvl/engineering.html). Offers links across all the engineering fields, including libraries, professional associations, journals, and discussion groups.
- *IEEE Spectrum* (http://www.spectrum.ieee.org). Offers information and links to all IEEE publications as well as a section called "Sites we like," which offers a wide variety of engineering and technology links.

Evaluating Internet Sources

Because of the wide accessibility of the Internet and because it often bypasses the careful review stages built into print publishing, you need to be especially concerned about the reliability of information you find on the Internet. Evaluate the usefulness and reliability of information on the Internet or the Web by the same standards you use to evaluate information from other sources.

The easiest way to ensure that information is valid is to obtain it from a reputable source. American businesses rely on and widely use the compilations of data from the Bureau of Labor Statistics, the Securities and Exchange Commission, and the Bureau of the Census. Likewise, the Internet (or Web) versions of established, reputable journals in medicine, management, engineering, computer software, and the like, merit the same level of trust as the printed versions. As you move away from established, reputable sites, exercise more caution. Be especially wary of unmoderated discussion groups on Usenet and other public bulletin board systems. Remember that anyone with access can publish on the Internet; thus, for many sources, there are no editorial checks and balances. Treat information obtained from such sources accordingly. (For information about citing Internet sources, see **documenting sources.**)

Writer's Checklist: Evaluating Internet Sources

Ask yourself these questions before you cite a Web site or other Internet forum.

- ☑ Is it accurate and up-to-date?
- ☑ Is the author qualified and reputable?
- ☑ Who maintains the site? A university? A government agency? A library?
- ☑ Is the sponsor of the site reputable?
- ☑ Is it objective? Are biases clearly stated?
- ☑ Are opinion pieces labeled as such?
- ☑ Is the scope of topics the site covers made clear?
- ☑ Are there any disclaimers about the scope or content of the information?

Writer's Checklist: Evaluating Internet Sources (continued)

These standards also apply to printed material, people you interview, and all other sources. One good source for Web-evaluation techniques is (http://www.science.widener.edu/~withers/webeval.htm).

interviewing for information

The **scope** of coverage you establish for your writing project may require information that is available only from individual persons: experts, customers, users of equipment, and those with opinions that may provide valuable insights. You may also find an interview helpful in preparing a **questionnaire.** When you reach this point, consider a personal interview.

The process of interviewing can be divided into three parts: (1) determining the proper person to interview, (2) preparing for the interview, and (3) conducting the interview.

Determining the Proper Person to Interview

Many times, your subject or your **purpose** in writing about the subject will logically point to the proper person to interview. If you were writing a proposal to a government agency, you might want to interview someone who has experience both in writing government proposals and with the agency involved. The following sources can help you determine the appropriate person to interview: (1) a search on the Internet, (2) local chapters of professional societies, and (3) the Yellow Pages of the local telephone directory. (See also **Internet research.**)

Preparing for the Interview

After determining whom you want to interview, it is best to request the interview by telephone or **email,** since letters are probably too slow to allow you to meet your deadline.

Before the interview, learn as much as possible about the person you are going to interview as well as the company or organization for which he or she works. When you make contact, whether by email or telephone, with the prospective interviewee to set up an interview, explain who you are, why you chose him or her for the interview, the subject of the interview, that you would like to arrange an interview at his or her convenience, and that you will allow him or her to review your draft.

After you have made the appointment, prepare a list of specific questions to ask the interviewee. The natural temptation for many interviewers is to ask vague, general questions rather than specific ones.

VAGUE   "What do you think of the Internet?"

Such a question is too general to elicit specific and useful information. Rather, ask more specific but open-ended questions that prompt interviewees to respond.

SPECIFIC   "Some local companies in your business are making extensive use of the Internet for marketing. How are you making use of the Internet?"

More specific questions prompt the interviewee to provide more useful information.

Writer's Checklist: Conducting an Interview

The following list summarizes effective techniques for conducting an interview.

☑ Arrive promptly for the interview and set the interviewee at ease.

☑ Be pleasant but purposeful. You are there to get information, so do not be timid about asking leading questions on the subject.

☑ Use your prepared list of questions. Begin with the least complex aspects of the topic to get the conversation started and then move to the more complex aspects.

☑ Do not get sidetracked. If the interviewee strays from the subject, be ready with a specific question to direct the conversation back on track.

☑ Avoid being rigid; if a prepared question is no longer suitable, move to the next question.

☑ Some answers prompt additional questions; ask them as they arise.

☑ Let your interviewee do most of the talking. Do not try to impress him or her with your own knowledge or opinions about the subject. Remember that the interviewee is the expert.

Writer's Checklist: Conducting an Interview (continued)

☑ Take only memory-jogging notes that will help you recall the conversation later. Concentrate on key facts and figures. Do not ask your interviewee to slow down so you can take detailed notes. To do so not only would be an imposition on the interviewee's time but might interrupt his or her train of thought.

☑ Use a tape recorder only if both you and your interviewee are comfortable with it. If you use a tape recorder, do not rely on it to do all your work—make sure you ask crucial questions.

☑ As the interview is reaching a close, take a few minutes to skim your notes. If time allows, ask the interviewee to clarify anything that is ambiguous. Do not, however, overstay your welcome.

☑ After thanking the interviewee, ask if you may call back if you need to clarify a point or two as you write up your interview notes.

☑ Immediately after leaving the interview, expand your memory-jogging notes to help you mentally review the interview. Do not postpone this step. No matter how good your memory is, you will forget some important points if you do not flesh out your notes at once.

interviewing for a job

A job interview may last for thirty minutes, or it may take several hours; it may be conducted by one person or by several, either at one time or in a series of interviews, by phone, by teleconferencing, or in person. Since it is impossible to know exactly what to expect, it is important that you be as well prepared as possible.

Before the Interview

The interview is not a one-way communication. It presents you with an opportunity to ask questions of your potential employer. In preparation, learn everything you can about the company before the interview. Use these questions as a guide.

- What kind of organization is it?
- How diversified is it?
- Is it locally owned?

- Is it a nonprofit organization?
- If it is public employment, at what level of government is it?
- Does it provide a service? If so, what kind?
- How large is the business? How large are its assets?
- Is the owner self-employed? Is it a subsidiary of a larger operation? Is it expanding?
- How long has it been in business?
- Where will you fit in?

You can obtain information from the Internet, current employees, company literature such as employee publications, and the business section of back issues of local newspapers (available in the library). You may be able to learn the company's size, sales volume, product line, credit rating, branch locations, subsidiary companies, new products and services, building programs, and other such information from its annual reports; publications such as *Moody's Industrials, Dun and Bradstreet, Standard and Poor's,* and *Thomas' Register;* and other business reference sources a librarian might suggest. What you cannot find through your own research, ask your interviewer. Now is your chance to make certain you are considering a healthy and growing company. It is also your chance to show interest in the company.

Try to anticipate the questions your interviewer might ask and prepare your answers in advance. Be sure you understand a question before answering it and avoid responding too quickly with a "canned" answer. Be prepared to discuss answers to the interviewer's questions in a natural and relaxed manner. Interviewers typically ask the following questions.

- What are your short-term and long-term occupational goals?
- What are your major strengths and weaknesses?
- Do you work better with others or alone?
- Why do you want to work for this employer?
- How do you spend your free time?
- What are your personal goals?
- Describe an accomplishment you are particularly proud of.
- Why are you leaving your current job?
- Why should I hire you?
- What salary do you expect?

Some of those questions are difficult. Give them careful thought and remember that there is no one "correct" answer.

Be sure you arrive for your interview at the appointed time. In fact, it is usually a good idea to arrive five or ten minutes early because you may be asked to fill out an application before you meet your interviewer. Always bring extra copies of your **résumé** and samples of your work (if applicable). Some of the people you meet may not have a copy of your résumé, and it contains much of the same information the application asks for: personal data, work experience, education. Read the application form before you fill it out and proofread it when you are finished. Not only does the application form provide the company with a record for its files, it also gives the company an opportunity to see how closely you follow directions and how thoroughly you complete a task.

During the Interview

The interview actually begins before you are seated: What you wear and how you act make a first impression. The way you dress does matter. In general, dress conservatively and avoid extremes. Be well groomed. For men, a recent haircut and a clean shave are essential; if you wear a beard, have a barber trim it. For women, basic clothes with simple jewelry look most businesslike. Avoid heavy perfume and extreme makeup. Remember, you have only one chance to make a good first impression.

Behavior. First, thank the interviewer for his or her time, express your pleasure at meeting him or her, and remain standing until you are offered a seat. Sit up straight (good posture suggests self-assurance), look directly at the interviewer, and try to appear relaxed and confident. Never chew gum. During the interview, you may find yourself feeling a little nervous. Use that nervous energy to your advantage by channeling it into alertness. Listen carefully and record important information in your memory. Do not attempt to take extensive notes during the interview, although it is acceptable to jot down a few facts and figures. (See also **listening.**)

Responses. When you answer questions, do not ramble or stray from the subject. Say only what you must to answer each question properly and then stop, but avoid giving just yes or no answers — they usually do not permit the interviewer to learn enough about you. Some interviewers allow a silence to fall just to see how you will react. The burden of conducting the interview is the interviewer's, not yours, and he or she may interpret your rush to fill a void in the conversation as a sign of insecurity. If such silences make you uncomfortable, be ready to ask an intelligent question about the company.

If the interviewer overlooks important points, bring them up. However, let the interviewer mention salary first, if possible. In the event you are forced to bring up the subject, put it in a straightforward question. Make sure you are aware of prevailing salaries in your field, so you will be better prepared to discuss salary. If you are a recent graduate, it is usually unwise to attempt to bargain. Many companies have inflexible starting salaries for beginners.

Interviewers look for a degree of self-confidence and an understanding of the field in which the applicant is applying for a job, as well as genuine interest in the field, the company, and the job. Less is expected of a beginner, but even a newcomer must show some self-confidence and command of the subject. One way to communicate your interest in the job and the company is to ask questions. Interviewers respond favorably to applicants who can communicate and present themselves well.

Conclusion.   At the conclusion of the interview, thank your interviewer for his or her time. Indicate that you are interested in the job (if true) and try to get an idea of when you can expect to hear from the company (do not press too hard). Reaffirm friendly contact with a firm handshake.

Sending Follow-up Communications

After you leave the interview, jot down the pertinent information you obtained—it may be helpful in comparing job offers. A day or two later, if the job is appealing, send the interviewer a note of thanks in a brief letter or email, saying that you find the job attractive and feel you can fill it well (Figure I–14 on page 320).

If you are offered a job you want, write a brief letter of acceptance as soon as possible, certainly within a week. The format for such a letter is simple. Begin by accepting the job you have been offered. Identify the job by title and state the exact salary so there will be no confusion on those two important points. The second paragraph might go into detail about moving dates and reporting for work. The details will vary, depending on the nature of the job offer. Conclude the letter with a statement that you are looking forward to working for your new employer. (See **acceptance letters.**)

Because you will probably have applied to more than one organization, you may need to write a letter of refusal if you receive more than one job offer. Be especially tactful and courteous because the employer you are refusing has spent time and effort interviewing you and may have counted on your accepting the job. It is

---

2647 Sitwell Road
Charlotte, NC 28210
March 17, 20--

Mr. F. E. Vallone
Manager of Human Resources
Information Systems, Inc.
3275 Commercial Park Drive
Raleigh, NC 27609

Dear Mr. Vallone:

Thank you for the informative and pleasant interview we had last Wednesday. Please extend my thanks to Mr. Wilson of the Media Group as well.

I came away from our meeting most favorably impressed with Information Systems. I find the position you are filling to be an attractive one and feel confident that my qualifications would enable me to perform the duties to everyone's advantage.

If I can answer any further questions, please let me know.

Sincerely yours,

Philip Ming

Philip Ming

FIGURE I–14. Follow-up Letter

also possible that you may apply for another job at that company in the future. Figure I–15 is an example of a job-refusal letter. It acknowledges the consideration given the applicant, offers a logical reason for the refusal, and concludes on a pleasant note. (See also **job search** and **application letters**.)

2647 Sitwell Road
Charlotte, NC 28210
March 26, 20--

Mr. F. E. Vallone
Manager of Human Resources
Information Systems, Inc.
3275 Commercial Park Drive
Raleigh, NC 27609

Dear Mr. Vallone:

I enjoyed talking with you about your opening for a technical writer, and I was gratified to receive your offer. Although I have given the offer serious thought, I have decided to accept a position as a copywriter with an advertising agency. I feel that the job I have chosen is better suited to my skills and long-term goals.

I appreciate your consideration and the time you spent with me. I wish you the best of luck in filling the position.

Sincerely,

Philip Ming

Philip Ming

FIGURE I–15. Letter of Refusal

introductions

The purpose of an introduction is to give **readers** enough general information about the subject to enable them to understand the detailed information in the body of the document. An introduction provides background details, such as why the document is necessary, and a frame of reference for the details contained in the body of the document. An introduction should accomplish the following:

- *State the subject.* The introduction should state the subject and provide background information, such as definition, history, or theory, to provide context for the reader.
- *State the purpose.* The statement of purpose should make readers aware of your goal as they read the document. It should also tell them why the document exists and whether your material provides a new perspective or clarifies an existing perspective.
- *State the scope.* By stating the scope of the document, you tell readers the amount of detail you plan to cover.
- *Preview the development of the subject.* In a longer work, such as an academic paper, a report, or a journal article, it may be helpful if you state how you plan to develop the subject. Providing such information allows readers to anticipate how the subject will be presented and gives them a basis for evaluating how you arrived at your **conclusions** or recommendations.

Consider writing the introduction last. Many writers find that only then do they have a full enough perspective on the subject to introduce it adequately.

Not every document needs an introduction. When your readers are already familiar with the major elements of your subject, you can save them time and better hold their interest by getting on to the details. If you do not need a full-scale introduction, turn to the entry **openings** for interesting ways to begin without an introduction. Figure I–16 on page 323 is an example from a **trade journal article.**

Introductions for Technical Manuals and Specifications

You may find that you need to write one kind of introduction for academic papers, reports, or journal articles and a different kind of

Subject and background

Toxic waste sites are viewed by the public as one of the most serious environmental hazards according to opinion polls (*New York Times* 1991). The Environmental Protection Agency's (EPA) current process for ranking sites uses the "best engineering judgment and does not consider socioeconomic factors" (Mitre Corporation 1981); the cost to physically clean the site is estimated, but the social costs from the existence of the site (the social benefits from the cleaning) are not calculated. The decline in house values has been well documented (for example, Kohlhase 1991; Michaels and Smith 1990), but the role of the EPA's actions in mitigating

Purpose

the costs has not been thoroughly examined. The EPA could potentially speed the market's adjustment process by disseminating information about the toxicity of the site and how effective cleaning is expected to be, thereby reducing adjustment costs. Social costs may also be eliminated by cleaning the site.

Method of development

One way to evaluate the social benefits obtained by removing a toxic site is to look at the impact of the site on adjacent house values over the period of adjustment. In theory, house values are determined by the discounted value of future rents that could be earned by that unit. If the existence of a toxic site causes rents to fall because the unit is less desirable, house prices close to the site will decrease. If cleaning the site completely removes all concern about locating near the site, the prices of neighboring houses should increase. By comparing the house values before and after the cleaning, the cost captured by house values can be measured.

Scope and method of development

This study uses a unique data set, consisting of information on the sales price, sales date, house and neighborhood characteristics, and distance from toxic sites, for over 2,000 houses in the Boston metropolitan area from 1975–92. The area under study (Woburn, Massachusetts) has two Superfund sites. Both were announced as sites by the EPA in the early 1980's, and the cleaning process has begun. Using regression techniques, the effect on house values of the EPA announcement of a discovered toxic site and of subsequent cleanup efforts can be measured.

K.A. Kiel, "Measuring the Impact of the Discovery and Cleaning of Identified Hazardous Waste Sites on House Values." *Land Economics* 71 (4).

FIGURE I–16. Journal Article Introduction

introduction for **technical manuals** or **specifications.** In writing an introduction for a technical manual or set of specifications, identify the topic and its primary purpose or function in the first sentence or two. Be specific but do not go into elaborate detail. Your introduction sets the stage for the entire document, and it should provide readers with a broad frame of reference and an understanding of the overall topic. Then you can introduce technical details in the body of the document.

How technical your introduction should be depends on your readers: What are their technical backgrounds? What kind of information are they seeking in the manual or specification? The topic should be introduced with a specific audience in mind—a computer user, for example, has different interests in an application program or utility routine than a programmer. Whether you need to provide explanations or definitions of terminology will depend on your intended audience. The following example is written for readers who understand such terms as "constructor" and "software modules."

- The System Constructor is a program that can be used to create operating systems for a specific range of microcomputer systems. The constructor selects requested operating software modules from an existing file of software modules and combines those modules with a previously compiled application program to create a functional operating system designed for a specific hardware configuration. It selects the requested software modules, establishes the necessary linkage between the modules, and generates the control tables for the system according to parameters specified at run time.

You may encounter a dilemma that is common in technical writing: Although you can't explain topic A until you have explained topic B, you can't explain topic B before explaining topic A. The solution is to explain both topics in broad, general terms in the introduction. Then, when you need to write a detailed explanation of topic A, you will be able to do so because your reader will know just enough about both topics to be able to understand your detailed explanation.

- The NEAT/3 programming language, which treats all peripheral units as file storage units, allows your program to perform data input or output operations depending on the specific unit. Peripheral units from which your program can only input data are referred to as *source units.* Units to which your program can only output data are referred to as *destination units* or as combination *source-destination units.*

Consider writing the introduction last. Many writers find this helpful because they feel that only then do they have a full enough perspective on the writing to introduce it adequately.

investigative reports

Although an investigative **report** may be written for a variety of reasons, it is most often produced in response to a request for information. You might be asked, for instance, to research the Web sites of a number of companies in a certain industry, to conduct an opinion survey among customers, or to study a number of different procedures for performing a specific operation. An investigative report gives a precise analysis of the topic and then offers the writer's conclusions and recommendations.

Open the report with a statement of its **purpose,** then define the **scope** of your investigation. If the report is on a survey of opinions, indicate the number of people and the geographical areas surveyed and the income categories, occupations, ages, sexes, and possibly even the racial backgrounds of those surveyed. (See also **questionnaires.**) Include any information pertinent in defining the extent of the investigation. Then report your findings and, if necessary, discuss their significance. End the report with your conclusions and, if appropriate, any recommendations. An example of an investigative report is shown in Figure I–17 on page 326.

irregardless / regardless

Irregardless is nonstandard English because it expresses a **double negative.** The prefix *ir-* renders the base word negative, but *regardless* is already negative, meaning "unmindful." Always use *regardless* or *irrespective.*

- *Regardless*
  ~~Irregardless~~ of the difficulties, we must increase the strength of the outer casing.

Memo

To: Noreen Rinaldo, Training Manager *CL*
From: Charles Lapinski, Senior Instructor
Date: February 14, 20--
Subject: Addison's Basic English Program

As you requested, I have investigated Addison Medical Instruments' (AMI's) Basic English Program to determine whether we might adopt a similar program. The purpose of AMI's program is to teach medical technologists outside the United States who do not speak or read English to understand procedures written in a special 800-word vocabulary called *Basic English*. This program eliminates the need for AMI to translate its documentation into a number of different languages. The Basic English Program does not attempt to teach the medical technologists to be fluent in English but, rather, to recognize the 800 basic words that appear in Addison's documentation.

The course does not train technologists. Students must know, in their own language, what a word like *hemostat* means; the course simply teaches them the English term for it. As prerequisites for the course, students must have a basic knowledge of their specialty, must be able to identify a part in an illustrated parts book, must have used AMI products for at least one year, and must be able to read and write in their own language.

Students are given the specially prepared instruction manual, an illustrated book of equipment with parts and their English names, and pocket references containing the 800 words of the Basic English vocabulary plus the English names of parts. Students can write the corresponding word in their language beside the English word and then use the pocket reference as a bilingual dictionary. The course consists of 30 two-hour lessons, each lesson introducing approximately 27 words. No effort is made to teach pronunciation; the course teaches only recognition of the 800 words, which include 450 nouns, 70 verbs, 180 adjectives and adverbs, and 100 articles, prepositions, conjunctions, and pronouns.

The 800-word vocabulary enables the writers of documentation to provide medical technologists with any information that might be required because the subject areas are strictly limited to usage, troubleshooting, safety, and the operation of AMI medical equipment. All nonessential words (such as *apple, father, mountain,* and so on) have been eliminated, as have most synonyms (for example, *under* appears, but *beneath* does not).

Conclusions

I see three possible ways in which we might be able to use some or all of the elements of AMI's Basic English Program: (1) in the preparation of all of our student manuals, (2) in the preparation of manuals for international students, or (3) as AMI uses the program.

I think it would be unnecessary to use the Basic English methods in the preparation of manuals for *all* of our students. Most of our students are English speakers to whom an unrestricted vocabulary presents no problem.

In conjunction with the preparation of student manuals for international students, the program might have more appeal. Students would take the Basic English course either before coming to this country to attend school or after arriving but before beginning their technical training.

As for our initiating a Basic English program similar to AMI's, we could create our own version of the Basic English vocabulary and write our instructional materials in it. Since our product lines are much broader than Addison's, however, we would need to create illustrated parts books for each of the different product lines.

FIGURE I–17. Investigative Report

it

ESL

The **pronoun** *it* has a number of uses. First, *it* can refer to a preceding **noun** that names an object or idea or to a baby or an animal whose sex is unknown or irrelevant. *It* should have a clear antecedent.

- Darwinism made an impact on nineteenth-century American thought. *It* even influenced economics.

It can also serve as an **expletive.**

- *It* is necessary to sand the surface before applying a finish.

Avoid overusing expletives, however.

- ~~It is~~ *We* seldom ~~that we~~ go.

(See also **its/it's** and **pronoun reference.**)

italics

Italics is a style of type used to denote **emphasis** and to distinguish foreign expressions, book titles, and certain other elements. Italic typeface is indicated by underlining in a handwritten draft or where italic font is not available (see also **email**). *This sentence is printed in italics.* You may need to italicize words that require special emphasis in a sentence.

- Contrary to projections, sales have *not* improved since we started the new procedure.

Do not overuse italics for emphasis, however.

- This will hurt *you* more than *me.*

- This will hurt you more than me!

Foreign Words and Phrases

Foreign words and phrases that have not been assimilated into the English language are italicized. (See also **foreign words in English.**)

- *sine qua non, coup de grâce, in res, in camera*

Foreign words that have been fully assimilated into the language, however, need not be italicized. A word may be considered assimilated if it appears in most standard dictionaries and is familiar to most readers.

- cliché, etiquette, vis-à-vis, de facto, siesta

Titles

Italicize the titles of books, periodicals, newspapers, movies, television programs, and paintings.

- The *New York Times* is one of the nation's oldest newspapers.

Abbreviations of such titles are italicized if their spelled-out forms would be italicized.

- *NYT* is one of the nation's oldest newspapers.

Titles of chapters, articles, and reports are placed in **quotation marks,** not italicized.

- "Clarity and Conciseness: The Writer's Tightrope" was an article in the *New York Times.*

Titles of holy books and legislative documents are not italicized.

- The Bible and the Magna Carta changed the history of Western civilization.

Titles of long poems and musical works are italicized, but titles of short poems, musical works, and songs are enclosed in quotation marks.

LONG POEM	Milton's *Paradise Lost*
LONG MUSICAL WORK	Handel's *Messiah*
SHORT POEM	T. S. Eliot's "The Love Song of J. Alfred Prufrock"
SONG	Elton John's "Candle in the Wind"

Proper Names

The names of ships, trains, and aircraft (but not the companies that own them) are italicized.

- They sailed to Africa on the Onassis *Clipper* but flew back on the United *New Yorker.*

Craft that are known by model or serial designations are exceptions and are not italicized.

- DC-7, Boeing 747

Words, Letters, and Figures

Words, letters, and figures discussed as such are italicized.

- The word *inflammable* is often misinterpreted.

- I need a new keyboard because the *s* and *6* keys on my old one do not function.

Subheads

Subheads in a report are sometimes italicized.

- There was no publications department as such, and the writing groups were duplicated at each location. Wellington, for example, had such a large number of publications groups that their publication efforts can only be described as disorganized. Their duplication of effort must have been enormous.
    Training Writers. We are certainly leading the way in developing first-line managers (or writing supervisors) who not only are professionally competent but can also train the writers under their direction and be responsible for writing quality as well.

(See also **headings** and **layout and design.**)

its / it's   ⒠⒮⒧

Its is the **possessive case** form of *it*; *it's* is a **contraction** of *it is*.

- *It's* important that the factory meet *its* quota.

Although pronouns normally form the possessive by the addition of an **apostrophe** and an *s*, the contraction of *it is* (*it's*) already uses that device. Therefore, the possessive form of the pronoun *it* is formed by adding only the *s*.

J

jargon

Jargon is a highly specialized technical slang that is unique to an occupational or professional group. Jargon is at first understood only by insiders; later, it becomes known more widely. For example, computer programmers coined the term *debugging* to describe the discovery and correction of errors in software. If all your readers are members of a particular occupational group, jargon may provide a timesaving and efficient means of communicating with them. If, however, you have any doubt that your entire reading audience is part of such a group, avoid using jargon.

When jargon becomes so specialized that it applies only to one company or a subgroup of an occupation, it is referred to as "shop talk." Obviously, shop talk is appropriate only for those familiar with its special vocabulary and should be reserved for speech or informal memos and email. (See also **affectation** and **gobbledygook.**)

job descriptions

Most large companies and many small ones specify, in formal *job descriptions,* the duties of and requirements for many of the jobs in the firm. Job descriptions fill several important functions: They provide information on which equitable salary scales can be based; they help management determine whether all responsibilities within a company are adequately covered; and they let both prospective and current employees know exactly what is expected of them. Together, all the job descriptions in a firm present a picture of the organization's structure.

Sometimes middle managers are given the task of writing job descriptions for their employees. In many organizations, though, employees are required to draft their own job descriptions, which supervisors then check and approve.

Although job description formats vary from organization to organization, they commonly contain the following sections:

- The *accountability* section identifies, by title only, the person to whom the employee reports.
- The *scope of responsibilities* section provides an overview of the primary and secondary functions of the job and states, if applicable, who reports to the employee.
- The *specific duties* section gives a detailed account of the specific duties of the job as concisely as possible.
- The personal *requirements* section lists the required or preferred education, training, experience, and licensing for the job.

Writer's Checklist: Writing Job Descriptions

If you have been asked to prepare a job description for your position, the following guidelines should help.

- ☑ Before attempting to write your job description, make a list of all the different tasks you do in a week or a month. Otherwise, you will almost certainly leave out some of your duties.

- ☑ Focus on content. Remember that you are describing your job, not yourself.

- ☑ List your duties in decreasing order of importance. Knowing how your various duties rank in importance makes it easier to set valid job qualifications.

- ☑ Begin each statement of a duty with a **verb** and be specific. Write "Greet and assist customers" rather than "Take care of customers."

- ☑ Review existing job descriptions that are considered well written.

The job description shown in Figure J–1 on page 332 is typical. It never mentions the person holding the job described; it focuses, instead, on the job and the qualifications required to fill the position.

Manager, Technical Publications
Dakota Electrical Corporation

Accountability

Reports directly to the Vice President, Customer Service.

Scope of Responsibilities

The Manager of Technical Publications plans, coordinates, and supervises the design and development of technical publications and documentation required in the support of the sale, installation, and maintenance of Dakota products. The manager is responsible for the administration and morale of the staff. The supervisor for instruction manuals and the supervisor for parts manuals report directly to the manager.

Specific Duties

Directs an organization presently composed of 20 people (including two supervisors), over 75 percent of whom are writing professionals and graphics artists.

Screens, selects, and hires qualified applicants for the department.

Prepares a formal program designed to orient writing trainees to the production of reproducible copy and graphic arts.

Evaluates the performance of departmental members and determines salary adjustments for all personnel in the department.

Plans documentation to support new and existing products.

Determines the need for subcontracted publications and acts as a purchasing agent when needed.

Offers editorial advice to supervisors.

Develops and manages an annual budget for the Technical Publications Department.

Cooperates with the Engineering, Parts, and Service Departments to provide the necessary repair and spare parts manuals upon the introduction of new equipment.

Serves as a liaison between technical specialists, the publications staff, and field engineers.

Recommends new and appropriate uses for the department within the company.

Keeps up with new technologies in printing, typesetting, art, and graphics and uses them to the advantage of Dakota Electrical Corporation where applicable.

Requirements

B.A. in technical communication desired.

Minimum of three years' professional writing experience and a general knowledge of graphics, production, and Web design.

Minimum of two years' management experience with a knowledge of the general principles of management.

Strong interpersonal skills.

FIGURE J–1. Job Description

job search

Whether you are trying to land your first job or you want to change careers entirely, begin by assessing your skills, interests, and abilities, perhaps through **brainstorming.** Next, ask yourself what your career goals and values are. For instance, do you prefer working independently or collaboratively? Does helping others interest you? How important are career stability and a certain standard of living? Finally, ask yourself what you would most like to be doing in the immediate future, two years from now, and five years from now.

Once you have reflected and brainstormed about the job that is right for you, a number of sources can help you locate the job you want:

- Internet resources
- Networking
- School placement services
- Letters of inquiry
- Advertisements in newspapers
- Trade and professional journals
- Private employment agencies

Internet Resources

Many traditional sources, such as newspaper and professional journal ads, are available on the **Internet,** but specific sources are also useful. For example, Internet discussion groups on topics in your job specialty can provide a useful way to keep up with trends and general employment conditions over time. More immediately, of course, some discussion groups post job openings.

Using the **World Wide Web** can enhance your search in a couple of ways. First, you can make use of general Web sites that give advice to college graduates about job seeking, résumé preparation, and similar topics. Several such sites, as this book goes to press, are the following:

- College Grad Job Hunter at (http://www.collegegrad.com/)
- America's Job Bank at (http://www.ajb.dni.us/)
- Monster.com at (http://www.monster.com/)
- Career Central at (http://www.careercentral.com/)
- CareerTech.com at (http://www.careertech.com/)

(Keep in mind that such sites come and go as well as vary in their usefulness.)

Second, you can go to sites created by businesses and organizations that may hire employees in your area. Business sites often not only list job openings but also provide instructions for applicants. Finding an appropriate employer on the Web may involve searching thousands of entries in numerous databases, many of which will be irrelevant to your search. Cyberspace can be as crowded as other settings for finding employment. Employment specialists also suggest that you spend time on the Web later in the evening or very early in the morning so you can focus on in-person contacts during working hours. (See also **Internet research.**)

Another option is to establish your own Web site and post your résumé to attract potential employers. If you post your résumé on the Web, keep in mind the advice given in the entry **résumés** concerning electronic versions. If you do not have a Web site or simply do not feel confident adapting your résumé for the Web, consider using a résumé-writing company that will put your résumé online. Before you use such services, however, browse the Web to learn what types of employers (for example, the computer industry) look at résumés on the Web. One free service is America's Job Bank, listed earlier. Be aware, however, that commercial résumé-writing companies often charge a monthly fee for their posting services in addition to various other fees.

Networking

Networking is communication with people who might provide useful advice or who might be able to connect you with potential jobs in your interest area. They may be people already working in your chosen field, contacts in professional organizations, professors, family members, friends, neighbors, members of the clergy—any of them might direct you to exactly the job lead you need. Use contacts to develop even more contacts.

School Placement Services

Your school's career-development center is another good place to begin a job search. Government, business, and industry recruiters often visit job-placement offices to interview prospective employees; recruiters also keep career counselors aware of their companys' current employment needs and submit job descriptions. Not only can career counselors help you in the brainstorming phase of your career selection, they can also put you in touch with important, current resources. They can identify where to begin your search and suggest ways to save you time. Career development centers often hold workshops on résumé preparation, which can help you develop a résumé for a particular field. Although career development centers have their own Web sites, nothing replaces talking to a career counselor about your interests.

Letters of Inquiry

If you would like to work for a particular firm, write and ask whether it has any openings for people with your qualifications. Normally, you should send the letter to either the director of human resources or the department head; for a small firm, however, write to the head of the firm. Your letter should present a general summary of your employment background or training. If you are responding to an advertisement, make the letter short and to the point; express interest in the job, but let your enclosed résumé do the rest.

Advertisements in Newspapers

Many employers advertise in the classified sections of newspapers and on their Internet sites. For the widest selection of help-wanted listings, look in the Sunday editions or the "help-wanted" Web pages of local and big-city newspapers. Occasionally, newspapers print special employment supplements that provide valuable information on résumé preparation, job fairs, and other facets of the job market. An item-by-item check is necessary, because many times a position can be listed under various classifications. A clinical medical technologist seeking a job, for example, might find the specialty listed under "Medical Technologist," "Clinical Medical Technologist," or "Laboratory Technologist." Depending on a hospital's or a pathologist's needs, the listing could be even more specific, such as "Blood Bank Technologist" or "Hematology Technologist." So try to read all areas that may be pertinent to your search.

As you read the ads, take notes on such things as salary ranges, job locations, job duties and responsibilities, and even the terminology used in the ads to describe the work. A knowledge of the words and expressions that are generally used to describe a particular type of work can be helpful when you prepare your résumé and letters of application. (See also **application letters.**)

Trade and Professional Journals

In many industries, associations publish periodicals of interest to people working in the industry. Such periodicals often contain listings of current job opportunities. If you were seeking a job in forestry, for example, you could check the job listings in the *Journal of Forestry*, published by the Society of American Foresters. To learn about the trade or professional associations for your occupation, consult the following references at a library: *Encyclopedia of Associations, Encyclopedia of Business Information Sources,* and *National Directory of Employment Services.* Many of the organizations have Web sites, where they often list open positions. (See also **library research.**)

Private Employment Agencies

Private employment agencies are profit-making organizations that are in business to help people find jobs — for a fee. Choose a private employment agency carefully. Some are well established and reputable, but others have questionable reputations. Check with your local Better Business Bureau as well as friends and acquaintances or your school's career office before you sign an agreement with a private employment agency.

Reputable private employment agencies provide you with job leads and help to organize your campaign to find the job you want. They may also provide useful information on the companies doing the hiring.

Often, the employer pays the agency's fee. Before signing a contract, be sure you understand who is paying the fee; if you have to pay, make sure you know exactly how much. As with any written agreement, read the fine print carefully.

Other Sources

Local, state, and federal government agencies offer many employment services. Local government agencies are listed in telephone and Web directories under the name of your city, county, or state. For information about jobs with the federal government, contact the U.S. Office of Personnel Management at (http://www.usajobs.opm.gov/).

Keep records with dated job ads, copies of letters of application and résumés, notes requesting interviews, and the names of important contacts. Such a collection can act as a future resource and as a tickler file. (See also **interviewing for a job** and **acceptance letters.**)

journal articles (*see* **trade journal articles**)

judicial / judicious

Judicial is a term that pertains only to law.

- The *judicial* branch is one of the three branches of the United States government.

Judicious refers to careful or wise judgment.

- We intend to convert to the new refining process by a series of *judicious* steps.

K

kind of / sort of

Kind of and *sort of* should be used in writing only to refer to a class or type of things.

- They used a special *kind of* metal in the process.

Do not use *kind of* or *sort of* to mean "rather," "somewhat," or "somehow."

- It was ~~kind of~~ a bad year for the firm.

know-how

An informal term for "special competence or knowledge," *know-how* should be avoided in formal writing **style**.

- He has a reputation for marketing ~~know-how.~~ *skill.*

L

laboratory reports

A laboratory report should state the reason the laboratory investigation was conducted and list the equipment and procedures used during the test, the problems encountered, the conclusions reached, and finally the recommendations made. (Similar reports that are less formal and smaller in scope are **test reports.**)

A laboratory report often emphasizes the equipment and procedures used in the investigation because those two factors can be critical in determining the accuracy of the data. Present the results of the laboratory investigation clearly and concisely. If your report requires **graphs** or **tables,** integrate them in your report, as described in the entry **illustrations.**

The example in Figure L–1 on pages 340–42 is a typical laboratory report.

lay / lie   ESL

Lay is a transitive **verb** (a verb that has a direct object to complete its meaning) that means "place" or "put."

- We will *lay* the foundation of the building one section at a time.

The past tense form of *lay* is *laid.*

- We *laid* the first section of the foundation on the 27th of June.

The perfect tense form of *lay* is also *laid.*

- Since June, we *have laid* all but two sections of the foundation.

Lay is frequently confused with *lie,* which is an intransitive verb (a verb that does not require an object to complete its meaning) that means "recline" or "remain."

PCB Exposure from Oil Combustion
Wayne County Professional Firefighters

Submitted to:
Mr. Philip Landowe
President, Wayne County Professional Firefighters Association
Wandell, IN 45602

Submitted by:
Analytical Laboratories, Incorporated
Mr. Arnold Thomas
Certified Industrial Hygienist
Mr. Gary Seabolm
Laboratory Manager
Environmental Analytical Services
1220 Pfeiffer Parkway
Indianapolis, IN 46223
February 28, 20--

L

INTRODUCTION

Waste oil used to train firefighters was suspected of containing polychlori-
nated biphenyls (PCBs). According to information provided by Mr. Philip
Landowe, President of the Wayne County Professional Firefighters Association, it has been standard practice in training firefighters to burn 20–100 gallons of oil in a diked area of approximately 25–50 m^3. Firefighters would then
extinguish the fire at close range. Exposure would last several minutes, and
the exercise would be repeated two or three times each day for one week.

Oil samples were collected from three holding tanks near the training area
in Englewood Park on November 11, 20--. To determine potential firefighter
exposure to PCBs, bulk oil analyses were conducted on each of the samples.
In addition, the oil was heated and burned to determine the degree to which
PCB is volatized from the oil, thus increasing the potential for firefighter exposure via inhalation.

TESTING PROCEDURES

Bulk oil samples were diluted with hexane, put through a cleanup step, and
analyzed in electron-capture gas chromatography. The oil from the underground tank that contained PCBs was then exposed to temperatures of
1008°C without ignition and 2008°C with ignition. Air was passed over the
enclosed sample during heating, and volatized PCB was trapped in an ab-

FIGURE L–1. Laboratory Report

sorbing medium. The absorbing medium was then extracted and analyzed for PCB released from the sample.

RESULTS

Bulk oil analyses are presented in Table 1. Only the sample from the underground tank contained detectable amounts of PCB. Aroclor 1260, containing 60 percent chlorine, was found to be present in this sample at 18 mg. Concentrations of 50 mg PCB in oil are considered hazardous. Stringent storage and disposal techniques are required for oil with PCB concentrations at these levels.

Results for the PCB volatilization study are presented in Table 2. At 1008°C, 1 mg PCB from a total of 18 mg PCB (5.6 percent) was released to the air. Lower levels were released at 2008°C and during ignition, probably as a result of decomposition. PCB is a mixture of chlorinated compounds varying in molecular weight; lightweight PCBs were released at all temperatures to a greater degree than the high-molecular-weight fractions.

TABLE 1. Bulk Oil Analysis

Source	Sample #	PCB Content (mg/g)
Underground tank (11' deep)	6062	18*
Circle tank (3' deep)	6063a	<1
	6063b	<1
Square pool (3' deep)	6064a	<1
	6064b	<1

*Aroclor 1260 is the PCB type. This sample was taken for volatilization study.

TABLE 2. PCB Volatilization Study
for the 11-Foot-Deep Underground Tank*

Outgassing Temp (°C)	Outgassing Time (min.)	Sample Outgassed (g)	PCB Total (mg)	PCB Outgassed (mg)
100	30	1	18	1
200	30	1	18	0.6
200 with ignition	30	1	18	0.2

*Bulk analysis of 18 mg/g PCB.

FIGURE L–1. Laboratory Report (*continued*)

DISCUSSION AND CONCLUSIONS

At a concentration of 18 mg/g, 100 gallons of oil would contain approximately 5.5 g of PCB. Of the 5.5 g of PCB, about 0.3 g would be released to the atmosphere under the worst conditions.

The American Conference of Governmental Industrial Hygienists has established a threshold limit value (TLV)* of 0.5 mg/m^3 air for a PCB containing 54 percent C1 as a time-weighted average over an 8-hour work shift and has stipulated that exposure over a 15-minute period should not exceed 1 mg/m^3. The 0.3 g of released PCB would have to be diluted to 600 m^3 air to result in a concentration of 0.5 mg/m^3 or less. Since the combustion of oil lasted several minutes, a dilution to more than 600 m^3 is likely; thus, exposure would be less than 0.5 mg/m^3. Since an important factor in determining exposure is time and the firefighters were exposed only for several minutes at intermittent intervals, adverse effects from long-term exposure to low-level concentrations of PCB should not be expected.

It should be stressed, however, that these conclusions are based solely on oil containing 18 mg/g PCB. If, on previous occasions, the PCB content of the oil was much higher, greater exposure could have occurred. PCB is a known liver toxin and has also been classified as a suspected carcinogen. Although the primary route of entry into the body is by inhalation, PCBs can be absorbed through the skin. PCBs can cause a skin condition known as *chloracne,* which results from a clogging of the pores. This condition, which is often associated with a secondary infection, should not occur at the PCB concentration found in this oil. A clinical test exists for determining if PCB has been absorbed by the liver.

In summary, because exposure to this oil was limited and because PCB concentrations in the oil were low, it is unlikely that exposure from inhalation would be sufficient to cause adverse health effects. However, we cannot rule out the possibility that excessive exposure may have occurred under certain circumstances, based on factors such as excessive skin contact and the possibility that oil with a higher-level PCB concentration could have been used earlier. The practice of using this oil should be terminated.

*The safe average concentration that most individuals can be exposed to in an 8-hour day.

FIGURE L–1. Laboratory Report (*continued*)

- Injured employees should *lie* down and remain still until the doctor arrives.

The past tense form of *lie* is *lay*. (This form causes the confusion between *lie* and *lay*.)

- The injured employee *lay* still for approximately five minutes.

The perfect tense form of *lie* is *lain*.

- The injured employee *had lain* still for approximately five minutes before the doctor arrived.

layout and design

DIRECTORY
Overview   343
Typography   344
Highlighting Devices   346
Page Design   348

L

Overview

The layout and design of a document are crucial to its success. A well-designed document makes even the most complex information look accessible and gives readers a favorable impression of the writer and the organization. To accomplish those goals, a design should offer a simple and uncluttered presentation of the topic; highlight structure, hierarchy, and order; help readers find information easily; and reinforce an organization's image.

Effective design is based on visual simplicity and harmony, such as using compatible fonts and the same highlighting device for similar items. Design should reveal hierarchy by signaling the difference between topics and subtopics, between primary and secondary information, and between general points and examples. Writers can achieve effective layout and design through their selection of fonts, their choice of devices to highlight information, and their arrangement of text and visual components on a page. Such visual cues make information easy to find. Finally, the design of a document should project the appropriate image of an organization. For example, if clients are paying a high price for consulting services, they may expect a sophisticated, polished design; if employees inside an

organization expect management to be frugal, they may accept—
even expect—economical and standard company design.

Typography

Typography refers to the style and arrangement of type on a printed
page. A complete set of all the letters, numbers, and symbols available in one typeface (or style) is called a *font*. The letters in a typeface have a number of distinctive characteristics, some of which are
shown in Figure L–2.

Typeface and Type Size.   For most on-the-job writing, select a
typeface primarily for its legibility. Avoid typefaces that may distract readers with contrasts in thickness or with odd-looking features, as is often the case with script and cursive typefaces. In addition, avoid typefaces that fade when printed or copied. Choose
popular typefaces with which readers are familiar, such as Times
Roman or the following typefaces:

- Baskerville
- **Bodoni**
- Century
- Garamond

- Gill Sans
- Helvetica
- Palatino
- **Univers**

Do not use more than two typefaces in a document. With so many
fonts available on most computers, a common mistake is to pepper
the document with many odd-looking typefaces, thereby creating
disharmony. To create a dramatic contrast between headlines and
text, as in a newsletter, use a typeface that is distinctively different.

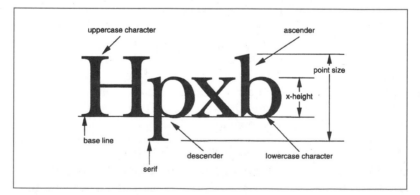

FIGURE L–2. Primary Components of Letter Characters

You can also use a noticeably different typeface *within* a graphic element. In any case, experiment before making final decisions. (Keep in mind that not all fonts have the same assortment of symbols and other characters.)

One way typefaces are characterized is by the presence or absence of *serifs*. Serifs, shown in Figure L–2 on page 344, are the small projections at the end of each stroke in a letter. Serif typefaces have projections; sans serif styles do not. (*Sans* is French for "without.") The text of this book is set in Plantin, a serif typeface. Although sans serif type has a modern look, serif type is easier to read, especially in the smaller sizes. Sans serif, however, works well for headings (like the ones in this book).

A font size that is too small will cause eyestrain and make the text look crammed and intimidating. Six-point type is too small for almost any application other than footnotes and classified ads. On the other hand, type that is too large uses more space than necessary, makes reading difficult and inefficient, and makes readers perceive words in parts rather than as wholes. Figure L–3 illustrates various type sizes.

Ideal point sizes for text on paper documents range from 8 to 13 points; 10- or 11-point type is most commonly used. For Web documents, you need to look carefully at both typeface and type size. Monitors display fonts in dramatically different sizes on the screen, even when the same typefaces are involved. Therefore, you should print out samples of a document to test your choice of sizes and typefaces.

The distance from which a document will be read should help determine type size. For example, instructions that will rest on a table at which the reader stands require a larger typeface than normal. Consider the age of your readers too. Visually impaired readers and some older adults may need larger type sizes.

6 pt. This size might be used for dating a source.

8 pt. This size might be used for footnotes.

10 pt. This size might be used for figure captions.

12 pt. This size might be used for main text.

14 pt. This size might be used for headings.

FIGURE L–3. Samples of 6- to 14-Point Type

Left- or Full-Justified Margins.   Left-justified margins are generally easier to read than full-justified margins because the uneven contour of the right margin (or "ragged" right) provides the eyes with more landmarks to identify. Left justification also is better if lines are short and justification causes your word-processing software to insert irregular-sized spaces between words, producing unwanted white space or unevenness in blocks of text.

However, because ragged-right margins look informal, full-justified text is more appropriate for publications aimed at a broad readership that expects a more formal, polished appearance. Further, full justification is often useful with multiple-column formats because the spaces between the columns (called *alleys* or *gutters*) need the definition that full justification provides.

Highlighting Devices

Writers use a number of methods to emphasize important words, passages, and sections within documents:

- Typographical devices
- Headings and captions
- Headers and footers
- Rules, icons, and color

When used thoughtfully, highlighting devices give a document a visual logic and organization. For example, rules and boxes can set off steps and illustrations from surrounding explanations. Consistency is important: Use the same technique to highlight a particular feature throughout your document.

Use typographical devices and special graphic effects in moderation; too many design devices clutter a page and interfere with comprehension.

Typographical Devices.   One method of achieving emphasis through typography is to use **capital letters.** HOWEVER, LONG STRETCHES OF ALL UPPERCASE LETTERS ARE DIFFICULT TO READ BECAUSE THEIR UNIFORM SIZE AND SHAPE DEPRIVE READERS OF IMPORTANT VISUAL CLUES AND THUS SLOW READING. Ascenders and descenders (see Figure L–2) make lowercase letters distinctive and easy to identify; therefore, a mixture of uppercase and lowercase letters is most readable. Use all uppercase letters only in short spans, such as in headings.

Also use **italics** sparingly. *Continuous italic type reduces legibility and thus slows readers.* Of course, italics are useful if your aim is to slow readers, as in cautions and warnings. Boldface, which is used for cross-references in this book, may be the best cuing device because it is visually different yet retains the customary shapes of letters and numbers. Of course, overuse of boldface type can be jarring to the eye.

Headings and Captions.   **Headings** (or heads) reveal the organization of a document and indicate the hierarchy within it. Headings help readers decide which sections they need to read. Captions are titles that highlight or describe illustrations or blocks of text. Captions often appear below or above figures and **tables** and in the left or right margins next to blocks of text.

Headings appear in many typeface variations (boldface being most common) and often use sans serif typefaces. The most common positions for headings and subheadings are centered, flush left, indented, or by themselves in a wide left margin. Insert an additional line of space above a heading to emphasize the logical division. Major section or chapter headings normally appear at the top of a new page. Never leave a heading as the final line on a page — the heading is disconnected from its text and thus ineffective. Instead, move the heading to the start of the next page.

Headers and Footers.   A header is identifying information carried at the top of each page; a footer contains similar information at the bottom of each page. Document pages may have headers or footers or both. Headers and footers carry such information as the topic or subtopic of a section, identifying numbers, the date the document was written, page numbers, the document name, or the section title. Although headers and footers are important reference devices, too much information in them can create visual clutter. (See also **headers and footers.**)

Rules, Icons, and Color.   Rules are vertical or horizontal lines used to divide one area of the page from another or to create boxes. Avoid the temptation to overuse rules; too many rules and boxed elements create a cluttered look that counteracts their initial purpose, which is simplicity and ease of reading.

An icon is a pictorial representation of an idea; it can be used to identify specific actions, objects, or sections of a document.

L

Commonly used icons include the small envelopes on Web pages to symbolize email links and national flags to symbolize different language versions of a document. To be effective, icons must be simple and intuitively recognizable or at least easy to define.

Color and screening can distinguish one part of a document from another or unify a series of documents. (Screening refers to shaded areas on a page.) Color and screening can set off sections within a document, highlight examples, or emphasize warnings. In tables, you can use screening to highlight column titles or sets of data to which you want to draw the reader's attention.

Page Design

Page design is the process of combining the various design elements on a page to make a coherent whole. The flexibility of your design is based on the capabilities of your word-processing software, how the document will be reproduced, and the budget. (High-quality offset color printing, for example, is far more expensive than black-and-white printing.)

L

Thumbnail Sketches.   Before you spend time positioning actual text and visuals on a page, you may want to create a *thumbnail sketch,* in which blocks indicate the placement of elements. You can go further by roughly assembling all the thumbnail pages, showing size, shape, form, and general style of a large document. Such a mock-up, called a *dummy,* allows you to see how a finished document will look. As you work with elements on the page, experiment with different layouts. Often what seems like a good layout in principle turns out to be unworkable in practice.

Columns.   As you design pages, consider the size and number of columns. A single-column format works well with larger typefaces, double spacing, and left-justified margins. For smaller typefaces and single-spaced lines, the two-column structure enhances legibility by keeping text columns narrow enough so readers need not scan back and forth across the width of the entire page for every line.

A word on a line by itself at the end of a column is called a *widow.* A single word carried over to the top of a column or page is called an *orphan.* Avoid both widows and orphans.

White Space.   White space visually frames information and breaks it into manageable chunks. For example, white space between paragraphs helps readers see the information in the paragraphs as units. Use extra white space between sections as a visual cue to signal that one section is ending and another is beginning. You need not have access to sophisticated equipment to make good use of white space. You can easily indent and skip lines for paragraphs, lists, and other blocks of material.

Lists.   Lists are an effective way to highlight words, phrases, and short sentences. Further, they are easy to read. Lists are particularly useful for certain types of information:

- Steps in sequence
- Materials or parts needed
- Items to remember
- Criteria for evaluation
- Concluding points
- Recommendations

Avoid both too many lists and too many items in lists, as described in the entry **lists.**

Illustrations.   Readers notice illustrations before they notice text, and they notice larger illustrations before they notice smaller ones. Thus, the size of an illustration suggests its relative importance. For **newsletter articles** and publications aimed at wide audiences, consider especially the proportion of the illustration to the text. Magazine designers often employ the "three-fifths" rule of thumb: Page layout is more dramatic and appealing when the major element (**photographs, drawings,** illustrations, and so on) occupies three-fifths rather than half the available space. The same principle can be used to enhance the visual appeal of a report.

Remember that clarity and usefulness take precedence over aesthetics in most technical documents. Illustrations can be gathered in one place (for example, at the end of a **report**), but placing them in the text closer to their accompanying explanations makes them more effective. Using illustrations in the text also provides visual relief. (For advice on the placement of illustrations, see the "Writer's Checklist" in the entry **illustrations.**)

leave / let

As a **verb,** *leave* should never be used in the sense of "allow" or "permit."

Let
• ~~Leave~~ me do it my way.
       ^

As a **noun,** however, *leave* can mean "permission to do something."

• Employees are granted a *leave* of absence if they have a chronic illness.

lend / loan

Both *lend* and *loan* can be used as **verbs,** but *lend* is more common.

• You can *lend* them the money if you wish.
• You can *loan* them the money if you wish.

Unlike *lend, loan* can be a **noun.**

• We made arrangements at the bank for a *loan.*

letters (*see* correspondence *and* email)

libel / liable / likely

The term *libel* refers to "anything circulated in writing or pictures that injures someone's good reputation." When someone's reputation is injured in speech, the term is *slander.*

• When an editorial charged our board of directors with bribing a representative, the board sued the newspaper for *libel.*
• If the mayor supports the bribery charge in her speech tonight, we will accuse her of *slander.*

The term *liable* means "legally subject to" or "responsible for."

• Employers are held *liable* for their employees' actions.

In technical writing, *liable* should retain its legal meaning. Where a condition of probability is intended, use *likely*.

likely
- Rita is ~~liable~~ to be promoted.
  ∧

library research

DIRECTORY
Overview   351
Reference Works   351
Catalogs, Indexes, and Statistical Sources: Online and Printed   353

Overview

The first step to using the library for research is to develop a search strategy. Your search strategy depends on the kind of information you are seeking. For example, if you need the latest data on a subject, you might need to check the Web for resources. If you need an overview of a subject, various reference works like encyclopedias are appropriate. If you need historical background, books and articles listed in a library's online catalogs may be your best option. Do not overlook the most important resources in the library—reference librarians. They are employed to help you find information and are most familiar with the resources available in any particular library. (For advice about taking notes during library research, see **note taking.**)

Reference Works

General reference works, such as encyclopedias, specialized dictionaries, handbooks, and manuals, are often good places to begin your research, particularly if you know little about your subject. Reference books can provide you with a brief but reliable overview of your subject and also direct you to more specialized sources. General information on a variety of subjects is also available on the **World Wide Web.** Although most libraries employ a reference librarian, you can search on your own for a listing of reference titles by typing keywords into your library's online search system.

Encyclopedias. Encyclopedias are comprehensive, multivolume collections of articles arranged alphabetically and often illustrated. Some encyclopedias cover a wide range of subjects, while others specialize in a particular subject. General encyclopedias provide overviews of many subjects that can be helpful to someone gathering general information on a topic. Many encyclopedia articles contain bibliographies that can lead you to additional information. Three of the best-known general encyclopedias are *Collier's Encyclopedia, Encyclopedia Americana,* and *Encyclopaedia Britannica.*

Subject encyclopedias provide detailed information on all aspects of a particular field of knowledge, such as *The Encyclopedia of Careers and Vocational Guidance,* edited by William Hopke. There are many more specialized encyclopedias than there are general encyclopedias.

Dictionaries. There are many good desk-size English-language **dictionaries,** including the *American Heritage Dictionary of the English Language* and the *Random House Webster's College Dictionary.* Unabridged dictionaries, which are larger and more comprehensive in their coverage, often contain basic terms from many specialized subjects. One respected unabridged English-language dictionary is Funk & Wagnall's *Standard Dictionary.*

For the meanings of words too specialized for a general dictionary, a subject dictionary is useful. Subject dictionaries define terms used in a particular field, such as business, geography, architecture, or consumer affairs. Definitions in subject dictionaries are generally more detailed and comprehensive than those found in general dictionaries. One example is *Dictionary of Personal Computing and the Internet,* edited by Simon Collin.

Although subject dictionaries are specialized and offer detailed definitions of field-specific terms, they are written in language that is straightforward enough to be understood by nonspecialists. Many general and specialized dictionaries are available on CD-ROM as well as on the Web.

Handbooks and Manuals. Handbooks and manuals are usually one-volume compilations of frequently used information in a particular field of knowledge. The information they offer includes brief definitions of terms or concepts, graphs, tables that display basic numerical data, maps, and the like. Handbooks and manuals offer a ready source of fundamental information about a subject, but they are usually intended for those who have basic knowledge, particularly in specialized fields.

Web Resources.   The Web offers a variety of Internet resources for research; however, as suggested in the entry **Internet research,** you must be careful to assess both the quality of the source and the accuracy of data. The following sources are basic and have links to other specific subject areas and resources on the Web:

- Reach It at (http://www.iTools.COM/research-it/)
- Library of Congress at (http://lcweb.loc.gov)
- Government Printing Office at (http://www.access.gpo.gov)

Catalogs, Indexes, and Statistical Sources: Online and Printed

When you are ready to consult more specific sources on your topic, you will need to use catalogs and indexes to find books and periodical articles. You may also need to consult statistical sources and atlases.

Most libraries have an online catalog, which lists the library's collection of books and other materials in an easily accessible database. An online catalog permits an automated search of a library's holdings and often provides more information than the traditional card catalog. In addition, online terminals are usually equipped with printers, so you can print out subject and title listings quickly and accurately. Finally, online systems tell you not only whether the library carries a particular item but also whether it is checked out, when it is due to be returned, and whether it is housed in a separate library branch or special collection. If the book or other item is available from another library, you can arrange to obtain it through an interlibrary loan.

Although online search methods vary somewhat among systems, generally you can find a book by looking up either the title or the author's name. If you are trying to find a book on a particular subject and do not have a title or author in mind, you can type in the subject, and a list of matching items will be displayed. If after selecting an item and reading its description, you decide the book merits further examination, you can locate it on the shelf by using the call number.

Whether you are searching online catalogs or using the Internet, you must choose keywords carefully. Choose the most specific term possible. Also try multiple terms (word processing AND grammar) and exclusionary terms (Macintosh NOT computer). The words and symbols to use in a search (for example, AND or + to combine terms and quotation marks to indicate a phrase) vary from one search engine to another. Read the directions provided by the engine you choose.

If you are looking for magazine or periodical articles on a particular subject, you can begin with one of the many indexes to publications. The *Magazine Index,* for example, indexes over four hundred popular and general-interest magazines. As with a book search, you narrow your search by typing specific terms until you find the article you want. In addition to being in a hard copy of the periodical, an article also may be available on CD-ROM, microfiche, or the periodical's Web site.

Finally, sources of statistical information and atlases are increasingly available on the World Wide Web as well as in print.

CD-ROMs.   CD-ROMs are the perfect medium for reference works such as dictionaries and encyclopedias. CD-ROM and **hypertext** technology permit easy full-text subject searches. Every word in a text (except conjunctions, prepositions, and articles) is indexed. Once you have found the information you need, you can print it for future use. Because CDs can store graphics, sound, and video, as well as text, they are the ideal medium for large reference works.

Most references are available on CD-ROM, including almanacs, encyclopedias, atlases, periodicals, general and specialized dictionaries, books of quotations, thesauruses, and numerous economic, demographic, geographic, political, legal, and statistical data collections. The U.S. Bureau of the Census alone makes scores of data collections available on CD-ROM in agriculture, housing, population, transportation, construction, retail and wholesale trade, and many other areas. Check with a reference librarian to learn the kinds of CD-ROM references available and for information about how to access the information using a computer.

Periodical Indexes, Bibliographies, and Abstracting Services.   Periodical indexes and **bibliographies** are lists of journal articles and books. Periodical indexes are devoted specifically to journal, magazine, and newspaper articles. (The term *periodical* is applied to publications that are issued at regular intervals—daily, weekly, monthly, and so on.) Bibliographies list books, periodicals, and other research materials published in a particular subject area, such as business, engineering, medicine, the humanities, or the social sciences. One example is *The St. Martin's Bibliography of Business and Technical Communication* by Gerald Alred. To find a bibliography using your library's online catalog, type in your topic, then add the word *bibliography*. Bibliographies of bibliographies provide listings in many areas. One of the most important resources is *Bibliographic Index: A Cumulative Bibliography of Bibliographies*, 1937– (*1937–* means that the index covers bibliographies published from 1937 through the present).

Abstracting services provide brief summaries of the sources cited, giving an idea of their content so you can judge whether they are relevant to your research. Some potentially useful indexes and abstracts are included in the following list. Check with a reference librarian to find out if a particular index or abstracting service is available through an online terminal or on a CD-ROM.

> *Applied Science and Technology Index,* 1958–. Alphabetical subject listing of articles from about 300 periodicals; issued monthly.
>
> *Biological and Agricultural Index,* 1964–. Alphabetical subject listing of biological and agricultural periodicals; issued monthly.
>
> *Engineering Index,* 1934–. Alphabetical subject listing containing brief abstracts; issued monthly.
>
> *Government Reports Announcements and Index,* 1965 . Semimonthly index of reports, arranged by subject, author, and report number.
>
> *Monthly Catalog of U.S. Government Publications,* 1895–. Unclassified publications of all federal agencies, listed by subject, author, and report number; issued monthly.
>
> *New York Times Index,* 1851–. Alphabetical list of subjects covered in *New York Times* articles; issued bimonthly.
>
> *Readers' Guide to Periodical Literature,* 1900–. Monthly index of about 200 general U.S. periodicals, arranged alphabetically by subject.

Safety Sciences Abstracts Journal, 1974–. Listing of literature on
industrial and occupational safety, transportation, and environmental and medical safety; issued quarterly.

To choose the most relevant indexes, bibliographies, and abstracts
for your research, consult a reference librarian or an appropriate
guide to reference books or materials, such as the following.

Balay, Robert. *Guide to Reference Books.* 11th ed. Gualala:
ALA, 1996.

Walford, A. J. *Walford's Guide to Reference Material.* Vol. 1: Science and Technology. Vol. 2: Social and Historical Sciences, Philosophy, and Religion. Vol. 3: Generalia, Languages, the Arts, and Literature. Lanham: UNIPUB, 1998.

Both of those guides list thousands of reference books, indexes, and
other items useful to researchers. They are annotated and arranged
so you can find a subject quickly.

After you have selected and located the index, bibliography, or
abstract that deals with your subject, consult the instructions in the
front of each work, the first volume of the series, or the help menu.
There you will find a key to the abbreviations and symbols used, an
explanation of how the information is arranged, a list of the specific
subjects covered, and, for periodical indexes, a list of the newspapers, magazines, and journals that are included.

Atlases and Statistical Sources. Atlases are classified into two
broad categories based on the type of information they present:
General atlases represent physical and political boundaries, and
thematic atlases represent special subjects, such as climate, population, natural resources, or agricultural products. Following are two
well-known general atlases.

Hammond Atlas of the World. 2nd ed. Maplewood: Hammond,
1998.

The Times Atlas of the World. 9th ed. Boston: Houghton Mifflin,
1995.

Statistical sources are collections of numerical data. They are
the best source for such information as the U.S. gross domestic
product (GDP), the consumer price index (CPI), or the demographic breakdown of the general population. You can find the answers to many statistical reference questions in almanacs and ency-

clopedias. However, for answers to more difficult or comprehensive questions, you may need to consult works devoted exclusively to statistical data, a selection of which follows.

> *American Statistics Index.* Washington: Congressional Information Service, 1978–. Monthly, quarterly, and annual supplements.

> United States Bureau of the Census. *County and City Data Book.* Washington: GPO, 1952–. Issued approximately every five years.

> United States Bureau of the Census. *Statistical Abstract of the United States.* Washington: GPO, 1879–. Annual.

> U.S. Census Bureau. (http://www.census.gov).

The *American Statistics Index* lists and summarizes all statistical publications issued by agencies of the U.S. government, including periodicals, reports, special surveys, and pamphlets. The *Statistical Abstract* includes statistics on social, political, and economic conditions in the United States. Compiled by the United States Bureau of the Census, the data include vital statistics and cover broad topics, such as population, demography, education, and public land. Some state and regional data are given, as well as selected global statistics. A wealth of statistical data is also available from the Census Bureau's Web site (new publications are added monthly).

-like

The suffix *-like* is sometimes added to **nouns** to make them into **adjectives.** The resulting compound word is hyphenated only if it is unusual or might not immediately be clear.

- childlike, lifelike, dictionary-like, computer-like
- Her new assistant works with machine-*like* efficiency.

like / as

To avoid confusion between *like* and *as,* remember that *like* is a **preposition** and *as* is a **conjunction.** Use *like* with a **noun** or **pronoun** that is not followed by a **verb.**

- The new supervisor behaves *like* a novice.

Use *as* before **clauses** (which contain verbs).

- She acted *as* though she owned the company.

- He responded *as* we expected he would.

Like may be used in elliptical constructions that omit the verb.

- She took to architecture *like* a bird to nest building.

(See also **figures of speech.**)

linking verbs (*see* verbs)

listening

Effective listening is a prerequisite for all types of communication. It enables the listener to understand the instructions of a teacher, the goals of a manager, and the needs and wants of customers. It also lays the foundation for productive interpersonal communication and cooperation.

There are three essential parts to every communication: a message, the sender of the message, and the receiver of the message. For communication to occur, all three parts must be synchronized. For example, a message that is blocked or distorted by noise (interference with reception or transmission) will not be received by the listener. "Noise" can be anything from a lack of attention on the part of the listener to cross-cultural differences to actual noise in the environment. Communication works best when both the speaker and the listener focus clearly on the content of the message and attempt to eliminate as much noise as possible.

Fallacies About Listening

Although listening is a crucial skill, it is the skill we concentrate on least in our education, know the least about, and take most for granted. That is probably why two fallacies about listening are widespread: (1) Hearing and listening are the same, and (2) words mean the same thing to everyone.

Most people assume that because they can hear they know how

to listen. In fact, *hearing* is passive whereas *listening* is active. Hearing voices in a crowd, a ringing telephone, or a door being slammed requires no analysis and no active involvement of any kind. We hear such sounds without choosing to listen to them—we have no choice but to hear them. Listening, however, requires effort and skill and involves a number of related activities, including interpreting the message and assessing its worth.

Although many people believe that words have absolute meanings, they can have multiple meanings, determined by the context in which they are used. Differences in meaning may be the result of differences in the speaker's and the listener's occupation, education, culture, sex, race, or other factors, such as the unique **idioms** of a language. (See also **biased language, English as a second language, jargon,** and **international correspondence.**)

Active Listening

To listen actively, you should (1) make a conscious decision to listen actively, (2) define your purpose for listening, (3) take specific actions to listen more efficiently, and (4) adapt to the situation.

Step 1: Make a Conscious Decision.   The first step to active listening is simply making up your mind to do so. Active listening requires a conscious effort, something that does not come naturally. To listen actively, "seek first to understand and then to be understood." That rule will help you make a conscious effort to interpret the speaker's message the way he or she intends.

Step 2: Define Your Purpose.   To listen actively, you must understand your purpose for listening. Knowing why you are listening can go a long way toward managing the most common listening problems: drifting attention, formulating your response while the speaker is still talking, and interrupting the speaker. To help you define your purpose for listening, ask yourself these questions:

- What kind of information do I hope to get from this exchange?
- How can I use this information?
- What kind of message do I want to send while I am listening? (Do I want to portray understanding, determination, flexibility, competence, or patience?)
- What kinds of "noise" might interfere with listening during the interaction—boredom, daydreaming, anger, impatience? How

can I keep noise from placing a barrier between me and the speaker?

Step 3: Take Specific Actions.   After you have consciously decided to become a listener and have examined your purpose for listening, you need to take specific actions to help you listen more actively. Becoming an active listener requires a willingness to become a responder rather than a reactor. A *responder* is a listener who slows down the communication to be certain that he or she is accurately receiving the message sent by the speaker. A *reactor,* on the other hand, does not check the meaning of the message and simply says the first thing that comes to mind based on superficial information. A reactor may never know if he or she has received an accurate version of the message. Take the following actions to help you become a responder and not a reactor.

- Make a conscious effort not to allow your preconceptions or personal biases to prevent impartial evaluation of the message. For example, do not dismiss the message because you dislike the speaker or are distracted by the speaker's appearance or mannerisms.
- Slow down the communication by asking for more information or by paraphrasing the message received before you offer thoughts, opinions, or recommendations. **Paraphrasing** not only lets the speaker know you are listening but also gives the speaker an opportunity to clear up any misunderstanding. Paraphrasing keeps you focused and helps you remember the discussion.
- Listen with empathy by putting yourself in the speaker's position. Empathy makes the speaker feel that you are trying to understand the communication. When people feel they are being listened to empathetically, they tend to respond with appreciation and cooperation, thereby improving the communication.
- To help you stay focused on what the speaker is saying, take notes while you are listening, especially during a longer presentation or lecture. **Note taking** not only communicates your attentiveness to the speaker, it reinforces the message and helps you remember it.

Step 4: Adapt to the Situation.   The requirements of active listening differ from one situation to another. For example, when you are listening to a lecture, you may be listening only for specific information. When someone stops you in the hall for a casual conversation, you may legitimately listen without giving the conversation your full attention. However, if you are on a team project where the success of the project depends on everyone's contribution, you need to be listening at the highest level so you can gather information as well as pick up on nuances the other speakers may be communicating. (See **also collaborative writing** and **oral presentations.**)

lists

Lists can save readers time by allowing them to see at a glance specific items, questions, or directions. Lists also help readers by breaking up complex statements and by allowing key ideas to stand out.

> Before we agree to hold the district meeting at the Brent Hotel, we should make sure the hotel facilities provide the following:
> - Service center with phones, faxes, Internet hookup, PCs, and copying services for the conference committee
> - Ground-floor exhibit area large enough for thirty 8-by-15-foot booths
> - Eight meeting rooms to accommodate 25 people each
> - Internet hookups, projection screens for presentations, and overhead projectors in each room
> - Ballroom and dining facilities for 250 people
>
> To confirm that the Brent Hotel is our best choice, we should tour the facilities during our stay in Kansas City.

Notice that all the list items in the example have **parallel structure.** In addition, all the items are balanced; that is, all points are relatively equal in importance and are of the same general length.

In an attempt to avoid writing paragraphs, some writers tend to overuse lists. Memos or reports that consist almost entirely of lists can be difficult to understand because the reader is forced to connect the separate items mentally to provide **coherence.** Do not expect a reader to deal with unexplained lists of ideas.

To ensure that your readers understand how a list fits with the surrounding sentences, provide adequate **transitions** before and after any list. If you do not want to indicate rank or sequence, which numbered lists suggest, use bullets or another typographical device.

Writer's Checklist: Using Lists

☑ List only comparable items.

☑ Use parallel structure throughout.

☑ Use only words, phrases, or short sentences.

☑ Use bullets when rank or sequence is not important.

☑ Do not include too many items in one list.

☑ Do not overuse lists.

☑ Provide adequate transitions before and after lists.

literature

In professional contexts, the word *literature* applies to a body of writing pertaining to a specific field, such as finance, insurance, or computers. We commonly speak, for example, of medical literature or campaign literature.

- Please send me any available *literature* on computerized translations of foreign languages.

literature reviews

A *literature review* is a summary **report** on the relevant **literature** (printed and electronic) available on a particular subject over a specified period of time. For example, a literature review might describe significant material published in the past five years on a technique for improving emergency diagnostic procedures, or it might describe all reports written in the past ten years on efforts to improve safety procedures at a particular company. A literature review tells readers what has been published on a particular subject and gives them an idea of what material they should read in full.

Some **trade journal articles** or theses begin with a brief literature review to bring the reader up to date on current research in the field. The writer then uses the review as background for his or her own discussion of the subject. Figure L–4 shows a two-paragraph literature review that serves as an **introduction** for a medical article on clefts in newborns. On the other hand, a literature review could be a whole document in itself. Fully developed literature reviews are good starting points for detailed **research.**

To prepare a literature review, you must begin by researching published material on your topic. Because your reader may begin research on the basis of your literature review, be especially careful to accurately cite all bibliographic information. As you review each source, note the scope of the book or article and judge its value to the reader. Save all printouts of computer-assisted searches — you may want to incorporate the sources in a bibliography. (See also **documenting sources, Internet research,** and **library research.**)

Assessment of attractiveness of newborn infants with unrepaired facial clefts by different groups of individuals was studied to investigate the validity of a surgeons' rule of thumb system for this based upon cleft severity.

Recent studies have attempted to construct objective rating scales to quantify the severity of facial disfigurement in children and young people (Eliason et al., 1991; Poole et al., 1991; Roberts-Harry et al., 1992; Tobiasen and Hiebert, 1993). The medical community increasingly recognizes the need for a holistic approach to congenital disfigurement, including such aspects as parents' reactions, feelings, and attitudes (Bull and Rumsey, 1988). Results of these studies suggest that there may be a mistaken assumption of consensus between plastic surgeons and people unfamiliar with facial clefts in how each group respectively rates attractiveness of facial appearance. The studies all refer, however, to repaired facial clefts and other anomalies; none has assessed unrepaired facial clefts in neonates. Tobiasen (1991) has concluded that it is premature to suggest that surgeons rate impairment differently from others. Surgeons commonly use a rule of thumb system to classify severity on the basis of dichotomous categorizations (I.e., bilateral or unilateral clefting along the axis of symmetry; complete or incomplete extension to the nasal alar, and with or without palate involvement). The proposed system used by plastic surgeons and derived from Freedlander et al. (1990) is shown in Table 1.

Source: Slade, Pauline. "Relationships Between Cleft Severity and Attractiveness of Newborns with Unrepaired Clefts." *Cleft Palate-Craniofacial Journal,* vol. 32, no. 4 (July 1995): 318–322.

FIGURE L–4. Literature Review in Introductory Paragraphs

Begin a literature review by defining the area to be covered and the types of works to be reviewed. For example, a literature review may be limited to articles and reports and not include any books. You can arrange your discussion chronologically, beginning with a description of the earliest relevant literature and progressing to the most recent (or vice versa). You can also subdivide the topic, discussing works in various subcategories of the topic.

Related to literature reviews are annotated **bibliographies,** which also give readers information about published material. Rather than discussing the literature in paragraphs, however, an annotated bibliography lists each bibliographic item and then briefly describes it. The description (or *annotation*) may include the purpose of the book, its scope, the main topics covered, its historical importance, and anything else the writer thinks the reader should know.

logic errors

In persuasive writing, logic (correct reasoning) is essential to convincing readers that your conclusions are valid (see **persuasion**). Errors in logic can undermine the point you are trying to communicate and destroy your credibility. Common logic errors you should watch for in your writing are lack of reason, sweeping generalizations, non sequiturs, false analogies, biased or suppressed evidence, opinions instead of facts, and loaded arguments.

Lack of Reason

When a statement is contrary to the reader's common sense, that statement is not reasonable. If, for example, you stated, "Los Angeles is a small town," readers might immediately question your logic. If, however, you stated, "Although Los Angeles is a large sprawling city, it is composed of many areas that are more like small towns," readers probably could accept the statement as reasonable.

Sweeping Generalizations

Sweeping generalizations are statements that are too large or all-inclusive to be supportable; they generally enlarge an observation about a small group to a generalization about an entire population.

• Engineers are poor writers.

That statement ignores any possibility that some engineers might write well. Such generalizations are not true, and using them will weaken your credibility.

Non Sequiturs

A statement that does not logically follow a previous statement is called a non sequitur.

• I arrived at work early today, so the weather is calm.

Common sense tells us that arriving early for work does not produce calm weather; thus, the second part of the sentence does not follow logically from the first. In your own writing, be careful that all points stand logically connected; non sequiturs cause difficulties for readers.

False Analogies

The term *false analogy* (also called *post hoc, ergo propter hoc,* Latin for "after this, therefore because of this") refers to the logical fallacy that because one event happened after another event, the first somehow caused the second.

• I didn't bring my umbrella today. No wonder it is now raining.

In on-the-job writing, such an error in reasoning usually happens when the writer hastily concludes that two events are related without examining the logical connection between them.

Biased or Suppressed Evidence

A conclusion reached as a result of self-serving data, questionable sources, suppressed evidence, or incomplete facts is both illogical and unethical. Suppose you are preparing a report on the acceptance of a new policy among employees. If you distribute **questionnaires** only to those employees who think the policy is effective, the resulting evidence will be biased. If you purposely ignore employees who do not believe the policy is effective, you will be suppressing evidence. Intentionally ignoring relevant data that might not support your position not only produces inaccurate data, but—perhaps more important—it is unethical. (See also **ethics in writing.**)

Opinions Instead of Facts

Distinguish between fact and opinion. Facts include verifiable data or statements, whereas opinions are personal conclusions that may or may not be based on facts. For example, it is verifiable that distilled water boils at 100 degrees centigrade; that it tastes better than tap water is an opinion. Distinguish the facts from your opinions so your readers can draw their own conclusions.

- Profits last year set a new record.
  [This sentence is stated as a fact that can be verified.]

- The employees believe that profits are not good.
  [The word *believe* identifies the statement as an opinion — statistics on profits may or may not verify the opinion as a fact.]

Loaded Arguments

When you include an opinion in a statement and then reach conclusions based on that statement, you are loading the argument. Consider the following opening for a memo:

- I have several suggestions to improve the poorly written policy manual. First, we should change . . .

By opening with the assumption that the manual is poorly written, the writer has loaded the statement to get readers to accept his or her arguments and conclusions. Be careful not to load arguments in your writing; conclusions reached with loaded statements are weak and can produce negative reactions in readers who detect the loading.

long variants

A long variant is an inflated version of a simpler word. Guard against inflating plain words by adding **prefixes** or **suffixes,** a practice that creates long variants. The following is a list of words and their inflated counterparts.

use	utilization (see **utilize**)
visit	visitation
priority	prioritization
orient	orientate
finish	finalize
commercial	commercialistic

(See also **gobbledygook** and **word choice.**)

loose / lose

Loose is an **adjective** meaning "not fastened" or "unrestrained."

* He found a *loose* wire.

Lose is a **verb** meaning "be deprived of" or "fail to win."

* Did you *lose* the operating instructions?

lowercase and uppercase letters   (ESL)

Lowercase letters are small letters, as distinguished from **capital letters** (known as uppercase letters). The terms were coined in the early history of printing, when printers kept the small letters in a "case," or tray, below the case where they kept the capital letters.

Avoid using all capital letters ("all caps") to emphasize important information. A passage set in all caps is hard to read, because the shapes of capital letters are not as distinctive as the shapes of lowercase letters or a combination of uppercase and lowercase letters. (See also **layout and design.**)

M

malapropisms

A malapropism is a word that sounds similar to the one intended but is ludicrously wrong in the context.

INCORRECT   No longer are our employees *sedimentary,* since the opening of our new fitness center.

CORRECT   No longer are our employees *sedentary,* since the opening of our new fitness center.

Intentional malapropisms are sometimes used in humorous writing; unintentional malapropisms can embarrass a writer. (See also **figures of speech.**)

male

The term *male* is usually restricted to scientific, legal, and medical contexts (a male patient, a male suspect). Keep in mind that the term sounds cold and impersonal. *Boy, man,* and *gentleman* are acceptable substitutes in other contexts, but they have connotations involving age, dignity, and social position. (See also **biased language** and **female.**)

maps

Maps are often used to show specific geographic features (roads, mountains, rivers, and the like), as in Figure M–1 on page 369. They can also illustrate geographic distributions of population, climate patterns, corporate branch offices, and so forth. A map might be useful, for example, in a report recommending several locations for a new corporate facility.

Place maps as close as possible to the portion of the text that

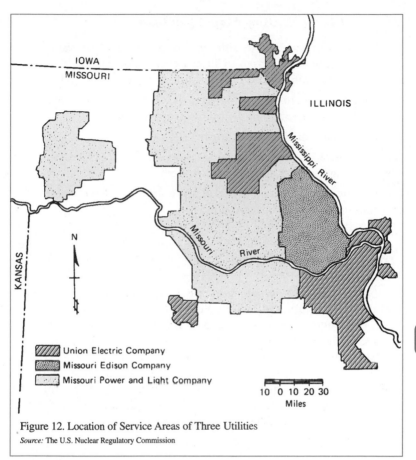

Figure 12. Location of Service Areas of Three Utilities

Source: The U.S. Nuclear Regulatory Commission

FIGURE M–1. Sample Map

refers to them and follow the other general guidelines discussed in **illustrations.**

Writer's Checklist: Using Maps

Bear these points in mind in creating and using maps:

☑ Label the map clearly and assign the map a figure number if it is one of a number of illustrations.

☑ Make sure that all boundaries in the map are clearly identified. Eliminate unnecessary boundaries.

Writer's Checklist: Using Maps (continued)

- ☑ Eliminate unnecessary information from the map. For example, if the purpose of the map is to show population centers, do not include mountain elevations, rivers, and the like.

- ☑ Include a scale of miles/kilometers or feet/meters to give your readers an indication of the map's proportions.

- ☑ Indicate which direction is north.

- ☑ Emphasize key features by using color, shading, dots, crosshatching, or other appropriate symbols.

- ☑ Include a key or legend that explains what the different colors, shadings, or symbols represent.

mathematical equations

You can accurately prepare material with mathematical equations and make it easy to read by following consistent standards throughout a document. Unless you need to follow a specific style manual or specifications, the following guidelines should serve you well.

Set short and simple equations, such as $x(y) = y^2 + 3y + 2$, as part of the running text rather than displaying them on separate lines, as long as an equation does not appear at the beginning of a sentence. If a document contains multiple equations, identify them with numbers, as the following example shows:

$$x(y) = y^2 + 3y + 2 \qquad\qquad (1)$$

Number displayed equations consecutively throughout the work. Place the equation number, in parentheses, at the right margin of the same line as the equation (or of the first line if the equation runs longer than one line). Leave at least four spaces between the equation and the equation number. Refer to displayed equations by number, for example, as "Equation 1" or "Eq. 1."

Positioning Displayed Equations

Equations that are set off from the text need to be surrounded by space. Triple-space between displayed equations and normal text. Double-space between one equation and another and between the lines of multiline equations. Count space above the equation from the uppermost character in the equation; count space below from the lowermost character.

Type displayed equations either at the left margin or indented five spaces from the left margin, depending on their length. When a series of short equations is displayed in sequence, align them on the equal signs.

$$p(x,y) = \sin(x + y) \tag{2}$$
$$p(x,y) = \sin x \cos y + \cos x \sin y \tag{3}$$
$$p(x_0, y_0) = \sin x_0 \cos y_0 + \cos x_0 \sin y_0 \tag{4}$$
$$q(x,y) = \cos(x + y) \tag{5}$$
$$= \cos x \cos y - \sin x \sin y$$
$$q(x_0, y_0) = \cos x_0 \cos y_0 - \sin x_0 \sin y_0 \tag{6}$$

Break an equation that requires two lines at the equal sign, carrying the equal sign over to the second portion of the equation.

$$\int_0^1 (f_n - \tfrac{n}{r}f_n)^2 \, r \, dr + 2n \int_0^1 f_n f_n dr \tag{7}$$
$$= \int_0^1 (f_n - \tfrac{n}{r}f_n)^2 \, r \, dr + n f_n^2(1)$$

If you cannot break an equation at the equal sign, break it at a plus or minus sign that is not in **parentheses** or **brackets.** Bring the plus or minus sign to the next line of the equation, which should be positioned to end near the right margin of the equation.

$$\varnothing(x, y, z) = (x^2 + y^2 + z^2)^{1/2} (x - y + z) (x + y - z)^2 \tag{8}$$
$$- [f(x,y,z) - 3x^2]$$

The next best place to break an equation is between parentheses or brackets that indicate multiplication of two major elements.

For equations that require more than two lines, start the first line at the left margin, end the last line at the right margin (or four spaces to the left of the equation's number), and center intermediate lines between the margins. Whenever possible, break equations at operational signs, parentheses, or brackets.

Omit punctuation after displayed equations, even when they end a sentence and even when a key list defining terms follows (for example, P = pressure, psf; V = volume, cu ft; T = temperature, °C). Punctuation may be used before an equation, however, depending on the grammatical construction.

The term $(n)_r$ may be written in a more familiar way by using the following algebraic device:

$$(n)_r = \frac{(n)(n-1)(n-2) \ldots (n-r+1)(n-r)(n-r-1) \ldots 3 \cdot 2 \cdot 1}{(n-r)(n-r-1) \ldots 3 \cdot 2 \cdot 1}$$
$$= \frac{n!}{(n-r)!} \tag{9}$$

maybe / may be   ESL

Maybe (one word) is an **adverb** meaning "perhaps."

- *Maybe* the legal staff can resolve this issue.

May be (two words) is a **verb** phrase.

- It *may be* necessary to ask an outside specialist.

media / medium

Media is the plural of *medium* and should always be used with a plural **verb.**

- Many communication *media* are available today.

- The Internet is a multifaceted *medium* that is popular with many people.

meetings

A meeting is a face-to-face exchange among a group of people who collaborate to produce better results than any one of the participants could have produced alone. Like other communication, such as writing and **presentations,** a successful meeting requires planning and preparation.

Planning a Meeting

For a meeting to be successful, it must be carefully planned. Planning consists of determining the focus of the meeting, deciding who should attend, and choosing the best time and place to hold the meeting. You also need to prepare an agenda for the meeting and determine who should take the minutes.

Determine the Purpose of the Meeting.   The first step in planning a meeting is to focus on the desired outcome. The following questions can help you determine the purpose of the meeting.

1. What do I want participants to *know* as a result of the meeting?
2. What do I want participants to *believe* after attending the meeting?
3. What do I want participants to *do* or *be able to do* as a result of attending the meeting?

For example, consider that you need to launch a sales campaign for one of your company's newest products, a scanner. You call a meeting of the sales staff to get their ideas for a successful promotional campaign. The answers to the pre-meeting questions for this particular meeting might be as follows:

1. I want the sales force to *know* that this is an outstanding scanner and that it will increase sales considerably.
2. I want the sales force to *believe* that this is the best scanner on the market and that our customers want it.
3. I want the sales force to *offer ideas* for the sales campaign.

Once you have focused your desired outcome, use the information to write a *purpose statement* for the meeting that answers the questions *what* and *why*.

- The purpose of this meeting is to gather ideas from the sales force [what] to create a successful sales campaign for our new scanner [why].

M

Decide Who Should Attend.   There is really no sense in having a meeting if not enough of the key people are present. The necessary people for the meeting about the scanner sales campaign might include the advertising manager, sales representatives who sell similar equipment, customer-service staff who might handle customers' questions, and service managers who might be involved with maintenance and repair. In the case that the meeting must be held without some key participants, email those people prior to the meeting for their contributions. Of course, the meeting minutes should be distributed to significant nonattendees.

Choose the Meeting Time.   The time of day and how long the meeting lasts can influence the outcome of the meeting. Consider the following when you are planning a meeting:

- People need Monday morning to focus on the week's work after the weekend.

- People need Friday afternoon to wrap up the week and take care of tasks that must be finished before the weekend.
- During the hour following lunch, most people tend to feel sleepy or sluggish, a condition called *postprandial letdown.*
- Be sure to plan for adequate breaks during long meetings to allow participants to check their messages, make phone calls, and refresh themselves.
- A meeting held during the last fifteen minutes of the day is sure to be quick—but it is likely that no one will remember what went on.

Choose the Meeting Location.   Having a meeting on your own premises gives you an advantage: You feel more comfortable, which along with your guests' newness to their surroundings may help you get your way. That advantage may backfire, however, by putting people on the defensive and introducing conflict into the meeting. Holding the meeting on another person's premises, on the other hand, can signal cooperation.

For balance, especially when people are meeting for the first time or are discussing sensitive issues, having a meeting at a neutral site may be the best solution. No one gains an advantage in off-site meetings, and attendees often feel freer to participate.

Establish the Agenda.   A tool for focusing the group, the agenda is an outline of what the meeting will address. Never begin a meeting without an agenda, even if it is only a handwritten list of topics. Ideally, the agenda should be distributed a day or two before the meeting, so the attendees have ample time to prepare. For a longer meeting in which participants are required to make a presentation, try to distribute the agenda a week or more in advance. If you are not able to distribute the agenda early, be sure to distribute it at the beginning of the meeting.

The agenda should list the attendees, the meeting time and place, and the topics you plan to discuss. If the meeting includes presentations, the agenda should list the time allotted for each speaker. Finally, the agenda should indicate an approximate length for the meeting so participants can plan the rest of their day. Figure M–2 on page 375 shows a typical agenda.

If the agenda is distributed in advance of the meeting, it should be accompanied by a **memo** or **email** informing people of the meeting. The cover memo should include the following information:

Sales Meeting Agenda

Purpose: To get input for a sales campaign for the new scanner
Date: January 27, 20--
Place: Conference Room 15-C
Time: 9:30 a.m.–11:00 a.m.
Attendees: New Products Advertising Manager, Equipment Sales Reps, Customer-Service Staff, and Service Managers

Topic	Presenter	Time
The Scanner	Bob Arbuckle	Presentation, 9:30–9:45
The Campaign	Maria Lopez	Presentation, 9:45–10:00
The Sales Strategy	Mary Winifred	Presentation, 10:00–10:15
Discussion	Led by Dave Crimes	Discussion, 10:15–11:00

FIGURE M–2. Meeting Agenda

- The purpose of the meeting
- Instructions on how to prepare
- The meeting start and stop times
- The date and place
- The names of the people invited

Figure M–3 on page 376 shows a cover memo to accompany an agenda.

Assign the Minute Taking.   Delegate the minute taking to someone other than the leader. It is difficult for the person leading a meeting to steer the meeting, coach participants, and take minutes at the same time. It is important to record major decisions that the group makes as well as any assignments for follow-up. Each assignment is usually given a date by which it must be completed. To avoid misunderstandings, the minute taker must record each assignment, the person responsible for it, and the date on which it is due.

Avoid delegating the minute taking to someone who is quiet and nonparticipative—the task will only make it harder for that person to participate in the meeting. On the other hand, the responsibility of meeting notes will help a talkative or overly dominant participant concentrate on what is going on. For a standing committee, it is best to rotate the responsibility of taking minutes. (See also **minutes of meetings** and **note taking**.)

Memo

TO: New Products Advertising Manager
Equipment Sales Representatives
Customer-Service Staff
Service Managers
FROM: Susan McLaughlin *SM*
DATE: May 7, 20--
SUBJECT: Planning Meeting

Purpose of the Meeting

The purpose of this meeting is to get your ideas for the upcoming introduction and sales campaign for our new scanner.

Time, Date, and Location

Date: May 11, 20--
Time: 9:30 a.m.–11:00 a.m.
Place: Conference Room E (go to the ground floor, take a right off the elevator, third door on the left)

Attendees

The groups listed above

Meeting Preparation

Everyone should be prepared to offer suggestions on the following items:

- Sales features of the new scanner
- Techniques for selling scanners
- Customer profile for potential scanner business
- FAQs—questions customers may ask
- Anticipated service needs of the scanner and attachments

FIGURE M-3. Memo to Accompany an Agenda

Conducting the Meeting

In addition to the minute taker, assign a person to the task of writing on a flip chart or otherwise recording information that needs to be viewed by everyone present. Alternatively, you can use a computer to gather and display information during a meeting, even project it on a screen.

During the meeting, keep to your agenda; however, allow room for differing views and foster an environment in which participants listen respectfully to one another. (See also **listening.**) To create such an environment, adopt the following principles:

- Consider the feelings, thoughts, ideas, and needs of others — do not let your own agenda blind you to other points of view.
- Help other participants feel valued and respected by listening to them and responding to what they say.
- Respond positively to the comments of others as best you can.
- Increase your willingness to accept opinions, perspectives, methods, and ways of doing things that are different from your own, particularly those from other cultures (see **global communication**).

Deal with Conflict.   Despite your best efforts, conflict is inevitable, whether because of personality differences or for other reasons. Conflict is potentially valuable; when managed positively, it can stimulate creative thinking by challenging complacency and showing ways to achieve goals more efficiently or economically. (For further advice on dealing with conflict and working with groups, see **collaborative writing.**)

Members of any group are likely to vary greatly in their personalities and attitudes, and you may encounter people who approach meetings differently. Consider the following tactics for the interruptive, negative, rambling, overly quiet, and territorial personality types.

- The *interruptive person* rarely lets anyone finish a sentence and intimidates the group's quieter members. The best way to deal with such a person is to tell him or her in a firm but nonhostile tone to let the others finish in the interest of getting everyone's input. By addressing the issue directly, you signal to the group the importance of putting common goals first.
- The *negative person* has difficulty accepting change and often considers a new idea or project from a negative point of view. Such negativity, if left unchecked, can demoralize the group as a whole and deflate all enthusiasm for new ideas. If the negative person brings up a valid point, ask for the group's suggestions to remedy the issue being raised. If the negative person's reactions are not valid or are outside the agenda, state the necessity of staying focused on the agenda and perhaps recommend a separate meeting to address those issues.

M

- The *rambling person* cannot collect his or her thoughts quickly enough to verbalize them and generally wastes valuable meeting time. You can be of significant help to this type of person by restating or clarifying his or her ideas. Be careful to avoid being demeaning by prefacing your clarification with "What you mean is . . ." Try to strike a balance between providing your own interpretation and drawing out the person's intended meaning.
- The *quiet person* may be timid or may just be deep in thought. To draw such people into a discussion, ask for their thoughts, being careful not to embarrass them. In some cases, you can have a quiet person jot down his or her thoughts and give them to you later.
- The *territorial person* fiercely defends his or her group against real or perceived threats and may refuse to cooperate with members of other departments, companies, and so on. Point out that although such concerns may be valid, everyone is working toward the same overall goal and it takes precedence.

Be aware that some conflict may be caused by cultural differences. For advice on dealing with different cultural perspectives, see **global communication.**

Close the Meeting.   In closing the meeting, review all decisions and assignments. Paraphrase each to help the group focus on what they have collectively agreed to do as well as to help the minute taker review his or her notes. This is the time to raise questions and clarify any misunderstandings. Be sure to set a date by which everyone at the meeting can expect to receive copies of the minutes. Finally, thank everyone for participating and close the meeting on a positive note.

memos

Memos, including memos sent via **email,** are the most common form of correspondence in an organization. In fact, many types of workplace writing described throughout this book are written in memo form. Among their many uses, memos announce policies, disseminate information, delegate responsibilities, instruct employees, and report results. They provide a record of decisions made and actions taken in an organization. They also play a key role in the management of many organizations because managers use memos to keep employees informed about company goals, to motivate employees to achieve those goals, and to build employee morale.

Writing a Memo

To produce a memo that effectively communicates your message, outline it first, even if you simply jot down points to be covered and then order them logically. With careful preparation, your memos will be both concise and adequately developed. Adequate development of your thoughts is crucial to a memo's clarity, as the following example indicates.

> ABRUPT   Be more careful on the loading dock.
>
> DEVELOPED   To prevent accidents on the loading dock, follow these procedures:
> 1. Check . . .
> 2. Load only . . .
> 3. Replace . . .

Although the original version is concise, it is not as clear and specific as the revision. Do not assume your **readers** will know what you mean. Readers who are pressed for time may misinterpret a vague memo.

Openings.   Although methods of development vary, a memo normally begins with a statement of the main idea. Consider the following example:

> MAIN IDEA   Because of our inability to serve our present and future clients efficiently, I recommend we hire an additional attorney.

When the reader is not familiar with the subject or with the background of a problem, provide an introductory background paragraph before stating the main point of the memo. Doing so is especially important in memos that serve as records of crucial information months (or even years) later. Generally, longer memos or those dealing with complex subjects benefit most from more thorough introductions. However, even when you are writing a short memo about a familiar subject, remind readers of the context. Readers have so much crossing their desks that they need a quick orientation. In the following examples, words that provide context are shown in italics.

- *As we discussed after yesterday's meeting,* we need to set new guidelines for . . .

- *As Maria recommended,* I reviewed the office reorganization plan. I like most of the features; however, the location of the receptionist and administrative assistant . . .

Do not state the main point first when (1) the reader is likely to be highly skeptical or (2) you are disagreeing with a person in a position of higher authority. In such cases, a more persuasive tactic is to state the problem first, then present the specific points supporting your final recommendation. (See also **persuasion.**)

Writing Style and Tone.   A carelessly prepared memo sends a garbled message that could baffle readers, cause a loss of time, produce costly errors, or even offend. Consider the unintended secondary message the following notice conveys:

- It has been decided that the office will be open the day after Thanksgiving.

The first part of the sentence ("It has been decided") not only sounds impersonal but also communicates an authoritarian, management-versus-employee tone: Somebody "decides"; you work. The passive voice also suggests that the decision maker does not want to say, "I have decided" and be identified (in any case, the office staff would undoubtedly know). One solution is to remove the first part of the sentence.

- The office will be open the day after Thanksgiving.

Even that statement sounds impersonal. The best solution would be to suggest both that there is a good reason for the decision and that employees are privy to (if not a part of) the decision-making process.

- Because we must meet the December 15 deadline to be eligible for the government contract, the office will be open the day after Thanksgiving.

By subordinating the bad news (the need to work on that day), the writer focuses on the reasoning behind the decision to work. Employees may not necessarily like the message, but they will at least understand that the decision is not arbitrary and is tied to an important deadline. (See also **correspondence** and **style.**)

Whether your memo is formal or informal depends entirely on your readers and your **purpose.** Is your reader a coworker, a superior, or a subordinate? A message to a coworker who is also a friend is likely to be more informal, while an internal proposal to several readers or to someone two or three levels higher in your organization is likely to use a more formal style. Consider the following versions of a statement:

TO AN EQUAL	I can't go along with the plan because I think it poses serious logistical problems. First, . . . [informal, casual, and forceful response written to an equal]
TO A SUPERIOR	The logistics of moving the department may pose serious problems. First, . . . [formal, impersonal, and cautious response to a superior]

A memo that gives instructions to a subordinate should be relatively formal, impersonal, and direct, unless you are trying to reassure or praise. Using an overly chatty, casual style in memos to your subordinates may make you seem either insincere or ineffectual. However, if you are too formal, sprinkling your writing with fancy words, you may seem stuffy and pompous. You may also be regarded as rigid and incapable of moving an organization ahead. (See also **affectation.**) When writing to subordinates, remember that managing does not mean dictating. An imperious tone—like false informality—will not make a memo an effective management tool. When you write a memo to a subordinate, adopt a positive yet reasonable tone, as in the following example.

- Because we must meet the December 15 deadline to be eligible for the government contract, the office will be open the day after Thanksgiving. I am also temporarily reassigning several members of the office staff as shown below.

Lists and Headings.   Using **lists** is an effective strategy to give your points impact in a memo. Lists can be read and their meaning grasped more quickly than a paragraph that says the same thing. Be careful, however, not to overuse lists. A memo that consists almost entirely of lists is difficult for readers to understand because they

PROFESSIONAL PUBLISHING SERVICES
MEMORANDUM

TO: Barbara Smith, Publications Manager
FROM: Hannah Kaufman, Vice President   *HK*
DATE: April 14, 20--
SUBJECT: Schedule for ACM Electronics Brochure

ACM Electronics has asked us to prepare a comprehensive brochure for its Milwaukee office by August 9, 20--. We have worked
with electronics firms in the past, so this job should be relatively
easy to prepare. My guess is that the job will take nearly two
months. Ted Harris has requested time and cost estimates for the
project. Fred Moore in production will prepare the cost estimates,
and I would like you to prepare a tentative schedule for the project.

Additional Personnel

In preparing the schedule, check the availability of the following:
    1. Production schedule for all staff writers
    2. Available freelance writers
    3. Dependable graphic designers
Ordinarily, we would not need to depend on outside personnel;
however, because our bid for the *Wall Street Journal* special project
is still under consideration, we could be short of staff in June and
July. Further, we have to consider vacations that have already been
approved.

Time Estimates

Please give me time estimates by April 19. A successful job done
on time will give us a good chance to obtain the contract to do ACM
Electronics' annual report for its stockholders' meeting this fall.

I know your staff can do the job.

cc: Ted Harris, President
     Fred Moore, Production Editor

FIGURE M–4. Typical Memo Format

must mentally connect the separate and disjointed items on the page. Further, lists lose their impact when they are overused.

Another attention-getting device, particularly in long memos, is **headings.** Headings divide material into manageable segments, call attention to main topics, and signal a shift in topic. Readers can scan the headings and read only the section or sections appropriate to their needs.

Format

Memos vary greatly in format and customs. Although there is no single standard, Figure M–4 on page 382 shows a typical 8½-by-11-inch format with a printed company name. Subject lines announce the topic; because they aid filing and later retrieval, they must be specific and accurate.

VAGUE	Subject: Tuition Reimbursement
VAGUE	Subject: Time-Management Seminar
SPECIFIC	Subject: Tuition Reimbursement for Time-Management Seminar

Capitalize all major words in a title, except articles, prepositions, and conjunctions with fewer than four letters unless they are the first or last word. Remember: The title in the subject line should not substitute for an opening that provides a context for the message.

The final step is signing or initialing a memo, a practice that lets readers know that you have approved its contents. Where you sign or initial the memo depends on the practice of your organization: Some writers sign at the end of a memo, others sign their initials next to their typed name. Follow the practice of your employer. Figure M–4 shows a typical placement of initials.

M

metaphors (*see* figures of speech)

methods of development

After you have completed your research, but before beginning your outline, ask yourself how you can most effectively "unfold" the topic for your **readers.** An appropriate method of development makes it easy for readers to understand your topic and moves the topic smoothly and logically from an **introduction** (or **opening**) to a **conclusion.**

A logical method of development of your subject satisfies the readers' need for shape and structure. As the writer, you must choose the method of development that best suits your subject, your readers, and your **purpose.** Below are the most common methods, each of which is discussed in further detail in its own entry.

- *Sequential.*   The **sequential method of development** emphasizes the order of elements. If you are writing a set of **instructions,** for example, your readers need the instructions in the order that will enable them to complete the task successfully.
- *Chronological.*   To emphasize the time element of the sequence, you would follow a **chronological method of development.** For example, a Federal Aviation Administration (FAA) report on an airplane crash might begin with takeoff and proceed sequentially to the crash.
- *Comparison.*   When writing about a new topic that is in many ways similar to another, more familiar topic, you might develop the new topic by using the **comparison method of development**.
- *Division and Classification.*   When describing physical objects or structures with component parts, you can use the **division and classification method of development** to explain each part's function and how all the parts work together.
- *Spatial.*   Use the **spatial method of development** to describe the physical appearance of an object from top to bottom, inside to outside, front to back, and so on.
- *Cause-and-Effect.*   The **cause-and-effect method of development** begins with either the cause or the effect. For example, if you were an FAA agent reporting on an airplane disaster, you might start your report with the cause of the crash and lead up to the crash itself. Conversely, you might start with the crash and trace the events back to the cause. You can also use this approach to develop a report that offers a solution to a problem, beginning with the problem and moving on to the solution, or vice versa.
- *General-to-Specific.*   The **general-to-specific method of development** proceeds from general to specific information. If

you are writing about the software for a new computer system, for example, you might begin with a general statement of the function of the total software package, then explain the functions of the larger routines in the software package, and finally deal with the functions of the various subroutines within the larger routines.

- *Specific-to-General.* The **specific-to-general method of development** begins with specific information and builds to a general conclusion. For example, you might describe a software problem in a minor application, leading to a larger, more global problem with the software.
- *Order of Importance.* To explain the functions of the departments in a company, you could use the **decreasing-order-of-importance method of development** to reflect their importance in the company: the executive department first and the temporary support staff last, with all other departments (sales, accounting, and so on) arranged in their relative order of importance. In another situation, you might reverse the order and use the **increasing-order-of-importance method of development.**

M

Methods of development often overlap — rarely does a writer rely on only one method. The important thing is to select one primary method of development, base your outline on it, and then subordinate any other methods to it. For example, in describing the organization of a company, you could use elements from three methods of development. You could divide the larger topic (the company) into departments, arrange the departments by their order of importance within the company, and present their operations sequentially.

Use the method of development appropriate to your writing task to create an outline, as discussed in the entry **outlining.**

minutes of meetings

Organizations and committees keep official records of their meetings; such records are known as *minutes.* If you attend many business meetings, you may be asked to write and distribute the minutes of a meeting. For continuing committees, most commonly seen in educational and governmental organizations, the minutes of the previous meeting are usually read aloud if they were not distributed to the members beforehand; the group then votes to accept the minutes as prepared or to revise specific items. Although most meetings are called to discuss a single topic and most committees are not permanent, minutes are still issued after each meeting. The bylaws policies and procedures of an organization may specify what must be included in the meeting minutes. (For advice on conducting meetings, see **meetings.**)

Writer's Checklist: Minutes of Meetings

In general, meeting minutes should include the following:

- ☑ The name of the group or committee holding the meeting
- ☑ The place, time, and date of the meeting
- ☑ The kind of meeting (a regular meeting or a special meeting called to discuss a specific subject or problem)
- ☑ The number of members present and, for committees or boards of ten or fewer members, their names
- ☑ A statement that the chairperson and the secretary were present or the names of any substitutes
- ☑ A statement that the minutes of the previous meeting were approved or revised
- ☑ A list of any reports that were read and approved
- ☑ All the main motions that were made, with statements as to whether they were carried, defeated, or tabled (vote postponed) and the names of those who made and seconded the motions (motions that were withdrawn are not mentioned)
- ☑ A full description of resolutions that were adopted and a simple statement of any that were rejected
- ☑ A record of all ballots with the number of votes cast for and against resolutions
- ☑ The time the meeting was adjourned (officially ended) and the place, time, and date of the next meeting
- ☑ The recording secretary's signature and typed name and, if desired, the signature of the chairperson

Because minutes are often used to settle disputes, they must be accurate, complete, and clear. When approved, minutes of meetings are official and can be used as evidence in legal proceedings.

NORTH TAMPA MEDICAL CENTER

Minutes of the Monthly Meeting
Medical Audit Committee

DATE: July 26, 20--

PRESENT: G. Miller (Chair), C. Bloom, J. Dades,
            K. Gilley, D. Ingoglia (Secretary), S. Ramirez,
            D. Rowan, C. Tsien, C. Voronski
ABSENT: R. Fautier, R. Wolf

Dr. Gail Miller called the meeting to order at 12:45 p.m. Dr. David Ingoglia made a motion that the June 1, 20--, minutes be approved as distributed. The motion was seconded and passed.

   The committee discussed and took action on the following topics.

(1) TOPIC: Meeting Time

   Discussion: The most convenient time for the committee to meet.
   Action taken: The committee decided to meet on the fourth Tuesday of every month, at 12:30 p.m.

FIGURE M–5. One Sample of Minutes

Keep your minutes brief and to the point. Give complete information on each topic, without rambling. Except for recording motions, which must be transcribed word for word, summarize what occurs and paraphrase discussions. Following a set format will help you keep the minutes concise. You might, for example, use the heading "Topic," followed by the subheadings "Discussion" and "Action Taken," for each major point discussed. Two possible formats for minutes of meetings are shown in Figure M–5 and Figure M–6 on page 388.

   Avoid abstractions and generalities; always be specific. Use names and titles and refer to people consistently—a lack of consistency in titles or names may reveal or unintentionally suggest a

WARETON MEDICAL CENTER
DEPARTMENT OF MEDICINE

Minutes of the Regular Meeting
Credentials Committee

DATE: April 18, 20--

PRESENT: M. Valden (Chairperson), R. Baron, M. Frank, J. Guern,
           L. Kingson, L. Kinslow (Secretary), S. Perry, B. Roman,
           J. Sorder, F. Sugihana

Dr. Mary Valden called the meeting to order at 8:40 p.m. The minutes of
the previous meeting were unanimously approved, with the following correction: the secretary of the Department of Medicine is to be changed from Dr.
Martino Alvarez to Dr. Barbara Golden.

Old Business

None.

New Business

The request by Dr. Henry Russell for staff privileges in the Department
of Medicine was discussed. Dr. James Guern made a motion that Dr. Russell
be granted staff privileges. Dr. Martin Frank seconded the motion, which
passed unanimously.

Similar requests by Dr. Ernest Hiram and Dr. Helen Redlands were discussed. Dr. Fred Sugihana made a motion that both physicians be granted all
staff privileges except respiratory-care privileges because the two physicians
had not had a sufficient number of respiratory cases. Dr. Steven Perry seconded the motion, which passed unanimously.

Dr. John Sorder and Dr. Barry Roman asked for a clarification of
general duties for active staff members with respiratory-care privileges.
Dr. Richard Baron stated that he would present a clarification at the next
scheduled staff meeting, on May 15.

Dr. Baron asked for a volunteer to fill the existing vacancy for Emergency Room duty. Dr. Guern volunteered. He and Dr. Baron will arrange a
duty schedule.

The meeting was adjourned at 9:15 p.m.; the next regular meeting is
scheduled for May 15, at 8:40 p.m.

Respectfully submitted,

Leslie Kinslow   *Mary Valden*

Leslie Kinslow   Mary Valden, M.D.
Medical Staff Secretary   Chairperson

FIGURE M-6. Another Sample of Minutes

deference to one person at the expense of another. Avoid adjectives and adverbs that suggest good or bad qualities, as in "Mr. Sturgess's *capable* assistant read the *comprehensive* report to the subcommittee." Minutes should be as objective and impartial as possible.

If a member of the committee is to follow up on something and report back to the committee at its next meeting, clearly state the person's name and the responsibility he or she has accepted.

If you have been assigned to take the minutes at a meeting, come to the meeting fully prepared. Bring more than one pen and plenty of paper. If convenient, record your notes on a laptop computer. Take memory-jogging notes during the meeting and expand them with the appropriate details immediately after the meeting, before you forget.

misplaced modifiers (*see* modifiers)

mixed constructions   ESL

A mixed construction produces a garbled meaning because elements of the sentences do not sensibly fit together. The problem may be a **grammar** error, a **logic error,** or both, as in the following example.

- I have a degree in mechanical engineering along with 12 years of experience sets a foundation for a strong background in analyzing problems and assessing solutions.

The writer begins with an independent clause (I have a degree in mechanical engineering), then follows with a phrase (along with 12 years of experience), and then provides a predicate (sets a foundation . . .) as though the independent clause were a subject in need of a predicate. Consider the following revision, which puts the elements together in a grammatical and logical complex sentence.

- I have a degree in mechanical engineering and 12 years of experience, which provide me with a strong background in analyzing problems and assessing solutions.

(See also **sentence construction.**)

MLA style (*see* documenting sources)

modifiers   ESL

Modifiers are words, phrases, or clauses that expand, limit, or make more precise the meaning of other elements in a sentence. Although we can create sentences without modifiers, we often need the detail and clarification they provide.

> **WITHOUT MODIFIERS**   Production decreased.
>
> **WITH MODIFIERS**   *Automobile* production decreased *rapidly*.

Most modifiers function as **adjectives** or **adverbs.** An adjective makes the meaning of a noun or pronoun more precise by pointing out one of its qualities or by imposing boundaries on it.

- *ten* automobiles
- *an* animal
- *this* crane
- *loud* machinery

An adverb modifies an adjective, another adverb, a **verb,** or an entire clause.

- Under test conditions, the brake pad showed *much* less wear than it did under actual conditions.
  [The adjective *less* is modified.]

- The brake pad wear was *very* much less than under actual conditions.
  [Another adverb, *much,* is modified.]

- The wrecking ball hit the wall of the building *hard.*
  [The verb *hit* is modified.]

- *Surprisingly,* the machine failed even after all the durability and performance tests it had passed.
  [A clause is modified.]

Adverbs are **intensifiers** when they increase the impact of adjectives (*very* fine, *too* high) or adverbs (*rather* quickly, *very* slowly). As a rule, be cautious using intensifiers; their overuse can lead to exaggeration and hence to inaccuracies.

Stacked (Jammed) Modifiers

Some writing is unclear or difficult to read because it contains stacked modifiers, that is, strings of modifiers preceding nouns.

STACKED   Your *staffing-level authorization reassessment* plan should result in a major improvement.

The noun *plan* is preceded by three long modifiers, a string that slows the reader down and makes the sentence awkward and clumsy. Stacked modifiers are often the result of an overuse of **jargon** or **vogue words.** Occasionally, they occur in a mistaken attempt to be concise by eliminating short prepositions or connectives — exactly the words that help to make sentences clear and readable. See how breaking up the stacked modifiers makes the example easier to read.

CLEAR   Your plan for the reassessment of staffing-level authorizations should result in a major improvement.

Misplaced Modifiers

A modifier is misplaced when it modifies the wrong word or phrase. A misplaced modifier can cause **ambiguity.**

- We *almost* lost all of the parts.

- We lost *almost* all of the parts.

The first sentence means that all of the parts were *almost* lost (but they were not), and the second sentence means that most of the parts (*almost all*) were in fact lost. If the intended meaning is not the stated meaning, the reader will be misled. Place modifiers as close as possible to the words they are intended to modify.

Likewise, place phrases near the words they modify. Note the two meanings possible when the phrase is shifted in the following sentences:

- The equipment *without the accessories* sold the best.
  [Different types of equipment were available, some with and some without accessories.]

- The equipment sold the best *without the accessories.*
  [One type of equipment was available, and the accessories were optional.]

Place clauses as close as possible to the words they modify.

REMOTE   We sent the brochure to four local firms *that had four-color art.*

CLOSE   We sent the brochure *that had four-color art* to four local firms.

M

Squinting Modifiers

A modifier "squints" when it can be interpreted as modifying either of two sentence elements simultaneously, thereby confusing readers about which is intended.

- We agreed *on the next day* to make the adjustments.

The meaning of the example is unclear; it could have either of the following two senses:

- We agreed *to make the adjustments on the next day.*
- *On the next day, we agreed* to make the adjustments.

A squinting modifier can sometimes be corrected simply by changing its position, but often it is better to rewrite the sentence:

- We agreed that *on the next day* we would make the adjustments. [The adjustments were to be made on the next day.]

- *On the next day,* we agreed that we would make the adjustments. [The agreement was made on the next day.]

M (See also **dangling modifiers.**)

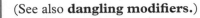

mood   ESL

The grammatical term *mood* refers to the verb functions that indicate whether the verb is intended to make a statement or ask a question, give a command, or express a hypothetical possibility.

The *indicative mood* refers to an action or a state that is conceived as fact.

- The setting *is* correct.
- *Is* the setting correct?

The *imperative mood* expresses a command, suggestion, request, or entreaty. In the imperative mood, the implied subject "you" is not expressed.

- *Install* the wiring today.

The *subjunctive mood* expresses something that is contrary to fact, conditional, hypothetical, or purely imaginative; it can also express a wish, a doubt, or a possibility. In the subjunctive mood, *were* is used instead of *was*. In clauses that speculate about the present or future, the base form (*be*) is used following certain verbs, such as *propose, request,* and *insist.*

- If we *were* to close the sale today, we would meet our monthly quota.

- The senior partner insisted that she [I, you, we, they] *be* in charge of the project.

The most common use of the subjunctive mood is to express clearly that the writer considers a condition to be contrary to fact. If the condition is not considered to be contrary to fact, the indicative is used.

SUBJUNCTIVE   If I *were* president of the firm, I would change several hiring policies.

INDICATIVE   If I *am* president of the firm, why don't I feel that I control every aspect of its policies?

M

ESL TIPS FOR DETERMINING MOOD

In written and especially in spoken English, there is an increasing tendency to use the indicative mood where the subjunctive traditionally has been used. Note the differences between traditional and contemporary usage in the following examples.

Traditional use of the subjunctive mood:
- I wish he *were* here now.
- If I *were* going to the conference, I would room with him.
- I requested that she *show* up on time.

Contemporary (informal) use of the indicative mood:
- I wish he *was* here now.
- If I *was* going to the conference, I would room with him.
- I requested that she *shows* up on time.

As a nonnative speaker of English, you are faced with a choice: Do you use the subjunctive and, consequently, in some circles sound sophisticated, intellectual, or even weird? Or do you use the indicative and in other circles sound uneducated? The answer might be to master both uses and be able to move freely between the different circles. In formal business and technical writing, however, it is best to use the more traditional expressions.

Ms. / Miss / Mrs.

Ms. is widely used in business and public life to address or refer to a woman, especially if her marital status is either unknown or irrelevant to the context. More traditionally, *Miss* is used to refer to an unmarried woman, and *Mrs.* is used to refer to a married woman. Some women may indicate a preference for *Ms., Miss,* or *Mrs.,* which you should honor. If a woman has an academic or professional title, use the appropriate form of address (*Doctor, Professor, Captain*) instead of *Ms., Miss,* or *Mrs.*

mutual / in common

When two or more persons (or things) have something *in common,* they share it or possess it jointly.

- What we have *in common* is our desire to make the company profitable.

- The fore and aft guidance assemblies have a power source *in common.*

Mutual may also mean "shared" (as in mutual friends), but it usually implies something given and received reciprocally and is used with reference to only two persons or parties.

- Melek mistrusts Roth, and I am afraid the mistrust is *mutual.* [Roth also mistrusts Melek.]

- Both business and government consider the policy to be of *mutual* benefit.

M

N

narration

Narration is the presentation of a series of events in a prescribed (usually chronological) sequence. Much narrative writing explains how something happened: a laboratory study, a site visit, an accident, the decisions in an important meeting. (See also **trip reports** and **trouble reports.**)

Effective narration rests on two key writing techniques: the careful, accurate sequencing of events and a consistent **point of view** on the part of the narrator. Narrative sequence and essential shifts in the sequence are signaled in three ways: chronology (clock and calendar time), transitional words pertaining to time (*before, after, next, first, while, then*), and verb tenses that indicate whether something has happened (past **tense**) or is under way (present tense). The point of view indicates the writer's relation to the information being narrated as reflected in the use of **person.** Narration usually expresses a first- or third-person point of view. First-person narration indicates that the writer is a participant ("this happened to me"), and third-person narration indicates that the writer is writing about what happened to someone or something else ("this happened to her, to them, or to it").

The narrative shown in Figure N–1 on pages 396–97 reconstructs the final hours of the flight of a small aircraft that attempted to land at the airport in Hailey, Idaho. Instead, the plane crashed into the side of a mountain, killing the pilot and the copilot. Because of the disastrous outcome of the flight, the investigators needed to "tell the story" in detail so that any lessons learned could be made available to other fliers. To do that, they had to recount the events as closely as possible. Thus, the chronology of the events is specified throughout the narrative in local (Hailey, Idaho) time.

To "tell the story," the narrator used the third-person point of view throughout, except for the first-person reports from witnesses.

History of the Flight*

At 0613 m.s.t.[1] on January 3, 1983, N805C, a Canadair Challenger owned and operated by the A.E. Staley Company departed Decatur, Illinois, on a flight to Friedman
Memorial Airport, Hailey, Idaho. The route of the flight was via Capitol, Illinois,
Omaha, Nebraska, Scotts Bluff, Nebraska, Riverton, Wyoming, Idaho Falls, Idaho,
direct 43°30′ north latitude, 114°17′ west longitude.

The en route portion of the flight was uneventful, and about 35 nmi east of Idaho
Falls N805C was cleared by the Salt Lake City, Utah, Air Route Traffic Control
Center (ARTCCS) to descend from 39,000 feet to 22,000 feet. N805C descended to
22,000 feet and the flightcrew then requested a descent to 17,000 feet. About 35 nmi
east of Sun Valley Airport, after being cleared, N805C descended to 17,000 feet.
About 0901, N805C's flightcrew canceled their flight plan and, shortly thereafter,
changed the transponder from 1311, the assigned discrete code, to 1200, the visual
flight rules (VFR) code. At 0901:07, the data analysis reduction tool (DART) radar
data showed a 1311 beacon code at 17,000 feet about 11 nmi east of the Sun Valley
Airport. At 0901:37, the DART radar data showed a 1200 VFR transponder beacon
code with no altitude readout about 2 nmi west of the 1311 beacon code that was
recorded at 0901:07.

At 0904:10, DART radar data recorded a 1200 code target at 13,500 feet almost
directly over the Sun Valley Airport. According to an employee of the airport's
fixed base operator, N805C's flightcrew called on the airport's UNICOM[2] frequency and requested a landing advisory and asked if a food order had been placed.
The flightcrew then stated that there would "be a quick turn," and placed a fuel request. This was the last transmission heard from N805C. The employee said that she
provided the latest altimeter setting to N805C, and "since we did not have the cloud
conditions in the area, I was glad when other pilots were able to give reports as they
saw things from the air."

The flightcrew of Cessna Citation, N13BT, which had landed at Sun Valley
about 0903, also heard N805C report "over the field." According to N13BT's pilot,
N805C reported over the field "sometime during our final approach or landing." According to the pilot, the weather at the airport when he landed "was 800 (feet) overcast with 10 miles visibility. The tops of the overcast fog bank was about 6,800
(feet) m.s.l." He said that the overcast was "solid northwest up the valley. Visibility
appeared lower (to the) northwest."

About 0908, Trans Western Flight 1301, a Convair 580, landed at Sun Valley
Airport. Flight 1301 had descended through a hole in the overcast about 15 nmi
southwest of Bellevue, Idaho, which is about 3 nmi southeast of the Sun Valley Airport. The first officer said that he gave position reports to the Sun Valley UNICOM
when the flight was 15 nmi from the airport, when it was 10 nmi from the airport
over Bellevue turning on final approach for runway 31, and when it was 1 mile from

---

*National Transportation Safety Board, "Aircraft Accident Report: A. E. Staley Manufacturing
Co., Inc., Canadair Challenger CL-600, N805C, Hailey, Idaho, January 3, 1983." National Technical Information Service, Springfield, Va., 1983.

[1]All times, unless otherwise noted, are mountain standard time based on the 24-hour clock.
[2]UNICOM. The non-government air/ground radio communications facility which may provide
airport advisory information at certain airports. The Sun Valley UNICOM did not record, nor
was it required to record or log, the time of radio communications.

the runway. The first officer said that he could see the visual approach slope indicator (VASI) lights for runway 30 during the landing approach. The captain and the first officer said that they neither saw N805C nor heard radio transmissions from N805C.

About 0900, a man who was driving his truck north on the highway between Bellevue and Hailey, Idaho, saw a twin engine, cream-colored jet, break through the clouds when he was about 2.5 miles north of Bellevue. He saw that the landing gear was down but he did not see any lights on the airplane. When the airplane appeared, "it was about 300 to 500 yards from the west hills adjacent to the airport and about 1,000 feet from the valley floor." The airplane was in a noseup attitude. The witness said that after the airplane descended below the clouds "and (the pilot) saw how close to the hills he was, he then started a sharp right turn." The airplane disappeared from his view into "low hanging clouds" over the northwest side of the hangar at the airport.

Between 0900 and 0930, another man, who was in the yard of his home in northeast Hailey, saw a jet airplane east of his home. The airplane was "white or silver with a blue tint." (N805C was painted white with blue and gold stripes along the length of the fuselage and tops of the wings.) The airplane was below the clouds, and he had "a good view of the airplane for about 10 to 15 seconds." He said that the airplane had a noseup attitude and "the wings were rocking up and down about 20°." The witness said that the clouds obscured all but the lower peaks of the mountains to the east and that after he lost sight of the airplane he thought it was "odd that the aircraft was under cloud cover."

Shortly after 0900, a woman who was located in an apartment in southeast Hailey, heard a jet airplane fly over "in a northerly direction." She thought that this was "odd because jets don't go over us heading north from the airport. The engines sounded very loud. . . ."

A fourth witness said that, between 0900 and 0940, she heard a jet airplane overfly her house in northeast Hailey. The woman was in the living room of her house when she heard the airplane and thought that "it must be low because of the loudness of the (engine) noise," and that "the sound of the jet did not trail off as they do as they fly farther away from you. The sound stopped less than 30 seconds from the time I first heard it." At the time, the clouds were resting on and hiding the tops of the mountains to the east.

About 1030, the chief pilot of the A. E. Staley Manufacturing Company, who was to board N805C at Sun Valley, arrived at the airport. Since the airplane was overdue, he instituted inquiries to several nearby airports to determine where the airplane had landed. At 1300, he asked air traffic control to make a full communications search. At 1400, after being told that the airplane had not been found, he requested an air search. While waiting for search and rescue teams to arrive, the chief pilot rented an airplane and at about 1700, found the accident site. The impact site, elevation about 6,510 feet, was about 2.2 nmi north of Sun Valley Airport at coordinates 43°32′50″ N latitude, 114°17′35″ W longitude.

FIGURE N–1. Example of a Narration (*continued*)

The words of each witness were quoted for their bearing on what happened, from that person's vantage point.

Investigators sequenced the events as precisely as possible from the beginning of the flight until the crash site was located by referencing verified clock times and approximating those that could not be verified. The verb tenses throughout indicate a past action: *departed, canceled, cleared, called, reported.*

Important but secondary information was either mentioned in a footnote or, like the color of the plane, inserted in parentheses. Information about the color of the plane was important to the narration because it verified a sighting by an eyewitness; otherwise, it would have been unnecessary.

Although narration often exists in combination with other **forms of discourse,** once a narrative is under way, it should not be interrupted by lengthy explanations or analyses. Explain only what is necessary for readers to follow the events.

nature

Nature, when used to mean "kind" or "sort," is vague. Avoid this usage in your writing. Say exactly what you mean.

- The ~~nature of~~ *exclusionary clause in* the contract caused the problem.

needless to say

Although the phrase *needless to say* sometimes occurs in speech and writing, it is redundant because it always precedes a remark that is stated despite the inference that nothing further need be said.

- ~~Needless to say, departmental~~ *Departmental* cutbacks have meant decreased efficiency.

- ~~Needless to say,~~ *Understandably,* departmental cutbacks have meant decreased efficiency.

neo-

The **prefix** *neo-* is derived from a Greek word meaning "new." It is hyphenated when used with a proper **noun**. Common nouns are not hyphenated unless the base word begins with the vowel *o*.

- neo-Fascism, neo-Darwinism, neo-impressionism
- neoclassicism, neonatal, neoplasm

new words

New words (also called *neologisms*) continually find their way into the language from a variety of sources. Some come from technology (*download*), scientific research (*in vitro*), and brand or trade names (*Xerox*). Others are **blend words** formed by combining existing words (*netiquette* from Inter*net* plus e*tiquette*), **acronyms** (*spool* from *s*imultaneous *p*eripheral *o*perations *onl*ine). Some words have come from other languages, such as *skiing* (Norwegian) and *whiskey* (Gaelic).

Business and technology are responsible for many new words. Some are necessary and unavoidable because they communicate new concepts concisely and accurately. However, it is best to avoid creating a new word if an existing word will do. If you do use a new word or expression you believe a reader may not know, be sure to define it the first time you use it. (See also **affectation, vogue words,** and **word choice.**)

newsletter articles

An employee newsletter is a publication that is designed primarily to keep employees informed about the company and its operations and policies. Many different types of newsletters are found in business and industry, from informal electronic newsletters to sophisticated, glossy publications. Employee newsletters increase employee or even customer understanding of the company, its purpose, and the industry; report recent breaking news about the company's divisions, policy, and achievements; and note employees' accomplishments (awards, anniversaries, and so on). If your company publishes a newsletter, you may be asked to contribute an article on a subject in your area of expertise.

Before you begin to write, consider the traditional *who, what, where, when,* and *why* of journalism (Who did it? What was done? Where was it done? When was it done? Why was it done?) and then add *how,* which may be of as much interest to your colleagues as any of the five *w*'s.

Next, determine whether the company has an official policy or position on your subject. If so, adhere to it as you prepare your article. If there is no company policy, determine as nearly as you can management's attitude toward your subject.

Gather several fairly recent issues of the company newsletter and study the **style** and **tone** of the writing and the approach used for various kinds of subjects. Understand those perspectives before you begin to work on your own article. Ask yourself the following questions about your subject: What is its significance to the company? What is its significance to my coworkers? The answers to those questions should help you establish the style, tone, and approach for your article and also heavily influence your conclusion.

Research for a newsletter article frequently consists of **interviewing.** Interview everyone concerned with your subject. Get all available information and all points of view. Be sure to give maximum credit to the maximum number of people.

Writer's Checklist: Newsletter Articles

Writing a newsletter article requires more imagination than writing a report. Because newsletters are usually not required reading, you do not have a captive audience. Use the following tips to achieve **emphasis.**

☑ Write an intriguing **title** to catch the **readers'** attention; **rhetorical questions** often work well.

☑ Include eye-catching **photographs** or **illustrations** that entice your audience to read your lead **paragraph.**

☑ Fashion a lead, or first paragraph, that will encourage further reading. The first paragraph generally makes the **transition** from the title to the substantive body of the article.

☑ Offer a well-developed presentation of your subject to hold the readers' interest all the way to the end of the article.

Author! Author!

It was a number of centuries back that the Roman satirist Persius said, "Your knowing is nothing, unless others know you know."

Today at Allen-Bradley some of our engineers are subscribing to Persius's philosophy. They contend it takes more than the selling of a product for a corporation to enjoy continuing success.

"We must sell our knowledge as well," stressed Don Fitzpatrick, Commercial Chief Engineer. "We have to get our customers to think of us as knowledge experts."

Fitzpatrick's point is well made. In today's high-technology business climate, it is essential to present a vanguard image as well as produce quality products. A corporation can be considered a "knowledge expert" when it "not only has a product for sale, but is the acknowledged leader in the application of that product," Fitzpatrick explained.

How does a corporation acquire an image as a "knowledge expert"?

Customer Roundtables, distributor schools, and participation by our engineers in professional societies, such as the Institute of Electrical and Electronics Engineers (IEEE), have helped Allen-Bradley to enhance its image in the industrial control market, noted Fitzpatrick. Yet, one subliminal selling tool that Fitzpatrick believes A-B engineers could sharpen to foster more influence is the writing and presentation of technical papers.

"I'm talking about technical writing. Not the kind of technical writing that explains our products or how to use them safely, but the kind of writing that is done as a method of sharing knowledge among companies who have similar concerns," Fitzpatrick explained. Such writings discuss new methods being used in the industry, or a company's difficulties working with an electrical phenomenon. They introduce state-of-the-art technology, new theories, or are tutorials on the developments in the industry to date, he added.

Fitzpatrick's Paper Cited

Fitzpatrick knows about technical papers. The former Purdue University professor has been preparing and presenting topics for Allen-Bradley Roundtables and distributor schools for several years. He also has been active in the presentation of papers for IEEE conferences. This past September in Houston, Fitzpatrick received a second-place award for a paper he presented at a Petrochemical Industrial Conference (PCIC) in San Diego. (PCIC is an organization of the Industry Application Society, which is part of the IEEE.) The award-winning paper, entitled "Transient Phenomena in the Motor Control Distribution Systems," also was published in IEEE's renowned annual publication "IAS Transactions." Only a few papers are accepted annually for presentation at professional conferences, noted Fitzpatrick. In turn, just a few of those presentations are published in IEEE-related "Transactions" catalogs. Occasionally a paper's topic also may generate an article in a trade publication. For these reasons preparing such papers can be plums—for both company recognition and the author's esteem among colleagues.

Cultivating peer recognition is as important as nurturing an expert image for your company in today's technological world, said Steven Bomba, Director of Advanced Technology Development. "In the area of advanced technology the only way to test if your work is current is to share it with people outside of the company," he said. "When you publish, you don't reveal proprietary information. You publish information that will be economically beneficial to your organization. Our employees should be anxious to test their sword against their competitors."

402   newsletter articles

Growing Fraternity Here

Bomba and Fitzpatrick noted that there is a "new spirit for Allen-Bradley" in the area of technical writing and presentation. The number of engineers participating as guest speakers at national engineering, application, and technical conferences is a small but growing fraternity. Participation by A-B personnel in these professional conferences spurs a "psyche tune-up" and is critical to "innovation, staying current, and invention. Without it, a company lacks a reference point. Encouraging your people to write these papers provides a measure of value and stimulation," Bomba contends.

Technical papers are always noncommercial. They don't mention specific products or the company's name. The papers are bylined, however, usually with an editor's note containing information about the writer's background and employment. "If you word a paper right, you can sell your company. You can make the paper work as a subliminal public relations tool for you. One payoff for the extra time it often takes to prepare a paper is the direct influence in the public eye that is gained. It is impossible to measure the exact influence, but the exposure generally generates positive reaction," Fitzpatrick notes.

Contact at Conferences

"Participation in conferences through writing of technical papers results in an employee becoming more active in professional organizations. That in turn leads to more business contacts. We're trying to broaden our market base, enter new industries. Roundtables, publishing, and conference programs are byproducts, or spinoffs, that work for us. They help us to sell our ability," said Fitzpatrick.

They also work for the engineers and technicians who get involved in the paper's preparation. Often a paper may be co-authored or researched as a team.

Engineers or technicians who take the time to research a possible topic, prepare an abstract for query, and finally write a paper upon abstract acceptance, are taking those small steps of knowledge that lead to leaps in industry technology.

"Getting published places the author in the elite 'Invisible College of Knowledge Workers,' which is a collection of technical people who transcend the boundaries of industry, cities, and nations," said Bomba. "Only 1 percent of degreed engineers write technical papers that are published," he added.

"Peer fear" may be the greatest barrier for an engineer to pick up the pen— or keyboard, as the case may be today. Walt Maslowski, Principal Engineer, Dept. 756, was an engineer for 12 years before he decided to give technical writing a whirl three years ago. He joined A-B in 1989. This past September he presented his second paper as an A-B employee at an Industry Application Society conference in Cincinnati. His latest paper, which addressed the techniques in controlling three-phase induction motor drives, received a second-place award from IAS. The article will be published in IEEE's next issue of IAS Transactions. "There was a lot of positive response to the paper, and motor drives recognition is an area that A-B has been pursuing for some time," Maslowski noted.

Many Benefits Intangible

"The fear of possibly writing substandard papers held me back from writing for years," Maslowski continued. "What I've discovered, though, is that when you do write a qualified paper, your technical expertise becomes known. Many of your personal gains are intangible, but you meet people, open contacts, and absorb more knowledge."

How does one decide to write a paper? "First you should review enough of a field to get a pulse for it. Read key

FIGURE N-2. Newsletter Article (*continued*)

papers. Review key results that people are talking about. Analyze the unsolved areas — or holes — of these writings and then figure out how those holes might relate to your job," suggested Dave Linton, a project engineer in Dept. 756. Linton presented a tutorial on computer software design at a Digital Equipment Computer User Society symposium in early November.

Another barrier for some engineers is getting an abstract accepted, since all papers are reviewed by published professionals, or "referees." "Making that jump from the inside to the outside is hard without some middle ground to test your ideas. Some companies active in publishing have internal review systems where your co-workers give you some input and encouragement. It would be nice if Allen-Bradley had a similar system of internal publishing," Linton suggested. Bomba, Fitzpatrick, and others involved in technical writing here, concurred.

"Industry does not make the 'publish or perish' threat that is prevalent in the academic world. At A-B we really encourage our employees to contribute to our industry: to get involved," said Bomba.

Being recognized in your field builds self-confidence and generates a sense of pride and accomplishment. Writing technical papers is a good way to let "others know you know."

Reprinted with permission from the employee publication of the Allen-Bradley Company, Milwaukee.

FIGURE N-2. Newsletter Article (*continued*)

Writer's Checklist: Newsletter Articles (continued)

☑ Write a **conclusion** that emphasizes the significance of your subject to your audience and stresses the points you want your readers to retain.

In preparing your newsletter article, you may find it helpful to follow the steps listed in the Checklist of the Writing Process (on the inside front cover). Figure N-2 on pages 401–403 is an article written for a newsletter (the original included photographs and boxed quotations). (See also **layout and design.**)

news releases (*see* **press releases**)

no doubt but

In the **phrase** *no doubt but,* the word *but* is redundant.

- There is no doubt ~~but~~ that he will be promoted.

nominalizations   (ESL)

We have a natural tendency to want to make our on-the-job writing sound formal, even impressive. One practice that contributes to that tendency is the use of nominalizations. A nominalization is a weak verb (*make, do, conduct, perform,* and so forth) combined with a noun, when the verb form of the noun would communicate the same idea more effectively and concisely.

- The quality assurance team will ~~perform an evaluation of~~ *evaluate* the new software.

You may occasionally have a legitimate use for a nominalization. For example, you might use a nominalization to slow down the pace of your writing. But if you use nominalizations just to make your writing sound more formal, the result will be **affectation.** (See also **technical writing style** and **conciseness/wordiness.**)

none   (ESL)

None can be considered either a singular or a plural **pronoun,** depending on the context.

- *None* of the material *has* been ordered.
  [Always use a singular **verb** with a singular **noun,** in this case, "material."]

- *None* of the clients *has* been called yet.
  [Use a singular verb even with a plural noun (*clients*) if the intended **emphasis** is on the idea of *not one.*]

- *None* of the clients *have* been called yet.
  [Use a plural verb with a plural noun if no emphasis is intended.]

For emphasis, substitute *no one* or *not one* for *none* and use a singular verb.

- I paid the full retail price for three of your firm's machines, ~~none~~ *not one* of which was worth the money.

(See also **agreement.**)

nor / or   (ESL)

Nor always follows *neither* in sentences with continuing negation.

- They will *neither* support *nor* approve the plan.

Likewise, *or* follows *either* in sentences.

- The firm will accept *either* a short-term *or* a long-term loan.

Two or more singular subjects joined by *or* or *nor* usually take a singular **verb.** But when one subject is singular and one is plural, the verb agrees with the subject nearer to it.

SINGULAR	Neither the *architect* nor the *client was* happy with the design.
PLURAL	Neither the *architect* nor the *clients were* happy with the design.
SINGULAR	Neither the *clients* nor the *architect was* happy with the design.

(See also **conjunctions.**)

notable / noticeable

Notable, meaning "worthy of notice," is sometimes confused with *noticeable,* meaning "readily observed."

- Her accomplishments are *notable*.

- The construction crew is making *noticeable* progress on the new building.

note taking

The purpose of note taking is to summarize and record information you extract during **research.** The great challenge in taking notes is to condense another writer's thoughts in your own words without distorting the original thinking or plagiarizing. As you extract information, let the purpose of your writing and your audience guide you. Resist the temptation to copy your source word for word as you take notes; instead, paraphrase the author's idea or concept in your own words. (See **paraphrasing.**) You must do more than just change a few words in the original passage; otherwise, you will be guilty of **plagiarism.**

On occasion, when your source concisely sums up a great deal of information or points to a trend or development important to your subject, you are justified in quoting the source verbatim and incorporating that **quotation** into your document. As a general rule, you will rarely need to quote anything longer than a paragraph. If you do quote something word for word from a source, be certain to enclose the material in **quotation marks** in your notes so you will not forget that it is a direct quotation. In your finished writing, be certain to give the source of your quotation. (See also **documenting sources.**)

When you are taking notes on abstract ideas, as opposed to factual data, be careful not to sacrifice **clarity** for brevity. You can be brief—if you are accurate—with statistics, but notes that express concepts can lose their meaning if they are too brief. The critical test of a note is whether a week later you will still know what the note means and from it are able to recall the significant ideas of the passage. If you are in doubt about whether to take a note, take it—it is easier to discard a note than to find the source again.

The mechanics of note taking are simple and, if followed conscientiously, can save you much unnecessary work. The guidelines in the following Writer's Checklist should help you take notes more effectively.

Writer's Checklist: Taking Notes

☑ Do not try to write everything down. Select the most important ideas and concepts to record as notes.

Writer's Checklist: Taking Notes (continued)

☑ To give proper credit, include the author, title, publisher, place and date of publication, and page number. (On subsequent notes from the same source, you need to include only the author and the page number.)

☑ Mark any notes you do not fully understand or think you might need to pursue further.

☑ Be sure to record all vital names, dates, and definitions.

☑ Create your own shorthand. Many words can be indicated by symbols or shortened forms like & for *and,* + for *plus, w* for *with, hst* for *history,* and *7* for *seven.*

☑ Photocopy pages for passages you intend to quote and highlight the quotations.

☑ Print out key sections from Web sites or CD-ROMs or download the information to a "notes" file.

☑ Check your notes for accuracy against the printed material before you move on to another source.

Consider the information in the following paragraph:

> Long before the existence of bacteria was suspected, techniques were in use for combating their influence in, for instance, the decomposition of meat. Salt and heat were known to be effective, and these do in fact kill bacteria or prevent them from multiplying. Salt acts by the osmotic effect of extracting water from the bacterial cell fluid. Bacteria are less easily destroyed by osmotic action than are animal cells because their cell walls are constructed in a totally different way, which makes them very much less permeable.

The paragraph says essentially three things:

1. Before the discovery of bacteria, salt and heat were used to combat the effects of bacteria.
2. Salt kills bacteria by extracting water from their cells by osmosis, hence its use in curing meat.
3. Bacteria are less affected by the osmotic effect of salt than are animal cells, because bacterial cell walls are less permeable.

If your readers' needs and your objective involved tracing the origin of the bacterial theory of disease, you might want to note that

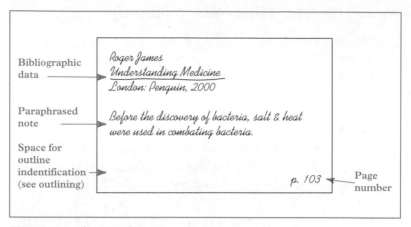

FIGURE N–3. First Note from a Source

salt was traditionally used to kill bacteria long before people realized what caused meat to spoil. It might not be necessary to your
topic to say anything about the relative permeability of bacterial cell
walls. Figure N–3 is an example of a first note taken from the paragraph on the use of salt as a preservative; Figure N–4 is an example
of a subsequent note.

Although you should record notes in a way you find efficient,
nothing replaces 3-by-5 index cards (as shown in Figures N–3 and
N–4) for some tasks. They are especially useful for **outlining** a
complex project.

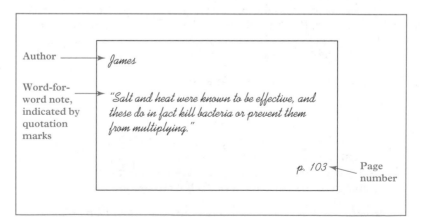

FIGURE N–4. Subsequent Note from a Source

nouns   ⓔSL

DIRECTORY
Overview   409
Types of Nouns   409
Noun Usage   410
Collective Nouns   410
Forming Plurals   411

Overview

A noun names a person, place, thing, concept, action, or quality.

Types of Nouns

The two basic types of nouns are proper nouns and common nouns. *Proper nouns,* which are capitalized, name specific people, places, and things. (See also **capital letters.**)

- Abraham Lincoln, New York, U.S. Army, Nobel Prize

Common nouns, which are not capitalized unless they begin sentences, name general classes or categories of persons, places, things, concepts, actions, and qualities. Common nouns include all other types of nouns except proper nouns.

- human, city, organization, award

Other types of nouns are collective nouns, abstract nouns, concrete nouns, count nouns, and mass nouns. *Collective nouns* are common nouns that indicate a group or collection. They are plural in meaning but singular in form. (See the subsection "Collective Nouns" for advice on using singular or plural forms with collective nouns.)

- audience, jury, brigade, staff, committee

Abstract nouns are common nouns that refer to things that cannot be discerned by the five senses.

- love, loyalty, pride, valor, peace, devotion

Concrete nouns are common nouns used to identify those things that can be discerned by the five senses.

- house, paper, keyboard, glue, nail, grease

Count nouns are concrete nouns that identify things that can be separated into countable units.

- desks, chisels, envelopes, engines, pencils

Mass nouns are concrete nouns that identify things that are a mass rather than individual units and that cannot be easily separated into countable units.

- electricity, water, sand, wood, air, uranium, gold, oil, wheat, cement

Noun Usage

Nouns function as subjects of **verbs,** objects of verbs and **prepositions, complements,** or **appositives.**

SUBJECTS	The *metal* bent as *pressure* was applied to it.
DIRECT OBJECT OF A VERB	The bricklayer cemented the *blocks* efficiently.
INDIRECT OBJECT OF A VERB	The company awarded our *department* a plaque for safety.
OBJECT OF A PREPOSITION	The event occurred within the *year.*
SUBJECTIVE COMPLEMENT	A dynamo is a *generator.*
OBJECTIVE COMPLEMENT	We elected the sales manager *chairperson.*
APPOSITIVE	George Thomas, the *treasurer,* gave his report last.

Words normally used as nouns can also be used as adjectives and adverbs.

ADJECTIVE	It is *company* policy.
ADVERB	He went *home.*

Collective Nouns

When a collective noun refers to a group as a whole, it takes a singular verb and pronoun.

- The staff *was* divided on the issue and could not reach *its* decision until May 15.

When a collective noun refers to individuals within a group, it takes a plural verb and pronoun.

- The staff *returned* to *their* offices after the conference.

A better way to emphasize the individuals on the staff would be to use the phrase *members of the staff*.

- The members of the staff returned to *their* offices after the conference.

Treat organization names and titles as singular.

- LRM Company *has* grown 200 percent in the last three years; *it* will move to a new facility in January.

Some collective nouns regularly take singular verbs (*crowd*); others do not (*people*).

- The crowd *was* growing impatient.

- Many people *were* able to watch the space shuttle land safely.

Forming Plurals

N

Most nouns form the plural by adding *s*.

- *Dolphins* are capable of communicating with people.

Nouns ending in *s, z, x, ch,* and *sh* form the plural by adding *es*.

- How many size *sixes* did we produce last month?

- The letter was sent to all the *churches*.

- Technology should not inhibit our individuality; it should fulfill our *wishes*.

Nouns that end in a consonant plus *y* form the plural by changing the *y* to *ies*.

- The store advertises prompt delivery but places a limit on the number of *deliveries* in one day.

Some nouns ending in *o* add *es* to form the plural, but others add only *s*.

- One tomato plant produced 12 *tomatoes*.

- We installed two *dynamos* in the plant.

Some nouns ending in *f* or *fe* add *s* to form the plural; others change the *f* or *fe* to *ves*.

- cliff/cliffs, fife/fifes, hoof/hooves, knife/knives

Some nouns require an internal change to form the plural.

- woman/women, man/men, mouse/mice, goose/geese

Some nouns do not change in the plural form.

- *Fish* swam lazily in the clear brook, while a few wild *deer* mingled with the *sheep* in a nearby meadow.

Hyphenated and open compound nouns form the plural in the main word.

- sons-in-law, high schools, editors-in-chief

Compound nouns written as one word add *s* to the end.

- Use seven *tablespoonfuls* of freshly ground coffee to make seven cups of coffee.

If you are unsure of the proper usage, check a **dictionary.** (See **possessive case** for a discussion of how nouns form the possessive case.)

nowhere near

The phrase *nowhere near* is colloquial and should be avoided in writing.

- His ability ~~is nowhere near~~ Katya's.
  does not approach ^

- His ability is ~~nowhere near~~ Katya's.
  not comparable to ^

- His ability is ~~nowhere near~~ Katya's.
  far inferior to ^

number

Number is the grammatical property of **nouns, pronouns,** and **verbs** that signifies whether one thing (singular) or more than one (plural) is being referred to. Nouns normally form the plural by simply adding *s* or *es* to their singular forms.

- Many new business *ventures* have failed in the past ten years.

- *Partners* in successful *businesses* are not always personal friends.

Some nouns require an internal change to form the plural.

- woman/women, man/men, goose/geese, mouse/mice

All pronouns except *you* change internally to form the plural.

- I/we, he/they, she/they, it/they

By adding an *s* or *es,* most verbs show the singular of the third **person,** present **tense,** indicative **mood.**

- he *stands,* she *works,* it *goes*

The verb *be* normally changes form to indicate the plural.

SINGULAR	I *am* ready to begin work.
PLURAL	We *are* ready to begin work.

(See also **agreement.**)

numbers

DIRECTORY
Overview   414
Numerals or Words   414
Plurals   414
Measurements   414
Fractions   415
Time   415
Dates   415
Addresses   416
Documents   416
Tips for Punctuating Numbers (ESL)   416

Overview

The standards for using numbers vary; however, unless you are following an organizational or professional style guide, observe the following guidelines.

Numerals or Words

Write numbers from zero to ten as words and numbers above ten as numerals. Spell out approximate numbers.

- We've had *over a thousand* requests on our help line.

   In most writing, do not spell out ordinal numbers, which express degree or sequence (42nd), unless they are single words (tenth, sixteenth). When several numbers appear in the same sentence or paragraph, write them the same way, regardless of other rules and guidelines.

- The company owned *150* trucks, employed *271* people, and rented *7* warehouses.

   Spell out numbers that begin a sentence, even if they would otherwise be written as figures.

- *One hundred and fifty* people attended the meeting.

If spelling out such a number seems awkward, rewrite the sentence so that the number does not appear at the beginning.

- The meeting was attended by *150* people.

Plurals

The plural of a written number is formed by adding *s* or *es* or by dropping *y* and adding *ies,* depending on the last letter, just as the plural of any other **noun** is formed.

- elevens, sixes, twenties

The plural of a numeral may be written either with *s* alone or with an apostrophe (*'s*).

- 5s or 5's, 12s or 12's

Measurements

Express units of measurement as numerals.

- 3 miles, 45 cubic feet, 9 meters, 27 cubic centimeters, 4 picas

When numbers run together in the same phrase, write one as a numeral and the other as a word.

- The order was for ~~12~~ *twelve* 6-inch pipes.

Generally give percentages as numerals and write out the word *percent*, except when the number is in a table.

- Approximately *85 percent* of the land has been sold.

Fractions

Express fractions as numerals when they are written with whole numbers.

- 27½ inches, 4¼ miles

Spell out fractions when they are expressed without a whole number.

- one-fourth, seven-eighths

Always write numbers with decimals as numerals.

- 5.21 meters

Time

Express hours and minutes as numerals when *a.m.* or *p.m.* follows.

- 11:30 a.m., 7:30 p.m.

Spell out time that is not followed by *a.m.* or *p.m.*

- four o'clock, eleven o'clock

Dates

The year and the day of the month should be written as numerals. Dates are usually written in a month-day-year sequence, in which the year may or may not be followed by a comma.

- *August 26, 2025,* is the payoff date for the loan.

Use the strictly numerical form for **dates** (8/26/25) in informal writing only.

ESL TIPS FOR PUNCTUATING NUMBERS

The rules for punctuating numbers in English are summarized as follows.

A comma separates numbers with five or more digits into groups of three, starting from the right.

- 57,890 cubic feet
- $187,291
- 5,289,112,001 atoms

In numbers with four digits, the comma is optional.

- 1,902 cases or 1902 cases

Do not use a comma in years, house numbers, zip codes, and page numbers.

- The Boeing 777 was first flown commercially by United Airlines in June *1995*.
- Autotech Industries is located at *92401* East Alameda Drive in Los Angeles.
- The zip code is *91601*.
- The citation is located on page *1204*.

Use a period to represent the decimal point.

- Their stock values increased at a monthly rate of *4.2* percent.
- The jackpot for last week's lottery was *$3,742,097.43*.

(See also **global communication** and **global graphics.**)

Addresses

Spell out numbered streets from one to ten unless space is at a premium.

- East Tenth Street

Write building numbers as numerals. The only exception is the building number *one*.

- 4862 East Monument Street
- One East Monument Street

Write highway numbers as numerals.

- U.S. 70, Ohio 271, I-94

Documents

In manuscripts, page numbers are written as numerals, but chapter and volume numbers may appear as numerals or words.

- Page 37
- Chapter 2 or Chapter Two
- Volume 1 or Volume One

Express figure and table numbers as numerals.

- Figure 4 and Table 3

Do not follow a word representing a number with a numeral in parentheses that represents the same number. Doing so is redundant.

- Send five ~~(5)~~ copies of the report.

numeral adjectives (*see* adjectives)

N

O

objective (*see* **purpose**)

objective complements (*see* **complements**)

objects (ESL)

There are three kinds of objects: direct objects, indirect objects, and objects of prepositions. All objects are **nouns** or noun equivalents (**pronouns, verbals,** and noun phrases and clauses).

Direct Objects

A direct object answers the question "What?" or "Whom?" about a verb and its subject.

- Harry prepared a *report*.

- I like *jogging*.

- I like *to jog*.

- I like *it*.

- I like *what I saw*.

- Sheila designed a new *circuit*.
  [*Circuit,* the direct object, answers the question, "Sheila designed *what?*"]

- George telephoned the *chief engineer*.
  [*Chief engineer,* the direct object, answers the question, "George telephoned *whom?*"]

A verb whose meaning is completed by a direct object is called a transitive **verb.**

Indirect Objects

An indirect object is a noun or noun equivalent that occurs with a direct object after certain kinds of transitive verbs, such as *give, wish, cause,* and *tell.* The indirect object answers the question "To whom or what?" or "For whom or what?" The indirect object always precedes the direct object.

- We sent the *general manager* a full report.
  [*Report* is the direct object; the indirect object, *general manager,* answers the question, "We sent a full report *to whom?*"]

- The general manager gave the *report* careful consideration.
  [*Consideration* is the direct object; the indirect object, *report,* answers the question, "The general manager gave careful consideration *to what?*"]

- The purchasing department bought *Sheila* a new printer.
  [*Printer* is the direct object; the indirect object, *Sheila,* answers the question, "The purchasing department bought a new printer *for whom?*"]

Objects of Prepositions

As the object of a **preposition,** a noun or pronoun combines with a preposition to form a prepositional phrase.

- *After the meeting,* the district managers adjourned *to the executive dining room.*

(See also **complements.**)

observance / observation

An *observance* is the "performance of a duty, custom, or law"; it is sometimes confused with *observation,* which is the "act of noticing or recording something."

- The *observance* of Veterans Day as a paid holiday varies from one organization to another.

- The laboratory technician made careful *observations* during the experiment.

OK / okay   (ESL)

The expression *okay* (also spelled *OK*) is common in informal writing but should be avoided in more formal documents, such as **reports.**

- Mr. Sturgess ~~gave his okay to~~ *approved* the project.

- The solution is ~~okay with~~ *acceptable to* the staff.

on account of

Avoid using the phrase *on account of* as a substitute for *because of.*

- He felt that he had lost his job ~~on account~~ *because* of the company's financial losses.

on / onto / upon

On is normally used as a preposition meaning "supported by," "attached to," or "located at."

- Install the phone *on* the wall.

Onto implies movement to a position on or movement up and on.

- The association members surged *onto* the platform to congratulate their new president.

Similarly, *on* stresses a position of rest, and *upon* emphasizes movement.

- A book lay *on* the table.
- She put a book *upon* the table.

on the grounds that / on the grounds of

The phrases *on the grounds that* and *on the grounds of* are wordy substitutes for *because*.

- She left ~~on the grounds that~~ *because* the NASA position offered a higher salary.

one (ESL)

When used as an indefinite **pronoun,** *one* may help you avoid repeating a noun.

- We need a new plan, not an old *one*.

One is often redundant in phrases in which it restates the noun, and it may take the proper emphasis away from the adjective.

- The computer program was not ~~a unique one.~~ *unique.*

One can also be used in place of a noun or personal pronoun in a statement such as the following:

- *One* cannot ignore *one's* physical condition.

Using *one* in that way is formal and impersonal; in any but the most formal writing, you are better advised to address your reader directly and personally as *you*.

- ~~One~~ *You* must be careful that the cost of ~~one's~~ *your* computer equipment does not exceed the use to which ~~one~~ *you* can put it.

(See also **point of view.**)

one of those . . . who

A dependent clause beginning with *who* or *that* and preceded by *one of those* takes a plural verb.

- She is *one of those* managers *who are* concerned about their writing.

- This is *one of those* policies *that make* no sense when you examine them closely.

In those two examples, *who* and *that* are subjects of dependent clauses; they refer to plural antecedents (*managers* and *policies*) and thus take plural verbs (*are* and *make*).

Because people so often use singular verbs in such constructions in everyday speech and informal writing, the rule confuses many writers. The principle behind the rule becomes clearer if *among* is substituted for *one of*. Compare the following examples with the preceding ones.

- She is *among* those managers *who are* concerned about their writing.
[You would not write, "She is among those managers who *is* concerned . . ."]

- This is *among* those policies *that make* no sense when examined closely.
[You would not write, "This is among those policies that *makes* no sense . . ."]

There is one exception to the plural-verb rule stated at the beginning of this entry: If the phrase *one of those* is preceded by *the only,* the verb in the following dependent clause should be singular.

- She is *the only one of those* managers *who is* concerned about her writing.
[The verb is singular because its subject, *who*, refers to a singular antecedent, *one*. If the sentence were reversed, it would read, "Of those managers, she is *the only one who is* concerned about her writing."]

- This is *the only one of those* policies *that makes* no sense when you examine it closely.
[If the sentence were reversed, it would read, "Of those policies, this is *the only one that makes* no sense when you examine it closely."]

only

In writing, the word *only* should be placed immediately before the word or phrase it modifies.

- We ~~only~~ lack ^*only*^ financial backing.

Incorrect placement of *only* can change the meaning of a sentence.

- *Only* he said that he was tired.
  [He alone said that he was tired.]

- He *only* said that he was tired.
  [He actually was not tired, although he said he was.]

- He said *only* that he was tired.
  [He said nothing except that he was tired.]

- He said that he was *only* tired.
  [He said that he was nothing except tired.]

openings

DIRECTORY
Overview   423
Statement of the Problem   425
Definition   425
Interesting Detail   425
Anecdote   426
Background   426
Quotation   426
Objective   427
Summary   427
Forecast   427
Scope   427

O

Overview

If your **readers** are already familiar with your subject, or if what you are writing is short, you may not need to begin your writing project with a full **introduction.** You may simply need to focus

your readers' attention with a brief opening. For many documents,
openings that immediately get to the point are adequate, as shown
in the following examples. (See also **correspondence, email,** and
reports.)

Correspondence

Mr. Charles B. Ignatowski
1720 Old Line Road
Thomasbury, WV 26401

Dear Mr. Ignatowski:

You will be happy to know that we have corrected the error in your
bank balance. The new balance shows . . .

Progress Report Letter

William Chang, M.D.
Phelps Building
9003 Shaw Avenue
Parksville, MD 21221

Dear Dr. Chang:

To date, 18 of the 20 specimens you submitted for analysis have
been examined. Our preliminary analysis indicates . . .

Longer Progress Report

PROGRESS REPORT ON REWIRING THE SPORTS ARENA
The rewiring program at the Sports Arena is continuing on schedule.
Although the costs of certain equipment are higher than our original
bid, we expect to complete the project without exceeding our budget,
because the speed with which the project is being completed will save
labor costs.

Work Completed
As of August 15, we have . . .

Email

Date: Fri, 26 Jan 1999 13:18:54 -0600 (CST)
From: Charles Spiros <charles.spiros@convex2.com>
To: Jane T. Meyers <meyers@execrw.org>
Cc: B. C. Sohns <bcs@execrw.org>
Subject: Budget Estimate for Fiscal 2004

Email (continued)

Parts/attachments:
1 Shown 13 lines Text
2   23 KB Application,""

Jane,
As I promised in my earlier email, I've attached the personnel budget estimates for fiscal year 2004.

Statement of the Problem

One way to give readers a perspective on your report is to present a brief account of the problem that led to the study or project being reported.

- Several weeks ago, a manager noticed an apparently simple problem in the software developed by Datacom Systems. He immediately reported his discovery to his supervisor. After an intensive investigation, we found that Datacom Systems . . .

Definition

Although a definition can be useful as an opening, do not define something with which the reader is familiar or provide a definition that is obviously a contrived opening (such as "Webster defines *technology* as . . ."). A definition should be used as an opening only if it offers insight into what follows.

- *Risk* is a loosely defined term. It is used here in the sense of physical risk as a qualitative combination of the probability of an event and the severity of the consequences of that event.

Interesting Detail

Often an interesting detail about your subject can be used to gain readers' attention and arouse their interest. Readers who work at a manufacturing company, for example, may be surprised by the diversity of products purchased by their company.

- From asbestos sheeting to zinc castings, from a chemical analysis of the water in Lake Maracaibo (to determine its suitability for use in steam injection units) to pistol blanks (for use in testing power charges), the purchasing department attends to the company's material needs. Approximately 15,000 requisitions, each containing from 1 to 14 separate items, are processed each year by this department. Every item or service that is bought . . .

Anecdote

An anecdote can also be used to catch your readers' attention and interest.

- In his poem "The Calf Path," Sam Walter Foss tells of a wandering, wobbly calf trying to find its way home at night through the lonesome woods. It made a crooked path, which was taken up the next day by a lone dog. Then "a bellwether sheep pursued the trail over vale and steep, drawing behind him the flock, too, as all good bellwethers do." At last the path became a country road; then a lane that bent and turned and turned again. The lane became a village street, and at last the main street of a flourishing city. The poet ends by saying, "A hundred thousand men were led by a calf, three centuries dead."

    Many companies today follow a "calf path" because they react to events rather than planning . . .

Background

The background or history of a topic may be interesting and put the topic in perspective for your readers. Consider the following example from a newsletter describing the process of oil drilling:

- From the bamboo poles the Chinese used when the pyramids were young to today's giant rigs drilling in a hundred feet of water, there has been a lot of progress in the search for oil. But whether four thousand years ago or today, in ancient China or a modern city, in twenty fathoms of water or on top of a mountain, the object of drilling is and has always been the same—to manufacture a hole in the ground, inch by inch.

Quotation

Occasionally, you can use a quotation to stimulate interest in your subject. To be effective, the quotation must be pertinent—not some loosely related remark selected from a book of quotations.

- Richard Smith, president of P. R. Smith Corporation, recently said, "I believe that the Web will have a greater impact on high-tech research than we might imagine." His statement represents a growing feeling among corporate leaders that . . .

Objective

In reporting on a project, you might open with a statement of the project's objective to give readers a basis for judging the results.

- The primary objective of the project was to develop new techniques to solve the problem of waste disposal. Our first step was to investigate . . .

Summary

You can provide a summary opening by describing in abbreviated form the results, conclusions, or recommendations of your article or report. Do not start the summary, however, by writing, "This report summarizes . . ."

CHANGE   This report summarizes the advantages offered by the photon as a means of examining the structural features of the atom.

TO   As a means of examining the structure of the atom, the photon offers several advantages.

Forecast

Sometimes you can use a forecast of a new development or trend to arouse the reader's interest.

- In the very near future, we may be able to use a handheld medical diagnostic device similar to those in science fiction to assess the physical condition of accident victims. This product and others are now being developed at The Seldi Group, Inc.

Scope

At times you may want to present the **scope** of your document in your opening. By providing the parameters of your material, the limitations of the subject, or the amount of detail to be presented, you enable your readers to determine whether they want to or need to read your document.

- This pamphlet provides a review of the requirements for obtaining an FAA pilot's license. It is not designed to be a textbook to prepare you to take the examination itself; rather, it gives you an idea of the steps you need to take and the costs involved.

(See also **conclusions.**)

oral presentations (*see* presentations)

oral / verbal

Oral refers to what is spoken.

- He offered an *oral* argument to the court.

Although it is sometimes used synonymously with *oral*, *verbal* literally means "in words" and can refer to what is spoken or written. To avoid possible confusion, do not use *verbal* if you can use *written* or *oral*.

- She offered a ~~verbal~~ *written* agreement to meet the deadline.
- She offered ~~a verbal~~ *only an oral* agreement to meet the deadline.

When you must refer to something both written and spoken, use both *written* and *oral* (rather than *verbal*) to make your meaning clear.

- He demanded either a *written* or an *oral* agreement before he would continue the project.

organization

Organization is essential to the success of any writing project, from a **formal report** to a Web page. Good organization is achieved through an outline created with a logical and appropriate **method of development** that suits your subject, your **readers,** and your **purpose.**

An outline gives shape and structure to the material you gathered during **research.** It enables you to emphasize your key points by placing them in the positions of greatest importance. By breaking your material into manageable parts, **outlining** also makes large or complex subjects easier for you to organize, and it ensures that your finished writing will move logically from idea to idea without omitting anything important. Finally, by forcing you to structure your thinking at an early stage, a good outline allows you to concentrate exclusively on writing when you begin the rough draft.

During organization, you must consider a **design and layout** that will be helpful to your reader and a **format** appropriate to your subject and purpose. If you intend to include **illustrations** with your writing, a good time to think about them is when you have completed your outline, especially if they need to be prepared by someone else while you are writing and revising the draft.

organizational charts

An organizational chart shows how the various components of an organization are related to one another. This type of visual aid is useful when you want to give your readers an overview of an organization or to display the lines of authority within it.

The title of each organizational component (office, section, division) is placed in a separate box. The boxes are then linked to a central authority, as shown in Figure O–1. If your readers need the information, include the name of the person and position title in each box.

As with all **illustrations,** place the organizational chart as close as possible to the text that refers to it.

O

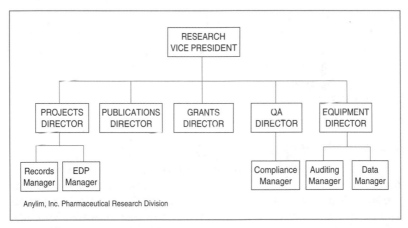

FIGURE O–1. Organizational Chart

orient / orientate

Orientate is merely a **long variant** of *orient,* meaning "locate in relation to something else." Use the shorter form because it is simpler.

- Let me ~~orientate~~ *orient* your group to our operation.

outlining

An outline is the skeleton of the document you are going to write; it lists the main topics of your subject and each subtopic. Outlining does many things to make your job easier and your writing better.

Advantages of Outlining

Outlining provides structure to your writing by ensuring that it has a beginning (**introduction** or **opening**), a middle (main body), and an end (**conclusion**). An outline gives your writing **coherence** and **transition** so that one part flows smoothly to the next without omitting anything important.

Like a road map, an outline indicates a starting point and keeps you moving logically so you do not get lost before you arrive at your conclusion. Using an outline offers many benefits: **Logic errors** are much easier to detect and correct in an outline than in a draft, and larger and more difficult subjects are easier to handle by breaking them into manageable parts. The less certain you are about your writing ability or about your subject, the fuller your outline should be. The parts of an outline are easily moved around, so you can see what arrangement of your ideas is most effective. Perhaps most important, creating a good outline frees you to concentrate on writing when you begin the rough draft.

Types of Outlines

Two types of outlines are most common: topic outlines and sentence outlines. A *topic outline* consists of short phrases that show the sequential order and relative importance of ideas; it provides order and establishes the relationships of topics to one another. Although a topic outline alone is generally not sufficient for a large or complex writing job, you can use it to structure the major and minor divisions of your topic in preparation for creating a sentence outline. An outline for a small job is not as detailed as one for a larger job, but it is just as important; a topic outline that lists your major and minor points can help greatly in a document as short as a letter, an **email,** or a **memo.** (See also **correspondence.**)

On a large writing project, create a topic outline first and then use it as a basis for creating a sentence outline. A *sentence outline* summarizes each idea in a complete sentence that may become the topic sentence for a paragraph in the rough draft. A sentence outline begins with a statement of the main idea that establishes the subject and then follows with a complete sentence for each idea in the major and minor divisions. If most of your notes can be shaped into topic sentences for paragraphs in your rough draft, you can be relatively sure that your document will be well organized.

Creating an Outline

When you are outlining large and difficult subjects with many pieces of information, the first step is to group related items and write them on note cards. Use an appropriate **method of development** to arrange items and label them with roman numerals. (See also **organization.**) For example, the major divisions for this discussion of outlining could be as follows:

  I. Advantages of outlining
 II. Types of outlines
III. Creating an outline

The second step is to establish your minor points by deciding on the minor divisions within each major division. Use a method of development to arrange minor divisions under the appropriate major divisions and label them with capital letters.

 II. Types of outlines
     A. Topic outlines
     B. Sentence outlines   } *Division and classification*
III. Creating an outline
     A. Establish major and minor divisions.
     B. Sort note cards by major and minor divisions. } *Sequential*
     C. Complete the sentence outline.

Of course, you will often need more than two levels of **headings.** If your subject is complicated, you may need three or four levels of heads to better organize all your ideas in proper relationship to one another. In that event, use the following numbering scheme:

I. First-level heading
  A. Second-level heading
    1. Third-level heading
      a. Fourth-level heading

The third step is to mark each detailed note card with the appropriate roman numeral and capital letter. Sort the note cards by major and minor division headings. (See also **note taking** and **research.**)

Organize the cards logically within each minor heading and mark each with the appropriate sequential Arabic number. Transfer your notes to a document, converting them to complete sentences. As you do, make sure that the subordination of minor division headings to major heads is logical. All headings, major and minor, within a division should be written in **parallel structure.** For example, all the second-level headings under "III. Creating an outline" are complete sentences in the active voice.

The outline samples shown to this point use a combination of numbers and letters to differentiate the various levels of information. But you could also use a decimal numbering system, such as the following, for your outline.

1 FIRST-LEVEL SECTION
  1.1 Second-level section
  1.2 Second-level section
    1.2.1 Third-level section
    1.2.2 Third-level section
      1.2.2.1 Fourth-level section
      1.2.2.2 Fourth-level section
  1.3 Second-level section
2 FIRST-LEVEL SECTION

This system should not go beyond the fourth level, because the numbers get too cumbersome beyond that point. In many documents, the decimal numbering system is carried over from the outline to the final version of the document for ease of cross-referencing sections.

Make certain your outline follows your method of development. Check that it develops your subject and does not stray into unrelated or only loosely related topics. Check your outline for completeness; scan it to see whether you need more information in any of your divisions and insert any information that is missing.

Treat **illustrations** as an integral part of your outline, noting in the outline approximately where each should appear. At each place, either include a rough sketch of the visual or write "illustration of . . ." As with other information in the outline, feel free to move, amend, or delete illustrations. Planning your **graphics** requirements from the beginning ensures their harmonious integration throughout all versions of the draft to the finished document.

You now have a complete sentence outline, and the most difficult part of the writing job is over. However, remember that an outline is not set in stone; it may need to change as you write the draft, but it should always be your point of departure and return.

outside of

In the phrase *outside of*, the word *of* is redundant.

- Place the rack outside ~~of~~ the incubator.

In addition, do not use *outside of* to mean "aside from" or "except for."

- *Except for*
  ~~Outside of~~ his frequent absences, Jim has a good work record.

O

over with

In the expression *over with*, the word *with* is redundant; moreover, the word *completed* often better expresses the thought.

- You may use the conference room when the managers' meeting is over ~~with~~.

- You may use the conference room when the managers' meeting is
  completed.
  ~~over with.~~

P

pace

Pace is the speed at which you present ideas to the **reader.** Your goal should be to achieve a pace that fits both your readers and your subject; at times you may need a fast pace, at other times a slow pace. The more knowledgeable the reader is, the faster your pace can be, but be careful not to lose control of it. In the first version of the following passage, facts are piled on top of each other at a rapid pace. In the second version, the same facts are spread out over two sentences and presented in a more easily assimilated manner, even though the length is no greater. In addition, the second version achieves a different and more desirable **emphasis.**

RAPID
: The hospital's generator is powered by a 90-horsepower engine, is designed to operate under normal conditions of temperature and humidity, produces 110 volts at 60 hertz, is designed for use under emergency conditions, and may be phased with other units of the same type to produce additional power when needed.

CONTROLLED
: The hospital's generator, which is powered by a 90-horsepower engine, produces 110 volts at 60 hertz under normal conditions of temperature and humidity. Designed especially for use under emergency conditions, this generator may be phased with other units of the same type to produce additional power when needed.

paragraphs

ESL

DIRECTORY
Overview   435
Topic Sentence   435

Paragraph Length   436
Writing Paragraphs   436
Unity and Coherence   437

Overview

A paragraph is a group of sentences that support and develop a single idea; think of it as an essay in miniature. The paragraph performs three functions: (1) It develops the unit of thought stated in the topic sentence; (2) it provides a logical break in the material; and (3) it creates a visual break on the page, which signals a new topic.

Topic Sentence

A topic sentence states the paragraph's main idea; the rest of the paragraph supports and develops that statement with carefully related details. The topic sentence is often the first sentence because it tells the **reader** what the paragraph is about.

- *The arithmetic of searching for oil is stark.* For all the scientific methods of detection, the only way the oil driller can actually know for sure that there is oil in the ground is to drill a well. The average cost of drilling an oil well is over $300,000, and drilling a single well may cost over $8,000,000. And once the well is drilled, the odds against its containing any oil at all are 8 to 1! Even after a field has been discovered, one out of every four holes drilled in developing the field is a dry hole because of the uncertainty of defining the limits of the producing formation. The oil driller can never know what Mark Twain once called "the calm confidence of a Christian with four aces in his hand."

On rare occasions, the topic sentence logically falls in the middle of a paragraph.

- It is perhaps natural that psychologists should awaken only slowly to the possibility that behavioral processes may be directly observed, or that they should only gradually put the older statistical and theoretical techniques in their proper perspective. But it is time to insist that science does not progress by carefully designed steps called "experiments," each of which has a well-defined beginning and end. *Science is a continuous and often a disorderly and accidental process.* We shall not do the young psychologist any favor if we agree to reconstruct our practices to fit the pattern

P

demanded by current scientific methodology. What the statistician means by the design of experiments is design which yields the kind of data to which *his* techniques are applicable. He does not mean the behavior of the scientist in his laboratory devising research for his own immediate and possibly inscrutable purposes.

—B. F. Skinner, "A Case History in Scientific Method"

The topic sentence is usually most effective early in the paragraph, but a paragraph can lead up to the topic sentence, which is sometimes done to achieve **emphasis.**

- Energy does far more than simply make our daily lives more comfortable and convenient. Suppose you wanted to stop—and reverse—the economic progress of this nation. What would be the surest and quickest way to do it? Find a way to cut off the nation's oil resources! Industrial plants would shut down; public utilities would stand idle; all forms of transportation would halt. The economy would plummet into the abyss of national economic ruin. *Our economy, in short, is energy-based.*

—*The Baker World* (Los Angeles: Baker Oil Tools)

Paragraph Length

Paragraph length should aid the reader's understanding of ideas. A series of short, undeveloped paragraphs can indicate poor organization of material and sacrifice unity by breaking a single idea into several pieces. A series of long paragraphs, on the other hand, can fail to provide the reader with manageable subdivisions of thought. A paragraph should be just long enough to deal adequately with the subject of its topic sentence. A new paragraph should begin whenever the subject changes significantly. Occasionally, a one-sentence paragraph is acceptable if it is used as a transition between larger paragraphs or in letters and memos, in which one-sentence **openings** and closings are appropriate.

Writing Paragraphs

Careful paragraphing reflects the writer's logical thinking and **organization.** Clear and orderly paragraphs help the reader follow the writer's thoughts more easily.

Outlining is the best guide to paragraphing. It is easy to group ideas into appropriate paragraphs when you follow a good working outline. (See also **outlining.**) Notice how the following topic outline plots the course of the subsequent paragraphs:

OUTLINE

I. Advantages of Chicago as location for new facility
- A. Transport infrastructure
    1. Rail
    2. Air
    3. Truck
    4. Sea (except in winter)
- B. Labor supply
    1. Engineering and scientific personnel
       a. Many similar companies in the area
       b. Several major universities
    2. Technical and manufacturing personnel
       a. Existing programs in community colleges
       b. Possible special programs designed for us

RESULTING PARAGRAPHS

Probably the greatest advantage of Chicago as a location for our new facility is its excellent transport facilities. The city is served by three major railroads. Both domestic and international air cargo service is available at O'Hare International Airport; Midway Airport's convenient location adds flexibility for domestic air cargo service. Chicago is a major hub of the trucking industry, and most of the nation's large freight carriers have terminals there. Finally, except in the winter months when the Great Lakes are frozen, Chicago is a seaport, accessible through the St. Lawrence Seaway.

Chicago's second advantage is its abundant labor force. An ample supply of engineering and scientific staff is ensured not only by the presence of many companies engaged in activities similar to ours but also by the presence of several major universities in the metropolitan area. Similarly, technicians and manufacturing personnel are in abundant supply. The colleges in the Chicago City College system, as well as half a dozen other two-year colleges in the outlying areas, produce graduates with associate's degrees in a wide variety of technical specialties appropriate to our needs. Moreover, three of the outlying colleges have expressed an interest in establishing special courses attuned specifically to our requirements.

Unity and Coherence

A good paragraph has **unity** and **coherence,** as well as adequate development. Unity is singleness of purpose, based on the topic sentence that states the core idea of the paragraph. When every sentence in the paragraph develops the core idea, the paragraph has unity.

Coherence is holding to one point of view, one attitude, one tense; it is the joining of sentences into a logical pattern. A careful choice of transitional words ties ideas together and thus contributes to coherence in a paragraph, as shown in the following example.

TOPIC SENTENCE	*Over the past several months, I have heard complaints about the Merit Award Program. Specifi-*
TRANSITION	*cally,* many employees feel that this program should be linked to annual *salary increases.*
TRANSITION	They believe that *salary increases* would provide a much better incentive than the current $500 to $700 cash awards for exceptional service. *In*
TRANSITION	*addition,* these *employees believe* that their supervisors consider the cash awards a satisfactory alternative to salary increases. Although I don't think this practice is widespread, the fact that
TRANSITION	the *employees believe* that it is justifies a reevaluation of the Merit Award Program.

Simple enumeration (*first, second, then, next,* and so on) also provides effective **transition** within paragraphs. Notice how the italicized words and phrases give coherence to the following paragraph.

* Most adjustable office chairs have nylon hub tubes that hold metal spindle rods. To ensure trouble-free operation, lubricate the spindle rods occasionally. *First,* loosen the set screw in the adjustable bell. *Then,* lift the chair from the base so that the entire spindle rod is accessible. *Next,* apply the lubricant to the spindle rod and the nylon washer, using the lubricant sparingly to prevent dripping. *When you have finished,* replace the chair and tighten the set screw.

parallel structure   ESL

Parallel sentence structure requires that sentence elements that are alike in function be alike in construction as well. In the following example, similar actions are stated in similar phrases.

* We need a supplementary workforce *to handle* peak-hour activity, *to free* full-time employees from routine duties, *to relieve* operators during lunch breaks, and *to replace* vacationing employees.

Parallel structure achieves an economy of words, clarifies meaning, expresses the equality of its ideas, and achieves **emphasis.** The technique assists **readers** because it allows them to anticipate the meaning of a sentence element on the basis of its parallel construction. When they recognize the similarity of word order or construction, readers know that the relationship between the new sentence element and the subject is the same as the relationship between the last sentence element and the subject. Because of that similarity, they can move from one idea to another more quickly and confidently.

Parallel structure can be achieved with words, **phrases,** or **clauses.**

WORDS	If you want to earn a satisfactory grade in the training program, you must be *punctual, courteous,* and *conscientious.*
PHRASES	If you want to earn a satisfactory grade in the training program, you must recognize the importance *of punctuality, of courtesy,* and *of conscientiousness.*
CLAUSES	If you want to earn a satisfactory grade in the training program, *you must arrive punctually, you must behave courteously,* and *you must study conscientiously.*

Correlative **conjunctions** (*either . . . or, neither . . . nor, not only . . . but also*) should always employ parallel structure. Both parts of the pairs should be followed immediately by the same grammatical form: two similar words, two similar phrases, or two similar clauses.

WORDS	Viruses carry either *DNA* or *RNA,* never both.
PHRASES	Clearly, neither *serological tests* nor *virus isolation studies* alone would have been adequate.
CLAUSES	Either *we must increase our operational efficiency* or *we must decrease our production goals.*

To make a parallel construction clear and effective, it is often best to repeat an **article,** a **pronoun,** a helping **verb,** a **preposition,** a subordinating conjunction, or the mark of an infinitive (*to*).

ARTICLE	The Babylonians had *a* rudimentary geometry and *a* rudimentary astronomy.
PRONOUN	*My* father and *my* teacher agreed that I was not really trying.
VERB	The driver *must* be careful to check the gauge and *must* move quickly when the light comes on.

PREPOSITION	New teams were being established *in* New York and *in* Los Angeles.
CONJUNCTION	He was promoted *because* he was industrious, *because* he was punctual, and *because* he was willing to put in extra effort.
INFINITIVE	*To* run and be elected is better than *to* run and be defeated.

Parallel structure is especially important in creating your outline, your **table of contents,** and your **headings,** because it lets your readers know the relative value of each item in your table of contents and each head in the body of your document. (See also **outlining.**)

Faulty Parallelism

Faulty parallelism results when joined elements are intended to serve equal grammatical functions but do not have equal grammatical form. Avoid that kind of partial parallelism. Make certain that each element in a series is similar in form and structure to all others in the same series. In work-related writing, **lists** often cause problems with parallel structure. When you use a list that consists of phrases or clauses, each phrase or clause in the list should begin with the same **part of speech.**

| NOT PARALLEL | The following recommendations were made regarding the Cost Containment Committee's position statement: |

1. *Stress* that this statement is for all departments.
2. *Start* the statement with, "If the company continues to grow, the following steps should be taken."
3. *The statement* should emphasize that it applies both to department managers and staff.
4. *Such strong words* as *obligation, owe,* and *must* should be replaced with words that are less harsh.

| PARALLEL | The following recommendations were made regarding the Cost Containment Committee's position statement: |

1. *Stress* that this statement is for all departments.
2. *Start* the statement with, "If the company continues to grow, the following steps must be taken."

3. *Emphasize* that it applies both to department managers and to staff.
4. *Replace* such strong words as *obligation, owe,* and *must* with words that are less harsh.

In the first list, the items are not grammatically parallel in structure: Items 1 and 2 begin with verbs, but items 3 and 4 do not. Notice how much more smoothly the corrected version reads, with all items beginning with verbs.

Faulty parallelism sometimes occurs because a writer tries to compare items that are not comparable.

NOT PARALLEL   The company offers special university training to help nonexempt employees move into professional careers like human resources, accounting, *customer representatives,* and *sales trainees.*
[Notice that occupations—*human resources* and *accounting*—are being compared to people—*customer representatives* and *sales trainees.*]

PARALLEL   The company offers special university training to help nonexempt employees move into professional careers like human resources, accounting, *customer service,* and *sales.*

paraphrasing

Paraphrasing is restating or rewriting in your own words the essential ideas from another writer's passage. Because the paraphrase does not quote the source word for word, **quotation marks** are not necessary. However, paraphrased material should be credited because the ideas are taken from someone else even though the words are not identical. The following example is an original passage that explains the concept of "object blur." The paraphrased version restates the passage in a form appropriate for a report.

ORIGINAL   One of the major visual cues used by pilots in maintaining precision ground reference during low-level flight is that of object blur. We are acquainted with the object-blur phenomenon experienced when driving an automobile. Objects in the foreground appear to be rushing toward us while objects in the background appear to recede slightly. There is a point in the observer's line of sight, however, at which objects appear to stand still for a moment, before once again rushing toward him with increasing angular velocity. The distance from the observer to this point where objects appear stationary is sometimes referred to as the "blur threshold" range.

PARAPHRASED   Object blur refers to the phenomenon by which observers in a moving vehicle report that foreground objects appear to rush at them, while background objects appear to recede. But objects at some point appear temporarily stationary. The distance separating observers from that point is sometimes called the "blur threshold" range.

Note that the paraphrased version includes only the essential information from the original passage. Strive to put the original ideas into your words without distorting them. (See also **note taking, quotations, plagiarism,** and **ethics in writing.**)

parentheses

Parentheses are used to enclose explanatory or digressive words, phrases, or sentences. The material in parentheses often clarifies a sentence or passage without altering its meaning. Parenthetical information may not be essential to a sentence — in fact, parentheses deemphasize the enclosed material — but it may be interesting or helpful to some readers. Parenthetical material applies to the word or phrase immediately preceding it.

- Aluminum is extracted from its ore (called bauxite) in three stages.

    Parenthetical material does not affect the punctuation of a sentence. If a parenthesis appears at the end of a sentence, the ending punctuation should appear after the parenthesis. A comma following a parenthetical word, phrase, or clause also appears outside the closing parenthesis.

- These oxygen-rich chemicals, such as potassium permanganate ($KMnO_4$) and potassium chromate ($KCrO_4$), were oxidizing agents (they added oxygen to a substance).

However, when a complete sentence within parentheses stands independently, the ending punctuation goes inside the final parenthesis.

- The new marketing approach appears to be a success; most of our regional managers report sales increases of 15 to 30 percent. (The only important exceptions are the Denver and Houston offices.)

Parentheses also are used to enclose numerals or letters that indicate sequence. Enclose the numeral or letter in two parentheses rather than using only one parenthesis.

- The following sections deal with (1) preparation, (2) research, (3) organization, (4) writing, and (5) revision.

In some **footnote** forms, parentheses enclose the publisher, place of publication, and date of publication. (See also **documenting sources.**)

- [1]W. P. Hall, *Handbook of Communication Methods* (New York: Stoddard, 2002): 9.

Use **brackets** to set off a parenthetical item that is already within parentheses.

- We should be sure to give Emanuel Foose (and his brother Emilio [1812–1882]) credit for his part in founding the institute.

Do not follow a spelled-out number with a numeral in parentheses representing the same number.

- Send five (5) copies of the report.

P

participial phrases (*see* **phrases**)

participles (*see* **verbs**)

parts of speech   ESL

The term *parts of speech* describes the class of words to which a particular word belongs according to its function in a sentence (naming, asserting, describing, joining, acting, modifying, exclaiming). Each of the parts of speech in the following list is described in its own entry. (See also **functional shift.**)

PART OF SPEECH	FUNCTION
Noun, pronoun	Naming/referring
Verb	Asserting
Adjective, adverb	Describing/modifying
Conjunction, preposition	Joining/linking
Interjection	Exclaiming

party

In legal language, *party* refers to an individual, a group, or an organization.

- The injured *party* brought suit against my client.

The term is inappropriate in all but legal writing; when you are referring to a person, use the word *person*.

- The ~~party~~ person whose file you requested is here now.

Party is, of course, appropriate when it refers to a group.

- Arrangements were made for the members of our *party* to have lunch after the tour.

passive voice (*see* **voice**)

per

When *per* is used to mean "for each," "by means of," "through," or "on account of," it is appropriate.

- per annum, per capita, per diem, per head

When it is used to mean "according to" (*per* your request, *per* your order), the expression is **jargon** and should be avoided. Equally incorrect is the phrase *as per*.

- As ~~per our discussion,~~ we discussed, I will send revised instructions.

percent / percentage

Percent (or *per cent*) is normally used instead of the symbol % (except in tables, where space is at a premium).

- Only 25 *percent* of the members attended the meeting.

Percentage, which is never used with numbers, indicates a general size.

- Only a small *percentage* of the engineers attended the meeting.

periods   ESL

A period usually indicates the end of a declarative or imperative sentence. Periods also link when used as leaders (for example, in a table of contents) and indicate omissions when used as **ellipses.** Periods may also end questions that are really polite requests and questions to which an affirmative response is assumed.

- Will you please send me the financial statement.

(See also **sentence construction.**)

Periods in Quotations

Use a **comma,** not a period, after a declarative sentence that is quoted in the context of another sentence.

- "There is every chance of success," she stated.

A period is conventionally placed inside **quotation marks.** (See also **quotations.**)

- He liked to think of himself as an "entrepreneur."

- He stated clearly, "My vote is yes."

Periods with Parentheses

If a sentence ends with a parenthesis, the period should follow the parenthesis.

- The institute was founded by Harry Denman (1902–1972).

If a whole sentence (beginning with an initial capital letter) is enclosed in parentheses, the period (or other end mark) should be placed inside the final parenthesis.

- The project director listed the problems her staff faced. (This was the third time she had complained to the board.)

Other Uses of Periods

Use periods after initials in names.

- Wilma T. Grant   J. P. Morgan

Use periods as decimal points with **numbers.**

- 109.2 degrees $540.26   6.9 percent

Use periods to indicate **abbreviations.**

- Ms.   Dr.   Inc.

When a sentence ends with an abbreviation that ends with a period, do not add another period.

- Please meet me at 3:30 p.m.

Use periods following the numerals in a numbered list.

- 1. Enter your name.
  2. Enter your address.
  3. Enter your telephone number.

Period Faults

The incorrect use of a period is sometimes referred to as a *period fault.* When a period is inserted prematurely, the result is a **sentence fragment.**

> FRAGMENT   After a long day at the office during which we finished the quarterly report. We left hurriedly for home.

> SENTENCE   After a long day at the office, during which we finished the quarterly report, we left hurriedly for home.

When two independent clauses are joined without any punctuation, the result is a *fused* or *run-on* sentence. Adding a period between the clauses is one way to correct a fused sentence.

> FUSED   Bill was late for ten days in a row Ms. Sturgess had to fire him.

CORRECT   Bill was late for ten days in a row. Ms. Sturgess had to fire him.

Other options are to add a comma and a coordinating **conjunction** (*and, but, for, or, nor, yet*) between the clauses, to add a **semicolon,** or to add a semicolon with a conjunctive **adverb** (such as *therefore* or *however*).

person   ESL

Person refers to the form of a personal **pronoun** that indicates whether the pronoun represents the speaker, the person spoken to, or the person or thing spoken about. A pronoun that represents the speaker is in the *first* person.

- *I* could not find the answer in the manual.

A pronoun that represents the person or people spoken to is in the *second* person.

- *You* are going to be a good supervisor.

A pronoun that represents the person or people spoken about is in the *third* person.

- *They* received the news quietly.

The following table shows first-, second-, and third-person pronouns.

PERSON	SINGULAR	PLURAL
First	I, me, my, mine	We, us, our, ours,
Second	You, your, yours	You, your, yours
Third	He, him, his, she, her, hers, it, its	They, them, their, theirs

(See also **number, case,** and **one.**)

personal / personnel

Personal is an **adjective** meaning "of or pertaining to an individual person."

- He left work early because of a *personal* problem.

Personnel is a **noun** meaning a "group of people engaged in a common job."

* All *personnel* should pick up their paychecks on Thursday.

Be careful not to use *personnel* when the word you need is *persons,* *people,* or a more descriptive word.

* The remaining ~~personnel~~ *employees* will be moved next Thursday.

personal pronouns (*see* pronouns)

persons / people

The word *persons* is used to refer to a specific number of people, often in legal or official contexts.

* Carpool lanes are restricted to vehicles carrying more than two *persons*.

In all other contexts, use *people*.

* We need to find more qualified *people* to fill the vacant positions.

persuasion

Persuasion attempts to convince the **reader** to adopt the writer's point of view. The range of persuasive writing required on the job varies widely. In one situation, you may need to reinforce ideas that your readers already have. In another situation, you may need to get your readers to change their current ideas. In yet another situation, you may need to persuade your readers to do something. You may find yourself pleading for safer working conditions, justifying the expense of a new program, writing a **proposal** for a large purchase, or writing a business plan to persuade a bank to authorize a line of credit. Persuasive writing can also take the form of **memos** and **email.**

   In persuasive writing, the way you present your ideas is as important as the ideas themselves. You must support your appeal with logic, a sound presentation of facts, statistics, and examples. Be careful to avoid ambiguity, not to wander from your main point, and above all never to make trivial, irrelevant, or false claims. You should also acknowledge any real or potentially conflicting opinions; doing so allows you to anticipate and overcome objections. In fact, by acknowledging negative details or opposing views, you not only gain credibility but also assume an ethical stance.

The memo in Figure P–1 was written by an MIS administrator to persuade staff to accept and participate in a change to a new computer system. Notice that not everything in this memo is presented in a positive light. Change brings disruption, and the writer acknowledges that fact.

InterOffice Memorandum

TO: Engineering Sales Staff
FROM: Bernadine Kovak, MIS Administrator *BK*
DATE: April 8, 20--
SUBJECT: Plans for Changeover to NRT/R4 System

As you all know, our workload has jumped by 30 percent in the past month. It has increased because our customer base and resulting technical support services have grown dramatically. This growth is a result, in part, of our recent merger with Datacom.

This growth has meant that we have all experienced the difficulty of providing our customers with up-to-date technical information when they need it. In the next few months, we anticipate that the workload will increase another 20 percent. Even a staff as experienced as ours cannot handle such a workload without help.

To cope with that expansion, in the next month we will be installing the NRT/R4 mainframe and QCS enterprise software with Web-based applications and global sales and service network. The system will speed processing dramatically as well as give us access to all relevant company-wide databases. It should enable us to access the information both we and our customers need.

The new system, unfortunately, will cause some disruption at first. We will need to transfer many of our existing programs and software applications to the new format. And all of us need to learn to navigate in the R4 and QCS environments. However, once we have made those adjustments, I believe we will welcome the changes.

I would like to put your knowledge and experience to work in getting the new system into operation. Let's meet in my office to discuss the improvements on Friday, April 12, at 1:00 p.m. I will have details of the plan to discuss with you. I'm also eager to get your comments, suggestions, and — most of all — your cooperation.

FIGURE P–1. Persuasive Memo

You also gain credibility—and thus persuasiveness—through the **readers'** impressions of you as a person. For that reason, the appearance, or **layout and design,** of a document (your **résumé,** for example) is important. (See also **ethics in writing, logic errors,** and **methods of development.**)

phenomenon / phenomena

A *phenomenon* is an observable thing, fact, or occurrence. Its plural form is *phenomena.*

- The natural *phenomenon* of earth tremors is a problem we must anticipate in designing the Los Angeles installation.

- The *phenomena* associated with the onset of Alzheimer's disease are not fully understood.

photographs

Photographs are the best way to show the surface of an object, record an event, or demonstrate the development of a phenomenon over a period of time. Photographs can be taken with a digital camera or scanned and then inserted into a document or Web page. Photographs are not always the best type of illustration, however. They cannot depict the internal workings of a mechanism or below-the-surface details of objects or structures. Such details are better represented in **drawings.**

If you take a photograph yourself, stand close enough to the subject so it fills the picture frame. To get precise and clear photographs, choose camera angles carefully. Select important details and the camera angles that will record them. To show relative size, place a familiar object—such as a ruler or a person—near the subject being photographed. For example, the photograph in Figure P–2 on page 451 shows a control device being held by a human hand to illustrate its relative size and features.

Treat photographs like other **illustrations,** as described in that entry. Give the photograph a figure number, "callouts" (labels) to identify key features in the photograph, and any other important information. Position the figure number and caption so the reader can view them and the photograph from the same orientation.

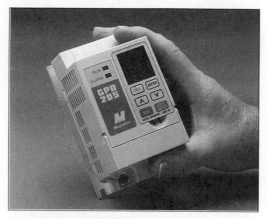

FIGURE P–2. Photo of Control Device
Photo courtesy of Ken Cook Company

phrases   (ESL)

DIRECTORY
Overview   451
Prepositional Phrases   452
Participial Phrases   453
Infinitive Phrases   454
Gerund Phrases   455
Verb Phrases   455
Noun Phrases   455

Overview

Below the level of the sentence, words are combined into two groups: **clauses** and phrases. Phrases are based on **nouns,** nonfinite **verb** forms, or verb combinations without subjects.

- She encouraged her staff *by her calm confidence.* [phrase]

A phrase may function as an **adjective,** an **adverb,** a noun, or a verb.

- The subjects *on the agenda* were all discussed.
  [phrase functioning as an adjective]

- We discussed the project *with great enthusiasm.*
  [phrase functioning as an adverb]

- *Working hard* is her way of life. [phrase functioning as a noun]

- The chief engineer *should have been notified.*
  [phrase functioning as a verb]

Even though phrases function as adjectives, adverbs, nouns, or verbs, they are normally named for the kind of word around which they are constructed: **preposition, participle, infinitive, gerund,** verb, or noun. A phrase that begins with a preposition is a *prepositional phrase;* a phrase that begins with a participle is a *participial phrase;* and so on.

Prepositional Phrases

A preposition is a word that shows relationship and combines with a noun or **pronoun** (its **object**) to form a modifying phrase. A prepositional phrase, then, consists of a preposition plus its object and the object's modifiers.

- *After the meeting,* the district managers adjourned *to the executive dining room.*

Prepositional phrases, because they normally modify nouns or verbs, usually function as adverbs or adjectives. A prepositional phrase may function as an adverb of motion.

- Turn the dial four degrees *to the left.*

A prepositional phrase may function as an adverb of manner.

- Answer customers' questions *in a courteous fashion.*

A prepositional phrase may function as an adverb of place.

- We ate lunch *in the company cafeteria.*

When functioning as adverbs, prepositional phrases may appear in different places.

- *In residential and farm wiring,* one of the wires must always be grounded.

- One of the wires must always be grounded *in residential and farm wiring.*

A prepositional phrase may function as an adjective. When functioning as adjectives, prepositional phrases follow the nouns they modify.

- Garbage *with a high protein content* can be processed into animal food.

Be careful when you use prepositional phrases, because separating a prepositional phrase from the noun it modifies can cause **ambiguity.**

AMBIGUOUS *The man* standing by the security guard *in the gray suit* is our president.

CLEAR *The man in the gray suit* who is standing by the security guard is our president.

(See also **modifiers.**)

Participial Phrases

A participle is any form of a verb that is used as an adjective. A participial phrase consists of a participle plus its object and its modifiers.

- The division *having the largest number of patents* will work with NASA.
- *Finding the problem resolved,* he went to the next item.
- *Having begun,* we felt that we had to see the project through.

The relationship between a participial phrase and the rest of the sentence must be clear to the reader. For that reason, every sentence containing a participial phrase must have a noun or pronoun that the participial phrase modifies; if it does not, the result is a dangling participial phrase.

Dangling Participial Phrases. A dangling participial phrase occurs when the noun or pronoun that the participial phrase is meant to modify is not stated but only implied in the sentence. (See also **dangling modifiers.**)

DANGLING *Being unhappy with the job,* his efficiency suffered. [His efficiency was not unhappy with the job; what the participial phrase really modifies — *he* — is not stated but merely implied.]

CORRECT   *Being unhappy with the job,* he grew less efficient.
                [Now what that participial phrase modifies—*he*—is
                explicitly stated.]

Misplaced Participial Phrases. A participial phrase is misplaced when it is too far from the noun or pronoun it is meant to
modify and so appears to modify something else. Such an error can
make the writer look ridiculous. (See also **modifiers.**)

MISPLACED   *Rolling around in the bottom of the vibration test cham-
                ber,* I found the missing bearings.

CORRECT   I found the missing bearings *rolling around in the bot-
                tom of the vibration test chamber.*

MISPLACED   We saw a large warehouse *driving down the highway.*

CORRECT   *Driving down the highway,* we saw a large warehouse.

Infinitive Phrases

An infinitive is the bare form of a verb (*go, run, talk*) without the restrictions imposed by **person** and **number;** an infinitive is generally preceded by the word *to* (which is usually a preposition but in
this use is called the sign, or mark, of the infinitive). An infinitive
phrase consists of the word *to* plus an infinitive and any objects or
modifiers.

• *To succeed in this field,* you must be willing *to assume responsibility.*

Do not confuse a prepositional phrase beginning with *to* with an infinitive phrase. In an infinitive phrase, *to* is followed by a **verb;** in a
prepositional phrase, *to* is followed by a **noun** or a **pronoun.**

PREPOSITIONAL PHRASE   We went *to the building site.*

INFINITIVE PHRASE   Our firm tries *to provide a comprehensive
                         training program.*

The implied subject of an introductory infinitive phrase should be
the same as the subject of the sentence. If it is not, the phrase is a
dangling modifier. In the following example, the implied subject of
the infinitive is *you* or *one,* not *practice.*

• To learn a new language, ~~practice is needed.~~ *you must practice.*

• To learn a new language, ~~practice is needed.~~ *one must practice.*

Gerund Phrases

A gerund phrase consists of a gerund plus any objects or modifiers and always functions as a noun.

SUBJECT   *Writing a technical manual* is a difficult task.
DIRECT OBJECT   She liked *designing the system.*

Verb Phrases

A verb phrase consists of a main verb and its helping verb.

- He *is* [helping verb] *working* [main verb] hard this summer.

Words can appear between the helping verb and the main verb of a verb phrase.

- He *is always working.*

The main verb is always the last verb in a verb phrase.

- You *will file* your tax return on time if you begin early.

- You *will have filed* your tax return on time if you begin early.

- You *are* not *filing* your tax return too early if you begin now.

Questions often begin with a verb phrase.

- *Will* he *audit* their account soon?

The adverb *not* may be appended to a helping verb in a verb phrase.

- He *did not work* today.

Noun Phrases

A noun phrase consists of a noun and its modifiers.

- *Many large companies* are concerned with product safety.

- Have *the two new employees* fill out *these forms.*

plagiarism   ESL

Plagiarism is the intentional use of someone else's exact words without **quotation marks** and appropriate credit or the use of someone else's unique ideas without acknowledgment. In publishing plagiarism is illegal; in other circumstances it is, at the least, unethical. Plagiarism is considered the theft of someone else's creative and intellectual property and is not accepted in business, science, journalism, academia, and many other fields. (For detailed guidance on quoting correctly, see **quotations.**)

You may quote or **paraphrase** the words and ideas of another if you document your source. (See also **documenting sources.**) Although you need not enclose paraphrased material in quotation marks, you must document its source. Paraphrased ideas are taken from someone else even though the words are not identical. Paraphrasing a passage without citing the source is permissible only when the information paraphrased is common knowledge in a field. Common knowledge refers to historical, scientific, geographical, technical, and other types of information on a topic widely known and readily available in handbooks, manuals, atlases, and other references. If you intend to publish or reproduce and distribute material in which you have included quotations from published works, you may have to obtain written permission to do so from the **copyright** holder.

In the workplace, employees often borrow from in-house manuals, booklets, reports, and other company sources in their writing. Borrowing passages from such sources is neither plagiarism nor a violation of copyright. The information, often referred to as "boilerplate," is simply being used in a different setting by the organization to which it already belongs. (See also **ethics in writing.**)

P

plurals (*see* nouns, pronouns, and verbs)

point of view

Point of view is the writer's relation to the information presented, as reflected in the use of grammatical **person.** The writer usually expresses point of view in first-, second-, or third-person personal **pronouns.** The use of the first person indicates that the writer is a participant or an observer: "This happened to *me*," "*I* saw that." The second and third person indicate that the writer is giving directions, **instructions,** or advice or writing about other people or something impersonal:

FIRST PERSON	*I* scrolled down to find the settings option.
SECOND PERSON	Scroll down to find the settings option and double-click. [*you* is understood]
THIRD PERSON	The report was written for *them*.

Consider the following sentence, written from an impersonal point of view.

- It is regrettable that your shipment of March 12 is not acceptable.

Now consider the same sentence written from the more personal first-person point of view.

- I regret that we cannot accept your March 12 shipment.

Although the meaning of both sentences is the same, the sentence with the personal point of view indicates that two people are involved in the communication.

Many people think they should avoid the pronoun *I* in their on-the-job writing. Such practice, however, leads to awkward sentences with people referring to themselves in the third person as *one* or as *the writer* instead of as *I*.

- ~~One~~ *I* can only conclude that the bond rate is too low.

- ~~The writer believes~~ *I believe* that this project will be completed by the end of June.

However, do not use the personal point of view when an impersonal point of view would be more appropriate or more effective. An impersonal point of view emphasizes the subject matter over the writer or the reader.

PERSONAL	I received objections to my proposal from several of you managers.
IMPERSONAL	Several managers have raised objections to the proposal.

In the example, it does not help to personalize the situation; in fact, the impersonal version may be more tactful.

Whether you adopt a personal or an impersonal point of view depends on the **purpose** and the **readers** of the document. For example, in an informal **email** to an associate, you would most likely adopt a personal point of view. But in a report to a large group, you would probably emphasize the subject by using an impersonal point of view.

- The evidence suggests that the absorption rate is too fast.

In **correspondence** on company stationery, use of the pronoun *we* may be interpreted as reflecting company policy, whereas *I* clearly reflects personal opinion. Which pronoun to use should be decided according to whether the matter discussed in the letter is a corporate or an individual concern.

- *I* understand your frustration with the price increase, but *we* must now include the import tax.

positive writing

Presenting positive information as though it were negative is a trap that writers in the workplace fall into easily because of the complexity of the information they must write about. It is a practice that confuses readers, however, and one that should be avoided.

NEGATIVE   If the error does not involve data transmission, the scan function will not be used.

POSITIVE   The scan function is used only if the error involves data transmission.

In the first sentence, the reader must reverse the two negatives to understand that it is only an exception that is being stated; the second sentence presents the exception in a straightforward manner.

Negative facts or conclusions should be stated negatively; stating a negative fact or conclusion positively is deceptive because it can mislead the reader. (See also **ethics in writing.**)

DECEPTIVE   In the first quarter of this year, employee exposure to airborne lead was within 10 percent of acceptable state health standards.

ACCURATE   In the first quarter of this year, employee exposure to airborne lead was 10 percent below acceptable state health standards.

Even if what you are saying is negative, do not use more negative words than necessary.

NEGATIVE   We are withholding your shipment until we receive
            your payment.

POSITIVE   We will forward your shipment as soon as we receive
            your payment.

(See also **double negatives.**)

possessive case   ESL

A **noun** or **pronoun** is in the possessive case when it represents a
person, place, or thing that possesses something. To make a singular noun possessive, add an **apostrophe** and *s*.

- The manufacturing *plant's* robotic inventory retrieval system is
  now in operation.

With plural nouns that end in *s*, show the possessive by placing an
apostrophe after the *s*.

- a managers' meeting

- the technicians' handbooks

The following list shows the relationships among singular, plural,
and possessive nouns that form their plurals by adding *s* or changing *y* to *ies*.

SINGULAR	employee	company
SINGULAR POSSESSIVE	employee's	company's
PLURAL	employees	companies
PLURAL POSSESSIVE	employees'	companies'

With plural nouns that do not end in *s*, show the possessive by
adding *'s* in both the plural and the singular forms.

SINGULAR	child	man
SINGULAR POSSESSIVE	child's	man's
PLURAL	children	men
PLURAL POSSESSIVE	children's	men's

Singular nouns that end in *s* form the possessive with either the addition of only an apostrophe or the addition of both an apostrophe
and an *s*.

- a *hostess'/hostess's* warm welcome

ESL   TIPS FOR INDICATING POSSESSION: 'S OR OF?

English expresses possession in two ways: apostrophe s ('s) and *of*.

Use *'s* with personal names, personal nouns, collective nouns, and animals. (Use just an apostrophe for plural nouns and multisyllable singular nouns that end with *s*.)

- Ms. *Corrales'* stock portfolio
- the *secretary's* lunch hour
- the *government's* pension plan
- the *dog's* tail
- the *cows'* milk

You can also use *'s* (or just an apostrophe) with some inanimate nouns: geographical and institutional names, nouns that refer to time, and nouns of special interest to human activity.

- the *company's* investors
- *today's* agenda
- a *week's* rest
- *business'* influence on politics

Use *of* with inanimate objects and to express measure.

- the title *of* the monthly report
- a cup *of* coffee
- the length *of* the memo

P

Singular nouns of one syllable that end in *s* always form the possessive by adding both an apostrophe and an *s*.

- The *boss's* desk was cluttered.

With coordinate nouns, the last noun takes the possessive form to show joint possession.

- *Michelson and Morely's* famous experiment on the velocity of light was completed in 1887.

To show individual possession with coordinate nouns, each noun should take the possessive form.

- The difference between *Thomasson's* and *Silson's* test results was statistically insignificant.

To form the possessive of a compound word, add *'s*.

- the *vice president's* car, the *pipeline's* diameter, the *editor-in-chief's* desk

When a noun ends in multiple consecutive *s* sounds, form the possessive by adding only an apostrophe.

- *Jesus'* disciples, *Moses'* journey

Do not use an apostrophe with possessive pronouns.

- yours, its, his, ours, whose, theirs

Several indefinite pronouns (*all, any, each, few, most, none,* and *some*) form the possessive case with the **preposition** *of*.

- Both dies were stored in the warehouse, and rust had ruined the surface *of each*.

Other indefinite pronouns, however, use an apostrophe.

- *Everyone's* contribution is welcome.

Only the possessive form of a pronoun should be used with a gerund.

- The safety officer insisted on *my* wearing protective clothing.

- *Our* monitoring was not affected by changing weather conditions.

P

(ESL) TIPS FOR INDICATING POSSESSION: PARTS OF THE BODY
AND PERSONAL BELONGINGS

English uses possessives to express possession of parts of the body and personal belongings.

- Ms. Winters broke *her* leg skiing.

- Mr. Sommers leaves *his* briefcase in the boardroom after every meeting.

English uses the definite article in prepositional phrases referring to the object and in passive construction referring to the subject.

- Dr. Meehan led me by *the* arm into the conference room.

- The stockroom clerk was struck on *the* leg by a box that fell from the shelf.

practicable / practical

Practicable means that something is possible or feasible. *Practical* means that something is both possible and useful.

* The program is *practical,* but considering the company's recent financial problems, is it *practicable?*

Practical, not *practicable,* is used to describe a person, implying common sense and a commitment to what works rather than to theory.

predicates (*see* sentence construction)

prefixes   ESL

A prefix is a letter or letters placed in front of a root word that changes the meaning of the root word. A prefix often causes the new word to mean the opposite of the root word. When a prefix ends with a vowel and the root word begins with the same vowel, the prefix may be separated from the root word with a **hyphen** (re-enter, co-operate, re-elect); the second vowel may be marked with a dieresis (reënter, coöperate, reëlect), although that practice is rare now; or the word may be written with neither hyphen nor dieresis (reenter, cooperate, reelect). The hyphen and the dieresis are visual aids that help the reader recognize that the two vowels are pronounced separately. (See also **diacritical marks.**)

Except between identical vowels, a hyphen rarely appears between a prefix and its root word. At times, however, a hyphen is necessary for clarity of meaning; for example, *reform* means "correct" or "improve," and *re-form* means "change the shape of."

preparation

The preparation stage of the writing process is essential. By determining your **readers'** needs, your **purpose,** and your **scope** of coverage, you understand the information you will need to gather during **research,** the next step. During preparation, you also need to consider the appropriate medium.

Writer's Checklist: Preparation

The following tips will help you prepare for any writing project.

- ☑ Determine who your readers are and learn certain key facts about them, such as their knowledge of your subject, their attitudes about the subject, and their needs relative to your subject.
- ☑ Determine your purpose: What exactly do you want your readers to know, believe, or be able to do when they have finished reading your document?
- ☑ Establish the scope of your document — the type and amount of detail you must include — by considering any external constraints (such as word limits for **trade journal articles** or the space limitations of **Web page design**) as well as by understanding your readers' needs and your purpose.
- ☑ Consider the appropriate medium for your message. (See also **selecting the medium.**)

---

prepositional phrases (*see* **phrases**)

prepositions   ESL

A preposition is a word that links a **noun** or **pronoun** (its **object**) to another sentence element by expressing such relationships as direction (*to, into, across, toward*), location (*at, in, on, under, over, beside, among, by, between, through*), time (*before, after, during, until, since*), or figurative location (*for, against, with*). Together, the preposition, its object, and the object's **modifiers** form a prepositional phrase that acts as a modifier.

Preposition Functions

The object of a preposition (the word or phrase following it) is always in the objective **case.** When the object is a compound noun, both nouns should be in the objective case. For example, the phrase "between you and *me*" is frequently and incorrectly written as "between you and *I*." *Me* is the objective form of the pronoun, and *I* is the subjective form.

Many words that function as prepositions also function as **adverbs.** If the word takes an object and functions as a connective, it is a preposition; if it has no object and functions as a modifier, it is an adverb.

PREPOSITIONS   The manager sat *behind* the desk *in* her office.

ADVERBS   The customer lagged *behind;* then he came *in* and sat down.

Certain verbs, adverbs, and adjectives are used with certain prepositions, such as "interested *in*," "aware *of*," "equated *with*," "adhere *to*," "capable *of*," "object *to*," and "infer *from*." A more detailed list of such usages appears in the entry **idioms.**

Prepositions at the End of a Sentence

A preposition at the end of a sentence can be an indication that the sentence is awkwardly constructed.

- ~~The~~ branch office ~~is where she was at~~ . *She was at the*

However, if a preposition falls naturally at the end of a sentence, leave it there.

- I don't remember which file name I saved it *under.*

Prepositions in Titles

When a preposition appears in a **title,** it is capitalized only if it is the first word in the title or if it has four letters or more.

- The article "Concerns About Distance Education" was reviewed recently in the newspaper column "In My Opinion."

(See also **capital letters.**)

Preposition Errors

Do not use redundant prepositions, such as "off *of*," "in back *of*," "inside *of*," and "at *about*." (See also **conciseness/wordiness.**)

ESL TIPS FOR USING PHRASAL VERBS (VERB + PREPOSITION)

Sometimes the words you learned as prepositions do not function as prepositions in a sentence. English creates many new verbs by combining verbs and prepositions. When a verb and a preposition are joined in that way, the preposition is called an *adverbial particle* or simply a *particle,* and it does not function as a preposition. The new verb (the verb + the preposition) is called a *phrasal verb,* a *two-word verb,* or sometimes a *prepositional verb.* The meaning of a phrasal verb is often not predictable based on the meanings of the words that make up the new verb. (For a list of common phrasal verbs, see **idioms**.)

In the following sentence, *around* functions as a preposition in the prepositional phrase *around the room.* The verb is *walk.*

- The chief engineer walked *around the room.*

However, in the next sentence, *after* does not function as a preposition; it functions as a participle in the phrasal verb *look after.* The meaning of *look after* ("supervise") is not transparent when you combine the meaning of *look* and the meaning of *after.*

- The chief engineer *looked after* his work crew.

Some phrasal verbs have *separate* particles. In the following examples, *up* can precede or follow the noun *word.*

- I looked *up* the word.

- I looked the word *up.*

However, in sentences where an object pronoun and a separate particle co-occur, the particle must follow the pronoun.

- I called up ~~him~~ *him* .
        ^

Some phrasal verbs have inseparable particles.

- I ran Jim ~~into~~ *into* at the water fountain.
        ^

Phrasal verbs can be confusing for nonnative speakers of English; unfortunately, memorization is the only way to master them. Two excellent resources are the *Longman Dictionary of Phrasal Verbs* and Dave's ESL café at (www.eslcafe.com/pv/).

P

EXACT	The client arrived at ~~about~~ four o'clock.
APPROXIMATE	The client arrived ~~at~~ about four o'clock.

Avoid unnecessarily adding the preposition *up* to verbs.

- Call ~~up~~ _{to} see if he is in his office.

Do not omit necessary prepositions.

- He was oblivious _{to} and not distracted by the view from his office window.

presentations

DIRECTORY
Overview   466
Determining Your Purpose   467
Analyzing Your Audience   467
Gathering Information   468
Structuring the Presentation   468
Using Visual Aids   471
Delivering a Presentation   475
Reaching Global Audiences   479

P

Overview

The steps required to prepare an effective presentation parallel the steps you follow to write a document. As with writing a document, you must determine your **purpose** and analyze your audience. You must find and gather the facts that will support your point of view and proposal and logically organize that information. Presentations do, however, differ from written documents in a number of important ways. They are intended for listeners, not readers. Because you are speaking rather than writing a memo or a report, your manner of delivery, the way you organize the material, and your supporting visual aids require as much attention as your content. (See also **readers.**)

Determining Your Purpose

Every presentation is given for a purpose, even if it is only to share information. To determine the purpose of your presentation, use the following questions as a guide.

- What do I want the audience to *know* when I have finished the presentation?
- What do I want the audience to *believe* when I have finished the presentation?
- What do I want the audience to *do* when I have finished the presentation?

Based on the answers to those questions, write a purpose statement that answers the questions *what* and *why*.

- The purpose of my presentation is to explain to my classmates the various tasks I performed last semester as a part-time volunteer at the Maplewood Adult Day-Care Center [*what*] so that other members of the class will want to become volunteers at Maplewood [*why*].
- The purpose of my presentation is to convince my company's chief information officer of the need to improve the appearance, content, and customer use of our company's Web site [*what*] so she will be persuaded to allocate additional funds for site development work in the next fiscal year [*why*].

Analyzing Your Audience

Once you have determined the desired end result of the presentation, you need to analyze your audience so you can tailor your presentation to their needs. Ask yourself these questions about your audience:

- What is your audience's level of experience or knowledge about your topic?
- What is the general educational level and age of your audience?
- What is the audience's attitude toward the topic you are speaking about, and — based on that attitude — what concerns, fears, or objections might the audience have?
- Are there subgroups in the audience that might have different concerns or needs?
- What questions might the audience ask about this topic?

Gathering Information

Now that you have focused the presentation, you need to find the facts that support your point of view or the action you propose. As you gather information, keep in mind that you should give the audience only the facts necessary to accomplish your goals; too much information will overwhelm them, and too little information will leave them with either a sketchy understanding of your topic or the feeling that you have not provided enough information to support the course of action you want them to take. (For detailed guidance about gathering information, see **research, library research,** and **Internet research.**)

Structuring the Presentation

When structuring the presentation, keep the focus on your audience. Listeners are freshest at the outset and refocus their attention as you wrap up your remarks. Take advantage of that pattern. Give them a brief overview of your presentation at the beginning, use the body to develop your ideas, and end with a summary of what you covered and, if appropriate, a call to action.

The Introduction. The **introduction** may include an **opening,** something designed to catch and focus the audience's attention. The following opening defines a problem.

- You have to write an important report, but you'd like to incorporate lengthy sections of an old report into your new one. The problem is that you don't have an electronic version of the old report. You will have to rekey many pages. You groan because that seems an incredible waste of time. Have I got a solution for you!

You could also use any of the following types of openings:

An attention-getting statement
- "As many as 50 million Americans have high blood pressure."

A rhetorical question
- "Would you be interested in a full-sized computer keyboard that is waterproof, is noiseless, and can be rolled up like a rubber mat?"

A personal experience
- "As I sat at my computer one morning last month deleting my eighth spam message of the day, I decided that it was time to find a solution to eliminate this time-waster."

An appropriate quotation
- According to researchers at the Massachusetts Institute of Technology, "Garlic and its cousin, the onion, confer major health benefits — including fighting cancer, infections, and heart disease."

(For additional examples, see **openings.**)

Following your opening, use the introduction to set the stage for the audience by giving them an overview of the presentation. The overview can include general or background information that the audience will need to understand the more detailed information in the body of your presentation. It can also be an overview of how you have organized the material.

- This presentation analyzes three different scanner models for us to consider purchasing. Based on a comparison of all three, I will recommend the one I believe best meets our needs. To do so, I'll discuss the following five points:
  1. Why we need a scanner [the problem]
  2. The basics of scanner technology [general information]
  3. The criteria I used to compare the three model scanners [comparison method of development]
  4. The scanner models I compared and why [possible solutions]
  5. The scanner I propose we buy [proposed solution]

The Body.   The body is where you present the evidence that will persuade the audience to agree with your conclusions and act on them. If there is a problem, demonstrate that it exists and offer a solution or range of possible solutions. For example, if your introduction stated that the problem is low profits, high costs, outdated technology, or high employee absenteeism, you could use the following approach.

1. Prove your point.
   - Gather the facts and data you need.
   - Present the information using easy-to-understand visual aids.
   - Determine whether you want the listeners to agree, change their minds, or do something.
   - Call for action.
2. Offer solutions.
   - "Increase profits by lowering production costs."
   - "Cut overhead to reduce costs or abolish specific programs or product lines."
   - "Replace outdated technology or upgrade existing technology."

- "Offer employees more flexibility in their work schedules or other incentives."

3. Anticipate questions ("How much will it cost?") and objections ("We're too busy now—when would we have time to learn the new software?") and incorporate the answers into your presentation.

The Closing.   The closing should achieve the goals of your presentation. If your purpose is to motivate the listeners to take action, ask them to do what you want them to do; if your purpose is to get your audience to think about something, summarize what you want them to think about. Many presenters make the mistake of not actually closing—they simply quit talking, shuffle papers, and then walk away.

Because your closing is what your audience is most likely to remember, it is the time to be strong and persuasive. Consider the following possible closing.

- Based on all the data, I believe that the Worthington scanner best suits our needs. It produces 3,000 units a month more than its closest competitor and creates electronic files we can use on our Web site and on paper. The Worthington is also compatible with our current computer network and includes staff training at our site. Although the initial cost is higher than the other two models, the additional capabilities, longer life-cycle for replacement parts, and lower maintenance costs make it a better value.

  Let's allocate the funds necessary for this scanner by the 15th of this month. Then we could be up and running before the first of next month and be well prepared for next quarter's customer presentations.

That closing brings the presentation full cycle and asks the audience to fulfill the purpose of the presentation—exactly what a closing should do.

Transitions.   Planned **transitions** should appear between the introduction and the body, between points in the body, and between the body and the closing. Transitions are simply a sentence or two to let the audience know that you are moving from one topic to the next. They also prevent a choppy presentation and provide the audience with assurance that you, the speaker, know where you are going and how to get there.

- Before getting into the specifics of each scanner I compared, I'd like to demonstrate how scanners work in general. That information will provide you with the background you'll need to compare the differences among the scanners and their capabilities discussed in this presentation.

It is also a good idea to pause for a moment after you have delivered a transitional line between topics to let your listeners shift gears with you. Remember, they do not know your plan.

Using Visual Aids

Well-planned visual aids not only add interest and emphasis to your presentation, they also clarify and simplify your message because they communicate clearly, quickly, and vividly. Charts, graphs, and illustrations greatly increase audience understanding and retention of information, especially for complex issues and technical information that could otherwise be misunderstood or overlooked. A bar graph, pie chart, diagram, or concise summary of key points can eliminate misunderstanding and save many words.

You can create and present visual aids in a variety of media, including flip charts, a whiteboard or chalkboard, overhead transparencies, 35mm slides, and computer presentation software.

Flip Charts.   Flip charts are large sheets of white paper bound like a tablet and fastened to the top of an easel. The presenter writes on the sheets with colored felt-tip pens, usually during the presentation. Flip charts are ideal for smaller groups in a conference room or classroom. To avoid the distraction of writing as you speak, you can prepare text and sketches ahead of time on a series of sheets and flip through them during your presentation. Flip charts are also an ideal medium for **brainstorming** with your audience. You can fill sheet after sheet with ideas, tape them around the walls for everyone to see, and use clean sheets to organize the ideas into an outline for follow-up work.

Whiteboard or Chalkboard.   The whiteboard or chalkboard common to classrooms is convenient for creating impromptu sketches and for jotting notes during your presentation. If your presentation requires extensive notes or complex drawings, create them before the presentation to minimize audience restlessness.

Before the presentation, make sure you have an ample supply of marking pens or chalk.

Overhead Transparencies.   Transparencies are page-size sheets of clear plastic on which you copy text and graphics using a copier or computer printer. During the presentation, you place the transparencies on an overhead projector and the images are projected onto a screen or a blank wall. An example of a bar graph on a transparency is shown in Figure P–3. The images can be anything from black text on clear film to multicolored computer images.

With overheads, you can create a series of overlays to explain a complex device or system, adding (or removing) the overlays one at a time. You can also lay a sheet of paper over a list of items on a transparency, uncovering one item at a time as you discuss it, to focus audience attention on each point in the sequence.

Slides.   Slides are 2-inch-by-2-inch 35mm film color transparencies that are inserted into a slide projector and projected onto a screen or a blank wall (Figure P–4 on page 473). Slides are especially useful if you need to include photographs in your presentation. Like overhead transparencies, they are best viewed in a darkened room. Slides also refer to the individual screens produced with presentation software.

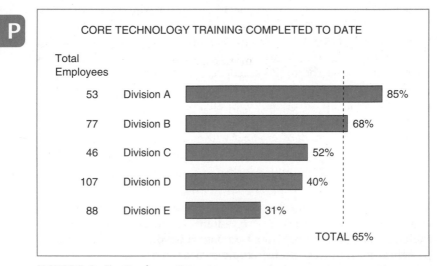

FIGURE P–3. Bar Graph on a Transparency

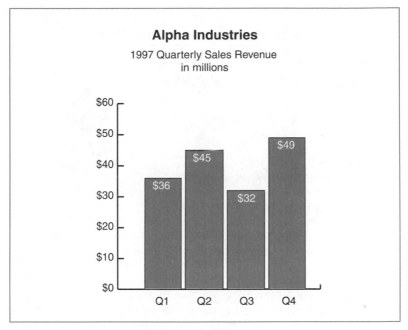

FIGURE P–4. Sample 35mm Slide

Presentation Software. Visual information in workplace presentations is frequently displayed on a computer monitor for a small audience or, for a larger audience, by a projector hooked to a laptop computer that displays the computer screen. The software used to create computer presentations, such as PowerPoint, Corel Presentations, and Freelance Graphics, lets you write most of a presentation with your word-processing program and develop charts and graphs with data from spreadsheet software. Those files then can be imported into one of the presentation packages and formatted with standard templates and other features that help you design effective visual aids. Enhancements include a selection of typefaces, background textures and colors, and clip art images. Beware, however, of using too many enhancements because they could distract viewers from your message. Images can also be printed out for use as overhead transparencies (or handouts), such as the bulleted list and clip art insert in Figure P–5 on page 474.

Be sure to integrate any computer visual aids with your presentation when you rehearse. Practice loading the presentation and

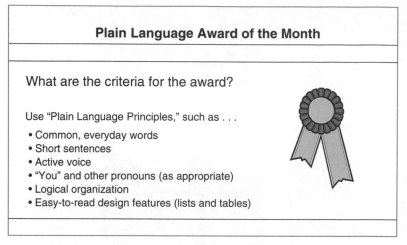

FIGURE P–5. Overhead Transparency with Bulleted List and Clip Art Insert

anticipate any technical difficulties that might arise. Should you encounter a technical snag during the presentation, stay calm and give yourself time to solve the problem. If you cannot, move on without the technology. As a backup, carry a printout of your electronic presentation in case there is a problem with the computer or the projection system. Carry an extra copy of your presentation on diskette for backup.

Writer's Checklist: Visual Aids

Keep in mind these tips when you are using visual aids.

☑ Use text sparingly. Instead of blocks of text, use bulleted or numbered lists, keeping them parallel in content and grammatical form. Use numbers if the sequence is important and bullets if it is not. (See **parallel structure.**)

☑ Limit the number of bulleted or numbered items to 5 or 6 per visual. Each visual should contain no more than 40 to 45 words. (See Figure P–5.) Any more will clutter the visual and force you to use a smaller type size that could impair the audience's ability to read it.

☑ Make your visuals consistent in type style, size, and spacing.

☑ Use a type size visible to members of the audience in the back of

Writer's Checklist: Visual Aids (continued)

the room. Type should be boldface and no smaller than 30 points. For headings, 45- or 50-point type works even better.

☑ Use graphs and charts rather than tables to show data trends. Use only one or two illustrations per visual. Otherwise, your presentation of the data will be cluttered and confusing.

☑ Make the contrast between your text and the background sharp. Use light backgrounds with dark lettering and avoid textured or fancy "wallpaper" backgrounds.

☑ Aim for about 20 visuals per presentation. Any more will tax the audience's concentration.

☑ Match the text of your presentation to the content of your visuals. Do not put one set of words or images on the screen and talk about the previous visual or, even worse, the next one.

☑ Do not read the text on your visual word for word. The members of the audience can read the visuals; they look to you to cover the salient points in detail.

Delivering a Presentation

Once you have outlined and drafted your presentation and prepared your visual aids, you are ready to think about practicing your presentation and delivery techniques.

Practice.   Begin by familiarizing yourself with the sequence of the material—major topics, notes, and visual aids—in your outline. Once you feel comfortable with the content, you are ready to practice the presentation itself.

PRACTICE ON YOUR FEET AND OUT LOUD.   Try to practice in the room where you will give the presentation. Practicing on-site helps you learn the idiosyncrasies of the room: the acoustics, the lighting, how the chairs will most likely be arranged, where the electrical outlets and switches are located, and so forth. Practicing out loud is more effective than just rehearsing mentally because you process information in your mind many times faster than you can possibly speak it. Rehearsing out loud makes clear exactly how long your presentation will take and highlights problems such as awkward transitions. You can also rehearse eliminating or reducing verbal tics, such as "um," "you know," and "like."

PRACTICE WITH YOUR VISUAL AIDS. Integrating your visual aids into your practice sessions will help your presentation go more smoothly. Operate the equipment (computer, slide or overhead projector) until you are comfortable with it. Even if things go wrong, being prepared and practiced will give you the confidence and poise to go on.

VIDEOTAPE YOUR PRACTICE SESSION. Videotape is an effective— and sometimes painful—way to catch what you are doing wrong. The tape will reveal from the audience's perspective how you present your material. If you do not have access to a video recorder, use an audiotape recorder to evaluate your vocal presentation. Another effective technique is to ask a friend or colleague to watch you and comment on your delivery.

Delivery Techniques That Work. Your delivery is both audible and visual. In addition to your words and message, your nonverbal communication affects your audience. To make an impression on your listeners and keep their attention, you must be animated. Your words will have more staying power when they are delivered with physical and vocal animation. If you want listeners to share your point of view, show enthusiasm for your topic. The most common delivery techniques include making eye contact, using movement and gestures, and varying voice inflection, pace, and projection.

EYE CONTACT. The best way to establish rapport with your audience is by eye contact. For smaller audiences, make eye contact with as many people as possible. In a large audience, directly address those people who seem most responsive to you in different parts of the audience. Address each person separately and focus your attention on him or her for several seconds before moving on. Doing that helps you establish rapport with your listeners by holding their attention. It also gives you important visual cues that let you know how your message is being received. Are people engaged and actively listening? Or are they looking around or staring at the floor? Such cues tell you that you may need to speed up or slow down the pace of your presentation.

MOVEMENT. Animate the presentation with physical movement. Simply take a step or two to one side after you have been talking for a minute or so. That type of movement is most effective at transitional points in your presentation between major topics or after pauses or emphases. Too much movement, however, can be distracting, so try not to pace.

Another way to integrate movement into your presentation is to walk to the screen and point to the visual aid as you discuss it. Touch the screen with the pointer and then turn back to the audience before beginning to speak (remember the three *t*'s: touch, turn, and talk). If you are using an overhead projector, place the pointer directly on the overhead so it casts a shadow that points to the appropriate item on the overhead.

GESTURES.   Gestures both animate your presentation and help communicate your message. Most people gesture naturally when they talk; nervousness, however, can inhibit gesturing during a presentation. Keep one hand free and above your waist and use that hand to gesture. It is OK to bring your hands together for effect, but don't lock them together; locking your hands into a rigid position looks unnatural and inhibits gesturing. Generally, the more relaxed you are, the more natural your gestures will be.

VOICE.   Your voice can be an effective tool in communicating your sincerity, enthusiasm, and command of your topic. Use it to your advantage to project your credibility. Vocal inflection is the rise and fall of your voice at different times, such as the way your voice naturally rises at the end of a question ("You want it *when?*"). Keep your audience's attention by using that inflection, just as you would in a conversation. A conversational delivery and eye contact also promote the feeling among members of the audience that you are addressing them directly. Do not fall into a monotone speech pattern that can hypnotize your listeners and make them drowsy. Vocal variety also allows you to highlight differences between key and subordinate points in your presentation.

PACE.   Pace is the speed at which you deliver your presentation. If you speak too fast, your words will run together, making it difficult for your audience to follow the presentation. If you speak too slowly, your listeners will become impatient and distracted. (See also **pace.**)

PROJECTION.   Most speakers think they are projecting more loudly than they are. Remember that your presentation is ineffective for anyone in the audience who cannot hear you. If listeners must strain to hear you, they may give up trying to listen. Correct projection problems by practicing out loud with someone listening from the back of the room.

Presentation Anxiety.   Everyone experiences nervousness before a presentation. Survey after survey reveals that for most people dread of public speaking ranks among their top five fears. Typical reactions to stress include shortness of breath, a racing heartbeat, trembling, perspiration, even nausea. Some people react by clearing their throats repeatedly, tugging at their clothing or earlobes, or moving continuously during the presentation. Instead of letting fear inhibit you, focus on channeling your nervous energy into a helpful stimulant. In other words, if you cannot eliminate your stress entirely, manage it. The best way to master anxiety is to know your topic thoroughly. Knowing what you are going to say and how you are going to say it will help you gain confidence and reduce anxiety as you become immersed in your subject.

Rehearsing your presentation will help. Do so alone and again, if possible, in front of one or more listeners. You may find it helpful to write out the presentation in full, put it aside, and rehearse using only brief notes. If you falter, refer to the written-out version. After a practice session, visualize yourself in front of your audience delivering your material point by point. Begin by saying to yourself, "My subject is important. I am ready. My listeners are here to listen to what I have to say." If you cannot remember every point you want to make during the presentation, review your notes or visual aids. They will trigger your memory both as you imagine the presentation and when you are actually giving it.

You can use a couple of techniques to reduce your fears before the presentation. Fill your lungs with a deep breath and hold it for a count of ten. (When your lungs are filled entirely, you should feel the lower part of your rib cage move outward and upward. Your breathing will be too shallow and ineffective if only the upper part of your chest moves as you inhale.) Then exhale. Repeat this process several times or until you feel your body begin to relax. Tensing and relaxing muscles is another effective stress reducer. Clench both fists tightly and count to ten while inhaling and then relax them as you exhale. Repeat this process several times until you feel the stress begin to diminish.

Writer's Checklist: Delivering a Presentation

When you need to make a presentation, the following guidelines will help you overcome nervousness and deliver an effective presentation.

☑ Practice your presentation thoroughly and, if possible, with visual aids and in front of listeners.

☑ Prepare a set of notes that will trigger your memory during the presentation.

☑ Visit the location of the presentation ahead of time to familiarize yourself with the surroundings.

☑ During the presentation, make as much eye contact as possible with your audience to establish rapport and maximize opportunities for audience feedback.

☑ Animate your delivery by integrating movement, gestures, and vocal inflection into your presentation. However, keep your movements and speech patterns natural.

☑ Do not read the text on your visuals word for word. Listeners can read — they look to you to explain the salient points in detail.

☑ Speak loudly and slowly enough to be understood.

Reaching Global Audiences

The prevalence of multinational corporations, the international subsidiaries of many companies, multinational trade agreements, the increasing diversity of the U.S. workforce, and even increases in immigration mean that the ability to reach audiences with varied cultural backgrounds will be essential in the years to come. The cross-cultural audiences for your presentations may include clients, business partners, colleagues, and current and potential employees and customers of varied backgrounds. For information and tips on communicating with cross-cultural audiences, see **global communication** and **international correspondence.**

press releases

Companies and organizations write press releases, or news releases, to announce new products and services, new policies, special events (such as branch openings, company anniversaries, mergers, and grand openings), management changes, and sponsorship of social-action and cultural programs. The purpose of a press release is both to inform the public about the company and its products and services and to promote a favorable image. Even the announcement of an unfavorable event, such as the closing of a division, should try to present the company in a good light. Large corporations and institutions usually have their own public relations staffs or use outside agencies. However, if you work for a small company without public relations resources, you may be called on to write a press release.

The press release should be clear, concise, and written with particular attention to the five *w*'s: *who, what, where, when,* and *why.* Write the release in **decreasing-order-of-importance method of development.** Put all critical information in the first paragraph, information of the next level of importance in the second paragraph, and so on. Editors must make your release fit the space they have available; if they must cut your release, they will delete the last paragraph first, and then the next to last, and so on. Make sure your facts are accurate and be careful to define any unfamiliar terms.

Releases are usually sent to local newspapers, television and radio stations, and other special groups, such as trade publications or professional associations. Send the release to a specific person whenever possible. Otherwise, try to address the news release to a particular editor, such as to the business editor or the education editor. (See also **newsletter articles.**)

Begin with the place and date of the announcement, as shown in the typical press release in Figure P–6 on page 481.

Writer's Checklist: Press Releases

Keep the following points in mind when you are writing a press release.

☑ Print the release on company letterhead and double- or triple-space for easy reading and editing.

☑ Leave the top third of the page blank for editors to mark any necessary headings on the copy.

☑ Give the name and phone number of someone in the company who can give further information.

☑ If the release is longer than one page, type *More* at the bottom of each page except the last.

☑ Type *-30-* or *End* at the bottom of the last page.

Bentley Plastics
3535 Michigan Avenue
Chicago, IL 60653
Email: bentleypr@bentley.org
Web: http://www.bentleyplas.org

FOR IMMEDIATE RELEASE

For More Information Call:
Marjorie Kohls
(312) 712-1946

CHICAGO, ILLINOIS, MAY 20, 20-- . . . Mark Williams has joined Bentley Plastics as vice president–marketing. He will direct the Illinois and Indiana district sales offices and coordinate overseas distribution through the company's Singapore office. Williams will travel extensively to the Far East while developing marketing channels for Bentley.

Formerly, Williams was director of services with International Marketing Associates, a consulting group in New York. While at IMA, he developed a computer-based marketing center that linked textile firms in the United States, Finland, and Great Britain. A graduate of the Columbia University Graduate School of Business, Williams is the author of *Marketing Dynamics,* an informal examination of psychological appeals to the buying public. The book has been used in marketing classrooms at several universities.

Bentley Plastics, headquartered in Chicago, produces polymer tubing and coils. The company's main plants are in Skokie, Illinois, and Gary, Indiana. Since 1978 Bentley has also been producing tubing for French distribution through the firm of Jourdan and Sons, Paris.

-30-

FIGURE P–6. Press Release

principal/principle

Principal, meaning an "amount of money on which interest is earned or paid" or a "chief official in a school or court proceeding," is sometimes confused with *principle,* which means "basic truth or belief."

- The bank will pay 6.5 percent on the *principal.*
- He sent a letter to the *principal* of the high school.
- She is a person of unwavering *principles.*

Principal is also an adjective, meaning "main" or "primary."

- My *principal* objection is that it will be too expensive.

process explanation

Many kinds of technical writing explain a process, an operation, or a procedure. The explanation involves describing in the appropriate order the steps that a specific mechanism or system uses to accomplish a certain result. For example, the process might be the steps necessary to design a product. In your **opening,** you might present a brief overview of the process or describe how the process works in relation to a larger whole.

In describing a process, transitional words and phrases serve to achieve **unity** within paragraphs, and **headings** mark the **transition** from one step to the next. **Illustrations,** like the one used in Figure P–7 on page 483, can help convey your message. (See also **instructions.**)

progress and activity reports

Progress reports and activity reports both document ongoing activities. They differ primarily in that progress reports are often used with major projects, whereas activity reports focus on the work and achievements of individual employees.

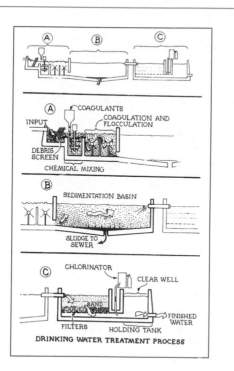

DRINKING WATER TREATMENT PROCESS

The Treatment Process

After it has been transported from its source to a local water system, most surface water must be processed in a treatment plant before it can be used. Some ground-water, on the other hand, is considered chemically and biologically pure enough to pass directly from a well into the distribution system that carries it to the home.

Although there are innumerable variations, surface water is usually treated as follows: First, it enters a storage lagoon where a chemical, usually copper sulfate, is added to control algae growth. From there, water passes through one or more screens that remove large debris. Next, a coagulant, such as alum, is mixed into the water to encourage the settling of suspended particles. The water flows slowly through one or more sedimentation basins so that larger particles settle to the bottom and can be removed. Water then passes through a filtration basin partially filled with sand and gravel where yet more suspended particles are removed.

At that point in the process, the Safe Drinking Water Act has mandated an addi-tional step for communities using surface water.... Water is to be filtered through activated carbon to remove any microscopic organic material and chemicals that have escaped the other processes. Activated carbon is extremely porous — one pound of the material can have a surface area of one acre. This honeycomb of minute pores attracts and traps pollutants through a process called adsorption....

FIGURE P-7. Illustration for a Process Description

Progress Reports

A progress report provides information about a project—its status, whether it is on schedule and within budget, and so on. Progress reports are often submitted by a contracting company to a client company. They are issued at regular intervals throughout the life of a project to state what has been done in the specified interval and what remains to be done before the project is completed. The progress report is used primarily with projects that involve many steps over a period of weeks, months, or even years. Progress reports help keep projects running smoothly by helping managers assign work, adjust schedules, allocate budgets, and schedule supplies and equipment.

All progress reports in a series or for a particular project should have the same **format.** Because progress reports are normally sent outside the company, they are often written in letter format. (See also **correspondence.**)

The **introduction** to the first progress report should identify the project, any methods and necessary materials or other expenditures, and the date by which the project is expected to be completed. Subsequent reports summarize the progress achieved since the preceding report as well as list the steps that remain to be taken.

The body of the progress report should describe the project's status, including details such as schedules and costs. The body also includes a statement of the work completed and perhaps an estimate of future progress.

The report ends with **conclusions** and recommendations about changes in the schedule, materials, techniques, and so on.

Figure P–8 on page 485 shows the initial progress report submitted by a construction company (Hobard Construction Company) to a client (the county administrator).

Activity Reports

Within an organization, employees often submit reports on the progress and status of ongoing projects. Managers may combine the activity reports (also called *status reports*) of several individuals or teams into larger activity reports and, in turn, submit those larger reports to their own managers. Activity reports help keep team members and managers aware of the progress and status of projects in an organization.

HOBARD CONSTRUCTION COMPANY
9032 Salem Avenue
Lubbock, TX 79409
(808) 769-0823

August 17, 20--

Walter M. Wazuski
County Administrator
109 Grand Avenue
Manchester, NH 03103

Dear Mr. Wazuski:

The renovation of the County Courthouse is moving forward on schedule and within budget. Although the cost of certain materials is higher than our original bid indicated, we expect to complete the project without exceeding the estimated costs because the speed with which the project is progressing will reduce overall labor expenses.

Costs
Materials used to date have cost $78,600, and labor costs have been $193,000 (including some subcontracted plumbing). Our estimate for the remainder of the materials is $59,000; remaining labor costs should not be in excess of $64,000.

Work Completed
As of August 15, we had finished the installation of the circuit-breaker panels and meters, of level-one service outlets, and of all subfloor wiring. The upgrading of the courtroom, the upgrading of the records-storage room, and the replacement of the air-conditioning units are in the preliminary stages.

Work Schedule
We have scheduled the upgrading of the courtroom to take place from August 25 to October 5, the upgrading of the records-storage room from October 6 to November 12, and the replacement of the air-conditioning units from November 15 to December 17. We see no difficulty in having the job finished by the scheduled date of December 23.

Sincerely yours,

Tran Nuguélen

TN2@hobard.com

FIGURE P–8. Initial Progress Report

INTEROFFICE MEMO

Date: June 5, 20--
To: Kathryn Hunter, Director of Engineering   *w7*
From: Wayne Tribinski, Manager, Applications Programs
Subject: Activity Report for May 20--

We are dealing with the following projects and problems, as of May 31.

Projects

Status of current projects

1. The *Problem-Tracking System* now contains both software and hardware problem-tracking capabilities. The system upgrade took place over the weekend of May 11 and 12 and was placed online on the 13th.
2. For the *Software Training Mailing Campaign*, we anticipate producing a set of labels for mailing software training information to customers by June 10.
3. The *Search Project* is on hold until the PL/I training has been completed, probably by the end of June.
4. The project to provide a database for the *Information Management System* has been expanded in scope to provide a database for all training activities. We are rescheduling the project to take the new scope into account.

Problems

Existing problems

The *Information Management System* has been delayed. The original schedule was based on the assumption that a systems analyst who was familiar with the system would work on this project. Instead, the project was assigned to a newly hired systems analyst who was inexperienced and required much more learning time than expected.

Bill Michaels, whose activity report is attached, is correcting a problem in the *CNG Software*. This correction may take a week.

Plans for Next Month

Plans for the coming months

- Complete the *Software Training Mailing Campaign.*
- Resume the *Search Project.*
- Restart the project to provide a database on information management with a schedule that reflects its new scope.
- Write a report to justify the addition of two software engineers to my department.
- Congratulate publicly the recipients of Meritorious Achievement Awards: Bill Thomasson and Nancy O'Rourke.

Current Staffing Level

Current staff: 11
Open requisitions: 0

FIGURE P–9. Activity Report

An activity report includes information about the status of all current projects, including any problems, actions taken to resolve those problems, and the author's plans for the next period.

Because the activity report is issued periodically (usually monthly) and contains material familiar to its readers, it normally needs no introduction or conclusion, although it may need an **opening** to provide context. The format for activity reports may vary from company to company or even among different parts of the same company. The following sections are typical and would be appropriate for most situations.

- *Current projects.* This section lists all the projects assigned to the report writer and summarizes their status.
- *Current problems.* This section details any problems in a project and explains the steps being taken to resolve them.
- *Plans for the next period.* This section projects what the writer expects to achieve on each project during the next reporting period.
- *Current staffing level.* This section, included by managers and project leaders, lists the number of employees assigned to the writer and correlates that number with the staffing level considered necessary for the projects assigned.

The activity report shown in Figure P–9 on page 486 was submitted by a manager of application programs (Wayne Tribinski) who supervises 11 employees. The reader of the report (Kathryn Hunter) is Tribinski's manager and the director of engineering.

pronoun reference   ESL

A **pronoun** should refer clearly to a specific antecedent. Avoid vague and uncertain references.

- We got the account after we wrote the proposal. ~~It was a big one.~~ *, which was a big one,*

For the sake of **coherence,** place pronouns as close as possible to their antecedents—the likelihood of an ambiguous reference increases with distance. (See also **ambiguity.**)

- The office building next to City Hall ~~was praised for its architectural design~~ .

  , praised for its architectural design, is

Avoid general and hidden references. A general (or broad) reference or one that has no real antecedent is a problem that often occurs when the word *this* is used by itself.

- He deals with personnel problems in his work. This helps him in his personal life.

  experience

A hidden reference is one that has only an implied antecedent.

- A high-lipid, low-carbohydrate diet is "ketogenic" because it favors ~~their~~ formation .

  the   *of ketone bodies*

Do not repeat an antecedent in parentheses following the pronoun. If you feel you must identify the pronoun's antecedent in that way, you need to rewrite the sentence.

> **AWKWARD**   The senior partner first met Bob Evans when he (Evans) was a trainee.
>
> **IMPROVED**   Bob Evans was a trainee when the senior partner first met him.

The use of the masculine personal pronoun to refer to both sexes implies sexual bias. Avoid the problem by substituting an article or changing the statement from singular to plural.

- Each employee is to have ~~his~~ annual medical exam by Friday.

  the

- ~~Each employee is~~ to have ~~his~~ annual medical ~~exam~~ by Friday.

  All employees are   *their*   *exams*

(See also **biased language**.)

---

pronouns   (ESL)

DIRECTORY
Overview   489
Case   490
Gender   492

Number 492
Person 493
Tips for Understanding Agreement: Possessive Pronouns
and Antecedents (ESL) 493

Overview

A pronoun is a word that is used as a substitute for a **noun** (the noun for which a pronoun substitutes is called the *antecedent*). Using pronouns in place of nouns relieves the monotony of repeating the same noun over and over. (See also **pronoun reference.**)

Personal pronouns refer to the person or people speaking (*I, me, my, mine; we, us, our, ours*), the person or people spoken to (*you, your, yours*), or the person, people, or thing(s) spoken of (*he, him, his; she, her, hers; it, its; they, them, their, theirs*). (See also **person** and **point of view.**)

- *I* wish *you* had told *me* that *she* was coming with *us*.

- If *their* figures are correct, *ours* must be in error.

Demonstrative pronouns (*this, these, that, those*) indicate or point out the thing being referred to.

- *This* is my desk.

- *These* are my coworkers.

- *That* will be a difficult job.

- *Those* are incorrect figures.

Relative pronouns (*who, whom, which, that*) perform a dual function: (1) They take the place of nouns, and (2) they connect and establish the relationship between a dependent **clause** and its main clause.

- The department manager decided *who* would be hired.

Interrogative pronouns (*who, whom, what, which*) are used only to ask questions.

- *What* is the trouble?

Indefinite pronouns specify a class or group of persons or things rather than a particular person or thing. *All, another, any, anyone, anything, both, each, either, everybody, few, many, most, much, neither, nobody, none, several, some,* and *such* are indefinite pronouns.

- Not *everyone* liked the new procedures; *some* even refused to follow them.

A *reflexive pronoun,* which always ends with the suffix -*self* or -*selves,* indicates that the subject of the sentence acts upon itself. (See also **sentence construction.**)

- The electrician accidentally shocked *herself.*

The reflexive pronouns are *myself, yourself, himself, herself, itself, oneself, ourselves, yourselves,* and *themselves. Myself* is not a substitute for *I* or *me* as a personal pronoun.

- Victor and ~~myself~~ *I* completed the report on time.

- The assignment was given to Ingrid and ~~myself~~ *me*.

Intensive pronouns are identical in form to the reflexive pronouns, but they perform a different function: to emphasize their antecedents.

- I *myself* asked the same question.

Reciprocal pronouns (*one another* and *each other*) indicate the relationship of one item to another. *Each other* is commonly used when referring to two persons or things and *one another* when referring to more than two.

- Salih and Kara work well with *each other.*
- The crew members work well with *one another.*

Case
Pronouns have forms to show the subjective, objective, and possessive cases, as the following chart shows.

SINGULAR	SUBJECTIVE	OBJECTIVE	POSSESSIVE
First person	I	Me	My, mine
Second person	You	You	Your, yours
Third person	He, she, it	Him, her, it	His, her, hers, its
PLURAL			
First person	We	Us	Our, ours
Second person	You	You	Your, yours
Third person	They	Them	Their, theirs

A pronoun that is used as the subject of a clause or sentence is in the subjective case (*I, we, he, she, it, you, they, who*). The subjective case is also used when the pronoun follows a linking **verb.**

- *She* is my boss.

- My boss is *she.*

A pronoun that is used as the object of a verb or **preposition** is in the objective case (*me, us, him, her, it, you, them, whom*).

- Ms. Davis hired Tom and *me.* [object of verb]

- Between you and *me,* she's wrong. [object of preposition]

A pronoun that is used to express ownership is in the possessive case (*my, mine, our, ours, his, hers, its, your, yours, their, theirs, whose*).

- He took *his* notes with him on the business trip.

- We took *our* notes with us on the business trip.

A pronoun appositive takes the case of its antecedents.

- Two systems analysts, Joe and *I,* were selected to represent the company.
  [*Joe and I* is in apposition to the subject, *systems analysts,* and must therefore be in the subjective case.]

- The systems analysts selected two members—Joe and *me.*
  [*Joe and me* is in apposition to *two members,* which is the object of the verb, *selected,* and therefore must be in the objective case.]

If you have difficulty determining the case of a compound pronoun, try using the pronoun singly.

- In his letter, Eldon mentioned *him* and *me*.

  In his letter, Eldon mentioned *him*.

  In his letter, Eldon mentioned *me*.

- *They* and *we* must discuss the terms of the merger.

  They must discuss the terms of the merger.

  We must discuss the terms of the merger.

When a pronoun modifies a noun, try it without the noun to determine its case. (See also **case**.)

- [*We/Us?*] pilots fly our own planes.

  We fly our own planes.

  [You would not write, "*Us* fly our own planes."]

- He addressed his remarks directly to [*we/us?*] technicians.

  He addressed his remarks directly to *us*.

  [You would not write, "He addressed his remarks directly to *we*."]

Gender

A pronoun must agree in gender with its antecedent. A problem sometimes occurs because the masculine pronoun has traditionally been used to refer to both sexes. To avoid the sexual bias implied in such usage, use *he or she* or the plural form of the pronoun, *they*.

- Each may stay or go as he *or she* chooses.

- ~~Each~~ *All* may stay or go as ~~he~~ *they* chooses.

If you use the plural form of the pronoun, be sure to change the indefinite pronoun *each* to its plural form, *all*. (See also **biased language.**)

Number

Number is a frequent problem with only a few indefinite pronouns (*each, either, neither,* and those ending with *-body* or *-one,* such as *anybody, anyone, everybody, everyone, nobody, no one, somebody, someone*) that are normally singular and so require singular verbs and are referred to by singular pronouns. (See also **number.**)

- As *each arrives* for the meeting and is seated, please hand him or her a copy of the confidential report. *Everyone* is to return the copy before *he or she* leaves. *No one* should be offended by these precautions when the importance of secrecy has been explained to *him or her*. I think *everybody* on the committee *understands* that *neither* of our major competitors *is* aware of the new process we have developed.

Person

Third-person personal pronouns usually have antecedents.

- Gina presented the report to the members of the board of directors. *She* [Gina] first read *it* [the report] to *them* [the directors] and then asked for questions.

First- and second-person personal pronouns do not normally require antecedents.

- *I* like my job.

- *You* were there at the time.

- *We* all worked hard on the project.

ESL TIPS FOR UNDERSTANDING AGREEMENT:
POSSESSIVE PRONOUNS AND ANTECEDENTS

In many languages, possessive pronouns agree in number and gender with the nouns they modify. In English, however, possessive pronouns agree in number and gender with their antecedents. Check your writing carefully for agreement between a possessive pronoun and the word, phrase, or clause it refers to.

- The woman brought ~~his~~ *her* brother a cup of soup.

  [*her* refers to woman]

- Robert sent ~~her~~ *his* mother flowers on Mother's Day.

  [*his* refers to Robert]

P

proofreaders' marks

Publishers have established **symbols,** called proofreaders' marks, that writers and editors use to communicate in the production of publications. Familiarity with those symbols makes it easy for you to communicate your changes to others. Figure P–10 on page 495 lists standard proofreaders' marks.

proofreading

Computer spellcheckers are both a boon and a bane to proofreading. They are a boon because they do so much work, but the downside is that they tend to make writers overconfident and thus a little sloppy in their proofreading. If a typographical error results in a legitimate English word (for example, *too* instead of *two*), the spellchecker will not flag the misspelling. Therefore, you still must proofread your work carefully. Proofread a printed copy as well as the computer display. (See also **spelling** and **proofreaders' marks.**)

Proofreading should be done in several stages. Read through the material four times: three times looking for specific things each time and a final time for content. The Writer's Checklist that follows offers tips for proofreading.

Writer's Checklist: Proofreading

During the first pass, ask yourself, "Does it look right?" Look for the following:

- ☑ Aesthetic placement of material on the page (see **layout and design**)
- ☑ Acceptable format (as for a letter or **memo;** see also **correspondence**)
- ☑ Correct spelling of names and places
- ☑ Accuracy of **numbers**
- ☑ Correct **usage**

During the second pass, ask, "Am I following the rules?" Check for these items:

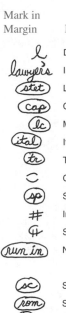

Mark in Margin	Instruction	Mark on Manuscript	Corrected Type
	Delete	the ~~lawyer's~~ Bible	the Bible
	Insert	the bible	the lawyer's bible
stet	Let stand	the ~~lawyer's~~ bible	the lawyer's bible
cap	Capitalize	the bible	the Bible
lc	Make lowercase	the Law	the law
ital	Italicize	the lawyer's bible	the *lawyer's* bible
tr	Transpose	the bible/lawyer's	the lawyer's bible
	Close space	the Bi ble	the Bible
sp	Spell out	2 bibles	two bibles
#	Insert space	the Bible	the Bible
	Start paragraph	The lawyer's . . .	The lawyer's . . .
run in	No paragraph	. . . marks. Below is a . . .	. . . marks. Below is a . . .
sc	Set in small capitals	the bible	the BIBLE
rom	Set in roman type	the *bible*	the bible
bf	Set in boldface	the bible	the **bible**
lf	Set in lightface	the **bible**	the bible
⊙	Insert period	The lawyers have their own bible	The lawyers have their own bible.
	Insert comma	However we cannot . . .	However, we cannot . . .
	Insert hyphens	half and half	half-and-half
	Insert colon	We need the following	We need the following:
	Insert semicolon	Use the law don't . . .	Use the law; don't . . .
	Insert apostrophe	Johns law book	John's law book
	Insert quotation marks	The law is law.	The "law" is law.
	Insert parentheses	John's law book	John's (law) book
	Insert brackets	(John Martin 1920–1962 went . . .)	(John Martin [1920–1962] went . . .)
	Insert en dash	1920 1962	1920–1962
	Insert em dash	Our goal victory	Our goal—victory
	Insert superior type	$3 = 9$	$3^2 = 9$
	Insert inferior type	HSO_4	H_2SO_4

FIGURE P–10. Proofreaders' Marks

Writer's Checklist: Proofreading (continued)

☑ Typographical errors
☑ Capitalization (see **capital letters**)
☑ Accurate **abbreviations** and **acronyms and initialisms**
☑ **Punctuation**
☑ General spelling
☑ Correct **grammar**

During the third pass, ask, "Is it complete?" Look for the following:

☑ Omissions
☑ Deletions
☑ Incomplete **World Wide Web** or **email** addresses and the like

When you read the document the final time for content, double-check to make certain that everything is there, that it is correct, and that it is in the right place.

proper nouns (*see* nouns)

proposals

P

DIRECTORY
Overview   496
Internal Proposals   497
Sales Proposals   501
Government Proposals   512

Overview

A proposal is a document written to persuade someone to follow a plan or course of action. (See also **persuasion**.) You may need to send a proposal to a colleague within your organization or to a potential client outside the organization.

Because a proposal offers a plan to fill a need, **readers** will evaluate your plan based on how well you answer their questions about what you are proposing to do, how you plan to do it, when you plan to do it, and how much it will cost.

To answer those questions satisfactorily, make certain that your proposal is written at your readers' level of knowledge. If you have more than one reader—and proposals often require more than one level of approval—take all of them into account. For example, if your primary reader is an expert on your subject but his or her supervisor, who must also approve the proposal, is not, provide an **executive summary** written in nontechnical language. You might also include a **glossary** of terms used in the body of the proposal or an **appendix** that explains highly detailed information in nontechnical language. If your primary reader is not an expert but his or her supervisor is, write the proposal with the nonexpert in mind and include an appendix that contains the technical details.

This entry discusses three typical types of proposals: internal proposals, sales proposals, and government proposals.

Internal Proposals

The purpose of an internal proposal is to suggest a change or an improvement within an organization. For example, an internal proposal can recommend one of the following:

- A change in the way something is being done
- That something new be done
- That funding be authorized for a large purchase

An internal proposal, often in **memo** format, is sent to a superior in the organization who has the authority to accept or reject the proposal.

In the **opening** of a proposal, you must establish that a problem exists that needs a solution. If the readers who will judge the proposal are not convinced that there is a problem, your proposal will be unsuccessful. Notice how the proposal opening in Figure P–11 on page 498 states the problem directly.

The body of a proposal should offer a practical solution to the problem. In building a case for a solution, be as specific as possible. When appropriate, include these typical sections:

- Breakdown of costs
- Information about equipment, material, and staff requirements
- Schedule for completing the task

Consider Figure P–12 on pages 498–99, which continues the proposal introduced in Figure P–11.

Date: May 11, 20--
To: Harriet V. Sullivan, President
From: Christine Thomas   *CT*
Subject: Advantages of Expanding HVS's Computer Capabilities

As we discussed earlier, our computer system has been in place since the founding of HVS four years ago and is currently operating at maximum capacity. Unfortunately, we are now in the position of turning new customers away and straining to meet the needs of our present customers. In addition, our system cannot run the latest accounting and financial-planning software. By purchasing a new system, we could attract new customers, provide better service to valued current customers, and enhance the range of services we provide.

FIGURE P–11. Proposal Opening

Last month, we turned away 13 new requests for service. At least six of these were equal to the largest accounts we currently serve. Furthermore, we cannot offer improved, faster service to our most profitable accounts (such as Talbot Trucking), nor can we attract new customers who need services we could only offer with an upgraded computer system.

Our present system is made up of stand-alone computers. They are not networked, so they cannot communicate with one another or with our customers. As a result, we are unable to exchange files electronically with our customers, a service more and more of them expect us to provide. Further, the staff at HVS needs to be able to track the status of customer work online and to exchange email messages both with customers and among ourselves.

The most efficient way to accomplish these goals is to upgrade our hardware and software and create an in-house network linked by modem to our present and future customers. This upgrade would allow us to respond to new service requests and attract new customers with our expanded capacity and means of communication.

Upgrading to such a system would give HVS the following additional advantages. It would

- Permit online access to up-to-the-minute market news, indexes, and financial trends for financial-planning services
- Provide access to the Internet and World Wide Web
- Allow HVS to create its own Web site to advertise, to respond to potential customers, and to communicate with professional organizations in accounting and financial planning
- Permit access to free software sites on the Web that allow specialized functions
- Increase work flow and reduce paperwork

FIGURE P–12. Proposal Body

The system that would achieve these goals has ample storage and processing capability, is easily upgraded, and is competitively priced. It would increase processing capacity fivefold and would enable us to use the full range of office-suite and specialized software on the market. The system's memory and processing speed would also accommodate more sophisticated software under development. It is made up of off-the-shelf components and could be upgraded one component at a time as needed.

The following details show not only system costs and annual operating expenses, but also savings that would pay for the new system in only two years. These figures are based on the ten service requests a month on average we have turned away over the past year. This business could generate profits of $45,000 per year, based on figures from Mr. Sadowski.

System Costs
- Hardware $37,995
- Software   9,750
- Installation and testing   16,995
- Training and maintenance   7,500
  Total costs $72,240

Annual Operating Expenses $2,680
Increased Profit Generated by System Upgrade
10 service requests × $375 per month × 12 months = $45,000/year

The purchase is a one-time investment. It requires no new staff, only training of current staff. (The attachment provides a listing of system components.)

Purchasing such a system would require that we shut down operations for three days for delivery, unpacking, installation, and testing. The third day would be set aside to train the staff at our site in the use of the specialized network software. If we plan the work before the next tax season and notify our customers at least a month in advance, I am sure that we can gain their cooperation, especially because the new system would permit us to serve them more efficiently.

FIGURE P–12. Proposal Body (*continued*)

P

The function of the conclusion in an internal proposal is to tie everything together, restate your recommendation, and close in a spirit of cooperation (offering to set up a meeting, supply additional information, or provide any other assistance that might be needed). Your conclusion should be brief, as in Figure P–13 on page 500.

A proposal to commit large sums of money is a common internal proposal. Although such internal proposals have various names, *capital appropriations proposal* (or *request*) is a common and descriptive name.

On the basis of these details, I recommend that we replace our present stand-alone computers with newer, networked models to provide electronic access to our customers and to the Internet. I will be happy to discuss this proposal with you and to provide additional information at your request.

FIGURE P–13. Proposal Conclusion

The introduction to a capital appropriations proposal should provide any background information your reader needs to make the decision in question. The introduction should also briefly describe any feasibility study you may have conducted, the conclusions you reached based on the study, and the decision you made on the basis of those conclusions.

The body of a capital appropriations proposal should identify all the equipment or property you are asking to acquire and state the justification for each expenditure. Be sure to list all the equipment you need to purchase so you do not have to prepare another proposal to purchase additional equipment later. Your proposal should include all the following items that are applicable to your particular request:

- A brief description of the item to be purchased
- The alternative choices you considered before deciding on the one you are recommending
- The specific products, services, or projects to be supported by the item
- The volume of usage that the item would receive over a specified period of time, such as the next two years
- If the company already owns some of the item, an explanation of why you require an additional purchase
- The life expectancy of the item and of the product, service, or project it is to support
- Any future products, services, or projects that the item could be used to support
- Whether a leasing arrangement is available and more economical than purchasing the item
- The impact on the company if the item is not acquired
- The latest feasible date of acquisition, including the impact that a 90-day delay would have
- The calculated return on investment or the internal rate of return

With your capital appropriations request, you should also submit a cover memo. The memo, which should be no more than a page, should give a brief description of the item and the reason it is needed.

Sales Proposals

The sales proposal, one of the major marketing tools in use in business and industry today, is a company's offer to provide specific goods or services to a potential buyer within a specified period of time and for a specified price. The primary purpose of a sales proposal is to demonstrate that the prospective customer's purchase of the seller's products or services will solve a problem, improve operations, or offer benefits. Indeed, more than selling equipment or services, the effective sales proposal sells solutions to problems.

Sales proposals vary greatly in size and sophistication. Some are a page or two written by one person; others are many pages written by several people; still others are hundreds of pages written by a team of professional proposal writers. A short sales proposal might bid for the construction of a single home; a sales proposal of moderate length might bid for the installation of a network of computer systems; and a very long sales proposal might be used to bid for the construction of a multimillion-dollar water purification system.

Your first task in writing a sales proposal is to find out exactly what your prospective customer needs. To do that, survey your potential customer's business and needs. You must then determine whether your organization can satisfy the customer's needs. Before preparing a sales proposal, you should try to find out who your principal competitors are. Then compare your company's strengths with those of competing firms, determine your advantages over your competitors, and emphasize those advantages in your proposal. For example, say a small software company is bidding for an Air Force contract at a local base. The proposal writer who believes that the company has better-qualified software engineers than its competitors might include the résumés of the key people who would be involved in the project, as a way of emphasizing that advantage.

Unsolicited and Solicited Proposals. Sales proposals may be either unsolicited or solicited. Unsolicited proposals are not as unusual as they sound: Companies often operate for years with a problem they have never recognized (unnecessarily high maintenance costs, for example, or poor inventory-control methods). You might prepare an unsolicited proposal if you were convinced that the potential customer could realize substantial benefits by adopting your solution to a problem. Of course, you would need to convince the customer of the need for what you are proposing and that your solution would be the best one.

Many unsolicited proposals are preceded by an inquiry to determine potential interest. Once you have received a positive response, you would conduct a detailed study of the prospective customer's needs to determine whether you can be of help and, if so, exactly how. You would then prepare your proposal on the basis of your study.

An unsolicited proposal should clearly identify the potential customer's problem. After stating the problem, the proposal must convince the customer that the problem needs to be solved. One way to do that is to emphasize the benefits the customer will realize from the solution being proposed.

The other type of sales proposal—the solicited sales proposal—is a response to a request for goods or services. Procuring organizations that would like competing companies to bid for a job commonly issue a *request for proposals* (*RFP*). An RFP is the means many companies and government agencies use to find the best method of doing a job and the most qualified company to do it. An RFP may be rigid in its specifications governing how the proposal should be organized and what it should contain, but it is normally flexible about the approaches the bidding firms may propose. Normally, the RFP simply defines the basic work that the firm needs.

The procuring organization generally publishes its RFP in one or more journals, in addition to sending it to certain companies that have good reputations for doing the kind of work needed. Some companies and government agencies even hold a conference for the competing firms at which they provide all pertinent information about the job being bid for.

Managers interested in responding to RFPs regularly scan the appropriate publications. Upon finding a project of interest, an executive in the sales department obtains all available information from the procuring company or agency. The data are then presented to management for a decision on whether the company is interested in the project. If the decision is positive, the technical staff is assigned to develop an approach to the work described in the RFP. The technical staff normally considers several alternatives, selecting the one that combines feasibility and a price that offers a profit for the bidder. The staff's concept is presented to higher management for a decision on whether the company wishes to present a proposal to the requesting organization. If the decision is to proceed, preparation of the proposal is the next step. When you respond to an RFP, pay close attention to any specifications in the request governing the preparation of the proposal and follow them carefully.

PROPOSAL
TO LANDSCAPE THE NEW CORPORATE HEADQUARTERS
OF THE
WATFORD VALVE CORPORATION

Submitted to: Ms. Tricia Olivera, Vice President
Submitted by: Jerwalted Nursery, Inc.
Date Submitted: February 1, 20--

Introduction states purpose and scope of proposal, indicates when project can be started and completed.

Jerwalted Nursery, Inc., proposes to landscape the new corporate headquarters of the Watford Valve Corporation, on 1600 Swanson Avenue, at a total cost of $14,871. The lot to be landscaped is approximately 600 feet wide and 700 feet deep. Landscaping will begin no later than April 30, 20--, and will be completed by May 31.

The following trees and plants will be planted, in the quantities and sizes given and at the prices specified.

Body lists products to be provided, cost per item.

    4 maple trees (not less than 7 ft.) @ $110 each—$440
    41 birch trees (not less than 7 ft.) @ $135 each—$5,535
    2 spruce trees (not less than 7 ft.) @ $175 each—$350
    20 juniper plants (not less than 18 in.) @ $15 each—$300
    60 hedges (not less than 18 in.) @ $12 each—$720
    200 potted plants (various kinds) @ $12 each—$2,400

                Total Cost of Plants = $ 9,745
                              Labor = $ 5,126
                         Total Cost = $14,871

Conclusion specifies time limit of proposal, expresses confidence, and looks forward to working with prospective customer.

All trees and plants will be guaranteed against defect or disease for a period of 90 days, the warranty period to begin June 1, 20--.

The prices quoted in this proposal will be valid until June 30, 20--.

Thank you for the opportunity to submit this proposal. Jerwalted Nursery has been in the landscaping and nursery business in the St. Louis area for thirty years, and our landscaping has won several awards and commendations, including a citation from the National Association of Architects. We are eager to put our skills and knowledge to work for you, and we are confident that you will be pleased with our work. If we can provide any additional information or assistance, please call.

FIGURE P–14. Short Sales Proposal

Simple Sales Proposals.   Even a short and uncomplicated sales proposal should be carefully planned. The introduction should state the purpose and scope of the proposal. It should indicate the dates on which you propose to begin and complete work on the project, any special benefits of your proposed approach, and the total cost of the project. The introduction could also refer to any previous positive association your company may have had with the potential customer.

The body of a sales proposal should itemize the products and services you are offering. It should include, if applicable, a discussion of the procedures you propose to use to perform the work and the materials that will be used. It should also present a schedule indicating when each stage of the project would be completed. Finally, the body of your proposal should include a breakdown of the costs of the project.

The conclusion should express your appreciation for the opportunity to submit the proposal and your confidence in your company's ability to do the job. You might add that you look forward to establishing good working relations with the customer and that you would be glad to provide any additional information that might be needed. Your conclusion could also review any advantages your company may have over its competitors. It should specify the time period during which your proposal can be considered a valid offer. If any supplemental materials, such as blueprints or price sheets, accompany the proposal, include a list of them at the end of the proposal. Figure P–14 on page 503 shows a typical short sales proposal.

Long Sales Proposals.   While the simple sales proposal is typically divided into the introduction, the body, and the conclusion, the long sales proposal contains more parts to accommodate the increased variety of information it represents. The long sales proposal may include some or all of the following sections.

- Transmittal (or **cover letter**)
- Title page
- Executive summary
- Product description
- Rationale
- Cost analysis
- Delivery schedule
- Site preparation

- Training requirements
- Statement of responsibilities
- Description of vendor
- Institutional sales pitch
- Conclusion
- Appendixes

The cover letter expresses your appreciation for the opportunity to submit your proposal and for any assistance you may have received in studying the customer's requirements. The letter should acknowledge any previous positive association with the customer. Then it should summarize the recommendations offered in the proposal and express your confidence that they will satisfy the customer's needs. Figure P–15 on pages 506–7 shows the cover letter for the proposal illustrated in Figures P–16 through P–23, a proposal that the Waters Corporation of Tampa, Florida, provide a computer system for the Cookson chain of retail stores.

A title page and an **executive summary** follow the cover letter. The title page contains the title of the proposal, the date, the company to which it is being submitted, and your company name. The executive summary is written to the executive who will ultimately accept or reject the proposal. It should summarize in nontechnical language how you plan to approach the work. Figure P–16 on page 508 shows the executive summary of the Waters Corporation proposal.

If your proposal offers products as well as services, it should include a general description of the products, as in Figure P–17 on pages 509–10. After the executive summary and the general description, explain exactly how you plan to do what you are proposing. This detailed section, called the rationale, will be read by specialists who can understand and evaluate your plan, so you can feel free to use technical language and discuss complicated concepts. Figure P–18 on page 511 shows one part of the detailed solution appearing in the Waters Corporation proposal, which included several other applications in addition to the payroll application. Notice that the rationale, like the discussion in an unsolicited sales proposal, begins with a statement of the customer's problem, follows with a statement of the solution, and concludes with a statement of the benefits to the customer.

A cost analysis and a delivery schedule are essential to any sales proposal. The cost analysis itemizes the estimated cost of all the

The Waters Corporation
17 North Waterloo Blvd.
Tampa, Florida 33607
Phone: (813) 919-1213   Fax: (813) 919-4411
http://www.waters.comp.com

September 1, 20--

Mr. John Yeung, General Manager
Cookson's Retail Stores, Inc.
101 Longuer Street
Savannah, Georgia 31499

Dear Mr. Yeung:

The Waters Corporation appreciates the opportunity to respond to Cookson's Request for Proposals dated July 26, 20--. We would like to thank Mr. Becklight, Director of your Management Information Systems Department, for his invaluable contributions to the study of your operations that we conducted before preparing our proposal.

It has been Waters's privilege to provide Cookson's with retail systems and equipment since your first store opened many years ago. Therefore, we have become very familiar with your requirements as they have evolved during the expansion you have experienced since that time. Waters's close working relationship with Cookson's has resulted in a clear understanding of Cookson's philosophy and needs.

Our proposal describes a Waters Interactive Terminal/Retail Processor System designed to meet Cookson's network and processing needs. It will provide all of your required capabilities, from the point-of-sale operational requirements at the store terminals to the host processor. The system uses the proven Retail III modular software, with its point-of-sale applications, and the superior Interactive Terminal, with its advanced capabilities and design. This system is easily installed without extensive customer reprogramming.

FIGURE P–15. Cover Letter

Mr. J. Yeung
Page 2
September 1, 20--

The Waters Interactive Terminal/Retail Processor System, which is compatible with much of Cookson's present equipment, not only will answer your present requirements but will provide the flexibility to add new features and products in the future. The system's unique hardware modularity, efficient microprocessor design, and flexible programming capability greatly reduce the risk of obsolescence.

Thank you for the opportunity to present this proposal. You may be sure that we will use all the resources available to the Waters Corporation to ensure the successful implementation of the new system.

Sincerely yours,

Janet A. Curtain

Janet A. Curtain
Executive Account Manager
General Merchandise Systems
(JCurtain@netcom.TF.com)

P

FIGURE P–15. Cover Letter (*continued*)

EXECUTIVE SUMMARY

The Waters 319 Interactive Terminal/615 Retail Processor System will provide your management with the tools necessary to manage people and equipment more profitably with procedures that will yield more cost-effective business controls for Cookson's.

The equipment and applications proposed for Cookson's were selected through the combined effort of Waters and Cookson's Management Information Systems Director, Mr. Becklight. The architecture of the system will respond to your current requirements and allow for future expansion.

The features and hardware in the system were determined from data acquired through the comprehensive survey we conducted at your stores in February of this year. The total of 71 Interactive Terminals proposed to service your four store locations is based on the number of terminals currently in use and on the average number of transactions processed during normal and peak periods. The planned remodeling of all four stores was also considered, and the suggested terminal placement has been incorporated into the working floor plan. The proposed equipment configuration and software applications have been simulated to determine system performance based on the volumes and anticipated growth rates of the Cookson stores.

The information from the survey was also used in the cost justification, which was checked and verified by your controller, Mr. Deitering. The cost-effectiveness of the Waters Interactive Terminal/Retail Processor System is apparent. Expected savings, such as the projected 46 percent reduction in sales audit expenses, are realistic projections based on Waters's experience with other installations of this type.

FIGURE P–10. Executive Summary

GENERAL SYSTEM DESCRIPTION

The point-of-sale system that Waters is proposing for Cookson's includes two primary Waters products. These are the 319 Interactive Terminal and the 615 Retail Processor.

Waters 319 Interactive Terminal

The primary component in the proposed retail system is the Interactive Terminal. It contains a full microprocessor, which gives it the flexibility that Cookson's has been looking for.

The 319 Interactive Terminal provides you with freedom in sequencing a transaction. You are not limited to a preset list of available steps or transactions. The terminal program can be adapted to provide unique transaction sets, each designed with a logical sequence of entry and processing to accomplish required tasks. In addition to sales transactions recorded on the selling floor, specialized transactions such as theater-ticket sales and payments can be designed for your customer-service area.

The 319 Interactive Terminal also functions as a credit authorization device, either by using its own floor limits or by transmitting a credit inquiry to the 615 Retail Processor for authorization.

Data-collection formats have been simplified so that transaction editing and formatting are much more easily accomplished. The IS manager has already been provided with documentation on these formats and has outlined all data-processing efforts that will be necessary to transmit the data to your current systems. These projections have been considered in the cost justification.

Waters 615 Retail Processor

The Waters 615 Retail Processor is a minicomputer system designed to support the Waters family of retail terminals. The

FIGURE P–17. General Description of Products

processor will reside in the computer room in your data center in Buffalo. Operators already on your staff will be trained to initiate and monitor its activities.

The 615 will collect data transmitted from the retail terminals, process credit, check authorization inquiries, maintain files to be accessed by the retail terminals, accumulate totals, maintain a message-routing network, and control the printing of various reports. The functions and level of control performed at the processor depend on the peripherals and software selected.

Software

The Retail III software used with the system has been thoroughly tested and is operational in many Waters customer installations.

The software provides the complete processing of the transaction, from the interaction with the operator on the sales floor through the data capture on cassette or disk in stores and in your data center.

Retail III provides a menu of modular applications for your selection. Parameters condition each of them to your hardware environment and operating requirements. The selection of hardware will be closely related to the selection of the software applications.

P

FIGURE P–17. General Description of Products (*continued*)

PAYROLL APPLICATION

Current Procedure

Your current system of reporting time requires each hourly employee to sign a time sheet; the time sheet is reviewed by the department manager and sent to the Payroll Department on Friday evening. Because the week ends on Saturday, the employee must show the scheduled hours for Saturday and not the actual hours; therefore, the department manager must adjust the reported hours on the time sheet for employees who do not report on the scheduled Saturday or who do not work the number of hours scheduled.

The Payroll Department employs a supervisor and three full-time clerks. To meet deadlines caused by an unbalanced work flow, an additional part-time clerk is used for 20 to 30 hours per week. The average wage for this clerk is $8.00 per hour.

Advantage of Waters's System

The 319 Interactive Terminal can be programmed for entry of payroll data for each employee on Monday mornings by department managers, with the data reflecting actual hours worked. This system would eliminate the need for manual batching, controlling, and data input. The Payroll Department estimates conservatively that this work consumes 40 hours per week.

Hours per week	40
Average wage (part-time clerk)	× 8.00
Weekly payroll cost	$320.00
Annual savings	$16,640

Elimination of the manual tasks of tabulating, batching, and controlling can save 0.25 hourly unit. Improved work flow resulting from timely data in the system without data-input processing will allow more efficient use of clerical hours. This would reduce payroll by the 0.50 hourly unit currently required to meet weekly check disbursement.

Eliminate manual tasks	0.25
Improve work flow	0.75
40-hour unit reduction	1.00
Hours per week	40
Average wage (full-time clerk)	9.00
Savings per week	$360.00
Annual savings	$18,720

TOTAL SAVINGS: $35,360

FIGURE P–18. Detailed Solution

products and services you are offering; the delivery schedule commits you to a specific timetable for providing those products and services. Figure P–19 on pages 513–15 shows the cost analysis and the delivery schedule of the Waters Corporation proposal.

If your recommendations include modifying the customer's physical facilities, you would include a site preparation description that details the required modifications. Figure P–20 on page 516 shows the site preparation section of the Waters proposal.

If the products and services you are proposing require training the customer's employees, your proposal should specify the required training and its cost. Figure P–21 on page 517 shows the training section of the Waters proposal.

To prevent misunderstandings about what you and your customer's responsibilities will be, you should draw up a statement of responsibilities (Figure P–22 on page 518), which usually appears toward the end of the proposal. Also toward the end of the proposal is a description of the vendor, which gives a description of your company, its history, and its present position in the industry, and an organizational sales pitch (Figure P–23 on page 519). Up to this point, the proposal has attempted to sell specific goods and services. The sales pitch, striking a somewhat different chord, is designed to sell the company and its general capability in the field. The sales pitch promotes the company and concludes the proposal on an upbeat note.

Some long sales proposals include a conclusion that summarizes the proposal's salient points and stresses your company's main strong points. Appendixes can be used to include statistical analysis, maps, charts, tables, and personal résumés.

Government Proposals

Government proposals are usually prepared as a result of either an invitation for bids or a request for proposals that has been issued by a government agency.

Invitation for Bids.   An invitation for bids (IFB) is an invitation by a government agency for companies or individuals to bid on a project that the agency has defined in detail. An IFB is inflexible; the rules are rigid and the terms are not open to negotiation. Any proposal prepared in response to an IFB must adhere strictly to its terms.

The IFB clearly defines the quantity and type of item that a government agency intends to purchase and is prepared at the

COST ANALYSIS

This section of our proposal provides detailed cost information for the Waters 319 Interactive Terminal and the 615 Waters Retail Processor. It then multiplies these major elements by the quantities required at each of your four locations.

319 Interactive Terminal

	Price	Maint. (1 yr.)
Terminal	$2,895	$167
Journal Printer	425	38
Receipt Printer	425	38
Forms Printer	525	38
Software	220	—
TOTALS	$4,490	$281

615 Retail Processor

	Price	Maint. (1 yr.)
Processor	$57,115	$5,787
CRT I/O Writer	2,000	324
Laser Printer	4,245	568
Software	12,480	—
TOTALS	$75,840	$6,679

The following breakdown itemizes the cost per store and the Data Center.

Store No. 1

Description	Qty.	Price	Maint. (1 yr.)
Terminals	16	$68,400	$4,496
Digital Cassette	1	1,300	147
Laser Printer	1	2,490	332
Software	16	3,520	—
TOTALS		$75,710	$4,975

FIGURE P–19. Cost Analysis and Delivery Schedule

Store No. 2

Description	Qty.	Price	Maint. (1 yr.)
Terminals	20	$85,400	$5,620
Digital Cassette	1	1,300	147
Laser Printer	1	2,490	332
Software	20	4,400	—
TOTALS		$93,590	$6,099

Store No. 3

Description	Qty.	Price	Maint. (1 yr.)
Terminals	17	$72,590	$4,777
Digital Cassette	1	1,300	147
Laser Printer	1	2,490	332
Software	17	3,740	—
TOTALS		$80,120	$5,256

Store No. 4

Description	Qty.	Price	Maint. (1 yr.)
Terminals	18	$76,860	$5,058
Digital Cassette	1	1,300	147
Laser Printer	1	2,490	332
Software	18	3,960	—
TOTALS		$84,610	$5,537

Data Center at Buffalo

Description	Qty.	Price	Maint. (1 yr.)
Processor	1	$57,115	$5,787
CRT I/O Writer	1	2,000	324
Laser Printer	1	4,245	568
Software	1	12,480	—
TOTALS		$75,840	$6,679

FIGURE P–19. Cost Analysis and Delivery Schedule (*continued*)

The following summarizes all costs.

Location	Hardware	Maint. (1 yr.)	Software
Store No. 1	$72,190	$4,975	$3,520
Store No. 2	89,190	6,099	4,400
Store No. 3	76,380	5,256	3,740
Store No. 4	80,650	5,537	3,960
Data Center	63,360	6,679	12,480
Subtotals	$381,770	$28,546	$28,100

TOTAL $438,416

DELIVERY SCHEDULE

Waters is normally able to deliver 319 Interactive Terminals and 615 Retail Processors within 30 days of the date of the contract. This can vary depending on the rate and size of incoming orders.

All the software recommended in this proposal is available for immediate delivery. We do not anticipate any difficulty in meeting your tentative delivery schedule.

FIGURE P-19. Cost Analysis and Delivery Schedule (*continued*)

SITE PREPARATION

Waters will work closely with Cookson's to ensure that each site is properly prepared prior to system installation. You will receive a copy of Waters's installation and wiring procedures manual, which lists the physical dimensions, service clearance, and weight of the system components in addition to the power, logic, communication-cable, and environmental requirements. Cookson's is responsible for all building alterations and electrical facility changes, including the purchase and installation of communication cables, connecting blocks, and receptacles.

Wiring

For the purpose of future site considerations, Waters's in-house wiring specifications for the system call for two twisted-pair wires and twenty-two shielded gauges. The length of communications wires must not exceed 2,500 feet.

As a guide for the power supply, we suggest that Cookson's consider the following.

1. The branch circuit (limited to 20 amps) should service no equipment other than 319 Interactive Terminals.
2. Each 20-amp branch circuit should support a maximum of three Interactive Terminals.
3. Each branch circuit must have three equal-size conductors — one hot leg, one neutral, and one insulated isolated ground.
4. Hubbell IG 5362 duplex outlets or the equivalent should be used to supply power to each terminal.
5. Computer-room wiring will have to be upgraded to support the 615 Retail Processor.

FIGURE P–20. Site-Preparation Section

TRAINING

To ensure a successful installation, Waters offers the following training course for your operators.

Interactive Terminal/Retail Processor Operations

Course number: 8256
Length: three days
Tuition: $500.00

This course provides the student with the skills, knowledge, and practice required to operate an Interactive Terminal/Retail Processor System. Online, clustered, and stand-alone environments are covered.

We recommend that students have a department-store background and that they have some knowledge of the system configuration with which they will be working.

P

FIGURE P–21. Training Section

RESPONSIBILITIES

On the basis of its years of experience in installing information-processing systems, Waters believes that a successful installation requires a clear understanding of certain responsibilities.

Generally, it is Waters's responsibility to provide its users with needed assistance during the installation so that live processing can begin as soon thereafter as is practical.

Waters's Responsibilities

- Provide operations documentation for each application that you acquire from Waters
- Provide forms and other supplies as ordered
- Provide specifications and technical guidance for proper site planning and installation
- Provide adviser assistance in the conversion from your present system to the new system

Customer's Responsibilities

- Identify an installation coordinator and system operator
- Provide supervisors and clerical personnel to perform conversion to the system
- Establish reasonable time schedules for implementation
- Ensure that the physical site requirements are met
- Provide personnel to be trained as operators and ensure that other employees are trained as necessary
- Assume the responsibility for implementing and operating the system

FIGURE P–22. Statement of Responsibilities

DESCRIPTION OF VENDOR

The Waters Corporation develops, manufactures, markets, installs, and services total business information-processing systems for selected markets. These markets are primarily in the retail, financial, commercial, industrial, health-care, education, and government sectors.

The Waters total system concept encompasses one of the broadest hardware and software product lines in the industry. Waters computers range from small business systems to powerful general-purpose processors. Waters computers are supported by a complete spectrum of terminals, peripherals, data-communication networks, and an extensive library of software products. Supplemental services and products include data centers, field service, systems engineering, and educational centers.

The Waters Corporation was founded in 1934 and presently has approximately 26,500 employees. The Waters headquarters is located at 17 North Waterloo Boulevard, Tampa, Florida, with district offices throughout the United States and Canada.

WHY WATERS?

Corporate Commitment to the Retail Industry

Waters's commitment to the retail industry is stronger than ever. We are continually striving to provide leadership in the design and implementation of new retail systems and applications that will ensure our users of a logical growth pattern.

Research and Development

Over the years, Waters has spent increasingly large sums on research and development efforts to ensure the availability of products and systems for the future. In 20--, our research and development expenditure for advanced systems design and technological innovations reached the $70-million level.

Leading Point-of-Sale Vendor

Waters is a leading point-of-sale vendor, having installed over 150,000 units. The knowledge and experience that Waters has gained over the years from these installations ensure well-coordinated and effective systems implementations.

FIGURE P–23. Description of Vendor and Organizational Sales Pitch

request of the agency that will use the item, such as the Department of Defense. An advertisement indicating that the government intends to purchase the item is announced in print and online publications, such as the *Commerce Business Daily* (http://www.cbdweb.com/) or the Contracts Reference Guide (http://www.hp.com/gov/federal/). The goods or services to be procured are defined in the IFB by references to performance standards called **specifications.** If, for instance, the item to be purchased is a truck, the important characteristics of that truck (height, weight, speed, carrying capacity, and so on) are listed in the machine specification. More than one specification may apply to a single purchase; that is, one specification may apply to the item, another to the manuals, another to welding procedures, and so forth. Bidders must be prepared to prove that their product will meet all requirements of all specifications.

A specification is restrictive, binding the bidder to produce an item that meets the exact requirements of the specification. For example, if a truck is the specified item, a company may not bid to furnish a cargo helicopter, even though the helicopter might do the job more efficiently than the truck. An IFB requires the bidder to furnish a specific product that meets the specification parameters; nothing else will be accepted or considered. An alternative suggestion will render the bid "nonresponsive," and it will be rejected.

Although the product will be tested, measured, and evaluated to see that it does, in fact, meet the requirements of the specification, price is the main criterion that enters into the selection of the vendor. However, an organization with superior engineering skills can often supply a product that meets the specification at a price well below that of a less sophisticated competitor. When a high degree of technological competence, specialized equipment, or other valid requirements make it necessary, the procuring agency may require the bidder to possess certain minimum qualifications in order to be considered; a paper clip manufacturer employing ten people, for instance, could not qualify to bid on a procurement of military aircraft.

Requests for Proposals.   In contrast to an IFB, a request for proposals (RFP) is flexible. It is open to negotiation and does not necessarily specify exactly what goods or services are required. Often, an RFP will define a problem and allow those who respond to suggest possible solutions. For example, if a procuring agency wanted to develop a way to make foot soldiers capable of covering difficult terrain at high speed, an RFP would normally be the means used to find the best method and select the most qualified vendor. In some instances, RFPs are presented in several stages, the first being development of a concept, the second being a prototype machine or device, and the third being the manufacture of the device selected.

The procuring agency defines the problem and publishes it as an RFP. Companies interested in government business scan the appropriate publications daily for RFPs. Upon finding a project of interest, the sales department obtains all relevant information from the procuring agency and presents the data to management for a decision on whether the company is interested in the project. If the decision is positive, the corporate technical and documentation staff are assigned the task of developing a "concept" to accomplish the task posed by the RFP. The technical staff normally considers several alternatives, selecting one that combines feasibility and price. The staff's concept is presented to management for a decision on whether the company wishes to present a proposal to the requesting government agency.

If management's decision is to proceed, preparation of a proposal is the next step. The proposal should articulate a clear-cut statement of the problem and the proposed solution. It should demonstrate through data that the company is financially and technologically sound, and it may include **résumés** of the people in charge and an **organizational chart** to show the chain of command. A discussion of the company's manufacturing capabilities and any other advantages is also advisable. However, in dealing with government agencies, keep in mind that unnecessarily elaborate and costly proposals may be construed as a lack of cost consciousness.

In many instances, the final cost to the government is determined by negotiation after the best overall concept has been selected. However, if your concept is one that provides a simple and inexpensive solution to the problem, cost is an advantage that you should emphasize. If your concept is expensive, explain the benefits your proposal brings to the government agency: advanced engineering, speed, increased capacity, and so forth.

The excerpt presented in Figure P–24 on pages 522–25 comes from a proposal that was prepared in response to an RFP. The rest of the proposal explained the design of the engine's cylinder heads, bearings, crankshaft, pistons, cylinder liners, connecting rod, and reversing mechanism. The proposal also contained various other sections, including budget estimates prepared by the company's financial analysts.

Advantages of the Proposed 5,000 H.P. Design

Statement of problem

The Navy Department has indicated to Burdlorn Manufacturing Company a need for main diesel propulsion engines to be used in conjunction with gas turbines on a program with the code name of CODAG. We have been advised that the power requirement for this engine is 5,000 horsepower, plus 10 percent overload. This engine must be contained within a space 24 feet long, 11 feet high, and 7 feet wide and must not exceed 60,000 pounds in weight. In addition, at a 5,000 horsepower operating level, it is not expected that the engine will require replacement or renewal of wearing parts or other major components more frequently than every 5,000 hours. The proposed design changes will result in a very lightweight engine (12 pounds/BHP) but one with durability characteristics that will not be equaled by any engine throughout the world in this compact high-horsepower output.

Brief statement of solution

5,000 Horsepower Engine

Company's technical capability

Burdlorn Manufacturing Company has completed shop tests on NOBS 72393, addenda 12 and 13. During these tests, the Navy engine serial GR-1047-0836 was operated at loads up to 4,950 brake horsepower (just 1 percent short of the 5,000 horsepower goal). As a result of these tests, we are satisfied that the 16-cylinder, 11″ bore, 12″ stroke V-engine, with a rating of 5,000 horsepower at 1,000 rpm, is well within practical limits. On the basis of an intensive review of the service history of the four engines installed on the LST-1176 and all of the research and development work at the Burdlorn factory, coupled with a substantial amount of analytical design work, Burdlorn Manufacturing Company recommends that certain redesigns be effected, based on the above information.

As a result of the recent tests at loads up to 4,950 horsepower that have been run in our factory, Burdlorn Manufacturing Company has successfully dealt with the thermodynamic problems of air/fuel requirements, airflow, heat transfer, and other related problems. In proposing design criteria, which are conservative to ensure the desired reliability, we have applied sufficient analysis to the individual components to make specific recommendations.

Advantages to be outlined

In order to summarize findings that we feel the Navy needs in order to evaluate our proposal, we will outline in numerical sequence the improvements and advantages based on our total experience with the 11″ bore, 12″ stroke, 16-cylinder V-engine.

1. Frame. Burdlorn Manufacturing Company recommends that the present 17 inches center distance from one cylinder to the adjacent cylinder be increased by 3 inches to provide 20 inches center distance from cylinder to cylinder in a given bank. The additional 3 inches per cylinder will mean an overall increase in engine length of 24 inches. This will mean an engine length of approximately 22 feet with the engine-driven auxiliary equipment. This will be well within the 24-foot maximum covered by the Navy specifications. The additional 3 inches will provide the following advantages.

FIGURE P–24. Government Proposal

a. The principal load-carrying member in the frame, which is the athwartship plate, will be increased in thickness from ½ inch to ¾ inch.

b. Additional clearance is provided for welding, and machining, particularly on the intermediate deck.

c. Additional length is provided for the main and crankpin journals. This will increase the bearing areas for the main and crankpin bearings and thus gain the advantage of reduced bearing loads.

Minor problem and proposed solution

The results of our experimental stress analysis have indicated a substantial bending movement in the top deck of the present design. Accordingly, Burdlorn Manufacturing Company proposes to increase the thickness of the top deck. The present design uses forgings. We are proposing to use controlled quality steel castings of substantially heavier construction to reduce the stress levels.

Burdlorn Manufacturing Company has done a substantial amount of stress analysis on a similar frame on our Model FS-13½″ × 16½″ V-engine. On the basis of this experience, we recommend a minor redesign of the main bearing saddles to create a more even stress distribution in transmitting the load through the athwartship plates to the main bearing saddles. These main bearing saddles would be made from steel casts such as our present 13½″ × 16½″ engines currently use.

To provide for reduction in stress levels in the sidesheets of the engine, we propose increasing the plate thickness from ½ inch to ⅝ inch. In addition, we have found through extensive testing of the Navy engine and our 13½″ × 16½″ V-engine and also by exhaustive photoelastic stress analysis techniques that the present "hourglass" design results in a stress concentration factor at the "waist" of the frame. Accordingly, we propose a modification in this shape that will reduce the stress concentration factor due to this shape effect (see Figure 9).

Introduction to illustration

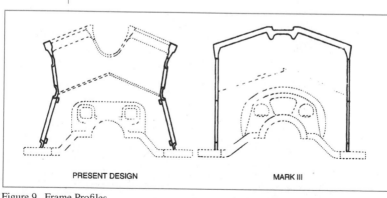

Figure 9. Frame Profiles

FIGURE P–24. Government Proposal (*continued*)

| Another problem and proposed solution with illustration | One problem experienced repeatedly on the LST-1176 was associated with the bosses for the handhole covers. Stress analysis work has shown a high stress concentration factor around these bosses because of unfused weld roots and also the increased localized stiffness caused by these bosses. Accordingly, we have designed handhole covers held in place by clamping action, which do not impose additional stress in the sidesheet due to the torquing down of the bolts against the bosses. We therefore gain the advantage of this direct means of eliminating a superimposed stress of 5,000 psi (see Figure 10). |

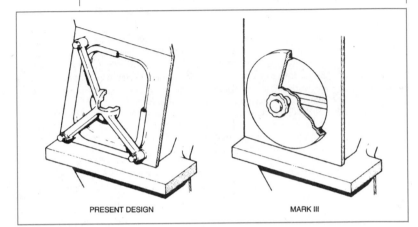

Figure 10. Side Cover Details

An additional problem experienced on earlier frames was cracking in the air manifold. The present design provides for the air header to be welded in as an integral stress member of the frame. A number of problems are associated with this type of design, some of which are stress concentration factors in the welding together of the top deck, the air header, and the athwartship plate. A second problem associated with this type of construction is the thermogradient experienced because the air manifold temperature is regulated to approximately 100°F., whereas other heat sinks at higher temperature levels create thermal stresses in the frame that are difficult to manage. Accordingly, Burdlorn proposes a separate and independent air header that will be bolted to the frame in such a manner as not to contribute to the overall stresses or stress concentration factors of the frame weldment.

2. Valves. During the 1,000 rpm test of the engine under Navy contract NOBS 72393, a wear rate of approximately .050 inch per 1,000 hours was experienced on the inlet valves. The wear rate on the exhaust valves was approximately .010 inch per 1,000 hours. A total wear of .100 inch would be considered acceptable for this size valve and insert. Accordingly, it will be appreciated that a wear life of more than 5,000 hours could be expected on the exhaust valve, but a much shorter life was experienced on the inlet valve. A careful analysis of all available information has revealed only two differences between the in-

take valve and the exhaust valve that would contribute to this difference in wear rates. These two differences are as follows:

a. The exhaust cam is driven through a hollow shaft that is 4 inches O.D. by 2½ inches I.D. The inlet cam is driven by a solid shaft that runs within the hollow exhaust camshaft and is 2½ inches in diameter. It will be recognized that the relative stiffness of these two shafts in torsion is in the ratio of 5½:1 as revealed by our calculations (see Figure 11).

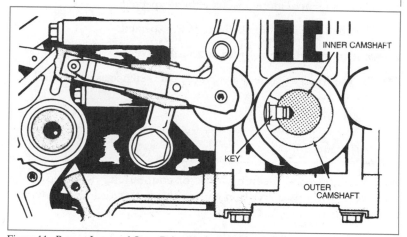

Figure 11. Present Inner and Outer Drive Shafts for Inlet and Exhaust Cams

b. The lift profiles of the intake and exhaust cams are identical; however, there is a greater dwell on the exhaust cam than there is on the inlet cam. The inlet cam has 15 degrees dwell, whereas the exhaust cam has 50 degrees dwell. The longer dwell on the exhaust cam, which cannot be incorporated in the inlet cam due to timing requirements, provides additional time in which the vibratory amplitudes in the valve train dampen or attentuate to a greater extent than the 15 degrees dwell provides on the inlet cam (see Figure 12). . . .

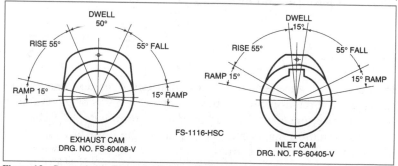

Figure 12. Comparison of Inlet and Exhaust Cam Profiles

FIGURE P-24. Government Proposal (*continued*)

proved / proven

Both *proved* and *proven* are acceptable past participles of *prove*, although *proved* is currently in wider use.

- They had *proved* more obstinate than expected.

- They had *proven* more obstinate than expected.

Proven is more commonly used as an **adjective.**

- She was hired because of her *proven* competence as a technical writer.

pseudo- / quasi-

As a prefix, *pseudo-*, meaning "false or counterfeit," is joined to the root word without a **hyphen** unless the root word begins with a capital letter.

- *pseudo*science [false science]

- *pseudo*-Freudian [pretending to agree with the psychoanalytic theories of Sigmund Freud]

Pseudo- is sometimes confused with *quasi-*, meaning "somewhat" or "partial." Unlike *semi-*, *quasi-* does not mean "half." *Quasi-* is usually hyphenated in combinations.

- *quasi*-scientific literature

(See also **bi-/semi-** and **prefixes.**)

punctuation ESL

Punctuation is a system of symbols that helps readers understand the structural relationships within a sentence. Marks of punctuation link, separate, enclose, indicate omissions, terminate, and classify. Most punctuation marks can perform more than one function. (See also **sentence construction.**)

Understanding punctuation is essential for writers because it enables them to communicate with clarity and precision. The use of punctuation is determined by grammatical conventions and the writer's intention. Think of punctuation as a substitute for the writer's facial expressions and vocal inflections. (See also **grammar.**)

Detailed information on each mark of punctuation is given in its own entry. The following are the 13 marks of punctuation.

Apostrophe	'
Brackets	[ ]
Colon	:
Comma	,
Dash	—
Exclamation mark	!
Hyphen	-
Parentheses	( )
Period	.
Question mark	?
Quotation marks	" "
Semicolon	;
Slash	/

(See also **abbreviations, acronyms and initialisms, capital letters, contractions, dates, ellipses, italics,** and **numbers.**)

purpose

What do you want your **readers** to know, believe, or be able to do when they have read your document? If you can answer that question, you have determined the purpose (or objective) of your writing project. Too often, beginning writers state their purposes in terms that are too broad to be practical. A purpose such as "to explain the fax modem feature of the IBM PC" is too general to be of any real help. In contrast, "to explain how to retrieve, send, print, and save faxes with the IBM PC" is a specific purpose that will help keep the writer on the right track. (See also **scope.**)

The writer's purpose is rarely simply to "explain" something. You must ask yourself, "Why do I need to explain it?" In answering that question, you may find that your purpose is also to persuade readers to change their attitude toward whatever you are explaining. (See also **persuasion.**)

Suppose a writer for a newsletter has been assigned to write an article about cardiopulmonary resuscitation (CPR). In answer to the question *what*, the writer could state the purpose as "to show the importance of CPR." To the question *why*, the writer might respond, "To encourage employees to sign up for evening CPR classes."

If you answer those two questions and put your answers in writing as your stated purpose, you will simplify your writing task and guarantee that you reach your goal. Try to state your purpose in a single sentence. If you find that you cannot, continue to work until you can. Remember that even a specific purpose is of no value unless you keep it in mind as you work. Be careful not to lose sight of your purpose as you become involved in the other steps of the writing process.

P

Q

The question mark (?) has several uses. Use a question mark to end a sentence that is a direct question.

* Where did you put the specifications?

Never use a question mark to end a sentence that is an indirect question.

* He asked me whether sales had increased this year?

Use a question mark to end a statement that has an interrogative meaning (a statement that is declarative in form but asks a question).

* The laboratory report is finished?

Use a question mark to end an interrogative clause within a declarative sentence.

* It was not until July (or was it August?) that we submitted the proposal.

When a directive is phrased as a question, a question mark is usually not used. However, a request (to a customer or a superior, for instance) almost always requires a question mark.

* Will you make sure that the machinery is operational by August 15. [directive]

* Will you email me if your entire shipment does not arrive by June 10? [request]

Question marks may follow a series of separate items within an interrogative sentence.

* Do you remember the date of the contract? Its terms? Whether you signed it?

Q

Retain the question mark in a title that is being cited, even though the sentence in which it appears has not ended.

- *Should Engineers Be Writers?* is the title of her book.

When used with quotations, the placement of the question mark is important. When the writer is asking a question, the question mark belongs outside the quotation marks.

- Did she say, "I don't think the project should continue"?

If the quotation itself is a question, the question mark goes inside the quotation marks.

- She asked, "When will we go?"

If both cases apply—the writer is asking a question and the quotation itself is a question—use a single question mark inside the quotation marks.

- Did she ask, "Will you go in my place?"

questionnaires

A questionnaire—a series of questions on a particular topic sent out to a number of people—serves the same purpose as an interview but does so on paper or by **email.** Questionnaires have several advantages over the personal interview as well as several disadvantages. A questionnaire allows you to gather information from many more people in a relatively brief time than you could by conducting personal interviews. It enables you to obtain simultaneous responses from people in different parts of the country or all over the world. Even people who live near you may be easier to reach by mail or email than in person. Those responding to a questionnaire do not face the pressure of composing thoughtful and complete answers to an interviewer—they have more time to complete a questionnaire, so they can think through their answers at a more leisurely pace. The questionnaire may yield more objective data because it reduces the possibility that the interviewer's tone of voice or facial expressions might influence an answer. Finally, the cost of writing, distributing, and tabulating a questionnaire is lower than the cost of conducting, recording, and tabulating numerous personal interviews.

Questionnaires have drawbacks, too. People who have strong opinions on a subject are more likely to respond to a questionnaire than those who do not, a factor that could slant the results. An interviewer can follow up on an answer with a pertinent question; at best, a questionnaire can be designed to let one question lead logically to another. Furthermore, sending out questionnaires and waiting for replies (especially if you use regular mail) takes considerably longer than conducting a personal interview. (See also **interviewing for information.**)

The advantages of a questionnaire will work in your favor only if the questionnaire is well designed. Your goal in designing a questionnaire is to obtain as much information as possible from your recipients with as little effort as possible on their part. The first rule to follow is to keep the questionnaire brief. The longer a questionnaire is, the less likely the recipient will be to complete and return it. Next, the questions should be easy to understand. A confusing question will yield confusing results, whereas a carefully worded question will be easy to answer. Ideally, recipients should be able to answer most questions with a "yes" or "no" or by checking or circling a choice among several options. Such answers require minimum effort on the part of the respondent and are easy to tabulate.

- Would you be willing to work a four-day workweek, ten hours a day, with every Friday off?

  Yes _____

  No _____

  No opinion _____

If you need more information than such questions produce, provide an appropriate range of answers, as in the following example.

- How many hours of overtime would you be willing to work each week?

  4 hours _____   10 hours _____

  6 hours _____   More than 10 hours _____

  8 hours _____   No overtime _____

Questions should be neutral; they should not be worded in such a way as to lead respondents to give a particular answer, which can result in inaccurate or skewed data.

Remember that you must eventually tabulate the answers; therefore, try to formulate questions with answers that can be easily counted and categorized. The easiest questions to tabulate are those for which the recipient does not have to compose an answer. Questions that require a comment take time to think about and write. As a result, they lessen your chances of obtaining a response, and any responses you do receive will be more difficult to interpret. Include a section on the questionnaire for additional comments, where recipients may clarify their overall attitude toward the subject. Include questions about the respondent's age, gender, education, occupation, and so on, if such information will be of value in interpreting the answers.

A questionnaire sent by mail or email must be accompanied by a letter explaining who you are, the purpose the questionnaire will serve, how the questionnaire will be used, the date by which you would like to receive a reply, and how and where you would like the completed questionnaire sent. If you send a paper copy, include your mailing address and a phone number; for an electronic questionnaire, include your email address. In general, the easier you make it for the recipients, the higher your response rate will be. (If you are using regular mail, include a stamped, self-addressed envelope.) If the information provided will be kept confidential, mention that in the letter.

Selecting the proper recipients for your questionnaire is crucial if you are to gather representative and usable data. To survey the opinions of all the employees in a small shop or a laboratory, you would simply send each worker a questionnaire. To survey the members of a club or professional society, you would mail questionnaires to those who are on a membership list. However, if you wanted to survey the opinions of large groups in the general population—for example, all medical technologists working in private laboratories or all independent garage owners—your task would not be so easy. Because you cannot include everybody in your survey, you need to choose a representative cross section. For example, you would want to include enough people from around the country (in other words, not from just one area, such as the Midwest), recipients of both genders, and people with an assortment of educational degrees. Only then could you make a generalized statement based on your findings from the sample. (The best sources of information on sampling techniques are marketing research and statistics texts.)

The sample questionnaire in Figure Q–1 on pages 533–35 was

October 18, 20--

To: All Company Employees

From: Nelson Barrett, Director *NB*
        Human Resources Department

Subject: Review of Flexible Working Hours Program

Please complete and return the questionnaire below regarding Luxwear Corporation's trial program of flexible working hours. Your answers will help us decide whether the program is worthwhile enough to continue permanently.

Return the completed questionnaire to Ken Rose, Mail Code 12B, by October 28. Your signature on the questionnaire is not necessary. Feel free to raise additional issues pertaining to the program. All responses will be given consideration.

If you want to discuss any item in the questionnaire, call Pam Peters in the Human Resources Department at extension 8812.

1. What kind of position do you occupy?

        Supervisory _____

        Nonsupervisory _____

2. Indicate to the nearest quarter of an hour your starting time under flextime.

    7:00 a.m. _____   8:15 a.m. _____

    7:15 a.m. _____   8:30 a.m. _____

    7:30 a.m. _____   8:45 a.m. _____

    7:45 a.m. _____   9:00 a.m. _____

    8:00 a.m. _____   Other (specify) _____

3. Where do you live?

    Talbot County _____   Greene County _____

    Montgomery County _____ Other (specify) _____

Q

FIGURE Q–1. Sample Questionnaire

4. How do you usually travel to work?

    Drive alone _____   Walk _____

    Taxi _____   Bus _____

    Train _____   Motorcycle _____

    Car pool _____   Other (specify) _____

    Bicycle _____

5. Has flextime affected your commuting time?

    Increase: Approx. number of minutes _____

    Decrease: Approx. number of minutes _____

    No change _____

6. If you drive alone or in a car pool, has flextime increased or decreased the amount of time it takes you to find a parking space?

    Increased _____ Decreased _____ No change _____

7. Has flextime had an effect on your productivity?

    a. Quality of work

        Increase _____ Decrease _____ No change _____

    b. Accuracy of work

        Increase _____ Decrease _____ No change _____

    c. Quiet time for uninterrupted work

        Increase _____ Decrease _____ No change _____

8. Have you had difficulty getting in touch with coworkers who are on different work schedules from yours?

    Yes _____   No _____

9. Have you had trouble scheduling meetings within flexible starting and quitting times?

    Yes _____   No _____

FIGURE Q-1. Sample Questionnaire (*continued*)

10. Has flextime affected the way you feel about your job?

Feel better about job   Feel worse about job

Slightly _____   Slightly _____

Considerably _____   Considerably _____

No change _____

11. How important is it for you to have flexibility in your working hours?

Very _____   Not very _____

Somewhat _____   Not at all _____

12. Has flextime allowed you more time to be with your family?

Yes _____   No _____

13. If you are responsible for the care of a young child or children, has flextime made it easier or more difficult for you to obtain baby-sitting or day-care services?

Easier _____   More difficult _____

No change _____

14. Do you recommend that the flextime program be made permanent?

Yes _____   No _____

15. Do you have suggestions for major changes in the program?

Yes (please specify) _____   No _____

Q

THANK YOU FOR YOUR ASSISTANCE

FIGURE Q–1. Sample Questionnaire (*continued*)

sent to employees in a large organization who had participated in a six-month program of flexible working hours. Under the program, employees worked a 40-hour, five-day week, with flexible starting and quitting times. Employees could start the workday between 7:00 and 10:00 a.m. and leave between 3:30 and 6:30 p.m., provided that they worked at least eight hours each day and took a half-hour lunch period midway through the day. (See also **research.**)

quid pro quo

Quid pro quo, which is Latin for "one thing for another," suggests mutual cooperation or "tit for tat" in a relationship between two groups or individuals. The term may be appropriate to business and legal contexts if you are sure your readers understand its meaning.

- It would be unwise to approve their request without insisting on a fair *quid pro quo.*

quotation marks   ESL

Quotation marks (" ") are used to enclose a direct quotation of spoken or written words. A direct quotation repeats the exact words of the original. Quotation marks have other special uses, but they should not be used for **emphasis.** A variety of guidelines govern the use of quotation marks.

Enclose in quotation marks anything that is quoted word for word (a direct quotation) from speech or written material.

- She said clearly, "I want the progress report by three o'clock."

- The report stated, "The potential in Florida for our franchise is as great as in California."

Do not enclose indirect quotations—usually introduced by the word *that*—in quotation marks. Indirect quotations are paraphrases of a speaker's words or ideas. (See also **paraphrasing.**)

- She said that she wanted the progress report by three o'clock.

- The report indicated that the potential for our franchise is as great in Florida as in California.

When you use quotation marks to indicate that you are quoting, do not make any changes in the quoted material unless you clearly indicate what you have done. For further information on using and incorporating quoted material, see **quotations.**

Use single quotation marks to enclose a quotation that appears within a quotation.

- John said, "Jane told me that she was going to 'stay with the project if it takes all year.'"

Use quotation marks to set off special words or terms only to point out that the term is used in context for a unique or special purpose (that is, in the sense of the term *so-called*).

- Typical of deductive analyses in real life are accident investigations: What chain of events caused the sinking of the "unsinkable" *Titanic* on its maiden voyage?

Slang, colloquial expressions, and attempts at humor, although infrequent in workplace writing, should seldom be set off by quotation marks.

- Our first six months amounted to a "~~shakedown cruise.~~" *shakedown cruise.*

Use quotation marks to enclose titles of reports, short stories, articles, essays, single episodes of radio and television programs, short musical works, paintings, and other artworks.

- Did you see the article "No-Fault Insurance and Your Motorcycle" in last Sunday's *Journal*?

Do not use quotation marks for titles of books and periodicals, which should appear in **italics.** Some titles, by convention, are not set off by quotation marks, underlining, or italics, although they are capitalized.

- Professional Writing [college course title], the Bible, the Constitution, Lincoln's Gettysburg Address, the Lands' End Catalog

  Commas and periods always go inside closing quotation marks.

- "Reading *Computer World* gives me the insider's view," he says, adding, "it's like a conversation with the top experts."

Semicolons and colons always go outside closing quotation marks.

- He said, "I will pay the full amount"; this statement surprised us.

ESL TIPS FOR USING QUOTATION MARKS AND PUNCTUATION

You may find that the correct use of punctuation and quotation marks in North American English frequently differs from that in your native culture.

> Style of quotation marks
>> "exceptional"   *not*   ,,exceptional" or <<exceptional>>
>
> Comma with quotation marks
>> "as a last resort,"   *not*   "as a last resort",
>
> Period with quotation marks
>> "to the bitter end."   *not*   "to the bitter end".
>
> Semicolon or colon with quotation marks
>> "there is no doubt";   *not*   "there is no doubt;"

All other punctuation follows the logic of the context: If the punctuation is a part of the material quoted, it goes inside the quotation marks; if the punctuation is not part of the material quoted, it goes outside the quotation marks.

Quotation marks can also be used as ditto marks. Instead of repeating a line of words or numbers directly beneath an identical set, place ditto marks directly beneath the word or set of words they represent. In formal writing, confine the use of ditto marks to **tables** and **lists**.

Q

- A is at a point equally distant from L and M.
  B   "   "   "   "   S and T.
  C   "   "   "   "   R and Q.

---

quotations   **ESL**

DIRECTORY
Overview   539
Direct Quotations   539
Indirect Quotations   539
Deletions or Omissions   539
Inserting Material into Quotations   540
Incorporating Quotations into Text   540

Overview

Using direct and indirect quotations is often an effective way to make or support a point. However, avoid the temptation to over-quote during the **note taking** phase of your **research;** concentrate on summarizing what you have read. When you do use a quotation, be sure to cite your source properly. If you do not, you will be guilty of **plagiarism.** (For specific details on MLA and APA citation styles, see **documenting sources.**)

Direct Quotations

A direct quotation is an exact, word-for-word copy of an original source. Choose direct quotations (which can be a word, a phrase, a sentence, or occasionally a paragraph) carefully and use them sparingly. Direct word-for-word quotations are enclosed in **quotation marks** and separated from the rest of the sentence by a **comma** or a **colon.**

- The economist stated, "Regulation cannot supply the dynamic stimulus that in other industries is supplied by competition."

When a quotation is divided, the material that interrupts the quotation is set off, before and after, by commas, and quotation marks are used around each part of the quotation.

- "Regulation," he said in a recent interview, "cannot supply the dynamic stimulus that in other industries is supplied by competition."

Indirect Quotations

An indirect quotation is a paraphrased version of an original text. It is usually introduced by the word *that* and is not set off from the rest of the sentence by punctuation marks. (See also **paraphrasing.**)

DIRECT   He said in a recent interview, "Regulation cannot supply the dynamic stimulus that in other industries is supplied by competition."

INDIRECT   In a recent interview he said that regulation does not stimulate the industry as well as competition does.

Deletions or Omissions

Deletions or omissions from quoted material are indicated by three ellipsis dots (. . .) within a sentence and by a period plus three ellipsis dots (. . . .) at the end of a sentence.

- "If monopolies could be made to respond . . . we would be able to enjoy the benefits of . . . large-scale efficiency. . . ."

If you are following the guidelines of the MLA (Modern Language Association), enclose the ellipsis dots in brackets.

- "If monopolies could be made to respond [. . .] we would be able to enjoy the benefits of [. . .] large-scale efficiency. [. . .]"

When a quoted passage begins in the middle of a sentence rather than at the beginning, ellipsis dots are not necessary; the fact that the first letter of the quoted material is not capitalized tells the reader that the quotation begins in midsentence.

- Rivero goes on to conclude that "coordination may lessen competition within a region."

Inserting Material into Quotations

When it is necessary to insert a clarifying comment within quoted material, use **brackets.**

- "The industry is organized as a relatively large, integrated system serving an extensive [geographic] area, with smaller systems existing as islands within the larger system's sphere of influence."

When quoted material contains an obvious error or might be questioned in some other way, the expression *sic* (Latin for "thus"), enclosed in brackets, follows the questionable material to indicate that the writer has quoted the material exactly as it appeared in the original.

Q

- The company considers the Baker Foundation to be a "guilt-edged [*sic*] investment."

Incorporating Quotations into Text

A quotation should make a significant contribution to the development of your subject by reinforcing a point you are making. When you work quoted material into a sentence or a passage, be sure it is related logically, grammatically, and syntactically to the rest of the sentence and to surrounding sentences.

Depending on the length of the quotations, there are two mechanical methods of handling them in text. In the MLA system of documentation, a quotation of three or fewer lines is incorporated into the text and enclosed in quotation marks. In the APA system, a quotation of fewer than 40 words is incorporated into the text and enclosed in quotation marks.

Material that is four lines or longer (MLA) or at least 40 words (APA) is usually inset; that is, it is set off from the body of the text by being indented from the left margin ten spaces (MLA) or five to seven spaces (APA). The quoted passage is spaced the same as the surrounding text and is not enclosed in quotation marks, as shown in Figure Q–2, which uses MLA style. If you are not following a specific style manual, you may block indent ten spaces from both the right and left margins for reports and other documents.

Notice in Figure Q–2 that the quotation blends with the content of the surrounding text, which makes use of **transitions** to introduce and then comment on the quotation. Later in the document, the following entry appears in the list of works cited as the source of the quotation in Figure Q–2.

* Alred, Gerald J. *The St. Martin's Bibliography of Business and Technical Communication.* New York: St. Martin's, 1997.

Do not rely too heavily on the use of quotations in the final version of your document. Overquoting can make your work appear derivative. Quote word for word only when your source concisely sums up a great deal of information or points to a trend or development important to your subject. Generally, avoid quoting anything that is more than one paragraph.

After reviewing a larger number of works in business and technical communication, Alred sees an inevitable connection between theory, practice, and pedagogy:

> Therefore, theory is necessary to prevent us from being overwhelmed by what is local, particular, and temporal. In turn, pedagogy both mediates practice and transforms our theory. Indeed, one reason I find this work rewarding is that I sense it puts me at the intersection of theory, practice, and pedagogy as they are involved with writing in the workplace. (ix–x)

The use of the Web today has reinforced this connection because it calls on the Web page designer to engage in a teaching function as well as reflect on the practice of Web design. For example, the widespread use of ...

Q

FIGURE Q–2. Long Quotation (MLA Style)

R

raise / rise

Both *raise* and *rise* mean "move to a higher position." However, *raise* is a transitive **verb** and always takes an **object** (*raise* crops), whereas *rise* is an intransitive verb and never takes an object (heat *rises*).

- *Raise* the lever to the ON position.

- Turn off the engine if the internal temperature *rises* above 110°.

re

Re (and its variant form, *in re*) is business and legal **jargon** meaning "in reference to" or "in the case of." Although *re* is sometimes used in **memos** and **email,** *subject* is a preferable term.

readers

The first rule of effective writing is to help your readers. If you overlook that commitment, your writing will not achieve its **purpose.** Further, your business or organization will lose time and money. Whether you are writing a letter, a **report,** or an **email,** you must first determine who the reader will be and then decide how to meet that reader's needs.

Your readers usually will be less familiar with your subject than you are. For example, when you are writing on a topic in your specialization for readers whose training is in other areas, you need to define specialized terms and explanations of principles that you, as a specialist, may take for granted. Even if you write a **trade journal**

article for others in your field, you must remember to explain new or unique uses of standard terms and principles.

When you write for many readers with similar backgrounds, try to visualize a typical member of that group and write to that reader. You might also make a list of characteristics (experience, training, and work habits, for example) of that reader to help you write at the appropriate level. This technique, used widely by professional writers, helps you to decide what you should or should not explain, according to the typical reader's needs.

When you have multiple audiences with varying backgrounds but cannot segment your document, determine your primary reading audience and address their needs. However, do not ignore your secondary audiences. Meet their needs to the best of your ability, as long as you do not sacrifice the needs of your primary audience. For example, recommendations, **executive summaries,** and **abstracts** can target executives who will be reading a report to understand the general implications of projects or technical systems. **Appendixes** containing **tables, graphs,** and raw technical data are geared to specialists who want to examine or use supporting data. The body of a report or proposal should be written to those readers with the most serious interest or who need to make decisions based on the details.

Many routine letters, **memos, email,** and short reports written for individual readers do not require such elaborate audience segmentation. Be sure to remember that person's exact needs as you write as well as consider how the reader will benefit from your message. (See also **correspondence, layout and design, point of view,** and **"you" viewpoint.**)

R

really

Really is an **adverb** meaning "actually" or "in fact." Although both *really* and *actually* are often used for **emphasis** or sarcasm in speech, such use should be avoided in writing. (See also **intensifiers.**)

- Did he ~~really~~ finish the report on time?

reason is because

The redundant phrase *the reason is because* is a colloquial expression that should be avoided in writing. In that phrase, the word *because* (which only repeats the notion of cause) should be replaced by *that*. You could also delete *the reason is* and use only *because*.

- Efficiency has increased 20 percent this year. The reason is
  that
  ~~because~~ our sales force has been more aggressive.
  ^

- Efficiency has increased 20 percent this year~~. The reason is~~ because our sales force has been more aggressive.

reciprocal pronouns (*see* pronouns)

reference books (*see* library research)

reference letters

Writing a reference letter (or letter of recommendation) can range from completing an admission form for a prospective student to composing a detailed description of professional accomplishments and personal characteristics for someone seeking employment.

To write an effective letter of recommendation, you must be familiar enough with the applicant's abilities or actual performance to offer an evaluation. Then you must truthfully and without embellishment communicate that evaluation to the inquirer. For the reference letter to achieve its purpose, you must specifically address the applicant's skills, abilities, knowledge, and personal characteristics in relation to the requested objective.

In a reference letter, always respond directly to the inquiry, being careful to address the specific questions asked. For the record, you must identify yourself: name, title or position, employer, and address. You could begin by stating the circumstances of your acquaintance and how long you have known the person for whom you are writing the letter. You should mention with as much substantiation as possible, one or two outstanding characteristics of the applicant. Organize the details in your letter in **decreasing order of importance.** Conclude with a brief summary of the applicant's qualifications and a clear statement of recommendation. Figure R–1 is a typical reference letter.

When you are asked to serve as a reference or to supply a letter of reference, be aware that applicants have a legal right to examine

---

IVY COLLEGE
DEPARTMENT OF BUSINESS
WEST LAFAYETTE, IN 47906
(691) 423-1719   (691) 423-2239 (FAX)
IVCO@IC.EDU (EMAIL)

January 14, 20--

Mr. Phillip Lester
Personnel Director
Thompson Enterprises
201 State Street
Springfield, IL 62705

Dear Mr. Lester:

How long writer has known applicant and the circumstances

As her employer and her former professor, I am happy to have the opportunity to recommend Kerry Hawkins. I've known Kerry for the last four years, first as a student in my class and for the last year as a research assistant.

Outstanding characteristics of applicant

Kerry is an excellent student, with above-average grades in our program. On the basis of a GPA of 3.6 (A = 4.0), Kerry was offered a research assistantship to work on a grant under my supervision. In every instance, Kerry completed her library search assignment within the time agreed upon. The material provided in the reports Kerry submitted met the requirements for my work and more. These reports were always well written. While working 15 hours a week on this project, Kerry has maintained a class load of 12 hours per semester.

Recommendation and summary of qualifications

I strongly recommend Kerry for her ability to work independently, to organize her time efficiently, and to write clearly and articulately. Please let me know if I can be of further service.

Sincerely yours,

Michael Paul

Michael Paul
Professor of Business

FIGURE R–1. Reference Letter

the materials in an organization's files that concern them, unless they sign a waiver. (See **correspondence** for letter format and general advice.)

reflexive pronouns (*see* **pronouns**)

refusal letters

When you receive a **complaint letter,** inquiry letter, or any letter to which you must give a negative reply or bad news, you may need to write a refusal letter, which could also take the form of a **memo** or an **email** message. (See also **correspondence** and **inquiries and responses.**)

Your letter should lead logically to the refusal. Often, stating the bad news in your **opening** will affect your reader negatively. The ideal refusal letter says "no" in such a way that you not only avoid antagonizing the reader but also maintain goodwill. To do that, you must convince the reader that your reasons for refusing are logical or understandable *before* you present the bad news. The following pattern is an effective way to deal with the problem.

1. In the opening, introduce the subject and establish a professional tone.
2. Review the facts leading to the bad news.
3. Give the bad news, based on the facts.
4. In the closing, establish or reestablish a positive relationship.

The primary purpose of the opening is to establish a professional tone and introduce the subject. If your refusal is in response to a complaint letter, do not begin by recalling the reader's disappointment ("We regret your dissatisfaction . . ."). You can express appreciation for your reader's time, effort, or interest, if appropriate, to soften the disappointment.

- The Screening Procedures Committee appreciates very much the time and effort you spent on your proposal for a new security-clearance procedure.

Next, analyze the circumstances of the situation sympathetically. Place yourself in the reader's position and try to see things from his or her point of view. Clearly establish the reasons you

WATASHAW ENGINEERING COMPANY
301 Industrial Lane
Decatur, IL 62525
Phone: (708) 222-3700
Fax: (708) 222-3707

March 26, 20--

Javier A. Lopez, President
TNCO Engineering Consultants
9001 Cummings Drive
St. Louis, MO 63129

Dear Mr. Lopez:

<table>
<tr><td>Buffer</td><td>I am honored to have been invited to address your regional meeting in St. Louis on May 17. That you would consider me as a potential contributor to such a gathering of experts is indeed flattering.</td></tr>
<tr><td>Review of facts and refusal</td><td>On checking my schedule, I find that I will be attending the annual meeting of our parent corporation's Board of Directors on that date. Therefore, as much as I would enjoy addressing your members, I must decline.</td></tr>
<tr><td>Goodwill close</td><td>I have been very favorably impressed over the years with your organization's contributions to the engineering profession, and I would welcome the opportunity to participate in a future meeting.</td></tr>
</table>

Sincerely,

Ralph P. Morgan

Ralph P. Morgan
Research Director

RPM/lcs

R

FIGURE R–2. Letter Refusing a Speaking Invitation

cannot do what the reader wants, even though you have not yet said you cannot do it. A good explanation should detail the reasons for your refusal so thoroughly that the reader will accept your refusal as a logical conclusion.

- We reviewed the potential effects of implementing your proposed security-clearance procedure on a company-wide basis. We asked the Systems and Procedures Department to review the data, surveyed industry practices, sought the views of senior management and department heads, and submitted the idea to our legal staff. As a result of this process, we have reached the following conclusions:
  - The cost savings you project are correct only if the procedure could be universally required.
  - The components of your procedure are legal, but most are not widely accepted by our industry.
  - Based on our survey, some components could alienate employees, who would perceive them as violating an individual's rights.
  - Enforcing company-wide use would prove costly and impractical.

Do not belabor the bad news—state your refusal quickly, clearly, and as positively as possible.

- For those reasons, the committee recommends that divisions continue their current security-screening procedures.

Close your letter or message in a way that reestablishes goodwill. You might provide an option, offer a friendly remark, assure the reader of your high opinion of his or her product or service, or merely wish the reader success. (See also **conclusions.**)

- Because some components of your procedure may apply in certain circumstances, we would like to feature your ideas in the next issue of *The Guardian.* I have asked the editor to contact you next week. On behalf of the committee, thank you for the thoughtful proposal.

The refusal letter in Figure R–2 on page 547 declines an invitation to speak.

regarding / with regard to

In regards to and *with regards to* are incorrect **idioms** for *in regard to* and *with regard to.* Both *as regards* and *regarding* are acceptable variants.

- In *regard* ~~regards~~ to the building contract, this question is pertinent.

- *Regarding* ~~With regards to~~ the building contract, this question is pertinent.

relative pronouns (*see* pronouns)

repetition

The deliberate use of repetition to build a sustained effect or to emphasize a feeling or idea can be a powerful device. (See also **emphasis.**)

- Similarly, atoms *come and go* in a molecule, but the molecule *remains;* molecules *come and go* in a cell, but the cell *remains;* cells *come and go* in a body, but the body *remains;* persons *come and go* in an organization, but the organization *remains.*
    —Kenneth Boulding, *Beyond Economics*

Repeating key words from a previous sentence or paragraph can also be used effectively to achieve **transition.**

- For many years, *oil* has been a major industrial energy source. However, *oil* supplies are limited, and other sources of energy must be developed.

Be consistent in the word or phrase you use to refer to something. In technical writing, it is generally better to repeat a word (so there will be no question in the reader's mind that you mean the same thing) than to use a synonym in order to avoid repetition. Your primary goal in technical writing is effective and precise communication rather than **elegant variation.**

MIXED	Several recent *analyses* support our conclusion. These *studies* cast doubts on the feasibility of long-range forecasting. The *reports,* however, are strictly theoretical.
CONSISTENT	Several recent theoretical *studies* support our conclusion. These *studies* cast doubts on the feasibility of long-range forecasting. They are, however, strictly theoretical.

Purposeless repetition, however, makes a sentence awkward and hides its key ideas.

- She said that the customer ~~said that the order was to be canceled.~~ *canceled the order.*

The needless repetition of ideas can be equally awkward.

REPETITIVE   In this modern world of ours today, the well-informed, knowledgeable employee will be well ahead of the competition.

ECONOMICAL   To succeed, today's employee must be well informed.

reports

A report is an organized presentation of factual information, often aimed at multiple audiences, that presents the results of, for example, an investigation, a trip, or a research project. For any report, assessing the **readers'** needs is important.

Formal reports present the results of projects that may require months of work and involve large sums of money. Such projects may be done either for your own organization or as a contractual requirement for another organization. Formal reports generally follow a precise **format** and include some or all of the report elements discussed in the entry **formal reports.**

Informal and short reports normally run from a few paragraphs to a few pages and include only the essential elements of a report: **introduction,** body, **conclusions,** and recommendations. Because of their brevity, informal reports are customarily written as letters (if written for someone outside the organization) or as **memos** (if written for someone inside the organization).

The introduction serves several functions: It announces the subject of the report, states the **purpose,** and, when appropriate, gives essential background information. The introduction may also summarize any conclusions, findings, or recommendations made in the report. Managers, supervisors, and clients find an **executive summary** useful because it gives them essential information at a glance and helps focus their thinking as they read the rest of the report. (See also **abstracts.**)

The body of the report should present a clearly organized account of the report's subject—the results of a test carried out, the status of a construction project, and so on. The amount of detail to include depends on the complexity of the subject and on your reader's familiarity with it.

The conclusion should summarize your findings and tell readers what you think the significance of those findings may be. In some reports, a final, separate section gives recommendations; in others, the conclusions and the recommendations are combined into one section. In the final section, you make suggestions for a course of action based on the data you have presented.

This handbook discusses many typical reports, which may be formal and informal and vary in length and complexity. The following is a list of relevant entries:

feasibility reports
investigative reports
laboratory reports
progress and activity reports
test reports
trip reports
trouble reports

(See also **proposals.**)

research

Research is the process of investigation, the discovery of facts. Research, however, must be preceded by **preparation,** especially consideration of your **readers, purpose,** and **scope.** Without adequate preparation, your research effort will not be focused.

On the job, your primary source of information is your own knowledge and experience. Therefore, before going elsewhere, interview yourself. That technique, commonly known as **brainstorming,** may also stimulate your thinking about additional places to seek information. In fact, many projects require information from a variety of sources. Researchers frequently distinguish between primary and secondary research, depending on the types of sources consulted and the method of gathering information.

Primary Research

Primary research refers to the gathering of raw data compiled from direct observation, surveys, experiments, **questionnaires,** interviews, audio- and videotape recordings, and the like. In fact, direct observation and hands-on experience are the only ways to obtain certain kinds of information, such as the behavior of people and animals, certain natural phenomena, mechanical processes, and the operation of tools and equipment.

If you are planning research that involves observation, choose your sites and times carefully. Be sure to obtain permission in advance when necessary—observations may be illegal, resented, or contrary to an organization's policy. Keep accurate, complete records that indicate date, time of day, duration of the observation, and so on. Save interpretations of your observations for future analysis. Be aware that research involving observations may be time-consuming, complicated, and expensive and that your presence may inadvertently influence the subjects you are observing.

If you have to write directions for some task, you might gather information by performing the task yourself. Afterward, you will in effect be interviewing yourself based on your experience. Brainstorm a rough outline of your experience. After you write the draft, have someone unfamiliar with the task operate the equipment or perform the procedure according to your written directions. If the tester has a problem, rewrite the troublesome passage until it is clear and easy to follow. (See also **usability testing.**)

Secondary Research

Secondary research refers to the gathering of information that has been analyzed, assessed, evaluated, compiled, or otherwise organized into accessible form. The forms or sources include books, articles, **reports,** Web documents, dissertations, operating and procedures manuals, brochures, and so forth. The entries **Internet research** and **library research** provide tactics for finding and evaluating sources.

As you seek information from such sources, keep the following guidelines in mind. The more recent the information, the better. **Trade journal articles** are essential sources of current-awareness information because books take longer to write, publish, and distribute than trade journal articles. Conference proceedings, an even better source of up-to-date information, contain papers presented at meetings of trade, industrial, and professional societies about recent research results or work in progress. Much of the information presented at conferences either will not be published elsewhere or will not appear for a year or more in a published journal. (See also **documenting sources, paraphrasing,** and **plagiarism.**)

When a resource seems useful, read it carefully and use **note taking** techniques for any information that falls within the scope of your research. If you think of additional questions about the topic as you read, jot them down. Some of your questions may eventually be answered in other research sources; those that remain unanswered can guide you to further research. For example, you may discover that you need to talk with an expert. Not only can someone skilled in the field answer many of your questions, he or she can also suggest further sources of information. To make the most of such discussions, see **interviewing for information** and **listening.**

resignation letters or memos

When you are planning to leave a job, for whatever reason, you usually write a resignation letter to your supervisor or to an appropriate person in the Human Resources Department. Start a resignation letter or memo on a positive note, regardless of the circumstances under which you are leaving. You might, for example, point out how you have benefited from working for the company or say something complimentary about how well the company is run. Or you might say something positive about the people with whom you have been associated. For strategies concerning negative messages, see **refusal letters.**

Then explain why you are leaving. Make your explanation objective and factual and avoid recriminations. Your resignation letter or memo will become part of your permanent file with the company; if it is angry and accusing, it could haunt you in the future when you need references.

Your letter or memo should give enough notice to allow your employer time to find a replacement. It might be no more than two weeks, or it might be enough time to enable you to train your replacement. Some organizations may ask for a notice equivalent to the number of weeks of vacation you receive. Check the policy of your employer before you begin your letter.

R

INTEROFFICE MEMORANDUM

To: W. R. Johnson, Director of Purchasing
From: J. L. Washburn, Purchasing Agent *JLW*
Date: January 7, 20--
Subject: Resignation from Barnside Appliances,
           effective January 21, 20--

Positive opening

My three years at Barnside Appliances have been an invaluable period of learning and professional development. I arrived as a novice, and I believe that today I am a professional—primarily as a result of the personal attention and tutoring I have received from my superiors and the fine example set by both my superiors and my peers.

Reason for leaving

I believe, however, that the time has come for me to move on to a larger company that can give me an opportunity to continue my professional development. Therefore, I have accepted a position with General Electric, where I am scheduled to begin on January 26. Thus, my last day at Barnside will be January 21. I will be happy to train my replacement during the next two weeks.

Positive closing

Many thanks for the experiences I have gained and best wishes for the future.

FIGURE R–3. Sample Resignation Memo (to Accept a Better Position)

INTEROFFICE MEMORANDUM

To: T. W. Haney, Vice President, Administration
From: L. R. Rupp, Executive Assistant *LRR*
Date: February 12, 20--
Subject: Resignation from Winterhaven, effective
           March 1, 20--

Positive opening

My five-year stay with the Winterhaven Company has been a very pleasant experience, and I believe that it has been mutually beneficial.

Reason for leaving

Because the recent restructuring of my job leaves no career path open to me, however, I have accepted a position with another company that I feel will offer me greater advancement opportunities. I am, therefore, submitting my resignation, to be effective on March 1, 20--.

Positive closing

I have enjoyed working with my coworkers at Winterhaven and wish the company success in the future.

FIGURE R–4. Sample Resignation Memo (Under Negative Conditions)

The sample resignation memo in Figure R–3 on page 554 is from an employee who is leaving to take a job offering greater opportunities. The memo of resignation in Figure R–4 on page 554 is written by an employee who is leaving under unhappy circumstances; notice that it opens and closes positively and that the reason for the resignation is stated without apparent anger or bitterness. (See also **correspondence** and **memos.**)

respective / respectively

Respective is an **adjective** that means "pertaining to two or more things regarded individually."

- The committee members prepared their *respective* reports.

Respectively is the **adverb** form of *respective,* meaning "singly, in the order designated."

- The first, second, and third prizes in the sales contest were awarded to Maria Juarez, Dan Wesp, and Simone Luce, *respectively.*

Respective and *respectively* are often unnecessary because the meaning of individuality is already clear.

- The committee members prepared their respective reports.

  Each ... *a report.*

R

restrictive and nonrestrictive elements (ESL)

Modifying **phrases** and **clauses** may be either restrictive or nonrestrictive. A nonrestrictive phrase or clause provides additional information about what it modifies, but it does not restrict the meaning of what it modifies. A nonrestrictive phrase or clause can be removed without changing the essential meaning of the sentence. It is, in effect, a parenthetical element set off by **commas** to show its loose relationship with the rest of the sentence.

NONRESTRICTIVE PHRASE	This instrument, *called a backscatter gauge*, fires beta particles at an object and counts the particles that bounce back.
NONRESTRICTIVE CLAUSE	The annual report, *which was distributed yesterday*, shows that sales increased 20 percent last year.

A restrictive phrase or clause limits, or restricts, the meaning of what it modifies. If it were removed, the essential meaning of the sentence would be changed. Because a restrictive phrase or clause is essential to the meaning of the sentence, it is never set off by commas.

RESTRICTIVE PHRASE	All employees *wishing to donate blood* may take Thursday afternoon off.
RESTRICTIVE CLAUSE	Companies *that adopt the plan* nearly always show profit increases.

It is important for writers to distinguish between nonrestrictive and restrictive elements. The same sentence can take on two entirely different meanings depending on whether a modifying element is set off by commas (because it is nonrestrictive) or not (because it is restrictive). A slip by the writer can not only mislead readers but also embarrass the writer.

MISLEADING	I think you will be positively impressed by our systems engineers who are located at our home office.
ACCURATE	I think you will be positively impressed by our systems engineers, who are located at our home office.

The first sentence suggests that readers may not be as positively impressed by systems engineers who are located elsewhere.

Use *which* to introduce nonrestrictive clauses and *that* to introduce restrictive clauses.

- After John left the restaurant, *which* is one of the finest in New York, he came directly to my office.

- Companies *that* diversify usually succeed.

résumés

DIRECTORY
Overview   557
Analyzing Your Background   557
Organizing the Résumé   558
Writing the Résumé   561
Electronic Résumés   563
Sample Résumés   571

Overview

A résumé is a summary of your qualifications and the key element in a **job search.** A résumé itemizes, in one or two pages, the qualifications that you can mention only briefly in your **application letter.** On the basis of the information in your résumé, prospective employers decide whether to ask you to come in for a personal interview. If you are invited to an interview, the interviewer can base specific questions on the data the résumé contains. (See also **interviewing for a job.**)

Résumés form the basis for a potential employer's first impression, so do not pinch pennies when you create a résumé. Use a high-quality printer and high-grade paper. Take the time to make sure your résumé is attractive, well organized, easy to read, and free of errors. Proofread your résumé carefully, verify the accuracy of the information, and have someone else review it. Experiment with the design to determine a layout that highlights your strengths. Try to keep the résumé to one page unless you have a great deal of experience. (See also **layout and design** and **proofreading.**)

Analyzing Your Background

In preparing to write your résumé, determine what kind of job you are seeking. Then ask yourself what information about you and your background would be most important to a prospective employer in the field you have chosen. On the basis of your answers, decide what sort of details you should include in your résumé and how you can most effectively present your qualifications. Brainstorm about yourself and your background. Ask yourself the following questions:

- What college or colleges did you attend? What degree(s) do you hold? What was your major field of study? What academic honors were you awarded?
- What extracurricular activities have contributed to your learning experience? Your leadership skills? Any collaborative experience?
- What jobs have you held? What were your principal and secondary duties in each of them? When and how long did you hold each job?
- What experience did you gain that would be of value in the kind of job you are seeking?

Use your answers as a starting point and let one question lead to another. (See also **brainstorming.**)

Organizing the Résumé

A number of different organizational patterns can be used effectively. A common one arranges information chronologically in the following topical categories:

> Heading
> Employment objective (optional)
> Education
> Employment experience
> Skills and activities
> References

Whether you place education or employment experience first depends on which would strengthen your résumé. If you are a recent graduate, you would probably list education first since you may not have much work experience. If you have years of related job experience, you would probably list job experience first since your interviewer will most likely be interested in the skills you gained at your previous jobs. In both cases, list the most recent education or job experience first, the next most recent experience second, and so on.

The Heading.   Create a heading that clearly shows your name, address, telephone and fax numbers, and email address. Do not include a date in the heading; if you do, you will have to change it every time you submit your résumé to a prospective employer. Centering your heading at the top of the page usually works best.

CONSUELA B. SANDOVAL
6819 Elm Street
Somerville, Massachusetts 02144
(617) 625-1552
cbsand@cpu.fairview.edu

Employment Objective.   If you have a clear employment objective, you can include it in your résumé. If you do, state not only your immediate employment objective but the direction you hope your career will take. An employment objective may be particularly useful in an electronic résumé because it serves both as a screening device and as an introduction to the material in the résumé.

- **EMPLOYMENT OBJECTIVE**
  To obtain a position that allows me to use my computer-science training to solve engineering problems with the potential to gain valuable management experience.

Education.   List the college or colleges you attended, the dates you attended each one, the degree or degrees you received, your major field of study, and any academic honors you earned. Mention the name of the high school you attended only if it was relatively recently, your résumé is sparse, or you want to call attention to awards you earned in high school or to related programs, internships, or study abroad.

- **EDUCATION**
  Bachelor of Science in Engineering (expected June 20--)
  Georgia Institute of Technology
  Cumulative Grade Point Average: 3.46 out of possible 4.0

  MAJOR COURSES
  Calculus I, II, III, IV
  Methods of Digital Computations
  Advanced Computer Techniques
  Special Computer Techniques
  Differential Equations
  Graphic Display
  Software Design

ACTIVITIES AND HONORS
Phi Chi Epsilon—Honor Society for Women in Business and
    Engineering
Society of Women Engineers—Secretary-Treasurer, Junior Year
American Institute of Industrial Engineers—Secretary, Junior
    Year
Engineering Science Club
Doris Harlow Scholarship recipient for two consecutive years
Dean's List six of eight semesters

Employment Experience.   List all your full-time jobs, starting
with the most recent and working backward. If you have had little
full-time work experience, list part-time and temporary jobs, including internships. Provide a concise description of your duties for
those jobs with duties similar to those of the job you are seeking; if a
job is not directly relevant, give only a job title and a brief description of duties that developed broad skills valued in the position you
are seeking. For example, if you were a lifeguard, focus on supervisory experience or even experience in averting disaster to highlight
management and decision making as well as crisis-control skills. If
you have been with one company for a number of years, highlight
your accomplishments and promotions during those years. List military service as a job; give the dates you served, your duty specialty,
and your rank at discharge. Discuss military duties only if they
apply to the job you are applying for.

- **EMPLOYMENT EXPERIENCE**
  COMPUTER SYSTEMS INTERNATIONAL, ATLANTA, GEORGIA
  September 20-- to Present
    As Assistant Training Director, assisted in preparing Profes-
    sional Training Program for the Design, Data Entry, and Engi-
    neering Departments.

  VACATIONLAND AMUSEMENT PARK, Toccoa, Georgia
  April 20-- to August 20--
    As Chief Lifeguard, trained and supervised three other life-
    guards.

Some résumés organize work experience by type rather than by
job chronology. Instead of listing the positions held in sequence, a
functional résumé lists jobs by the functions performed in all jobs.

If you were preparing a functional résumé, you might group your experience and skills under categories such as "Management," "Project Development," "Training," "Sales," and so on. Organization by function is useful for applicants who want to stress certain skills important to the prospective employer or industry or who have been employed at only one job and want to demonstrate the diversity of their experience in that position. It is also useful for anyone with gaps in his or her résumé caused by unemployment or illness. Although functional arrangement can be effective, prospective employers know that it is sometimes used to cover weaknesses, so use it only when you feel it is to your advantage and be prepared to explain any gaps it may reveal.

Skills and Activities.   The skills and activities category usually comes near the end of the résumé. Include items such as fluency in a foreign language, writing and editing abilities, specialized technical knowledge (such as knowledge of specific computer systems or desktop-publishing programs), student or community activities, professional or club memberships, and published works. Be selective: Do not duplicate information given in other categories and include only activities and skills that support your employment objective. Provide a heading for this category that fits its contents, depending on which skills or activities you want to emphasize, such as "Skills and Activities," "Professional Affiliations," or "Publications and Memberships."

References.   You can include references as part of the résumé or provide a statement on the résumé that references will be provided upon request. Either way, do not give anyone as a reference without first obtaining his or her permission.

Writing the Résumé

When writing your résumé, use action **verbs** (for example, "managed" rather than "was the manager") and state ideas concisely. Even though the résumé is about you, do not overuse "I."

- *Promoted*
  I was promoted to Section Leader in June 20--.

Be truthful in your résumé. If you give false information and are found out, the consequences could be serious. In fact, the truthfulness of your résumé reflects not only your own ethical stance but also the integrity with which you would represent the organization. (See also **ethics in writing**.)

Avoid listing the salary you desire in the résumé. On the one hand, you may price yourself out of a job you want if the salary you list is higher than a potential employer is willing to pay. On the other hand, if you list a low salary, any offer you receive could be less than it might otherwise have been.

If you are returning to the workplace after an absence, most career experts say that it is important to acknowledge the gap in your career rather than trying to hide it. That is particularly true if, for example, you are reentering the workforce because you have devoted a full-time period to care for children or dependent adults. Do not undervalue such work. Although unpaid, it often provides experience that develops important time-management, problem-solving, organizational, and interpersonal skills. The following examples illustrate how you might reflect such experiences in a résumé.

- **Primary Child-Care Provider, 20-- to 20--**
  Furnished full-time care to three preschool children in home environment. Instructed in crafts, beginning scholastic skills, time management, basics of nutrition, and swimming. Organized activities, managed household, and served as block-watch captain.

- **Home Caregiver, 20-- to 20--**
  Provided 60 hours per week in-home care to Alzheimer's patient. Coordinated medical care, developed exercise programs, completed and processed complex medical forms, administered medications, organized budget, and managed home environment.

If you have participated in volunteer work during such a period, list that experience. Volunteer work often results in the same experience as does full-time employed work, a fact that your résumé should reflect, as in the following example.

- **School Association Coordinator, 20-- to 20--**
  Managed special activities of the Briarwood High School Parent-Teacher Association. Planned and coordinated meetings, scheduled events, and supervised fund-drive operations. Raised $70,000 toward refurbishing the school auditorium.

Electronic Résumés

In addition to the traditional paper résumé, you can submit a résumé on disk, through **email** to a potential employer, or to a commercial database service or Web site that functions as an electronic employment agency. Or you can display and periodically update an electronic résumé at your own Web site. If you plan to post your résumé on the Web, keep the following points in mind.

- Follow the general advice for **Web page design,** such as viewing your résumé on several browsers to see how it looks.
- Do not list your home address or phone number; instead include an email link ("mailto") at the top of the résumé.
- Format the résumé as an ASCII text file so it will scroll correctly and be universally readable.
- Just below your name, you may wish to provide a series of hypertext links to such important categories as "experience" and "education."
- Use a counter to keep track of the number of times your résumé Web page has been visited.

A non-Web application of electronic résumés, particularly in large companies, is used to facilitate the screening of many applicants. Although an applicant's paper résumé can be scanned into computer files, employers appreciate receiving an electronic version that facilitates loading the information into their résumé database.

Electronic non-Web résumés differ from paper ones in a number of ways; one significant difference is the abundant use of nouns rather than action verbs. The use of nouns as keywords (sometimes called *descriptors*) is important because potential employers use them to screen candidates for specific qualifications and job descriptions. Keywords that produce a hit in the search process are critical to the success of an electronic résumé. One way to determine what keywords are appropriate to include in your résumé is to read job-vacancy postings and note terms used for job classifications that match your interests and qualifications. (Note: If you plan to post your résumé on the Web, do not include your home address or telephone number.)

The organization and design of electronic résumés also vary somewhat from those of paper ones. The main heading with your name and email address should be the first lines on the electronic

R

résumé, with all lines centered. Immediately follow the main heading with a section of keywords, as shown in the following example:

CAROL ANN WALKER
1436 W. Schantz Avenue
Dayton, Ohio 45401
(513) 339-2712
caw@hbk.com

Keywords: Financial Planner. Research Analyst. Banking Intern. Executive Curriculum. Ph.D. (in progress) The Wharton School. University of Pennsylvania. Computer Model Development. Articles published in *Finance Journal*. Written and oral communication skills. Grant writing for Foundation and Government Funding

Following the section of keywords (which can number as many as 50), you can include the fairly standard sections shown earlier in this entry, such as "Employment Objective," "Education," "Employment," and "Skills and Activities." However, as in the keyword section, use nouns rather than verbs (for example, *designer* and *management* rather than *designed* and *managed*).

If a résumé is to be sent as an attachment or included on a Web page as a PDF file, feel free to use formatting (boldface, italics, and underlining); it will remain in place. If you submit a résumé on a disk or on paper to be scanned, the document needs to be universally readable in any application. Therefore, as suggested earlier, either format the résumé as an ASCII file on your disk or use ASCII-compatible features on the paper copy. For example, do not use underlining, italics, or boldface type. Avoid decorative, uncommon, or otherwise fancy typefaces; use simple font styles (sans serif, for example) and sizes between 10 and 14 points. For résumés to be scanned, use white space generously; scanners use it to recognize that one topic has ended and another has begun. While it is best to keep the length of a paper résumé to one page, do not worry about limiting an electronic résumé to a single page. Do keep the résumé as simple, clear, and concise as possible. Use white or beige paper and do not fold the document when you mail it; during scanning, a folded line can produce a misreading.

DAVID B. EDWARDS
6819 Locustview Drive
Topeka, Kansas 66614
(913) 233-1552
dedwards@cpu.fairview.edu

EDUCATION

Fairview Community College, Topeka, Kansas
Associate's Degree, Computer Science, June 20--
Dean's Honor List Award — six quarters

RELEVANT COURSE WORK
    Operating Systems Design   Computer Graphics
    Database Management   Data Structures
    Introduction to Cybernetics   Technical Writing

EMPLOYMENT EXPERIENCE

COMPUTER CONSULTANT September 20---June 20--, Fairview Community College Computer Center: Advised and trained novice computer users; wrote and maintained Unipro operating system documentation.

TUTOR January–June 20--, Fairview Community College: Assisted students in mathematics and computer programming.

SKILLS AND ACTIVITIES

UNIPRO OPERATING SYSTEM: Thorough knowledge of its word-processing, text-editing, and file-formatting programs.

WRITING AND EDITING SKILLS: Experience in documenting computer programs for beginning programmers and users.

FAIRVIEW COMMUNITY MICROCOMPUTER USERS' GROUP: Cofounder and editor of monthly newsletter; listserv manager.

FURTHER INFORMATION

References, college transcripts, a portfolio of computer programs, and writing samples available upon request.

FIGURE R–5. Sample Résumé for Recent College Graduate

ROBERT MANDILLO
7761 Shalamar Drive
Dayton, Ohio 45424

Home: (513) 255-4137   Fax: (513) 255-3117
Business: (513) 543-3337   mand@juno.com

EMPLOYMENT EXPERIENCE

MANAGER, ENGINEERING DRAFTING DEPARTMENT — March 1995 to
    Present
Wright-Patterson Air Force Base, Dayton, Ohio

Supervise 17 Drafting Mechanics in support of the engineering design staff. Develop, evaluate, and improve materials
and equipment for the design and construction of exhibits.
Write specifications, negotiate with vendors, and initiate
procurement activities for exhibit design support.

SUPERVISOR, GRAPHICS ILLUSTRATORS — May 1983 to February
    1995
Henderson Advertising Agency, Cincinnati, Ohio

Supervised five Illustrators and four Drafting Mechanics after
promotion from Graphics Technician; analyzed and approved
work-order requirements; selected appropriate media and
techniques for orders; rendered illustrations in pencil and ink;
converted department to CAD system.

EDUCATION

Bachelor of Science in Mechanical Engineering Technology, 19--
Edison State College, Wooster, Ohio

Associate's Degree in Mechanical Drafting, 19--
Wooster Community College, Wooster, Ohio

PROFESSIONAL AFFILIATIONS

National Association of Mechanical Engineers and Drafting
Mechanics

REFERENCES

References, letters of recommendation, and a portfolio of original designs and drawings available upon request.

FIGURE R-6. Sample Résumé for Applicant with Extensive Work Record

CAROL ANN WALKER
1436 W. Schantz Avenue
Laurel, Pennsylvania 17322
(717) 339-2712
caw@hbk.com

EMPLOYMENT OBJECTIVE

Obtaining a position in financial research, leading to a management
position in corporate finance.

EDUCATION

Bachelor of Science in Business Administration
    (expected June 20--)
Indiana University
Emphasis: Finance   Minor: Technical Communication
Dean's List: 3.88 grade point average out of possible 4.0
Senior Honor Society, 20--

EMPLOYMENT EXPERIENCE

FIRST BANK, INC., of Bloomington, Indiana, 20--
Research Assistant, Summer and Fall Quarters
    Assisted manager of corporate planning and developed long-
    range planning models.

MARTIN FINANCIAL RESEARCH SERVICES, Bloomington, Indiana, 20--
    to 20--
Financial Audit Intern
    Developed a design concept for in-house financial audits and
    provided research assistance to staff.

SKILLS AND ACTIVITIES

Associate Editor, Business School Alumni Newsletter
    Wrote articles on financial planning with computer models; sur-
    veyed business periodicals for potential articles; edited submis-
    sions.

President, Women's Transit Program
    Coordinated activities to provide safe nighttime transportation to
    and from residence halls and campus buildings.

REFERENCES

Available upon request.

FIGURE R–7. Student Résumé

CAROL ANN WALKER
1436 W. Schantz Avenue
Laurel, Pennsylvania 17322
(717) 339-2712
caw@hbk.com

EMPLOYMENT EXPERIENCE

KERFHEIMER CORPORATION, Dayton, Ohio
November 20-- to Present

Senior Financial Analyst

Report to Senior Vice President for Corporate Financial Planning.
Develop manufacturing cost estimates totaling $30 million annually
for mining and construction equipment with Department of Defense.

Financial Analyst

Developed $50-million funding estimates for major Department of
Defense contracts for troop carriers and digging and earth-moving
machines.

Researched funding options, recommending those with most favorable
rates and terms.

Promoted to senior financial analyst: 20--

FIRST BANK, INC., Bloomington, Indiana
September 20-- to August 20--

Planning Analyst

Developed successful computer models for short- and long-range
planning.

EDUCATION

Ph.D. in Finance: expected, June 20--
The Wharton School of the University of Pennsylvania

M.S. in Business Administration, 20--
University of Wisconsin–Milwaukee
"Executive Curriculum" for employees identified as promising by their
employers.

B.S. in Business Administration (*magna cum laude*), 20--
Indiana University
Emphasis: Finance   Minor: Technical Communication

FIGURE R–8. Advanced Résumé Organized Chronologically by Job (first page)

Carol Ann Walker   Page 2

SKILLS AND ACTIVITIES

Published "Developing Computer Models for Financial Planning," *Midwest Finance Journal* (Vol. 34, No. 2, 20--), pp. 126–136.

Association for Corporate Financial Planning, Senior Member.

REFERENCES

References and a portfolio of financial plans are available upon request.

FIGURE R-8. Advanced Résumé Organized Chronologically by Job (*continued*)

CAROL ANN WALKER
1436 W. Schantz Avenue
Laurel, Pennsylvania 17322
(717) 339-2712
caw@hbk.com

MAJOR ACCOMPLISHMENTS

FINANCIAL PLANNING
- Researched funding options to achieve a 23% return on investment.
- Developed long-range funding requirements for over $1 billion in government and military contracts.
- Developed a computer model for long- and short-range planning that saved 65% in proposal-preparation time.
- Received the Financial Planner of the Year Award from the Association of Financial Planners, a national organization composed of both practitioners and academics.

CAPITAL ACQUISITION
- Developed strategies to acquire over $1 billion at 3% below market rate.
- Secured over $100 million through private and government research grants.
- Developed computer models for capital acquisition that enabled the company to decrease its long-term debt during several major building expansions.

FIGURE R-9. Advanced Résumé Organized by Function

Carol Ann Walker

RESEARCH AND ANALYSIS
- Researched and developed computer models applied to practical problems of corporate finance.
- Functioned primarily as a researcher at two different firms for over 11 years.
- Published research in financial journals while pursuing an advanced degree at the Wharton School.

EDUCATION

Ph.D. in Finance: expected, June 20--
The Wharton School of the University of Pennsylvania

M.S. in Business Administration, 20--
University of Wisconsin–Milwaukee
"Executive Curriculum" for employees identified as promising by their employers.

B.S. in Business Administration (*magna cum laude*), 20--
Indiana University
Emphasis: Finance   Minor: Technical Communication

EMPLOYMENT EXPERIENCE

KERFHEIMER CORPORATION, Dayton, Ohio
November 20-- to Present
Senior Financial Analyst
Financial Analyst

FIRST BANK, INC., Bloomington, Indiana
September 20-- to August 20--
Planning Analyst

PUBLICATIONS AND MEMBERSHIPS

Published "Developing Computer Models for Financial Planning,"
Midwest Finance Journal (Vol. 34, No. 2, 20--), pp. 126–136.

Association for Corporate Financial Planning, Senior Member.

REFERENCES

References and a portfolio of financial plans are available upon request.

FIGURE R-9. Advanced Résumé Organized by Function (*continued*)

Sample Résumés

This section includes a number of sample résumés that are formatted for paper, email attachments, or PDF files. Keep in mind that they are intended to stimulate your thinking about your own résumé, which must be tailored to your own job search. Examine as many résumés as possible and select the format that best suits your goals.

Figures R–5 and R–6 on pages 565–66 are résumés by a recent community college graduate and an applicant who has been employed for many years, respectively. The writer of the résumé in Figure R–5 has only limited work experience and therefore puts the education section first and gives fairly detailed information about his college work. In the résumé in Figure R–6, work experience appears first because the applicant has many years of full-time employment—his experience is much more important to a prospective employer than his educational data.

The résumés in Figures R–7, R–8, and R–9 are all for the same person. Figure R–7 on page 567 is the applicant's first résumé, when she was a student; the résumés in Figure R–8 on pages 568–69 and Figure R–9 on pages 569–70 reflect work experience later in her career (one is organized by job, the other by function).

revision

R

The more natural a work of writing seems to the reader, the more effort the writer has probably put into its revision. If possible, put your draft away for a day or two before you begin to revise it. If that is not possible, take a break from your writing and try to do something else before you revise. Without such a cooling period, you are too close to the draft to evaluate it objectively.

When you return to revise your draft, read and evaluate it with deliberation and objectivity and from the point of view of your **readers.** Be determined to find and correct faults—be honest. Do not try to revise everything at once. Read through your rough draft several times, each time searching for and correcting a different set of problems. The following Writer's Checklist provides tips for revision.

Writer's Checklist: Revision

Check your draft for the following:

☑ *Appropriate introduction.* Check to see that your introduction frames the rest of the document. (See **introductions** and **openings.**)

☑ *Completeness.* Your writing should give readers exactly what they need but not overwhelm them.

☑ *Accuracy.* No matter how careful and painstaking you may have been in conducting your research, compiling your notes, and creating your outline, you could easily have made errors when transferring your thoughts from the outline to the rough draft. Look for any inaccuracies that may have crept into your draft.

☑ *Unity and coherence.* Check to see that sentences and ideas are closely tied together and contribute directly to the main idea expressed in the topic sentence of the **paragraph.** Provide **transitions** where they are missing and strengthen those that are weak. (See also **coherence.**)

☑ *Consistent usage.* Make sure you have not called the same item by one term on one page and an alternative term on another page.

☑ *Conciseness.* Tighten your writing so it says exactly what you mean. Prune unnecessary words, phrases, sentences, even paragraphs. (See **conciseness/wordiness.**)

☑ *Awkwardness.* Look for awkward passive **voice** constructions; the active voice makes your writing more direct. (See also **awkwardness.**)

☑ *Word choice.* Delete or replace vague or pretentious words and unnecessary intensifiers. Check for **affectation** and unclear **pronoun references.** (See also **word choice.**)

☑ *Ethical language.* Check for **ethics in writing,** including language that might be interpreted as implying bias against a particular group. (See also **biased language** and **sexism in language.**)

☑ *Jargon.* Unless you are sure that *all* your readers will understand any **jargon** you have used, eliminate it.

☑ *Clichés.* Replace **clichés** with fresh figures of speech or direct statements.

☑ *Grammar.* Check your draft for possible grammatical errors. (See also **grammar.**)

☑ *Typographical errors.* Check your final draft for typographical errors with your spellchecker and by **proofreading** it.

rhetorical questions

A rhetorical question is a question to which a specific answer is neither needed nor expected. The question is often intended to make readers think about the subject from a different perspective; the writer then answers the question in the article or essay.

* Is space exploration worth the cost?

* Does advertising lower consumer prices?

The answer to a rhetorical question may not be a simple yes or no; it might be a detailed explanation of, for example, the pros and cons of the value of space exploration or the effect of advertising on consumer prices.

The rhetorical question can be an effective **opening,** and it is often used as a **title.** By its nature, it is somewhat informal and therefore should be used judiciously in technical writing. For example, a rhetorical question would not be an appropriate opening for a report or memo addressed to a busy superior in your company. When you do use a rhetorical question, be sure it is not trivial, obvious, or forced. More than any other writing device, the rhetorical question requires that you know your **readers.**

run-on sentences   ESL

A run-on sentence, sometimes called a fused sentence, is two or more sentences without punctuation to separate them. The term is also sometimes applied to a pair of independent **clauses** separated by only a **comma,** although this variation is usually called a **comma splice.** Run-on sentences can be corrected, as shown in the following examples, by (1) making two sentences, (2) joining the two clauses with a **semicolon** (if they are closely related), (3) joining the two clauses with a comma and a coordinating **conjunction,** or (4) subordinating one clause to the other.

* The new manager instituted several new procedures ~~some~~ were impractical.

  . Some

- The new manager instituted several new procedures *;* some were impractical.

- The new manager instituted several new procedures *, but* some were impractical.

- The new manager instituted several new procedures *,* some *of which* were impractical.

(See also **sentence construction** and **sentence faults.**)

S

sales proposals (*see* **proposals**)

same

When used as a **pronoun,** *same* is awkward and outdated.

- We received your proposal, and we will respond to ~~same~~ *it* next week.

schematic diagrams

The schematic diagram, used primarily for electrical and mechanical systems, attempts to portray a system's operation with lines and symbols rather than through a physical likeness. The schematic diagram emphasizes the relationships among the parts without giving precise proportions.

The schematic diagram in Figure S–1 on page 576 depicts the process of trapping particulates from coal-fired power plants. Note that the particulate filter is depicted both symbolically and as an enlarged drawing to show relevant structural details.

For guidelines on how to use illustrations and incorporate them into your text, see the entry **illustrations.**

scope

Scope is the depth and breadth of detail needed to cover a subject. The **readers'** needs and your **purpose** are primary in determining the type and amount of detail to include in a document. Your scope

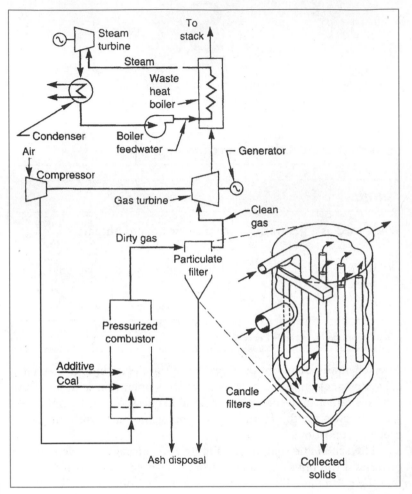

FIGURE S–1. Schematic Diagram

will also be affected by the type of document you are writing as well
as the medium you select for your message. If you do not determine
your scope of coverage in the planning stage of your writing project,
you will not know how much or what kind of information to include. Design the scope to satisfy your purpose and the needs of
your readers. (See also **selecting the medium.**)

search engines

Search engines are computer programs that help you locate information in and across electronic databases. In their simplest form, they compare user requests with indexed database information and list the available files. The term *search engine* often refers to software used through the **World Wide Web** or at a dedicated terminal, such as in a library, but it can also refer to the search capabilities of dictionary, encyclopedia, or reference manual CD-ROMs.

It is difficult to make definitive statements about the available search engines because of the rapid changes in both the amount and the type of information on the Web, as well as the ways in which information is displayed and organized. Even the most popular and stable search engine sites are upgraded fairly often to keep up with the growth of the Internet and the pace of software development. (See also **Internet.**) Because the Web has no standard for page design, uniform content, uniform identifiers or keywords, or other markers that would make a single kind of search engine sufficient, different search engines have been created. Pay close attention to the subtle differences among search engines, search in more than one search environment, and vary keywords and phrases to maximize opportunities for locating the information you seek.

The key differences among search engines are in the size of the database a search engine can access and the flexibility of the searches it conducts. Depending on the indexing, Internet search engines usually return a list of hits arranged from most relevant to least relevant. For example, a search on AltaVista for "Bedford/St. Martin's" yielded the screen in Figure S–2 on page 578, which shows the first 8 of 337,098 hits. At the bottom of the screen, there usually are links to more pages that meet the search criteria, from most relevant to least.

Most search engines use variables called Boolean operators—the words AND, OR, and NOT—to refine searches; some include additional variables that identify the exact position of parts of strings of text (such as ADJ, for *adjacent*). Some search engines allow another refinement called *truncation,* which allows users to search for a portion of a word and receive hits on all possible variants, including both **noun** and **verb** forms. A few engines search by keyword or exact title alone. Those are the least useful.

S

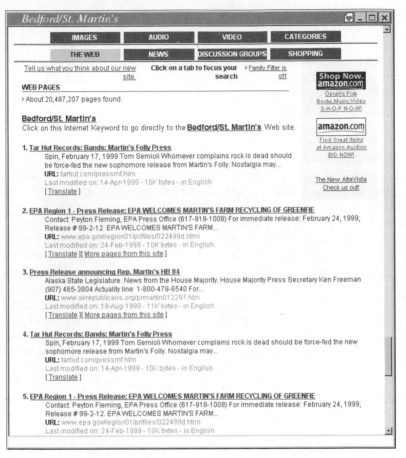

FIGURE S–2. Results of a Search Engine Query

The current generation of search engines allows creative and individually tailored approaches, including automated regular searches for particular kinds of information. For example, users interested in stock quotes or news stories about international business can tailor their searches or pay for search engines to seek out new information regularly in those areas. Some search engines, such as Northern Lights, provide searches of both the World Wide Web and their own proprietary database of articles not available on the Web. Other new search engines offer visual representations of data in the form of clickable image maps, in which information is represented visually by shapes and colors rather than, for example, alphabetized indexes. Users navigate through three-dimensional graphics spaces by moving the mouse. A new search option, meta-searching software, allows you to search multiple indexes, sites, and search engines at the same time, thus replicating your search as if you searched in several different places manually. With meta-searching, you could, for example, make a single request and search four or five regional libraries, AltaVista, and Yahoo! at the same time. (See also **Internet research.**)

Experienced users develop strategies that they employ across search engines. For example, a search using the same keywords in a variety of search engines but also variations on those keywords will provide the greatest number of hits across search platforms. By altering the searched terms or phrases to reflect the actual subject headings found in an initial broad search, users can progressively narrow the number of hits returned by the search engine.

The following search engines are the ones most widely used on the Web.

- *AltaVista* (http://www.altavista.com) offers the largest, most comprehensive search of Web pages and usenet newsgroups. To initiate searches, you can enter words or phrases. When you enter phrases, enclose them in double quotation marks: "marketing strategy" or "just in time inventory." Be careful about using capital letters. The word *cook* will give you a list of Web sites with recipes, but *Cook* will return sites listing proper names and historical figures. The same confusion can occur with *Turkey* and *turkey*.
- *Excite* (http://www.excite.com) also does keyword searches of Web pages, usenet groups, and classified ads. Again, to make your search productive, be as descriptive as possible with your search terms. For example, if you are looking for information about nuclear power and enter the term *nuclear,* the search will also yield listings for *nuclear family, nuclear medicine,* and *nuclear winter.*
- *Infoseek* (http://www.infoseek.com) searches Web sites, usenet groups, FTP sites, and gopher sites. It also offers national directories that include personal and company phone numbers, addresses, and email addresses. Search results are displayed in order of relevance, with those most frequently or most closely

S

matching your search terms listed first. Infoseek supports searches for phrases enclosed in quotation marks and recognizes capitalized words as proper nouns (*cook/Cook*). Like an index tool, Infoseek also offers broad subject category searches.

- *Lycos* (http://www.lycos.com) searches Web sites, FTP sites, and gopher sites and offers keyword and subject searching. You can customize searches by indicating if the match should be "strong," "close," "good," "fair," or "loose." "Strong" returns the fewest results, "loose" the most.

Other Web search engines include the following sites:

Yahoo! (http://www.yahoo.com)
HotBot (http://www.hotbot.com)
WebCrawler (http://www.webcrawler.com)

For reviews of search engines, search the Web sites of major computer-industry magazines.

selecting the medium

Early in your **preparation,** you must select the most appropriate medium for communicating your message. You can choose from a wide array of possibilities, from recent technologies like **email, fax,** voice mail, and video conferencing to more traditional means like letters, **memos,** telephone calls, and face-to-face meetings.

The most important considerations in selecting the appropriate medium are the audience (see **readers**) and the **purpose** of the communication. For example, if you need to collaborate with someone to solve a problem or if you need to establish rapport with someone, written exchanges (even by email) could be a waste of time and far less efficient than a phone call or a face-to-face meeting. On the other hand, if you need precise wording or you need to provide a record of a complex message, communicate in writing. Following are descriptions of typical means of communicating on the job today: letters on organizational stationery, memos, email messages, faxes, telephone calls, voice-mail messages, face-to-face meetings, and video conferencing. Understanding their primary characteristics will help you select the most appropriate medium for certain audiences and purposes.

Letters on Organizational Stationery

Letters are often the most appropriate choice for initial contacts with new business associates or customers and for other formal communications. Letters written on your organization's letterhead communicate formality, respect, and authority. A letter can travel much more quickly than before through express delivery services. Express delivery of letters (or other documents) should be preceded by a phone call, a voice-mail message, or an email message. (See also **correspondence.**)

Memos

Memos on printed company stationery are appropriate for communication among members of the same organization, even when offices are geographically separated. They have many of the same characteristics as letters, such as formality and authority, but they are used for a wider variety of functions, from reminders to short reports. The use of memos must follow an organization's protocol.

Email Messages

Email is used to send information, elicit discussions, collect opinions, and transmit many other kinds of messages quickly to those inside the organization as well as those outside who have access to the **Internet.** Email can be a less formal means of communication than either letters or memos; however, it is rapidly becoming *the* medium through which documents are transmitted. Because email recipients can print copies of messages they receive and easily forward them to others, always write your email with care and reread the message carefully before you send it.

Faxes

A fax (for "facsimile transmission") is most useful when speed is essential and when the information — a drawing or contract, for example — must be viewed in its original form. Faxes are also useful when the recipient does not have access to email or the programs to view attachments or when the material has not yet been converted into electronic form. Note that faxed correspondence often seems less official than a traditional letter, in part because the recipient does not receive the stationery. Of course, if formality is important, the original, official paper copy can be sent separately by overnight or regular mail.

Telephone Calls

The range of information exchanged through telephone calls is virtually unlimited—everything from a call of less than a minute to confirm a meeting time to a call lasting an hour or more to negotiate or clarify the conditions of a contract. Because phone calls enable participants to interpret tone of voice, they facilitate resolving misunderstandings, although they do not provide the visual cues that can be observed during face-to-face meetings.

A conference call among three or more participants is a less expensive alternative to a face-to-face meeting requiring travel. To ensure maximum efficiency, the person setting up the call works from an agenda shared by all the participants and directs the discussion as though he or she were leading a meeting.

Voice-Mail Messages

Telephone answering systems (voice mail) allow callers to record messages when the person called is not available. When you leave a voice-mail message, give your name and phone number, the date and time of the call, and a brief message ("Call me about the deadline for the new project" or "I got the package, so you don't need to call the distributor"). If the message is complicated or contains numerous details, use another medium, such as an email message or a letter, to ensure that the information is communicated accurately. If you want to discuss a subject, let the recipient know the subject so he or she can prepare a response before returning your call.

Face-to-Face Meetings

Face-to-face meetings are most appropriate for initial or early contacts with associates and clients with whom you intend to develop an important, long-term relationship. Meetings are also the best medium for solving a technical problem that needs input from more than two people. For instructions on how to conduct meetings as well as record discussions and decisions, see **meetings** and **minutes of meetings.**

Video Conferencing

Two-way and three-way video conferencing is becoming an increasingly common medium for business communication. Video conferences are particularly useful for meetings when travel is impractical or too expensive. Unlike telephone conference calls, video conferences have the advantage of allowing participants to see as well as to hear one another. Video conferences work best with participants who are at ease in front of a camera.

semicolons

The semicolon (;) links independent **clauses** or other sentence elements of equal weight and grammatical rank, especially **phrases** in a series that have **commas** in them. The semicolon indicates a greater pause between clauses than a comma, but not as great a pause as a **period.**

When the independent clauses of a compound sentence are not joined by a comma and a **conjunction,** they are linked by a semicolon.

* No one applied for the position; the job was too difficult.

Make sure, however, that such clauses balance or contrast with each other. The relationship between the two statements should be so clear that further explanation is not necessary.

* The new Web page was very successful; every division reported increased online sales.

Do not use a semicolon between a dependent clause and its main clause. Remember that elements joined by semicolons must be of equal grammatical rank or weight.

* No one applied for the ~~position;~~ *position,* even though it was heavily advertised.

With Strong Connectives

In complicated sentences, a semicolon may be used before transitional words or phrases (*that is, for example, namely*) that introduce examples or further explanation.

* The study group was aware of her position on the issue; that is, federal funds should not be used for the housing project.

(See also **transition.**)

A semicolon should also be used before conjunctive **adverbs** (such as *therefore, moreover, consequently, furthermore, indeed, in fact, however*) that connect independent clauses.

S

- I won't finish today; *moreover,* I doubt that I will finish this week.

The semicolon in the example shows that *moreover* belongs to the second clause.

For Clarity in Long Sentences

Use a semicolon between two independent clauses connected by a coordinating conjunction (*and, but, for, or, nor, yet*) if the clauses are long and contain other **punctuation.**

- In most cases, these individuals are corporate executives, bankers, Wall Street lawyers; *but* they do not, as the economic determinists seem to believe, simply push the button of their economic power to affect fields remote from economics.

A semicolon may also be used if any items in a series contain commas.

- Among those present were John Howard, president of the Omega Paper Company; Carol Delgado, president of Environex Corporation; and Larry Stanley, president of Stanley Papers.

Do not use semicolons to enclose a parenthetical element that contains commas. Use **parentheses** or **dashes** for that purpose.

INCORRECT   All affected job classifications; receptionists, secretaries, transcriptionists, and clerks; will be upgraded this month.

CORRECT   All affected job classifications (receptionists, secretaries, transcriptionists, and clerks) will be upgraded this month.

Do not use a semicolon as a mark of anticipation or enumeration. Use a **colon** for that purpose.

- Three decontamination methods are under ~~consideration;~~ *consideration:* a zeolite-resin system, an evaporation and resin system, and a filtration and storage system.

The semicolon always appears outside closing **quotation marks.**

- The attorney said, "You must be accurate"; her client replied, "I will."

sentence construction ⓔⓢⓛ

DIRECTORY
Overview   585
Subjects   585
Tips for Understanding the Subject of a Sentence ⓔⓢⓛ   586
Predicates   586
Sentence Types   587
Constructing Effective Sentences   589
Tips for Understanding the Minimum Requirements for a Sentence ⓔⓢⓛ   590

Overview

A sentence is the most fundamental and versatile tool available to the writer. Sentences generally flow from a subject, to a **verb,** to any objects, **complements,** or **modifiers,** but can be ordered in a variety of ways to achieve **emphasis.** When shifting word order for emphasis, however, be aware that word order can make a great difference in the meaning of a sentence.

- He was *only* the accountant.

- He was the *only* accountant.

Subjects

The most basic components of sentences are subjects and predicates. The subject of a sentence is a **noun** or a **pronoun** (and its **modifiers**) about which the predicate of the sentence makes a statement. Although a subject may appear anywhere in a sentence, it most often appears at the beginning.

- *To increase sales* is our goal.

- *The wiring* is defective.

Grammatically, a subject must agree with its **verb** in **number.**

- These *departments have* much in common.

- This *department has* several functions.

The subject is the actor in active-**voice** sentences.

- The *webmaster* reported a record number of hits in November.

S

> **ESL** TIPS FOR UNDERSTANDING THE SUBJECT OF A SENTENCE
>
> In English, every sentence, except commands, must have an explicit subject.
>
> *He established*
> - *Ozzie* worked fast. ~~Established~~ the parameters for the project.
>   ^
>
> In commands, the subject *you* is understood and is used only for emphasis.
>
> - (*You*) Show up at the airport at 6:30 tomorrow morning.
>
> - *You* do your homework, young man. [parent to child]
>
>   If you move the subject from its normal position (subject-verb-object), English often requires you to replace the subject with an expletive (*there, it*). In this construction, the verb agrees with the subject that follows it.
>
> - *There are* two files on the desk.
>   [The subject is *files.*]
>
> - *It is* presumptuous for me to speak for Jim.
>   [The subject is *to speak for Jim.*]
>
> Time, distance, weather, temperature, and environmental expressions use *it* as their subject.
>
> - *It* is ten o'clock.
>
> - *It* is ten miles down the road.
>
> - *It* never snows in Florida.
>
> - *It* is very hot in Jorge's office.
>
> - *It* gets very stuffy in here very quickly.

S

A compound subject has two or more **substantives** as the subject of a verb.

- *The president* and *the treasurer* agreed to begin the audit.

Predicates

The predicate is the part of a sentence that contains the main verb and any other words used to complete the thought of the sentence (the verb's modifiers and **complements**). The principal part of the predicate is the verb, just as a noun (or noun substitute) is the principal part of the subject.

- Bill *piloted the airplane.*

The *simple predicate* is the verb (or verb phrase) alone; the *complete predicate* is the verb and its modifiers and complements. A *compound predicate* consists of two or more verbs with the same subject. It is an important device for **conciseness** in writing.

- The company *tried* but *did not succeed* in that field.

A *predicate nominative* is a noun construction that follows a linking verb and renames the subject.

NOUN	She is my *attorney.*
NOUN CLAUSE	His excuse was *that he had been sick.*

Sentence Types

Sentences are classified according to *structure* (simple, compound, complex, compound-complex); *intention* (declarative, interrogative, imperative, exclamatory); and *stylistic use* (loose, periodic, minor).

Structure. A *simple sentence* consists of one independent clause. At its most basic, the simple sentence contains only a subject and a predicate.

- Profits [subject] rose [predicate].
- The storm [subject] finally ended [predicate].

A *compound sentence* consists of two or more independent clauses connected by a **comma** and a coordinating **conjunction,** by a **semicolon,** or by a semicolon and a conjunctive **adverb.**

- Drilling is the only way to collect samples of the layers of sediment below the ocean floor, *but* it is by no means the only way to gather information about these strata.
  [comma and coordinating conjunction]

- There is little similarity between the chemical composition of seawater and that of river water; the various elements are present in entirely different proportions. [semicolon]

- It was 500 miles to the site; *therefore,* we made arrangements to fly.
  [semicolon and conjunctive adverb]

A *complex sentence* contains one independent clause and at least one dependent clause that expresses a subordinate idea.

- The generator will shut off automatically [independent clause] if the temperature rises above a specified point [dependent clause].

S

A *compound-complex sentence* consists of two or more independent clauses plus at least one dependent clause.

- Productivity is central to controlling inflation [independent clause], because when productivity rises [dependent clause], employers can raise wages without raising prices [independent clause].

Intention.   A *declarative sentence* conveys information or makes a factual statement.

- The motor powers the conveyor belt.

An *interrogative sentence* asks a direct question.

- Does the conveyor belt run constantly?

An *imperative sentence* issues a command.

- Start the generator.

An *exclamatory sentence* is an emphatic expression of feeling, fact, or opinion. It is a declarative sentence that is stated with great feeling.

- The files were deleted!

Stylistic Use.   A *loose sentence* makes its major point at the beginning and then adds subordinate phrases and clauses that develop or modify that major point. A loose sentence might seem to end at one or more points before it actually does end, as the periods in brackets illustrate in the following sentence.

- It went up[.], a great ball of fire about a mile in diameter[.], changing colors as it kept shooting upward[.], an elemental force freed from its bonds[.] after being chained for billions of years.

A *periodic sentence* delays its main ideas until the end by presenting subordinate ideas or modifiers first.

- During the last century, the attitude of the American citizen toward automation underwent a profound change.

A *minor sentence* is an incomplete sentence. It makes sense in its context because the missing element is clearly implied by the preceding sentence.

- In view of these facts, is automation really useful? *Or economical?*

Constructing Effective Sentences

The subject-verb-object pattern is effective because it is most familiar to readers. In "The company dismissed Joe," we know the subject and the object by their positions relative to the verb. The
knowledge that the usual sentence order is subject-verb-object
helps readers interpret what they read.

An inverted sentence places the elements in other than normal
order.

- A better job I never had. [direct object-subject-verb]

- More optimistic I have never been.
  [subjective complement-subject-linking verb]

Inverted sentence order is used in questions and exclamations and
also to achieve **emphasis.**

- Have you a pencil? [verb-subject-complement]

- A sorry sight we presented! [complement-subject-verb]

Use uncomplicated sentences to state complex ideas. If readers
have to cope with a complicated sentence in addition to a complex
idea, they are likely to become confused.

CONFUSING   When you are purchasing parts, remember that al-
             though an increase in the cost of aluminum forces
             all the vendors to increase their prices, some vendors
             will have a supply of aluminum purchased at the old
             price, and they may be willing to sell parts to you at
             the old price in order to get your business.

SIMPLIFIED   Although an increase in the cost of aluminum forces
             all vendors to increase their prices, some vendors
             will have a supply of aluminum purchased at the old
             price. When you are purchasing aluminum parts, re-
             member that vendors may be willing to sell you the
             parts at the old price in order to get your business.

Just as simpler sentences make complex ideas more digestible, a
complex sentence construction makes a series of simple ideas
smoother and less choppy.

CHOPPY   The company was founded ten years ago. It now
             employs 50 people. It has branch offices in two
             other states.

SMOOTH	The company, which was founded ten years ago, now employs 50 people and has branch offices in two other states.

Avoid loading sentences with a number of thoughts carelessly tacked together. Such sentences are monotonous and hard to read because all the ideas seem to be of equal importance. Rather, distinguish the relative importance of sentence elements with **subordination.**

LOADED	We started the program three years ago, there were only three members on the staff, and each member was responsible for a separate state, but it was not an efficient operation.
SUBORDINATED	When we started the program three years ago, there were only three members on the staff, each having responsibility for a separate state; however, that arrangement was not efficient.

Express coordinate or equivalent ideas in similar form. The very construction of a sentence helps the reader grasp the similarity of its components, as illustrated in **parallel structure.** (See also **garbled sentences.**)

ESL TIPS FOR UNDERSTANDING THE MINIMUM REQUIREMENTS FOR A SENTENCE

- A sentence must start with a capital letter.
- A sentence must end with a period, a question mark, or an exclamation point.
- A sentence must have a subject.
- A sentence must have a verb.
- A sentence must conform to subject-verb-object word order (or inverted word order for questions or emphasis).
- A sentence must express an idea that can stand on its own (called the main or independent clause).

---

sentence faults   **ESL**

A number of problems can create sentence faults, including faulty **subordination, clauses** with no subjects, rambling sentences, and omitted **verbs.**

Faulty Subordination

Faulty subordination occurs when a grammatically subordinate element, such as a dependent clause, actually contains the main idea of the sentence or when a subordinate element is so long or detailed that it dominates or obscures the main idea. Both of the following sentences appear logical; to determine which one is appropriate, the writer needs to decide which of two ideas should be emphasized.

- Although the new filing system saves money, many of the staff are unhappy with it.

- The new filing system saves money, although many of the staff are unhappy with it.

If the main point is that *many of the staff are unhappy,* the first sentence is correct. If the writer's main point is that *the new filing system saves money,* the second sentence is correct.

The other major problem with subordination is loading so much detail into a subordinate element that the main point is crushed by the sheer size and weight of the subordinate information.

LOADED   Because the noise level in the assembly area on a typical shift is as loud as an alarm clock ringing three feet away, employees often develop hearing problems.

CONCISE   Because the noise level in the assembly area is so high, employees often develop hearing problems.

Writers sometimes inappropriately assume a subject that is not stated in a clause.

CONFUSING   Your application program can request to end the session after the next command.
[Request *who* or *what* to end the session?]

CLEAR   Your application program can request *the host program* to end the session after the next command.

The assertion that a sentence's predicate makes about its subject must be logical. "Mr. Wilson's *job* is a sales representative" is not logical, but "*Mr. Wilson* is a sales representative" is. "Jim's *height* is six feet tall" is not logical, but "*Jim* is six feet tall" is.

Sentences that contain more information than the reader can comfortably absorb in one reading are known as *rambling sentences.*

The obvious remedy for a rambling sentence is to divide it into two or more sentences. When you do that, put the main message of the rambling sentence into the first of the revised sentences.

RAMBLING   The payment to which a subcontractor is entitled should be made promptly in order that in the event of a subsequent contractual dispute we, as general contractors, may not be held in default of our contract by virtue of nonpayment.

DIRECT   Pay subcontractors promptly. Then if a contractual dispute occurs, we cannot be held in default of our contract because of nonpayment.

Do not omit a required verb.

written

- I never have ⌃ and probably never will write the annual report.

(See also **run-on sentences** and **sentence fragments.**)

sentence fragments   (ESL)

A sentence fragment is an incomplete grammatical unit that is punctuated as a sentence.

SENTENCE   He quit his job.

FRAGMENT   And quit his job.

A sentence fragment lacks either a subject or a **verb** or is a subordinate **clause** or **phrase.** Sentence fragments are often introduced by relative **pronouns** (*who, which, that*) or subordinating **conjunctions** (such as *although, because, if, when,* and *while*).

S

- The new manager instituted several new procedures. ~~Although~~ *, although* ⌃ she didn't train her staff first.

A sentence must contain a finite verb; verbals do not function as verbs. The following examples are sentence fragments because their **verbals** (*providing, to work, waiting*) cannot function as finite verbs.

FRAGMENTS	*Providing* all employees with disability insurance.
	To work a 40-hour week.
	The customer *waiting* to see you.
SENTENCES	The company must provide all employees with disability insurance.
	All employees are expected to work a 40-hour week.
	The customer is waiting to see you.

Explanatory phrases beginning with *such as, for example,* and similar terms often lead writers to create sentence fragments.

- The staff wants additional benefits. For example, the use of company automobiles. *, such as*

- The staff wants additional benefits. For example, the use of company automobiles. *one possible benefit is*

A hopelessly snarled fragment simply has to be rewritten. To rewrite such a fragment, pull the main points out of the fragment, list them in the proper sequence, and then rewrite the sentence. (See also **garbled sentences.**)

FRAGMENT	Removing the protection cap and the piston secured in the housing by means of the spring placed between the piston and the housing.
MAIN POINTS	1. Remove the protection cap.
	2. The piston is held in the housing by a spring.
	3. The spring is connected to the piston at one end and the housing at the other.
	4. To remove the piston, disconnect the spring.
SENTENCE	Removing the protection cap lets you remove the piston by disconnecting a spring that connects to the piston at one end and the housing at the other.

S

(See also **sentence construction, sentence faults,** and **run-on sentences.**)

sentence types (*see* sentence construction)

sentence variety

Sentences can be long or short; loose or periodic; simple, compound, complex, or compound-complex; declarative, interrogative, exclamatory, or imperative; even elliptical. There is never a legitimate excuse for letting your sentences become tiresomely alike. However, the best time to achieve sentence variety is during **revision.**

Sentence Length

Because a long series of sentences of the same length is monotonous, varying sentence length makes writing more interesting to the reader. For example, avoid stringing together a number of short independent **clauses.** Either connect them with subordinating connectives, thereby making some dependent clauses, or make some clauses into separate sentences. (See also **subordination.**)

STRING   The river is 60 miles long, and it averages 50 yards in width, and its depth averages 8 feet.

IMPROVED   The river, which is 60 miles long and averages 50 yards in width, has an average depth of 8 feet.

IMPROVED   The river is 60 miles long. It averages 50 yards in width and 8 feet in depth.

You can often effectively combine short sentences by converting **verbs** into **adjectives.**

- The steeplejack ~~fainted. He~~ *fainting* collapsed on the scaffolding.

Although too many short sentences make your writing sound choppy and immature, a short sentence can be effective following a long one.

- During the past two decades, many changes have occurred in American life, the extent, durability, and significance of which no one has yet measured. *No one can.*

In general terms, short sentences are good for emphatic, memorable statements. Long sentences are good for detailed explanations and support. There is nothing inherently wrong with a long sentence or even with a complicated one, as long as its meaning is clear and direct. Sentence length becomes an element of **style** when varied for **emphasis** or contrast; a conspicuously short or long sentence can be used to good effect.

Word Order

When a series of sentences all begin in exactly the same way (usually with an article and a noun) the result is likely to be monotonous. You can make your sentences more interesting by occasionally starting with a modifying word, phrase, or clause. However, overuse of this technique itself can be monotonous, so use it in moderation.

SINGLE MODIFIER   *Exhausted,* the project director slumped into a chair.

PHRASE   *To salvage the project,* she presented constructive alternatives when current policies failed to produce results.

CLAUSE   *Because we now know the result of the survey,* we can proceed with certainty.

Inverted sentence order can be an effective way to achieve variety, but do not overdo it.

- Then occurred the event that gained us the contract.

For variety, you can alter normal sentence order by inserting a phrase or clause.

- Titanium fills the gap, *both in weight and in strength,* between aluminum and steel.

The technique of inserting a phrase or clause is good for emphasis, providing detail, breaking monotony, and regulating **pace.**

Loose and Periodic Sentences

A loose sentence makes its major point at the beginning and then adds subordinate phrases and clauses that develop or modify the point. A loose sentence might seem to end at one or more points before it actually does end, as the periods in brackets illustrate in the following example.

S

LOOSE   It went up[.], a great ball of fire about a mile in diameter[.], an elemental force freed from its bonds[.] after being chained for billions of years.

A periodic sentence delays its main idea until the end by presenting **modifiers** or subordinate ideas first, thus holding the readers' interest until the end.

PERIODIC   During the last century, the attitude of the American citizen toward automation underwent a profound change.

Experiment with shifts from loose sentences to periodic sentences in your own writing, especially during revision. Avoid the singsong monotony of a long series of loose sentences, particularly a series containing coordinate clauses joined by **conjunctions.** Subordinating some thoughts to others makes your sentences more interesting.

LOOSE   The auditorium was filled to capacity, *and* the chairman of the board came onto the stage. The meeting started at eight o'clock, *and* the president made his report of the company's operations during the past year. The stockholder audience was obviously unhappy, *but* the members of the board of directors were all reelected.

PERIODIC   By eight o'clock, *when* the chairman of the board came onto the stage and the meeting began, the auditorium was filled to capacity. *Although* the stockholder audience was obviously unhappy with the president's report of the company's operations during the past year, the members of the board of directors were all reelected.

(See also **sentence construction.**)

sequential method of development

The sequential, or step-by-step, **method of development** is especially effective for explaining a process or describing a mechanism in operation. It is also the logical method for writing **instructions,** as shown in Figure S–3 on page 597.

PROCESSING FILM

Developing.   In total darkness, load the film on the spindle and enclose it in the developing tank. Be careful not to allow the film to touch the tank walls or other film. Add the developing solution, turn the lights on, and set the timer for seven minutes. Agitate for five seconds initially and then every half minute.

Stopping.   When the timer sounds, drain the developing solution from the tank and add the stop bath. Agitate continuously for 30 seconds.

Fixing.   Drain the stop bath and add the fixing solution. Allow the film to remain in the fixing solution for two to four minutes. Agitate for five seconds initially and then every half minute.

Washing.   Remove the tank top and wash the film for at least 30 seconds under running water.

Drying.   Suspend the film from a hanger to dry. It is generally advisable to place a drip pan below the rack. Sponge the film gently to remove excess water. Allow the film to dry completely.

FIGURE S–3. Sequential Method of Development

The main advantage of the sequential method of development is that it is easy to follow because the steps correspond to the elements of the process or operation being described. The disadvantages are that it can become monotonous and does not lend itself well to achieving **emphasis.**

Practically all methods of development have elements of sequence. The **chronological method of development,** for example, is also sequential: To describe a trip chronologically, from beginning to end, is also to describe it sequentially.

service   S

When used as a **verb,** *service* means "keep up or maintain" as well as "repair."

- Our company will *service* your equipment.

If you mean "providing a more general benefit," use *serve.*

- Our company ~~services~~ *serves* the northwest area of the state.

set / sit

Sit is an intransitive **verb;** it does not, therefore, require an **object.** Its past **tense** is *sat.*

- I *sit* by a window in the office.
- We *sat* around the conference table.

Set is usually a transitive verb, meaning "put or place," "establish," or "harden." Its past tense is *set.*

- Please *set* the trophy on the shelf.
- The jeweler *set* the stone beautifully.
- Can we *set* a date for the tests?
- The high temperature *sets* the epoxy quickly.

Set is occasionally intransitive.

- The sun *sets* a little earlier each day.
- The glue *set* in 45 minutes.

sexism in language   ESL

Sexism is defined as attitudes, conditions, or behaviors that promote stereotyping of social roles based on a person's sex. Although it may not be used intentionally, sexist language can suggest as well as perpetuate sexist attitudes or practices that are unethical, if not illegal, in business. For discussions of sexist language and how to avoid it, see the following entries:

agreement of pronouns and antecedents
biased language
chair/chairperson
correspondence
ethics in writing
everybody/everyone
female
gender
he/she

S

male

Ms./Miss/Mrs.

shall / will

Although traditionally *shall* was used to express the future tense with *I* and *we*, *will* is now generally accepted with all persons. *Shall* is commonly used today only in questions requesting an opinion or a preference rather than a prediction (compare "Shall we go?" to "Will we go?") and in statements expressing determination ("I shall return").

sic

Latin for "thus," *sic* is used in **quotations** to indicate that the writer has quoted the material exactly as it appears in the original source. It is most often used when the original material contains an obvious error or might be questioned in some other way. *Sic* is placed in **brackets.**

- In the textbook *Basic Astronomy,* the author notes that the "earth does not revolve around the son [*sic*] at a constant rate."

similes (*see* **figures of speech**)

simple sentences (*see* **sentence construction**)

-size / -sized

As **modifiers,** the **suffixes** *-size* and *-sized* are more common to advertising copy than to general writing.

- king-size bed, economy-sized packaging

However, they usually are redundant unless they are part of the name of a product and generally should not be used. (See also **conciseness/wordiness.**)

slashes

The slash (/) performs punctuating duties by separating and show-ing omission. The slash is called a variety of names, including *slant line, virgule, bar, solidus,* and *shilling.*

The slash is often used to separate parts of addresses in contin-uous writing.

- The return address on the envelope was Ms. Rose Howard/Klein-lindener Str. 62/Giessen/D-35394/Germany.

The slash can indicate alternative items.

- David's telephone number is 549-2278/2335.

The slash often indicates omitted words and letters.

- miles/hour for "miles per hour"

- w/o for "without"

In fractions, the slash separates the numerator from the denom-inator.

- 2/3 [2 of 3 parts], 3/4 [3 of 4 parts], 27/32 [27 of 32 parts]

The slash also separates items in the URL (Uniform Resource Locator) address for sites on the **World Wide Web.**

- http://www.bedfordstmartins.com/

In informal writing, the slash separates day from month and month from year in **dates.**

- 12/29/02

Do not use this form for **international correspondence,** since the order of the items varies.

so / so that / such

So is often vague and should be avoided if another word would be more precise.

- *Because she*
  ~~She~~ reads faster, ~~so~~ she finished before I did.

Another problem occurs with the phrase *so that,* which should never be replaced with *so* or *such that.*

- The report should be written ~~such that~~ it can be copied.

so that

Such, an **adjective** meaning "of this or that kind," should never be used as a **pronoun.**

- Our company does not need on-site attorneys, and we do not anticipate using ~~such.~~

any.

some

When *some* functions as an indefinite **pronoun** for a plural count **noun** or as an indefinite **adjective** modifying a plural count noun, use a plural **verb.**

- *Some* of us *are* prepared to work overtime.
- *Some* people *are* more productive than others.

Some is singular, however, when used with mass nouns.

- *Some* sand *has* trickled through the crack.
- *Some* oil *was* spilled on the highway.
- *Some* stationery *is* sold through our outlet stores.
- Most of the water evaporated, but *some remains.*

some / somewhat

Some, an **adjective** or a **pronoun** meaning "an undetermined quantity" or "certain unspecified persons," should not replace the **adverb** *somewhat,* which means "to some extent."

- His writing has improved ~~some.~~

somewhat.

some time / sometime / sometimes

Some time refers to a duration of time.

* We waited for *some time* before calling the customer.

Sometime refers to an unknown or unspecified time.

* We will visit with you *sometime*.

Sometimes refers to occasional occurrences at unspecified times.

* He *sometimes* visits the branch offices.

spatial method of development

In a spatial sequence, you describe an object or a process according to the physical arrangement of its features. Depending on the subject, you describe its features from bottom to top, side to side, east to west, outside to inside, and so on. Descriptions of this kind rely mainly on dimension (height, width, length), direction (up, down, north, south), shape (rectangular, square, semicircular), and proportion (one-half, two-thirds). Features are described in relation to one another.

* One end is raised six to eight inches higher than the other end to permit the rain to run off.

Features are also described in relation to their surroundings.

* The lot is located on the east bank of the Kingman River.

The spatial **method of development** is commonly used in descriptions of laboratory equipment, **proposals** for landscape work, construction-site **progress and activity reports,** and, in combination with a step-by-step sequence, in many types of **instructions.** The following description for a house inspection relies on a bottom-to-top, clockwise (south to west to north to east) sequence, beginning with the front door.

Interior of a Two-Story, Five-Room House
The front door faces south and opens into a hallway 7 feet deep and 10 feet wide. At the end of the hallway is the stairwell, which

begins on the right (east) side of the hallway, rises five steps to a landing, and reverses direction at the left (west) side of the hallway. To the left (west) of the hallway is the dining room, which measures 15 feet along its southern exposure and 10 feet along its western exposure. Directly to the north of the dining room is the kitchen, which measures 10 feet along its western exposure and 15 feet along its northern exposure. To the east of the kitchen, along the northern side of the house, is a bathroom that measures 10 feet (west to east) by 5 feet. Parallel to the bathroom is a passageway with the same dimensions as the bathroom and leading from the kitchen to the living room. The living room, which measures 15 feet (west to east) by 20 feet (north to south), occupies the entire eastern end of the floor.

On the second floor, at the top of the stairs is an L-shaped hallway, 5 feet wide. The base of the "L," over the front door, is 15 feet long. The vertical arm of the "L" is 13 feet long. To the west of the hall is the southwest bedroom, which measures 10 feet along its southern exposure and 8 feet along its western exposure. Directly to the north, over the kitchen, is the northwest bedroom, which measures 12 feet along its western exposure and 10 feet along its northern exposure. To the east, at the end of the hall, is a bathroom, which is 5 feet wide along the northern side of the house and 7 feet long. To the east of that bathroom and also along the northern side of the house is the master bathroom, which is 10 feet square and is entered from the master bedroom, which is directly over the living room. Like the living room, the master bedroom measures 15 feet along the northern and southern exposures of the house and 20 feet along the eastern exposure.

specie / species

Specie means "coined money" or "in coin."

- Paper currency was virtually worthless, and creditors began to demand payment in *specie*.

Species means a category of animals, plants, or things having some of the same characteristics or qualities. *Species* is the correct spelling for both the singular and the plural.

- The wolf is a member of the canine *species*.

- Many animal *species* are represented in the Arctic.

S

specific-to-general method of development

The specific-to-general **method of development** begins with a specific statement and builds to a general **conclusion.** It is similar to the **increasing-order-of-importance method of development** in that it carefully builds its case, often with examples and analogies in addition to facts or statistics, and does not actually make its point until the end. For example, if your subject were highway safety, you might begin with a specific highway accident and then go on to generalize about how details of the accident were common enough to many similar accidents that recommendations could be made to reduce the probability of such accidents. The following is an example of the specific-to-general method of development.

* Recently, a government agency studied the use of passenger-side air bags in 4,500 accidents involving nearly 7,200 front-seat passengers of the vehicles involved. Nearly all the accidents occurred on routes that had a speed limit of at least 40 mph. Only 20 percent of the adult front-seat passengers were riding in vehicles equipped with passenger-side air bags. Those riding in vehicles not equipped with passenger-side air bags were more than twice as likely to be killed as passengers riding in vehicles that were so equipped.

    A conservative estimate is that 40 percent of the adult front-seat passenger-vehicle deaths could be prevented if all vehicles came equipped with passenger-side air bags. Children, however, should always ride in the backseat because other studies have indicated that a child can be killed by the deployment of an air bag. If you are an adult front-seat passenger in an accident, your chances of survival are far greater if the vehicle in which you are riding is equipped with a passenger-side air bag.

S

specifications

A specification is a detailed and exact statement of particulars, including a statement that prescribes materials, dimensions, and quality of something to be built, installed, or manufactured.

There are two broad categories of specifications — *industrial specifications* and *government specifications* — and both require precision. A specification must be written so clearly that no one could misinterpret any statement contained in it; therefore, do not imply or suggest—state *explicitly* what is needed. Because of the stringent requirements for completeness and exactness of detail in specifications, careful **research** and **preparation** are especially important before you begin to write, as is careful **revision** after you have completed the draft. (See also **clarity** and **ambiguity.**)

Industrial Specifications

Industrial specifications are used in areas like computer software, in which there are no engineering drawings or other means of documentation. An industrial specification is a permanent document, whose purpose is twofold: (1) to document the item being devised so it can be maintained by someone other than the person who designed it and (2) to provide detailed technical information on the item being devised to all those in the company who need it (including engineers, technical writers, technical instructors, and possibly salespeople and purchasing agents).

The industrial specification describes a planned project, a newly completed project, or an old project. The specification for a planned project describes how it will be implemented, while the specification for a newly completed project describes how it was implemented. The specification for an old project describes the project as it exists after it has been operational long enough for all the problems to have been discovered and corrected. All three types of industrial specifications contain detailed technical descriptions of all aspects of the item: what was done, how it was done, what is required to use the item, how it is used, its function, who would use it, and so on.

Government Specifications

Government agencies are required by law to contract for equipment strictly according to definitions provided in formal specifications. A government specification is a precise definition of exactly what the contractor is to provide. In addition to a technical description of the device to be purchased, the specification normally includes an estimated cost, an estimated delivery date, and standards for the design, manufacture, quality, testing, training of government employees, governing codes, inspection, and delivery of the item.

In addition, government specifications contain details on the following:

- Scope of the project
- Documents the contractor is required to furnish with the device
- Required product characteristics and functional performance of the device
- Required tests, test equipment, and test procedures
- Required preparations for delivery
- Notes
- **Appendixes**

Government specifications are often used to prescribe the content and the deadline for government **proposals** submitted by vendors or companies that want to bid on a project.

You can request detailed information on military and non-military government specifications from the U.S. Government Printing Office (GPO) or at the GPO Web site (http://www.access.gpo.gov/).

spelling   ESL

The use of a computer spellchecker helps enormously with spelling problems; however, it will not catch all mistakes. It cannot detect a spelling error if the error results in a valid word; for example, if you mean *to* but inadvertently type *too,* the spellchecker will not detect the error. So, you still must check your document carefully. (See also **proofreading.**)

Writer's Checklist: Spelling

The following system will help you catch spelling errors.

☑ Keep your dictionary handy and use it regularly. If you are unsure about the spelling of a word, do not rely on memory or guess-work—consult the dictionary. When you look up a word, focus on both its spelling and its meaning. (See also **dictionaries.**)

☑ After you have looked in the dictionary for the spelling of the word, write the word from memory several times. Then check the

Writer's Checklist: Spelling (continued)

accuracy of your spelling. If you have misspelled the word, repeat this step. If you do not follow through by writing the word from memory, you lose the chance of retaining it for future use. Practice is essential.

☑ Keep a list of the words you commonly misspell and work regularly at whittling it down. Do not load the list with exotic words; many of us would stumble over *asphyxiation* or *pterodactyl*. Concentrate instead on more commonly used words like *calendar, maintenance,* and *unnecessary*. Keep frequently used words on your list until you have learned to spell them.

☑ Proofread all your writing for misspellings.

spin-off

Spin-off usually refers to benefits that come about in one area (for example, insulating materials) as the result of achievements in another area (for example, space technology research). It also means a divestiture by a corporation of a division or subsidiary by issuing to stockholders shares in a new company set up to continue the operations of the division or subsidiary. Used in the latter sense, *spin-off* is **jargon** that you should not use unless you are certain that all your readers understand it.

strata / stratum

Strata is the plural form of *stratum,* meaning a "layer of material."

- The land's *strata* are exposed by erosion.
- Each *stratum* is clearly visible in the cliff.

S

style   ESL

A dictionary definition of style is "the way in which something is said or done, as distinguished from its substance." Writers' styles are determined by the way writers think and transfer their thoughts to paper—the way they use words, sentences, images, **figures of speech,** and so on.

A writer's style is the way his or her language functions in particular situations. For example, an email to a friend would be relaxed, even chatty in tone, whereas a job application letter would be more restrained and formal. Obviously, the style appropriate to the one would not be appropriate to the other. In both situations, the **readers** and the **purpose** determine the manner or style the writer adopts. Beyond an individual's personal style, various kinds of writing have distinct stylistic traits, such as **technical writing style.**

Standard English can be divided into two broad categories of style—formal and informal—according to how it functions in certain situations. Understanding the distinction between formal and informal writing styles helps writers use the appropriate style in the appropriate place. We must recognize, however, that no clear-cut line divides the two categories and that some writing may call for a combination of the two.

Formal Writing Style

A formal writing style can perhaps best be defined by pointing to certain material that is clearly formal, such as scholarly and scientific articles in professional journals, lectures read at meetings of professional societies, and legal documents. Material written in a formal style is usually the work of a specialist writing to other specialists or writing that embodies laws or regulations. As a result, the vocabulary is specialized and precise. The writer's tone is impersonal and objective because the subject matter looms larger in the writing than does the author's personality (see **point of view**). Unlike an informal writing style, a formal writing style does not use **contractions,** slang, or dialect (see **English, varieties of**). Because the material generally examines complex ideas, sentences may be elaborate (see **sentence construction**).

Formal writing need not be dull and lifeless. By using such techniques as the active **voice** whenever possible, **sentence variety,** and **subordination,** a writer can make formal writing lively and interesting, especially if the subject matter is inherently interesting to the reader.

- Although a knowledge of the morphological chemical constitution of cells is necessary to the proper understanding of living things, in the final analysis it is the activities of their cells that distinguish organisms from all other objects in the world. Many of these activities differ greatly among the various types of living things, but some of the basic sorts are shared by all, at least in their essentials. It is these fundamental actions with which we are concerned here. They fall into two major groups—those that are characteristic of the cell in the *steady state,* that is, in the normally functioning cell not engaged in reproducing itself, and those that occur during the process of *cellular reproduction.*

—Lawrence S. Dillon, *The Principles of Life Sciences*

Whether you should use a formal style in a particular instance depends on your readers and purpose. When writers attempt to force a formal style when it should not be used, their writing is likely to fall victim to **affectation, awkwardness,** and **gobbledygook.**

Informal Writing Style

An informal writing style is a relaxed and colloquial way of writing standard English. It is the style found in most private letters and in some technical **correspondence, memos, email,** nonfiction books of general interest, and mass-circulation magazines. There is less distance between the writer and the reader because the tone is more personal than in a formal writing style. Contractions and elliptical constructions are commonplace. Consider the following passage, written in an informal style, from a nonfiction book.

* Business, like art and science, has been revealed and conceived through the intellect and imagination of people, and it develops or declines because of the intellect and imagination of people.

  In fact, there is no business; there are only people. Business exists only *among* people and *for* people.

  Seems simple enough, and it applies to every aspect of business, but not enough businesspeople seem to get it.

  Reading the economic forecasts and the indicators and the ratios and the rates of this or that, someone from another planet might actually believe that there really are invisible hands at work in the marketplace.

  It's easy to forget what the measurements are measuring. Every number—from productivity rates to salaries—is just a device contrived by people to measure the results of the enterprise of other people. For managers, the most important job is not measurement but motivation. And you can't motivate numbers.

  —James A. Autry, *Love and Profit: The Art of Caring Leadership*

As the example illustrates, the vocabulary of an informal writing style is made up of generally familiar rather than unfamiliar words and expressions, although slang and dialect are usually avoided. An informal style approximates the cadence and structure of spoken English while conforming to the grammatical conventions of written English.

Writers who consciously attempt to create a style usually defeat the purpose. Attempting to impress readers with a flashy writing style can lead to affectation; attempting to impress them with scientific objectivity can produce a style that is dull and lifeless. Technical writing need be neither affected nor dull. It can and should be simple, clear, direct, even interesting—the key is to master basic writing skills and always to keep your readers in mind. What will be both informative and interesting to your audience? When that question is uppermost in your mind as you apply the steps of the writing process, you will achieve an interesting and informative writing style.

Writer's Checklist: Style

The following guidelines will help you produce a brisk, interesting style. Concentrate on them as you revise your rough draft.

- ☑ Use the active voice—not exclusively but as much as possible without becoming awkward or illogical.

- ☑ Use **parallel structure** whenever a sentence presents two or more thoughts of equal importance.

- ☑ Use a variety of sentence structures to avoid the monotony of a singsong style.

- ☑ Avoid stating positive thoughts in negative terms (write "40 percent responded" instead of "60 percent failed to respond"). (See also **positive writing** and **ethics in writing.**)

- ☑ Concentrate on achieving the proper balance between **emphasis** and subordination.

S

---

subjective complements (*see* **complements**)

---

subjects of sentences (*see* **sentence construction**)

---

subordinating conjunctions (*see* **conjunctions**)

subordination (ESL)

Subordination is a technique that writers use to show, by the structure of a sentence, the appropriate relationship between ideas of unequal importance by subordinating the less important ideas to the more important ideas.

- Beta Corporation now employs 500 people. It was founded just three years ago. [The two ideas are equally important.]

- Beta Corporation, *which now employs 500 people,* was founded just three years ago. [The number of employees is subordinated.]

- Beta Corporation, *which was founded just three years ago,* now employs 500 people. [The founding date is subordinated.]

Effective subordination can be used to achieve s**entence variety, conciseness,** and **emphasis.** For example, consider the sentence, "The city manager's report was carefully illustrated, and it covered five pages." See how it might be rewritten, using subordination, in any of the following ways:

DEPENDENT CLAUSE	The city manager's report, *which covered five pages,* was carefully illustrated.
PHRASE	The city manager's report, *covering five pages,* was carefully illustrated.
SINGLE MODIFIER	The city manager's *five-page* report was carefully illustrated.

Use a coordinating **conjunction** (*and, but, for, or, so, yet*) to concede that an opposite or balancing fact is true; however, a subordinating connective can often make the point more smoothly.

- *Although their*
  ~~Their~~ bank has a lower interest rate on loans, ~~but~~ ours provides a
  ^
  wider range of essential services.

The relationship between a conditional statement and a statement of consequences is clearer if the condition is expressed as a subordinate **clause.**

- *Because the*
  ~~The~~ bill was incorrect, ~~and~~ the customer was angry.
  ^

Subordinating connectives (*such as, because, if, while, when, though*) achieve subordination effectively.

S

- A buildup of deposits is impossible *because* the apex seals are constantly sweeping the inside chrome surface of the rotor housing.

Relative **pronouns** (*who, whom, which, that*) can be used effectively to combine related ideas that would be less smooth as independent clauses or sentences.

- The generator is the most common source of electric current. It ^, *which* uses mechanical energy to produce electricity. ^

Avoid overlapping subordinate constructions, in which each construction depends on a preceding one. In such a construction, the relationship between a relative pronoun and its antecedent often is not clear.

OVERLAPPING	Shock, *which* often accompanies severe injuries and infections, hemorrhages, burns, heat exhaustion, heart attacks, food or chemical poisoning, and some strokes, is a failure of the circulation, *which* is marked by a fall in blood pressure *that* initially affects the skin (*which* explains pallor) and later the vital organs, such as the kidneys and the brain.
CLARIFIED	Shock often accompanies severe injuries and infections, hemorrhages, burns, heat exhaustion, heart attacks, food or chemical poisoning, and some strokes. It is a failure of the circulation, initially to the skin (this explains pallor) and later to vital organs such as the kidneys and the brain.

substantives   ESL

A substantive is a word or a group of words that functions in a sentence as a subject. A substantive may be a **noun,** a **pronoun,** or a **verbal** (gerund or infinitive), or it may be a **phrase** or a **clause** that is used as a noun.

NOUN	The *report* is due today.
PRONOUN	*We* must finish the project on schedule.
GERUND	*Drilling* is expensive.
INFINITIVE	*To succeed* will require hard work.

| NOUN PHRASE | *Several local college graduates* applied for the job. |
| NOUN CLAUSE | *What I think* is unimportant. |

suffixes (ESL)

A suffix is a letter or letters added to the end of a word to change its meaning in some way. Suffixes can change the part of speech of a word.

NO SUFFIX	The market survey was *thorough*. [adjective]
SUFFIX	The *thoroughness* should be obvious to all. [noun]
NO SUFFIX	The report was issued at just the right *time*. [noun]
SUFFIX	The report was very *timely*. [adjective]

surveys (*see* questionnaires)

sweeping generalizations (*see* logic errors)

symbols

From highway signs to mathematical equations, people communicate with written symbols. When a symbol seems appropriate in your writing, either be certain that your **readers** understand its meaning or place the symbol in **parentheses** following the spelled-out term the first time it appears. Never use a symbol when readers would more readily understand the full term. (See also **abbreviations, numbers,** and **global communication.**)

S

synonyms (ESL)

A *synonym* is a word that means nearly the same thing as another word does.

- purchase, acquire, buy
- seller, vendor, supplier

The dictionary definitions of synonyms are very similar, but the connotations may differ. (A *seller* may be the same thing as a *supplier,* but the term *supplier* does not suggest a commercial transaction as strongly as *seller* does.)

Do not try to impress your reader by finding fancy or obscure synonyms in a thesaurus; the result is likely to be **affectation.** (See also **connotation/denotation** and **antonyms.**)

syntax (ESL)

Syntax refers to the way that words, phrases, and clauses are combined to form sentences. In English, the most common structure is the subject-verb-object pattern. For more information about the word order of sentences, see **sentence construction, sentence faults, sentence fragments,** and **sentence variety.**

(ESL) TIPS FOR DETERMINING WORD ORDER

In English, word order is an important signaler of the meaning in a sentence. In the sentences "Jessica called Derby" and "Derby called Jessica," only the word order conveys who is doing the calling and who is getting the call. Here are some helpful generalizations for determining word order in English sentences.

Declarative sentences follow the subject-verb-object/complement word order.

- Edward arrived late this morning.

Interrogative sentences add a question word and invert the subject and the first auxiliary verb (question word-first auxiliary verb or *be*-subject-rest of predicate).

- Will Edward arrive on time?

  Adjectives precede nouns.

- the book
- seven files
- contented coworkers
- angry chairman

(continued)

ESL TIPS FOR DETERMINING WORD ORDER (*continued*)

(See Tips for Using Adjectives on page 24 for advice on ordering adjectives when you have more than one adjective modifying a noun.)

Adverbs generally can be ordered depending on the other components of the sentence. An adverb is often placed at the end of the sentence.

- Edward arrived *late*.

An adverb should never separate a verb from its object.

- Edward phoned ~~immediately~~ his boss . *immediately*

When adverbs of time, manner, and location are present, there are two tendencies:

- *Tomorrow,* Edward will arrive *late at his office*.
  [adverb of time at beginning; adverbs of manner, location at end]

- Edward will arrive *late at his office tomorrow*.
  [adverbs of manner, location, time at end]

Frequency adverbs generally follow *be* or the first auxiliary verb and precede the main verb.

- Vacations are *always* too short.

- Vacations have *always* been assigned on the basis of seniority.

- Vacations *always* fly by too quickly.

S

T

table of contents

A table of contents is a list of the major **headings** in a report or the chapters in a book in their order of appearance and with their corresponding page numbers. A table of contents is typically included in a document longer than ten pages. Because it appears at the front of a work, a table of contents previews what is in the work and allows **readers** to assess the work's usefulness. It also helps those looking for specific information to locate sections quickly and easily.

When you are creating a table of contents, use the major headings and subheadings of your document exactly as they appear in the text, as shown in Figure T–1 on page 617.

To punctuate a table of contents, you can place a series of spaced periods, called *leaders*, between the heading and the page number on which the heading appears. For guidance on the placement of the table of contents in a report, see **formal reports.**

tables

A table is useful for showing numerous specific, related facts or statistics in a small space. A table can present data more concisely than text, and it is more accurate than graphic presentations because it provides facts that a graph cannot convey. A table facilitates comparisons among data because of the arrangement of the data into rows and columns. Do not rely on a table (or any **illustration**) as the only method of presenting significant information; overall trends are more easily conveyed in charts and **graphs.**

Table Elements

Tables typically include the elements shown in Figure T–2 on page 617.

Table of Contents

Abstract iii
Foreword vii
Executive Summary 1
Part I: A Condensed Report of Observations 3
   Publications 4
   Organization 6
   Operation 8
   Human Resources 10
Part II: Conclusions and Recommendations 13
Appendix A: Observations in Detail 15
   Publications 16
   Organization 23
   Operation 41
   Staff 56
Appendix B: Individual Reports 85
   Memo from the Chair 86
   Report by Elaine Browser 88
   Report by Gerald Watson 92
   Report by Gunther Braun 96

FIGURE T–1. Table of Contents

Table number → Table 1.   Estimated Emissions from Electric Power Generation (tons per gigawatthour) ← Table title

Fuel	Sulphur Dioxide	Nitrogen Oxides	Parti-culate Matter	Carbon Dioxide	Volatile Organic Compounds
Eastern coal	1.74	2.90	0.10	1,000	0.06
Western coal	0.81	2.20	0.06	1,039	0.09
Gas	0.003	0.57	0.02	640	0.05
Biomass	0.06	1.25	0.11	0*	0.61
Oil	0.51	0.63	0.02	840	0.03
Wind	0	0	0	0	0
Geothermal	0	0	0	0	0
Hydro	0	0	0	0	0
Solar	0	0	0	0	0
Nuclear	0	0	0	0	0

*Net emissions.

SOURCE: Department of Energy

Labels: Boxhead, Column headings, Stub, Body, Rule, Footnote, Source line

FIGURE T–2. Elements of a Typical Table

T

Table Number.   Table numbers are usually Arabic and should be assigned sequentially to the tables throughout the text.

Table Title.   The title, which is normally placed just above the table, should describe concisely what the table represents.

Boxhead.   The boxhead contains the column headings, which should be brief but descriptive. Units of measurement, where necessary, should be either specified as part of the heading or enclosed in parentheses beneath the heading. Standard **abbreviations** and **symbols** are acceptable. Avoid vertical lettering whenever possible.

Stub.   The left vertical column of a table is called the stub. It lists the items about which information is given in the body of the table.

Body.   The body comprises the data below the column headings and to the right of the stub. Within the body, arrange columns so that the items to be compared appear in adjacent rows and columns. Where no information exists for a specific item, substitute a row of dots or a dash to acknowledge the gap.

Rules.   Rules are the lines that separate the table into its various parts. Horizontal rules are placed below the title, below the body of the table, and between the column headings and the body of the table. Tables should not be closed at the sides. The columns within the table may be separated by vertical rules if such lines aid clarity.

Footnotes.   Footnotes are used for explanations of individual items in the table. Symbols (such as * and †) or lowercase letters (sometimes in parentheses) rather than numbers are ordinarily used to key table **footnotes** because numbers might be mistaken for numerical data or could be confused with the numbering system for text footnotes.

Source Line.   The source line identifies where the data originated. When a source line is appropriate, it appears below the table. Many organizations place the source line below the footnotes. (See also **copyright** and **plagiarism.**)

Continued Tables.   When a table must be divided so it can be continued on another page, repeat the column headings and give the table number at the head of each new page a "continued" label (for example, "Table 3, *continued*").

Informal Tables

To list relatively few items that would be easier for the reader to grasp in tabular form, you can use an informal table, as long as you introduce it properly.

- Dear Customer:
  To order replacement parts, use the following part numbers and prices:

Part	Part Number	Price ($)
Diverter valve	2-912	12.50
Gasket kit	2-776	0.95
Adapter	3-212	0.90
Hose assembly	7-211	3.50

Although informal tables do not need titles or table numbers to identify them, they do require column headings that accurately describe the information listed.

technical information letters (*see* correspondence)

technical manuals

Technical manuals help technical specialists and customers use and maintain products. They are usually written by professional technical writers, although in smaller companies, engineers and technicians may write them. Technical manuals may be produced as either paper or online documents. For printed manuals, packaging is important. They should be sized for easy use; for example, printed manuals are often loose-leaf to enable easy revision and updating. (See also **layout and design.**)

Many companies consider good technical manuals to be an important marketing tool; if a manual makes a product easy to operate or repair, consumers are more likely to buy the product. This entry describes typical technical manuals and the readers and objectives they serve.

User Manuals

User manuals are aimed at skilled or unskilled consumers of equipment. User manuals often provide **instructions** that enable users of the product to set up, operate, and maintain the product. User manuals also typically include safety precautions, troubleshooting charts and guides, and manufacturer contact information.

Tutorials

Tutorials are self-study guides for users of a product or system. Whether packaged with user manuals, packaged as part of the product, or published separately, tutorials walk the novice user through the operation of the product or system.

Training Manuals

Training manuals are used to train individuals in some procedure or skill, such as operating equipment, flying an airplane, or processing an insurance claim. In many technical and vocational fields, training manuals are the primary teaching device. They are often accompanied by videos, CD-ROMs, and other audiovisual material.

Operators' Manuals

Written for skilled operators of construction, manufacturing, computer, or military equipment, operators' manuals contain essential instructions and safety warnings. They are often published in a convenient format that allows operators to use them at a work site.

Service Manuals

Service manuals help trained technicians repair equipment or systems, usually at the customer's location. Such manuals often contain elaborate troubleshooting guides for locating technical problems.

Special-Purpose Manuals

A number of manuals are intended for limited purposes.

Program reference manuals provide computer programmers with definitions, syntax rules, and other technical information about specific programming languages, such as FORTRAN and COBOL.

Overhaul manuals are guides for rebuilding items at the factory. Overhaul manuals are often used in the heavy-equipment industry, in which rebuilding a huge crane, for example, is much less costly than manufacturing a new one.

Handling and setup manuals are guides for skilled and unskilled employees in the safe handling, installation, or setup of various kinds of equipment.

Safety manuals are guides for operators of potentially dangerous equipment and illustrate safe operating practices.

Writer's Checklist: Technical Manuals

☑ Pay close attention to the **readers'** needs and their level of technical knowledge of the product you are documenting.

☑ Pay attention also to **organization** and **outlining** because of the complexity of the products and systems for which manuals are written.

☑ Use **illustrations,** such as exploded-view **drawings, flowcharts, photographs,** and **tables,** as appropriate.

☑ Include parts lists, troubleshooting guides, warning statements (as shown in the entry **instructions**), **schematic diagrams,** standard **symbols** for potential dangers, and **indexes** where appropriate.

☑ To ensure that manuals are helpful and accurate, submit them to technical, legal, and peer reviews.

☑ Consider **usability testing** of manuals in which typical product users perform tasks by using the manuals.

☑ Carefully review the entries in this book on **process explanation, proofreading, revision,** and **technical writing style.**

technical writing style

The goal of technical writing is to enable **readers** to use a technology or understand a process or concept. Because the subject matter is more important than the writer's voice, technical writing style uses an objective, not subjective, **tone.** The writing style is direct and utilitarian, emphasizing exactness and clarity rather than elegance or allusiveness. A technical writer employs figurative language only when a **figure of speech** would facilitate understanding.

Technical writing is often—but not always—aimed at readers who are not experts in the subject, such as consumers or employees learning to operate unfamiliar equipment. Because the audience is often inexperienced and the procedures described may involve hazardous material or equipment, clarity becomes an ethical as well as a stylistic concern. (See also **ethics in writing.**) Figure T–3 is an excerpt from a **technical manual** that instructs eye care specialists about operating testing equipment. As the figure illustrates, **layout and design** enhance clarity by giving the reader both text and visual cues.

Technical writing style may use a technical vocabulary appropriate for the reader, as Figure T–3 shows (*saccade, monocular elec-*

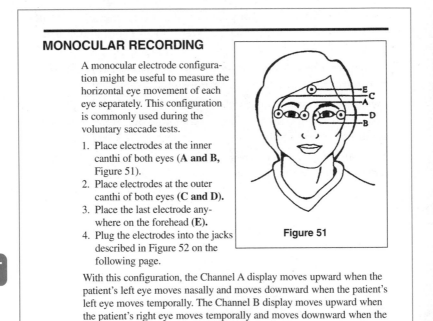

FIGURE T–3. Example of Technical Writing

trode), but it avoids **affectation.** Good technical writing also avoids overusing the passive **voice.** (See also **illustrations, instructions, organization, process explanation,** and **style.**)

telegraphic style   ESL

Telegraphic style condenses writing by omitting **articles, pronouns, conjunctions,** and transitional expressions. Although **conciseness** is important in writing, especially in **instructions,** writers sometimes make their sentences too brief by omitting words. Telegraphic style forces **readers** to supply the missing words mentally, thus slowing their progress. Compare the following two passages and notice how much easier the revised version reads (the added words are italicized).

TELEGRAPHIC	Take following action when treating serious burns. Remove loose clothing on or near burn. Cover injury with clean dressing and wash area around burn. Secure dressing with tape. Separate fingers/toes with gauze/cloth to prevent sticking. Do not apply medication unless doctor prescribes.
CLEAR	Take *the* following action when treating *a* serious burn. Remove *any* loose clothing on or near *the* burn. Cover *the* injury with *a* clean dressing and wash *the* area around *the* burn. *Then* secure *the* dressing with tape. Separate *the* fingers *or* toes with gauze *or* cloth to prevent *them from* sticking *together.* Do not apply medication unless *a* doctor prescribes *it.*

Telegraphic style can also produce **ambiguity,** as the following example demonstrates.

AMBIGUOUS	Grasp knob and adjust lever before raising boom. [Does this sentence mean that the reader should adjust the lever or also grasp an "adjust lever"?]
CLEAR	Grasp the knob and adjust the lever before raising the boom.

Although you may save yourself work by writing telegraphically, your readers will have to work that much harder to decipher your writing. (See also **transition.**)

tenant / tenet

A *tenant* is one who holds or temporarily occupies a property owned by another person.

- The *tenants* were upset by the increased rent.

A *tenet* is an opinion or belief held by a person, an organization, or a system.

- The idea that competition will produce adequate goods and services for a society is a central *tenet* of capitalism.

tense   (ESL)

DIRECTORY
Overview   624
Past Tense   625
Past Perfect Tense   625
Present Tense   625
Present Perfect Tense   626
Future Tense   626
Future Perfect Tense   626
Tips for Using the Progressive Form (ESL)   627
Shift in Tense   628

Overview

Tense is the grammatical term for **verb** forms that indicate time distinctions. There are six tenses in English: past, past perfect, present, present perfect, future, and future perfect. Each tense also has a corresponding progressive form.

TENSE	BASIC FORM	PROGRESSIVE FORM
Past	I began	I was beginning
Past perfect	I had begun	I had been beginning
Present	I begin	I am beginning
Present perfect	I have begun	I have been beginning
Future	I will begin	I will be beginning
Future perfect	I will have begun	I will have been beginning

Perfect tenses allow you to express a prior action or condition that continues in a present, past, or future time.

PRESENT PERFECT	*I have begun* to write the annual report and will continue for the rest of the month.
PAST PERFECT	*I had begun* to read the manual when the lights went out.
FUTURE PERFECT	*I will have begun* this project by the time funds are allocated.

Progressive tenses allow you to describe some ongoing action or condition in the present, past, or future.

PRESENT PROGRESSIVE	*I am beginning* to be concerned that we will not meet the deadline.
PAST PROGRESSIVE	*I was beginning* to think we would not finish by the deadline.
FUTURE PROGRESSIVE	*I will be requesting* a leave of absence when this project is finished.

Past Tense

The simple past tense indicates that an action took place in its entirety in the past. The past tense is usually formed by adding *-d* or *-ed* to the root form of the verb.

* We *closed* the office early yesterday.

Past Perfect Tense

The past perfect tense (sometimes called *pluperfect*) indicates that one past event preceded another. It is formed by combining the helping verb *had* with the past participle form of the main verb.

* He *had finished* by the time I arrived.

Present Tense

The simple present tense represents action occurring in the present, without any indication of time duration.

* I *use* the beaker.

A general truth is always expressed in the present tense.

* He learned that the saying "time *heals* all wounds" is true.

T

The present tense can be used to present actions or conditions that have no time restrictions.

- Water *boils* at 212 degrees Fahrenheit.

The present tense can be used to indicate habitual action.

- I *pass* the paint shop on the way to my department every day.

The present tense can be used as a "historical present" to make things that occurred in the past more vivid.

- "Commission Calls for Development of Better Information Technology" [news headline]

Present Perfect Tense

The present perfect tense describes something from the recent past that has a bearing on the present—a period of time before the present but after the simple past. The present perfect tense is formed by combining a form of the helping verb *have* with the past principle form of the main verb.

- He *has retired,* but he visits the office frequently.

- We *have finished* the draft and are ready to begin revising it.

Future Tense

The simple future tense indicates a time that will occur after the present. It uses the helping verb *will* (or *shall*) plus the main verb.

- I *will finish* the job tomorrow.

Do not use the future tense needlessly; doing so is unnecessarily complex.

- This system ~~will be~~ *is* explained on page 3.

- When you press this button, the hoist ~~will move~~ *moves* the plate into position.

Future Perfect Tense

The future perfect tense indicates action that will have been completed at a future time. It is formed by linking the helping verbs *will have* to the past participle form of the main verb.

(ESL) TIPS FOR USING THE PROGRESSIVE FORM

English uses the progressive form (also known as the continuous form), particularly the present progressive, much more frequently than do many other languages. Here are some guidelines to help you practice using the progressive form.

The progressive form of the verb is composed of two features: a form of the helping verb *be* and the *-ing* form of the base verb.

PRESENT PROGRESSIVE	I *am rewriting* the memo.
PAST PROGRESSIVE	I *was rewriting* the memo for several days.
FUTURE PROGRESSIVE	I *will be rewriting* that memo forever.

The present progressive is used in three ways:

1. To refer to an action that is in progress at the moment of speaking or writing
   - The conference chair *is* constantly *interrupting* the speakers.
   - The parliamentarian *is wearing* an Oscar de la Renta suit.
2. To highlight that a state or action is not permanent
   - The office temp *is helping* us for a few weeks.
3. To express future plans
   - The summer intern *is leaving* to return to school this Friday.

The past progressive is used to refer to a continuing action or condition in the past, usually with specified limits.

   - I *was failing* calculus until I got glasses.

The future progressive is used to refer to a continuous action or condition in the future.

   - We *will be monitoring* his condition all night.

Verbs that express mental activity are generally not used in the progressive:

appear	believe	belong	hate	have
hear	know	like	love	need
own	prefer	resemble	see	seem
smell	taste	think	understand	want

   believe
   - I ~~am believing~~ the defendant's testimony.
     ^

T

- She *will have driven* the test car 40 miles by the time she returns.

Shift in Tense

Be consistent in your use of tense. The only legitimate shift in tense records a real change in time. When you choose a tense, stay with that tense. Illogical shifts in tense will only confuse your readers.

- Before he installed the printed circuit, the technician ~~cleans~~ *cleaned* the contacts.

---

test reports

The test report differs from the more formal **laboratory report** in both size and scope. Considerably smaller, less formal, and more routine than the laboratory report, the test report can be a **memo** or a formal business letter, depending on its recipient. Either way, the report should have a subject line at the beginning to identify the test being discussed. (See also **correspondence.**)

The **opening** of a test report should state the test's purpose, unless it is obvious. The body of the report presents the data. A report on the tensile strength of metal, for example, would include the readings from the test equipment. If the procedure used to conduct the test is relevant, describe it as well. State the results of the test and, if necessary, interpret them. Conclude the report with any recommendations made as a result of the test. Figure T–4 on page 629 shows a typical test report.

---

that   ESL

Do not delete *that* from a sentence if it is necessary for the reader's understanding.

- Some designers fail to recognize sufficiently *that* the human beings who operate the equipment constitute an important safety system.

Avoid the unnecessary repetition of *that.*

Biospherics, Inc.

4928 Wyaconda Road
Rockville, MD 20852

Phone: 301-598-9011
Fax: 301-598-9570

September 9, 20--

Mr. Leon Hite, Administrator
The Angle Company, Inc.
1869 Slauson Boulevard
Waynesville, VA 23927

Subject: Monitoring Airborne Asbestos at the Route 66 Site

Dear Mr. Hite:

On August 30, Biospherics, Inc., performed asbestos-in-air monitoring at your Route 66 construction site, near Front Royal, Virginia. Six people and three construction areas were monitored.

All monitoring and analyses were performed in accordance with "Occupational Exposure to Asbestos," U.S. Department of Health and Human Services, Public Health Service, National Institute for Occupational Safety and Health, 1999. Each worker or area was fitted with a battery-powered personal sampler pump operating at a flow rate of approximately two liters per minute. The airborne asbestos was collected on a 37-mm Millipore-type AA filter mounted in an open-face filter holder. Samples were collected over an 8-hour period.

In all cases, the workers and areas monitored were exposed to levels of asbestos fibers well below the standard set by OSHA. The highest exposure found was that of a driller exposed to 0.21 fibers per cubic centimeter. The driller's sample was analyzed by scanning electron microscopy followed by energy dispersive X-ray techniques that identify the chemical nature of each fiber, to identify the fibers as asbestos or other fiber types. Results from these analyses show that the fibers present were tremolite asbestos. No nonasbestos fibers were found.

If you need more details, please let me know.

Yours truly,

Gary Geirelach
Chemist

Email: biospher@juno.org

FIGURE T–4. Test Report

T

WORDY   You will note *that,* as you assume greater responsibility and as your years of service with the company increase, *that* your benefits will increase accordingly.

CONCISE   You will note *that* your benefits will increase as you assume greater responsibility and your years of service with the company increase.

(See also **conciseness/wordiness.**)

that / which / who   ESL

That and *which* refer to animals and things; *who* refers to people.

- Companies *that* fund basic research must not expect immediate results.

- The jet stream, *which* is approximately 8 miles above the earth, blows at an average of 64 miles per hour.

- Diane Stoltzfus, *who* is retiring tomorrow, has worked for the company for 20 years.

The word *that* is often overused. However, do not eliminate it if doing so would cause **ambiguity** or problems with **pace.** (See also **that** and **who/whom.**)

- On the file specifications input to the compiler for any chained file,
  the user must ensure ‸*that* the number of sectors per main file section is
  a multiple of the number of sectors per bucket.

Use *which*, not *that*, with nonrestrictive clauses (clauses that do not change the meaning of the basic sentence).

NONRESTRICTIVE	After John left the restaurant, *which* is one of the best in New York, he came directly to my office.
RESTRICTIVE	Companies *that* diversify usually succeed.

(See also **restrictive and nonrestrictive elements.**)

ESL TIPS FOR OMITTING A RELATIVE PRONOUN

In English, the relative pronoun in a restrictive clause may be deleted when it is not the subject of the relative clause. (See also **restrictive and nonrestrictive elements.**)

   Deletion is not permitted when the relative pronoun is the subject of the relative clause:

- He went to the stockroom ‸*where* the files are kept.

   Deletion is permitted when the relative pronoun is not the subject of the relative clause:

- The stockroom ~~that~~ we always use is on the second floor.

T

there / their / they're   Ⓔˢᴸ

There is an **expletive** (a word that fills the position of another word, phrase, or clause) or an **adverb.**

EXPLETIVE   *There* were more than 1,500 people at the conference.

ADVERB   More than 1,500 people were *there.*

Their is the possessive form of *they.*

• Our employees are expected to keep *their* records current.

They're is a contraction of *they are.*

• If *they're* right, we should change the design.

thesaurus

A thesaurus is a book or electronic file of words and their **synonyms** and **antonyms,** arranged or retrievable by categories. Thoughtfully used, a thesaurus can help you with **word choice** during the **revision** phase of the writing process. However, the variety of words may tempt you to choose inappropriate or obscure synonyms just because they are available. Use a thesaurus only to clarify your meaning, not to impress your **reader.** Never use a word unless you are sure of its meanings; **connotations** of the word that might be unknown to you could mislead your reader. (See also **affectation.**)

titles

The title of a report or other document should both state its topic and indicate its **scope** and **purpose,** as in the following.

• "Effects of 60 Hertz Electric Fields on Embryo Chick Development, Growth, and Behavior"

A title should be concise, but do not make your title so short that it is not specific. For example, the title "Electric Fields and Living Organisms" announces the topic of a report, but it does not answer important questions such as "What is the relationship between electric fields and living organisms?" and "What stage of the organism's life cycle does the report discuss?" Titles are important because many readers decide to read a document, such as a **trade journal article,** based on the title you provide.

For guidelines on how to capitalize titles and when to use **italics** and **quotation marks,** see those entries and **capital letters.**

Writer's Checklist: Titles

The following guidelines will help you produce useful and specific titles.

☑ Avoid titles with redundancies that begin with "Notes on," "Studies on," "A Report on." However, works like annual reports or feasibility studies should be identified as such in the title because that information specifies the purpose and the scope of the report.

☑ Use the active **voice** — not exclusively but as much as possible without becoming awkward or illogical.

☑ Do not put titles in sentence form except for titles that ask a question.

☑ Do not indicate dates in the title of a periodic or progress report; put them in a subtitle ("Quarterly Report on the Hospital Admission Rates: January–March 20--").

☑ Avoid including **abbreviations, acronyms and initialisms,** chemical formulas, and the like in your title unless the work is addressed exclusively to specialists in the field.

☑ For multivolume publications, repeat the title on each volume. The distinction between the volumes is made by the volume number (Volume 1, 2, 3, and so on), and sometimes each volume has a different subtitle. For example, Volume 1 could be the executive summary; Volume 2, the main report; and any subsequent volumes, lengthy appendixes.

to / too / two   ⓔⓢⓛ

To, too, and *two* are confused only because they sound alike. *To* is used as a **preposition** or to mark an infinitive (see **verbs**).

- Send the report *to* the district manager. [preposition]

- I do not wish *to* go! [mark of the infinitive]

Too is an **adverb** meaning "excessively" or "also."

- The price was *too* high. ["excessively"]
- I, *too,* thought it was high. ["also"]

Two is a number.

- Only *two* buildings have been built this fiscal year.

tone   (ESL)

Tone is the writer's attitude toward the subject and his or her **readers.** The tone may be casual or serious, enthusiastic or skeptical, friendly or hostile, personal or formal. In **correspondence,** tone is particularly important because the message represents direct communication between two people. Moreover, a positive tone helps establish rapport between your organization and the public.

In workplace writing, the tone may range widely, depending on the **purpose,** situation, context, audience, and even medium. For example, in an **email** message to be read only by an associate who is also a friend, your tone might be casual.

- Your proposal to Smith and Kline is super. If we get the contract, I owe you lunch! I've marked a couple of places where we could be overly optimistic on the schedule. It's on its way by snail mail. See what you think.

In a **memo** to your manager or superior, however, your tone would be more formal and respectful.

- I think your proposal to Smith and Kline is excellent. I have marked a couple of places for you to consider to ensure that we are not committing ourselves to a schedule we might not be able to keep. If I can help in any other way, please let me know.

In a message that serves as a report to numerous readers, the tone would be professional, without casual language that could be misinterpreted.

- The Smith and Kline proposal appears complete and thorough, based on our department's evaluation. Several small revisions, however, would ensure that the company is not committing itself to an unrealistic schedule. These are marked on the copy of the report being circulated.

The choice of words, the **introduction** or the **opening,** and even the **title** contribute to the overall tone of your document. For instance, a title such as "Ecological Consequences of Diminishing Water Resources in California" clearly sets a different tone from "What Happens When We've Pumped California Dry?" The first title would be appropriate for a report; the second title would be appropriate for a newsletter or popular magazine article. The important consideration is to make sure that your tone is best suited to your purpose and your readers. (See also **style.**)

topic sentences (*see* **paragraphs**)

topics

On the job, the topic of a writing project is usually determined by need. In a college writing course, on the other hand, you may have to select your own topic. If you do, keep the following points in mind.

Writer's Checklist: Topic Selection

- ☑ Select a topic that interests you so that you will maintain your enthusiasm.
- ☑ Select a topic you can **research** adequately with the facilities available to you.
- ☑ Limit your topic so its **scope** is small enough to handle in the time you are given. A topic like "Environmentalism," for example, would be too broad. On the other hand, keep the topic broad enough so you will have enough to write about, for example, "Water Purification Options for Cities."
- ☑ Select a topic for which you can make adequate **preparation** to ensure a good final **report.**

toward / towards

Both *toward* and *towards* are acceptable variant spellings of the preposition meaning "in the direction of." *Toward* is more common in the United States, while *towards* is more common in Great Britain.

- We walked *toward* the Golden Gate Bridge.

- They were headed *towards* London Bridge.

trade journal articles

DIRECTORY
Overview   635
Planning the Article   635
Gathering the Data   636
Organizing the Draft   637
Preparing Sections of the Article   637
Preparing the Manuscript   639
Obtaining Publication Clearance   639

Overview

A trade journal article is an article written on a specific subject for a professional periodical. Such periodicals, commonly known as trade journals (or professional or scholarly journals), are often the official publications of professional societies. *Technical Communication*, for example, is an official voice of the Society for Technical Communication. Other professional publications include *Electrical Engineering Review, Chemical Engineering, Nucleonics Week,* and hundreds of similar titles. Professional staff people, such as engineers, scientists, educators, and legal professionals, regularly contribute articles to trade journals.

From time to time in your career, you may wish to write a trade journal article that would be of interest to others in your field. Such an article makes your work more widely known, provides publicity for your employer, gives you a sense of satisfaction, and may even improve your chances for professional advancement.

Planning the Article

When you are thinking about writing an article for a trade journal, consider the following questions:

- Is your work or your knowledge of the subject original? If not, what is there about your approach that justifies publication?
- Will the significance of the article justify the time and effort needed to write it?

- What parts of your work, project, or study are most appropriate to include in the article?

To help you answer those questions, learn as much as possible about the periodical or periodicals to which you plan to submit an article and consult your colleagues for advice. Once you have decided on several journals, consider the following factors about each one.

- the professional interests of its readership
- the size of its readership
- the professional reputation of the journal
- the appropriateness of your article to the journal's goals, as stated on its masthead page or in a mission statement on its Web site
- the frequency with which its articles are cited in other journals

Next, read back issues of appropriate journals to find out information like the amount and kind of details that the articles include, their length, and the typical writing style.

If your subject involves a particular project, begin work on your article when the project is in progress. That allows you to write the draft in manageable increments and record the details of the project while they are fresh in your mind. It also makes the writing integral to the project and may even reveal any weaknesses in the design or details, such as the need for more data.

As you plan your article, decide whether to invite one or more coauthors to join you. Coauthors can add strength and substance to a paper, but they also can add complications. To minimize potential problems, one of you should take the responsibility of being the primary author, that is, the person who makes assignments and sets up a schedule. The primary author must also ensure that the finished paper reads as smoothly as if it had been written by one person. (See **collaborative writing.**)

Gathering the Data

As you gather information, take notes from all the sources of primary and secondary **research** available to you, such as the following:

- published material on the subject
- your own experience

- notebooks recorded during your research
- survey results
- **progress reports**
- performance records and test data

Begin your research with a careful literature review to establish what has been published about your topic. A review of the relevant information in your field can be insurance against writing an article that has already been published. (Some articles, in fact, begin with a literature review.) As you compile that information, record your references in full; include all the information you need to document the source. (See also **Internet research** and **documenting sources.**)

Organizing the Draft

Some trade journals use a prescribed **organization** for the major sections. The following organization is common in scientific journals: **introduction,** materials and methods, results, discussion, and sources cited. If the major organization is not prescribed, choose and arrange the various sections of the draft in a way that shows your results to best advantage.

The best guarantee of a logically organized article is a good **outline.** In addition to shaping your information in a logical order, outlining helps you organize your thinking. If you have coauthors, it is essential that the writing team work from a common, well-developed outline; otherwise, coordinating the various writers' work will be impossible, and the parts will not fit together logically as a whole.

Once you have written your outline, consider it to be flexible; you may well change and improve your organization as you write the rough draft.

Preparing Sections of the Article

As you are preparing an article, pay particular attention to a number of key sections and elements: the **abstract,** the introduction, the **conclusion, illustrations** and **tables,** and **headings.**

Abstract.   Although it will appear at the beginning of the article, you can write the abstract only after you have finished writing the body of the manuscript. Be sure to follow any instructions provided by the journal on writing abstracts and review abstracts previously published in that journal. Prepare your abstract carefully—it will be the basis on which many other researchers decide whether to read your article in full. Abstracts are often published independently in abstract journals and are a source of terms (called *keywords*) used to index, by subject, the original article for computerized information-retrieval systems.

Introduction.   The purpose of an introduction is to give your readers enough general information about your topic for them to be able to understand the detailed information in the body of the article. The introduction should discuss these aspects of the article:

- the purpose of the article
- a definition of the problem examined
- the **scope** of the article
- the rationale for your approach to the problem or project and the reasons you rejected alternative approaches
- previous work in the field, including other approaches described in previously published articles

Above all, your introduction should emphasize what is new and different about your approach, especially if you are not dealing with a new concept. It should also demonstrate the overall significance of your project or approach by explaining how it fills a need, solves a current problem, or offers a useful application.

Conclusion.   The conclusion section pulls together your results and interprets them in light of the purpose of your study and the methods used to conduct it. Your conclusion must grow out of the evidence for the findings in the body of the article. The conclusion must also be consistent with the scope of information presented in the introduction.

T

Illustrations and Tables.   Use illustrations and tables wherever they are appropriate, but design each for a specific purpose: to describe a function, to show an external appearance, to show internal construction, to display statistical data, or to indicate trends.

Consider working your rough illustrations into your outline so you can decide on your illustrations as you write the manuscript. Used effectively and appropriately, illustrative material can clarify information and reinforce the point you are making in the article.

Headings. Headings are important for an article because they break the text into manageable portions. They also allow journal readers to understand the development of your topic and pinpoint sections of particular interest to them. The main headings in your outline will often become the headings in your final manuscript. Check the use of headings when you review back issues of the journal.

References. The references section (often titled "Works Cited") of the article lists the sources you used in the article. The specific format for listing sources varies from field to field. Usually the journal to which you submit an article will specify the form the editors require for citing sources. (For a detailed discussion of using and citing sources, see **documenting sources** and **quotations.**)

Preparing the Manuscript

Some trade journals recommend a particular style guide like the *Chicago Manual of Style* or offer a style sheet with detailed guidelines on style and format. Such style sheets often include specific instructions about how to format the manuscript, how many copies to submit, and how to handle **abbreviations, symbols, mathematical equations,** and the like. The following guidelines are typical.

- Double-space the manuscript, leaving one-inch margins all around, and number each page.
- Provide specific, accurate, and self-explanatory captions for all figures and tables. Add *callouts* (labels) to those illustrations that need them.
- Provide clear and accurately worded labels for drawings and other illustrations.
- Place any mathematical equations on separate lines in the text and number them consecutively.
- Include only high-quality **photographs** or scanned images. If you submit an original photograph, check with the editor about any special requirements. Identify each photograph by writing lightly in pencil on the back. If necessary, indicate which edge of the photograph is the top.

Obtaining Publication Clearance

After you have finalized your article, submit a copy to your employer for review before sending it to the journal. A review will make sure that you have not inadvertently revealed any proprietary information. Likewise, secure permission ahead of time to print information for which someone else holds the **copyright.** (See also **plagiarism.**)

transition (ESL)

Transition is the means of achieving a smooth flow of ideas from sentence to sentence, paragraph to paragraph, and subject to subject. Transition is a two-way indicator of what has been said and what will be said; it provides a means of linking ideas to clarify the relationship between them. You can achieve transition with a word, a phrase, a sentence, or even a paragraph. Without the guideposts of transition, **readers** can lose their way.

Transition can be obvious.

- *Having considered* the technical problems outlined in this proposal, *we turn now* to the question of adequate staffing.

Or it can be subtle.

- *Even if* the technical problems could be solved, there *still remains* the problem of adequate staffing.

Either way, you now have your readers' attention fastened on the problem of adequate staffing, exactly what you set out to do.

Certain words and phrases are inherently transitional. Consider the following terms and their functions.

- Result: *therefore, as a result, consequently, thus, hence*
- Example: *for example, for instance, specifically, as an illustration*
- Comparison: *similarly, likewise*
- Contrast: *but, yet, still, however, nevertheless, on the other hand*
- Addition: *moreover, furthermore, also, too, besides, in addition*
- Time: *now, later, meanwhile, since then, after that, before that time*
- Sequence: *first, second, third, then, next, finally*

Within a **paragraph,** such transitional expressions clarify and smooth the movement from idea to idea. Conversely, the lack of transitional devices can make for disjointed reading. Consider the following passage, which lacks adequate transition.

CHOPPY   People had always hoped to fly. Until 1903 it was only a dream. It was thought by some that human beings were not meant to fly. The Wright brothers launched the world's first heavier-than-air flying machine. The airplane has become a part of our everyday life.

Now read the same passage with words and phrases of transition added — notice how much more smoothly the thoughts flow.

SMOOTH   People had always hoped to fly, *but* until 1903 it was only a dream. *Before that time,* it was thought by some that human beings were not meant to fly. *However,* in 1903 the Wright brothers launched the world's first heavier-than-air flying machine. *Since then,* the airplane has become a part of our everyday life.

Transition between Sentences

In addition to using transitional words and phrases, you can achieve effective transition between sentences by repeating key words or ideas from preceding sentences and by using pronouns that refer to antecedents in previous sentences. Consider the following short paragraph, which employs both those means.

* Representative of many American university towns is Millville. *This Midwestern town,* formerly a *sleepy farming community,* is today the home of a large and bustling *academic community.* Attracting students from all over the Midwest, *this university* has grown rapidly in the last ten years. *This same decade* has seen a physical expansion of the campus. The state, recognizing *this expansion,* has provided additional funds for the acquisition of land adjacent to the university.

Another device for achieving transition is enumeration.

* The recommendation rests upon *three conditions. First,* the department staff must be expanded to a sufficient size to handle the increased workload. *Second,* sufficient time must be provided for the training of the new members of the staff. *Third,* a sufficient number of qualified applicants must be available.

Transition between Paragraphs

All the means discussed so far for achieving transition between sentences can also be effective for transition between paragraphs. For paragraphs, however, longer transitional elements are often required. One technique is to use an opening sentence that summarizes the preceding paragraph and then moves on to a new paragraph.

- One property of material considered for manufacturing processes is hardness. Hardness is the internal resistance of the material to the forcing apart or closing together of its molecules. Another property is ductility, the characteristic of the material that permits it to be drawn into a wire. The smaller the diameter of the wire into which the material can be drawn, the greater the ductility. Material also may possess malleability, the property that makes it capable of being rolled or hammered into thin sheets of various shapes. Engineers, in selecting materials to employ in manufacturing, must consider these properties before deciding on the most desirable for use in production.
  The requirements of hardness, ductility, and malleability account for the high cost of such materials. . . .

Another technique is to ask a question at the end of one paragraph and answer it at the beginning of the next.

- New technology has always been feared because it has at times displaced some jobs. But the all-important fact that is often overlooked is that it invariably creates many more jobs than it eliminates. Historically, the vast number of people employed in the American automobile industry as compared with the number of people who had been employed in the harness-and-carriage-making business is a classic example. Almost always, the jobs that have been eliminated by technological advances have been menial, unskilled jobs, and those who have been displaced have been forced to increase their skills, which resulted in better and higher-paying jobs for them. *In view of these facts, is new technology really bad?*
  Certainly technology has given us unparalleled access to information and created many new roles for employees. . . .

A purely transitional paragraph may be inserted to aid readability.

- The problem of poor management was a key factor that has caused the weak performance of the company.
  Two other setbacks to the company's fortunes that year also marked the company's decline: the loss of many skilled workers through the early retirement program and the intensification of the devastating rate of employee turnover.
  The early retirement program caused the failure. . . .

If you provide logical **organization** and you have prepared an **outline,** your transitional needs will easily be satisfied, and your writing will have **unity** and **coherence**. During **revision,** look for places where transition is missing and add it. Look for places where it is weak and strengthen it.

transmittal letters (*see* **cover letters**)

trip reports

Many employees must prepare reports of the business trips they take. A trip report both provides a permanent record of a business trip and its accomplishments and enables many employees to benefit from the information that one employee has gained.

A trip report is normally written in **memo** format and is often addressed to an immediate superior. The subject line gives the destination and the dates of the trip. The body of the report explains why you made the trip, whom you visited, and what you accomplished. The report should devote a brief section to each major event and may include a head for each section (you need not give equal space to each event; elaborate on the more important events). Follow the body of the report with the appropriate conclusions and recommendations. A typical trip report is shown in Figure T–5.

INTEROFFICE MEMO

TO: Roberto Camacho, Manager Customer Services
FROM: James D. Kerson, Maintenance Specialist   *JDK*
DATE: January 13, 20--
SUBJECT: Trip to Smith Electric Co., Huntington, West Virginia,
           January 3–5, 20--

I visited the Smith Electric Company in Huntington, West Virginia, to determine the cause of a recurring failure in a Model 247 printer and to fix it.

Problem
The printer stopped printing periodically for no apparent reason. Repeated efforts to bring it back online eventually succeeded, but the problem recurred at irregular intervals. Neither customer personnel operating the printer nor the local maintenance specialist was able to solve the problem.

FIGURE T–5. Trip Report

> *Investigation*
> On January 3, I met with Ms. Ruth Bernardi, the Office Manager, who explained the problem. My troubleshooting did not reveal the cause of the problem then or on January 4.
>
> Only when I tested the logic cable did I find that it contained a broken wire. I replaced the logic cable and then ran all the normal printer test patterns to make sure no other problems existed. All patterns were positive, so I turned the printer over to the customer.
>
> *Conclusion*
> There are over twelve thousand of these printers in the field, and to my knowledge, this is the first occurrence of a bad cable. Therefore, I do not believe the logic cable problem found at Smith Electric Company warrants further investigation.

FIGURE T–5. Trip Report (*continued*)

trite language

Trite language is made up of words, phrases, or ideas that have been used so often they are stale.

TRITE	*It may interest you to know* that all the folks in the branch office are *hale and hearty.* I should finish my report *quick as a wink,* and we should *clean up* on it.
IMPROVED	Everyone here at the branch office is well. I should finish my project within a week, and I'm sure it will prove profitable for us.

Trite language suggests that the writer is thoughtless in his or her **word choice** or is not thinking carefully about the topic but is relying instead on the thoughts and words of others. (See also **clichés.**)

trouble reports

The trouble report is used to report an accident, an equipment failure, a health emergency, and so on. The report assesses the causes of the problem and suggests changes necessary to prevent its recurrence. Because it is an internal document, the trouble report normally follows a simple **memo** format. Usually trouble reports are not large enough in either size or scope to require the format of a **formal report.**

In the subject line of the memo, state the precise problem you are reporting. Then begin your description of the problem in the body of your report. What happened? Where did it occur? When did it occur? Was anybody hurt? Was there any property damage? Was there a work stoppage? Because insurance claims, worker's compensation awards, and, in some instances, lawsuits may hinge on the information contained in a trouble report, be sure to include precise times, dates, locations, treatment of injuries, names of witnesses, and any other crucial information. Give a detailed analysis of what caused the problem. Be thorough and accurate in your analysis and support any judgments or conclusions with facts. Be objective; always use a neutral **tone** and avoid assigning blame. If you speculate about the cause of the problem, make it clear that you are speculating. (See also **ethics in writing.**)

In your **conclusion,** state what has been done, what is being done, or what will be done to correct the conditions that led to the problem. That may include training in safety practices, improved equipment, protective clothing (for example, shoes or goggles), and so on.

The report shown in Figure T–6 describes an accident involving personal injury. Notice the careful use of language and the factual details.

MEMO

To: Dianne Bell
From: Elaine Susuki   *ES*
Date: August 19, 20--
Subject: John Markley Accident Review Summary

An Accident Review was conducted on Friday, August 16, 20--, at the Reed Service Center. The attendees were as follows:

John Markley, Injured Employee
Harry Hartsock, Union Representative
Carl Timmerinski, Employee's Supervisor
Kalo Katarian, Manager
Marie Sonora, Safety Officer, Service Operations

FIGURE T–6. Trouble Report

Date of Accident: August 7, 20--

Days of Lost Time: 2

Accident Summary
On August 7, John Markley stopped by a rewiring job on German Road. Chico Ruiz was working there, stringing new wire, and John needed to check with Chico about the materials he wanted for framing a pole. Some tree trimming had been done in the area, and John offered to help remove the debris by loading it into the pickup truck he was driving. While John was loading branches into the bed of the truck, a piece broke off in his hand and struck his eye. John then backed his truck into a tree, shattering the rear window of the truck cab with his head.

Accident Details
1. John's right eye was struck by a piece of tree branch. John had just undergone laser surgery on his right eye on Monday, August 5, to reattach his retina.
2. John immediately covered his right eye with his hand, and Chico Ruiz gave him a paper towel with ice to cover his eye and help ease the pain.
3. After the initial pain subsided, John began to back up his truck to return to the Service Center. Chico reminded John about the pole trailer parked behind his truck and then returned to the crews he was supervising. John continued backing up, without seeing the tree behind him because his visibility was blocked by the debris in the bed of his truck.
4. As the truck hit the tree, John's head struck the back window of the truck, shattering the glass. He was not wearing a safety helmet because he was inside the truck cab.
5. John returned to the Service Center to report the accident to his supervisor. Because he had pieces of glass inside his clothes and on his neck, he decided to go home to shower and change clothes. He also used eyedrops prescribed for him after his surgery to thoroughly wash his eye.
6. The next day, August 8, John went to Downtown Worker's Care because he was experiencing headaches. He was diagnosed with a bruised eyeball and eyelid. The headaches were caused by the impact of his head against the rear window of the pickup truck.
7. On Monday, August 12, John returned to his eye surgeon. Although bruised, his eye was not damaged.

To prevent a recurrence of such an accident, the Safety Department will require the following actions in the future.

- When working near or moving debris, such as tree limbs or branches, employees must wear safety eyewear with side shields.
- When an accident occurs, employees must consider the possibility of shock to the injured victim. If no one can leave the job site to care for the injured person, assistance must be requested from the Service Center.
- Employees must always conduct a "circle of safety" check around any vehicle before moving it.

FIGURE T–6. Trouble Report (*continued*)

try and

The phrase *try and* is colloquial for *try to*. Unless you are writing a casual personal letter, use *try to*.

- Please try ~~and~~ finish the report on time.
  to

T

U

unity

Unity is singleness of purpose and treatment, the cohesive element that holds a document together; it means that everything in the document is essentially about one idea.

To achieve unity, the writer must select one **topic** and then treat it with singleness of **purpose,** without digressing into unrelated or loosely related paths. The logical sequence provided by a good **outline** is essential to achieving unity. An outline enables the writer to lay out the most direct route from **introduction** to **conclusion.** After you have completed your outline, check it to see that each part relates to your subject.

Effective **transition** is also a prime contributor to unity. Be certain that your transitional terms make each part of the relationship clear to what precedes it. (See also **coherence.**) Notice, for example, how neatly the sentences in the following paragraph are made to fit together by the italicized words and phrases of transition:

- Any company that operates internationally today faces a host of difficulties. Keeping up with technological change is a challenge. Most countries are struggling with economic problems *as well. In addition,* many monetary uncertainties and growing economic nationalism are working against multinational companies. *Yet* ample business is available in most developed countries if you have the right products, services, and marketing organization. To maintain the growth that Futura, Inc., has achieved globally, we recently restructured our international operations into four major trading areas. *This* reorganization will improve the services and support that the corporation can provide to its subsidiaries around the world. *At the same time,* the reorganization establishes firm management control, ensuring consistent policies around the world. *So* you might say the problems of doing business abroad will be more difficult this year, but we are better organized to meet those problems.

up

Adding the word *up* to **verbs** often creates a redundant phrase.

* Next open ~~up~~ the exhaust valve.

* He wrote ~~up~~ the report.

(See also **conciseness/wordiness.**)

usability testing

In documentation, *usability* refers to whether **readers** can use a document to easily fulfill their goals or accomplish tasks. For example, the usability of a **technical manual** might refer to the ability of readers to perform a procedure accurately and smoothly with the aid of the **instructions.** The usability of a brochure might refer to the ability of readers to scan it quickly and understand the contents in order to make an informed decision (such as whether to order a product or use a service).

To develop usable documents, you need to know how potential members of your target audience typically read and use those types of documents to fulfill goals (such as making a decision or comparing services) or perform tasks (such as sending a fax or assembling office furniture). Most users are busy people — they expect documentation to increase, not hinder, their productivity. If documents are consistent, predictable, and easy to use, users are more likely to read them, learn more quickly, and successfully apply what they learn.

Ideally, usability is built into documents or products from the beginning of their life cycle during the early planning, development, and design phases. To ensure that happens, it is important to focus on users throughout the evolution of a document. Usability engineering, or user-centered design (UCD), refers to the process of designing usable products and systems, with the user at the center of the process. Everything involved with a product or a service, including its goals, context, and environment, are approached from the user's viewpoint. With a user-centered approach, the team of engineers and technical writers interacts with potential customers to determine how well the product, system, or document meets their needs and what needs, if any, are unmet.

U

The benefits of a user-focused approach to document design and development include reducing the learning curve, allowing more functionality with less effort, and increasing productivity. At the same time, by focusing development on users, companies also reap benefits in reduced costs and increased customer satisfaction in the following ways:

- The documents are functional and more likely to be used.
- Product and document changes can be made before they become expensive.
- Efficient document development and training are facilitated.
- The need for updates and maintenance releases is minimized.
- More documents (and products) are sold.
- The organization's reputation is enhanced.

Usability testing is a common way to involve users throughout a document's evolution, for example, by testing periodic drafts on users and then revising the document in response to the test results. The process typically begins with the establishment of specific, quantitative, and measurable goals for documents. Next, the documents are designed to fulfill those goals. Finally, tests must be conducted with representative users to detect problems and determine whether the established goals have been achieved.

Usability testing is a collaborative process that involves teams of skilled usability specialists, interface designers, and technical writers. Usability testing has three main goals:

- It seeks to create a document that is easy to use and allows users to accomplish the tasks outlined in that document.
- It aims to detect potential problems for users as well as guide designers in resolving such issues. That process minimizes the risk of releasing ineffective documents and poorly designed products.
- Because it improves the process by which documents are designed and developed, companies can avoid repeating mistakes when developing future documents.

Test participants typically are members of the actual target audience for the document. If the participants encounter problems with the document, it is likely that in the future, actual users will experience similar problems. For example, if page 15 of a tax form is unclear to test participants, it is likely to be confusing to most taxpayers later on and thus should be revised. By testing documents on representative audience members, document designers can predict whether and where problems exist and revise the documents accordingly.

To detect user problems, usability tests can employ one or more of the following methods.

- *User testing.* To determine accuracy, precision, and **clarity** of writing, testers observe and record the actions of test participants who perform real tasks using the document.
- *Protocols.* Test participants make comments aloud as they read documents and perform document tasks, to reveal their thought processes, attitudes, and reasons for decision making.
- *Comprehension tests.* These tests determine whether users understand and can recall document features.
- *Surveys and interviews.* Testers interview users both before and after they read documents to determine their comprehension and attitudes and the clarity of the documents.
- *Review.* If test participants have trouble navigating a document or quickly locating specific items, the organization as well as **layout and design** are revised.

Analyzing the results of those test methods can help document designers detect problems and determine whether the document **purposes** have been met. Furthermore, usability tests can provide both specific and general guidance on how to revise the drafts to resolve problems by the time the document is distributed for use by the target audience. (See also **questionnaires** and **interviewing for information.**)

usage

Usage describes the choices we make among the various words and constructions available in our language. The lines between standard English and nonstandard English and between formal and informal English are determined by those choices. Your guideline in any situation requiring such choices should be appropriateness: Is the word or expression you use appropriate to your **readers** and your subject? When it is, you are practicing good usage.

This book contains many entries that focus on various usages (they appear in the Index, in italics). Good **dictionaries** are also an invaluable aid in your selection of the right word.

---

utilize

Do not use *utilize* as a **long variant** of *use,* which is the general word for "employ for some purpose." *Use* will almost always be clearer and less pretentious. (See **affectation.**)

- You can ~~utilize~~ *use* the company credit card for the trip.

V

vague words

A vague word is one that is imprecise in the context in which it is used. Some words encompass such a broad range of meanings that there is no focus for their definition. Words such as *real, nice, important, good, bad, contact, thing,* and *fine* are often called "omnibus words" because they can mean everything to everybody. In speech, we sometimes use words that are less than precise, but our vocal inflections and the context of our conversation make their meanings clear. Because you cannot rely on vocal inflections when you are writing, avoid using vague words. Be concrete and specific. (See also **abstract words/concrete words.**)

VAGUE	It was a *good* meeting.
SPECIFIC	The meeting resolved three questions: pay scales, fringe benefits, and workloads.

verb phrases (*see* phrases)

verbals   ESL

Verbals are derived from **verbs** and function as **nouns, adjectives,** and **adverbs.** There are three types of verbals: gerunds, infinitives, and participles.

Gerunds

A gerund is a verbal ending in *-ing* that is used as a noun.

* *Smelting* is a technique used to extract metal from ore.

A gerund can be used as a subject, a direct **object,** the object of a **preposition,** a subjective **complement,** or an **appositive.**

V

SUBJECT	*Estimating* is an important managerial skill.
DIRECT OBJECT	I find *estimating* difficult.
OBJECT OF PREPOSITION	We were unprepared for their *coming*.
SUBJECTIVE COMPLEMENT	Seeing is *believing*.
APPOSITIVE	My primary departmental function, *programming*, occupies about two-thirds of my time on the job.

Only the possessive form of a noun or **pronoun** should precede a gerund.

- *John's* working has not affected his grades.

- *His* working has not affected his grades.

Infinitives

An infinitive is the bare, or uninflected, form of a verb (for example, *go, run, fall, talk, dress, shout*) without the restrictions imposed by **person** and **number.** Along with the gerund and the participle, it is one of the nonfinite verb forms. The infinitive is generally preceded by the word *to*, which, although not an inherent part of the infinitive, is considered to be the sign of an infinitive.

- It is time *to go* to work.

- We met in the conference room *to talk* about the new project.

An infinitive is a verbal and can function as a noun, an adjective, or an adverb.

NOUN	*To expand* is not the only objective.
ADJECTIVE	These are the instructions *to follow*.
ADVERB	The company struggled *to survive*.

The infinitive can reflect two **tenses:** the present and (with a helping verb) the present perfect.

- *to go* [present tense]

- *to have gone* [present perfect tense]

The most common mistake made with infinitives is using the present perfect tense when the simple present tense is sufficient.

- I should not have tried to ~~have gone~~ *go* so early.

Infinitives formed with the root form of transitive verbs can express both active and (with a helping verb) passive **voice.**

- *to hit* [present tense, active voice]

- *to have hit* [present perfect tense, active voice]

- *to be hit* [present tense, passive voice]

- *to have been hit* [present perfect tense, passive voice]

A split infinitive is one in which an adverb is placed between the sign of the infinitive, *to,* and the infinitive itself. Because they make up a grammatical unit, the infinitive and its sign are better left intact than separated by an intervening adverb.

- To ~~initially~~ build the table in the file, you could input transaction records containing the data necessary to construct the record and table.

However, it may occasionally be better to split an infinitive than to allow a **sentence** to become awkward, ambiguous, or incoherent.

AMBIGUOUS	She agreed immediately *to deliver* the toxic materials. [could be interpreted to mean that she agreed immediately]
CLEAR	She agreed *to* immediately *deliver* the toxic materials. [no longer ambiguous]

Participles

A participle is a verb form that functions as an adjective.

- The *waiting* driver raced his engine.

- Here are the *revised* estimates.

- The *completed* report lay on his desk.

- *Rising* costs reduced our profit margin.

V

A participle cannot be used as the verb of a sentence. Inexperienced writers sometimes make that mistake, and the result is a **sentence fragment.**

- The committee chairperson was responsible. *, his* ~~His~~ vote being the decisive one.

- The committee chairperson was responsible. His vote ~~being~~ *was* the decisive one.

Present participles end in *-ing.*

- *Declining* sales forced us to close one branch office.

Past participles end in *-ed, -t, -en, -n,* or *-d.*

- What are the *estimated* costs?
- Repair the *bent* lever.
- Here is the *broken* calculator.
- What are the metal's *known* properties?
- The story, *told* many times before, was still interesting.

The perfect participle is formed with the present participle of the helping verb *have* plus the past participle of the main verb.

- *Having gotten* [perfect participle] a large bonus, the *smiling* [present participle], *contented* [past participle] sales representative worked harder than ever.

verbs   ESL

DIRECTORY
Overview   657
Types of Verbs   657
Forms of Verbs   658
Tips for Using Modal Verbs ESL   659
Properties of Verbs   660
Conjugation of Verbs   661
Tips for Avoiding Shifts in Voice, Mood, or Tense ESL   661

V

Overview

A verb is a word or group of words that describes an action ("The copier *jammed* at the beginning of the job"), states the way in which something or someone is affected by an action ("He *was disappointed* that the proposal was rejected"), or affirms a state of existence ("She *is* a district manager now").

Types of Verbs

Verbs are either transitive or intransitive. A *transitive verb* requires a direct object to complete its meaning.

- They *laid* the foundation on October 24.
  [The word *foundation* is the direct object of the transitive verb *laid*.]

- Rosalie Anderson *wrote* the treasurer a letter.
  [The word *letter* is the direct object of the transitive verb *wrote*.]

An *intransitive verb* does not require an object to complete its meaning. It makes a full assertion about the subject without assistance (although it may have modifiers).

- The engine *ran*.

- The engine *ran* smoothly and quietly.

A *linking verb* is an intransitive verb that links a complement to the subject. When the complement is a noun or a pronoun, it refers to the same person or thing as the noun or pronoun that is the subject.

- The winch *is* rusted.
  [*Rusted* is an adjective modifying *winch*.]

- A calculator *remains* a useful tool.
  [*A useful tool* is a subjective complement renaming *calculator*.]

Some intransitive verbs, such as *be, become, seem,* and *appear,* are almost always linking verbs. A number of others, such as *look, sound, taste, smell,* and *feel,* can function as either linking verbs or simple intransitive verbs. If you are unsure about whether one of those verbs is a linking verb, try substituting *seem;* if the sentence still makes sense, the verb is probably a linking verb.

- Their antennae *feel* delicately for their prey.
  [simple intransitive verb because *seem* cannot be substituted]

- Their antennae *feel* delicate.
  [linking verb because *seem* can be substituted]

Forms of Verbs

Verbs are described as being either *finite* or *nonfinite*.

Finite Verbs.   A finite verb is the main verb of a clause or sentence. It makes an assertion about its subject and often serves as the only verb in its clause or sentence.

- The telephone *rang* and the receptionist *answered* it.

A helping verb (sometimes called an *auxiliary verb*) is used in a verb **phrase** to help indicate **mood, tense,** and **voice.**

- The work *had* begun.

- I *am* going.

- I *was* going.

- I *will* go.

- I *should have* gone.

- I *must* go.

Phrases that function as helping verbs are often made up of combinations with the sign of the infinitive, *to* (for example, *am going to, is about to, has to,* and *ought to*).

The helping verb always precedes the main verb, although other words may intervene.

- Machines *will* never completely *replace* people.

Nonfinite Verbs.   Nonfinite verbs are **verbals,** which, although they are derived from verbs, actually function as nouns, adjectives, or adverbs.

When the *-ing* form of a verb is used as a noun, it is called a *gerund*.

GERUNDS   *Seeing* is *believing*.

(ESL) TIPS FOR USING MODAL VERBS

English uses primary auxiliaries (*have, be,* and *do*) to mark the main verb for tense, aspect, and mood. Modal auxiliaries (*can, may, would*) never function as main verbs; they add meaning but not tense to the sentence. Only one modal may occur in any verb phrase. The main modals used in English are as follows.

- *Can*

    He *can* type fast. [ability]

    Bill, you *can* still improve. [possibility]
- *Could*

    He *could* type fast before he broke his wrist. [past ability]

    Bill, you *could* still improve. [possibility]
- *May*

    Juanita *may* show up for the meeting. [possibility]

    You *may* come and go as you please. [permission]
- *Might*

    Juanita *might* show up for the meeting. [possibility]

    Might I go home early today? (very formal) [permission]
- *Must*

    We *must* finish this report by the end of the week. [necessity]

    You *must* see his new office. [recommendation]

    You *must* be hungry; you haven't eaten all day. [inference]
- *Should*

    You *should* apologize immediately. [advisability]

    Prentiss *should* be here any minute. [expectation]
- *Will*

    Greg *will* finish as soon as he can. [intention]
- *Would*

    Would you excuse me? [permission]

    He *would* review his work incessantly when he first started working here. [habitual past]

    That *would* be a good guess. [probability]

When several auxiliaries occur simultaneously, they must be in the following order:

(1) modal + (2) perfect + (3) progressive + (4) participle

    ①   ②   ③   ④

- He may have been being defrauded for several years.

V

An *infinitive,* which is the root form of a verb (usually preceded by *to*), can be used as a noun, an adverb, or an adjective.

- He hates *to complain.* [noun, direct object of *hates*]

- The valve closes *to stop* the flow. [adverb, modifies *closes*]

- This is the proposal *to select.* [adjective, modifies *proposal*]

A *participle* is a verb form used as an adjective.

- His *closing* statement was *convincing.*

- The *rejected* proposal was ours.

Properties of Verbs

Verbs must agree in person with personal pronouns functioning as subjects, and verbs must agree in tense and number with their subjects. Verbs must also be in the appropriate voice.

Person is the grammatical term for the form of a personal pronoun that indicates whether the pronoun refers to the speaker, the person spoken to, or the person (or thing) spoken about. Verbs change their forms to agree in person with their subjects.

- I *see* [first person] a yellow tint, but she *sees* [third person] a yellow-green hue.

- I *am* [first person] convinced, but you *are* [second person] not convinced.

Tense refers to verb forms that indicate time distinctions. There are six tenses: present, past, future, present perfect, past perfect, and future perfect. (See **tense.**)

Number refers to the two forms of a verb that indicate whether the subject of a verb is singular or plural.

SINGULAR   The machine *was* in good operating condition.

PLURAL   The machines *were* in good operating condition.

Most verbs show the singular of the third person, present tense, indicative mood by adding *s* or *es.*

- he *stands,* she *works,* it *goes*

The verb *to be* normally changes its form to indicate the plural.

SINGULAR   I *am* ready to begin work.

PLURAL   We *are* ready to begin work.

If you are in doubt about the plural form of a word, look it up in a **dictionary.** Most dictionaries give the plural if it is formed in any way other than by adding *s* or *es*.

Voice refers to the two forms of a verb that indicate whether the subject of the verb acts or receives the action. If the subject of the verb acts, the verb is in the *active voice;* if it receives the action, the verb is in the *passive voice.* (See also **voice.**)

ACTIVE   The aerosol bomb *propels* the liquid as a mist.

PASSIVE   The liquid *is propelled* as a mist by the aerosol bomb.

Conjugation of Verbs

The conjugation of a verb arranges all forms of the verb so that the differences caused by the changing of the tense, number, person, and voice are readily apparent. Figure V–1 on pages 662–63 shows the conjugation of the verb *drive.*

ESL TIPS FOR AVOIDING SHIFTS IN VOICE, MOOD, OR TENSE

To achieve clarity in your prose, it is important to maintain consistency and avoid shifts. A shift occurs when there is an abrupt change in voice, mood, or tense. Pay special attention when you edit your writing to check for the following types of shifts.

Voice

- The captain permits his crew to go ashore, but ~~they are not~~ *he does not permit* ~~permitted~~ *them* to go downtown.

  [The entire sentence is now in the active voice.]

Mood

- Reboot your computer~~,~~ and ~~you should~~ empty the cache too.

  [The entire sentence is now in the imperative mood.]

Tense

- I was working quickly, and suddenly a box ~~falls~~ *fell* off the conveyor belt and ~~breaks~~ *broke* my foot.

  [The entire sentence is now in the past tense.]

V

TENSE	NUMBER	PERSON	ACTIVE VOICE	PASSIVE VOICE
Present	Singular	1st	I drive	I am driven
		2nd	You drive	You are driven
		3rd	He drives	He is driven
	Plural	1st	We drive	We are driven
		2nd	You drive	You are driven
		3rd	They drive	They are driven
Progressive present	Singular	1st	I am driving	I am being driven
		2nd	You are driving	You are being driven
		3rd	He is driving	He is being driven
	Plural	1st	We are driving	We are being driven
		2nd	You are driving	You are being driven
		3rd	They are driving	They are being driven
Past	Singular	1st	I drove	I was driven
		2nd	You drove	You were driven
		3rd	He drove	He was driven
	Plural	1st	We drove	We were driven
		2nd	You drove	You were driven
		3rd	They drove	They were driven
Progressive past	Singular	1st	I was driving	I was being driven
		2nd	You were driving	You were being driven
		3rd	He was driving	He was being driven
	Plural	1st	We were driving	We were being driven
		2nd	You were driving	You were being driven
		3rd	They were driving	They were being driven
Future	Singular	1st	I will drive	I will be driven
		2nd	You will drive	You will be driven
		3rd	He will drive	He will be driven
	Plural	1st	We will drive	We will be driven
		2nd	You will drive	You will be driven
		3rd	They will drive	They will be driven
Progressive future	Singular	1st	I will be driving	I will have been driven
		2nd	You will be driving	You will have been driven
		3rd	He will be driving	He will have been driven
	Plural	1st	We will be driving	We will have been driven
		2nd	You will be driving	You will have been driven
		3rd	They will be driving	They will have been driven

FIGURE V–1. Verb Chart

TENSE	NUMBER	PERSON	ACTIVE VOICE	PASSIVE VOICE
Present perfect	Singular	1st	I have driven	I have been driven
		2nd	You have driven	You have been driven
		3rd	He has driven	He has been driven
	Plural	1st	We have driven	We have been driven
		2nd	You have driven	You have been driven
		3rd	They have driven	They have been driven
Past perfect	Singular	1st	I had driven	I had been driven
		2nd	You had driven	You had been driven
		3rd	He had driven	He had been driven
	Plural	1st	We had driven	We had been driven
		2nd	You had driven	You had been driven
		3rd	They had driven	They had been driven
Future perfect	Singular	1st	I will have driven	I will have been driven
		2nd	You will have driven	You will have been driven
		3rd	He will have driven	He will have been driven
	Plural	1st	We will have driven	We will have been driven
		2nd	You will have driven	You will have been driven
		3rd	They will have driven	They will have been driven

FIGURE V 1. Verb Chart (*continued*)

very

The temptation to overuse **intensifiers** like *very* is great. Evaluate your use of them carefully. When you do use intensifiers, clarify their meaning.

- Bicycle manufacturers had a *very* good year: Sales across the country were up 43 percent over the previous year.

In many sentences, however, the word can simply be deleted.

- The board was ~~very~~ angry about the newspaper report.

via

Via is Latin for "by way of."

- The equipment is being shipped to Los Angeles *via* Chicago.

Use *via* only for routing instructions.

- His project was funded ~~via~~ *as the result of* the recent legislation.

vogue words

Vogue words are words that suddenly become popular and, because of an intense period of overuse, lose their freshness and preciseness. They may become popular through their association with science, technology, or even sports. We include them in our vocabulary because they seem to give force and vitality to our language. Ordinarily, vogue words sound pretentious in everyday writing. (See also **word choice.**)

interface [as a verb]	*deliverables*	*dialog* [as a verb]
impact [as a verb]	*newbie*	*cutting edge*
skill sets	sales*wise*	right*size*

Obviously, some vogue words are appropriate in the right context. It is when they are used outside that context that imprecision becomes a problem.

- An *interface* was established between the computer and the satellite hardware.
  [*Interface* is appropriately used as a noun.]

- We must ~~interface~~ *cooperate* with the Human Resources Department.

  [*Interface* is inappropriately used as a verb; *cooperate* is more precise.]

voice

In grammar, voice indicates the relation of the subject to the action of the **verb.** When the verb is in the *active voice,* the subject acts; when it is in the *passive voice,* the subject is acted upon.

ACTIVE   David Cohen *wrote* the advertising copy.

PASSIVE   The advertising copy *was written* by David Cohen.

The two sentences say the same thing, but each has a different **emphasis:** In the first sentence, the emphasis is on *David Cohen;* in the second sentence, the focus is on *the advertising copy.* Because they are wordy and indirect, passive-voice sentences often are more difficult for **readers** to understand.

PASSIVE   Things *are seen* by the normal human eye in three dimensions: length, width, and depth.

ACTIVE   The human eye *sees* things in three dimensions: length, width, and depth.

Notice how much stronger and more forceful the active sentence is. Always use the active voice unless you have a good reason to use the passive.

Problems with the Passive Voice

Impeding Clarity.   The passive voice often impedes **clarity** because it is confusing, especially in **instructions,** which should always use the active voice.

PASSIVE   Plates B and C *should be marked* for revision.
[Are they already marked?]

ACTIVE   You *should mark* plates B and C for revision.

ACTIVE   *Mark* plates B and C for revision.

Passive voice can also lead to the use of **dangling modifiers.**

PASSIVE   Hurrying to complete the work, the cables *were connected* improperly.
[Who was hurrying, the cables?]

ACTIVE   Hurrying to complete the work, the technician improperly *connected* the cables.
[Here, *hurrying to complete the work* properly modifies *technician.*]

Burying Subjects.   One difficulty with passive sentences is that they can bury the performer of the action in **expletives** and prepositional **phrases.**

PASSIVE   It *was reported by* Marketing that the new model is defective.

ACTIVE   Marketing *reported* that the new model is defective.

Sometimes writers using the passive voice fail to name the performer, information that might be missed.

PASSIVE   The problem *was discovered* yesterday.

ACTIVE   The Maintenance Department *discovered* the problem yesterday.

Be careful not to use the passive voice to hide information from the reader, evade responsibility, or obscure an issue. (See also **ethics in writing.**)

- Several mistakes were made.
  [*Who* made the mistakes?]

- It has been decided.
  [*Who* has decided?]

- The product will be inspected.
  [*Who* will inspect the product?]

Causing Wordiness.   Passive-voice sentences are wordy because they always use a helping verb in addition to the main verb, as well as an extra **preposition** if they identify the doer of the action specified by the main verb. The passive-voice version is also indirect because it puts the doer of the action behind the verb instead of in front of it.

PASSIVE   Changes in policy *are resented by* employees.

ACTIVE   Employees *resent* changes in policy.

The active-voice version takes one verb (*resent*); the passive-voice version takes two verbs (*are resented*) and an extra preposition (*by*). The passive-voice version is also indirect because it puts the doer of the action behind the verb instead of in front of it. (See also **conciseness/wordiness.**)

Using the Passive Voice

There are instances when the passive voice is effective or even necessary. Indeed, for reasons of tact and diplomacy, you might need to use the passive voice to *avoid* identifying the doer of the action.

ACTIVE   Your sales force did not meet the quota last month.

PASSIVE   The quota was not met last month.

When the performer of the action is either unknown or unimportant, use the passive voice.

- The copper mine *was discovered* in 1929.

- Fifty-six barrels *were processed* in two hours.

When the performer of the action is less important than the receiver of that action, the passive voice is sometimes more appropriate.

- Ann Bryant *was presented* with an award by the president.

When you are explaining an operation in which the reader is not actively involved or when you are explaining a process or a procedure, the passive voice may be more appropriate.

- Area strip mining *is used* in regions of flat-to-gently rolling terrain, like that found in the Midwest and West. Depending on applicable reclamation laws, the topsoil *may be removed* from the area to be mined, *stored,* and later *reapplied* as surface material during reclamation of the mined land. After the removal of the topsoil, a trench *is cut* through the overburden to expose the upper surface of the coal to be mined. The length of the cut generally corresponds to the length of the property or of the deposit. The overburden from the first cut *is placed* on the unmined land adjacent to the cut. After the first cut *has been completed,* the coal *is removed,* and a second cut *is made* parallel to the first.

In the example, anyone—it really does not matter who—could be the doer of the action. Therefore, it would be pointless for the writer to put the passage in the active voice. Do not, however, simply assume that any such explanation should be in the passive voice. Ask yourself, "Would it be of any advantage to the reader to know the doer of the action?" If the answer is no, use the passive voice. But if the answer is yes, use the active voice. The following example uses the active voice to explain an operation:

- In the operation of an internal combustion engine, an explosion in the combustion chamber *forces* the pistons down in the cylinders. The movement of the pistons in the cylinders *turns* the crankshaft. The crankshaft, in turn, *operates* the differential. The differential *turns* the rear axles, which *turn* the wheels and *move* the automobile.

Whether you use the passive or the active voice, be careful not to shift voices in a sentence.

- Ms. McDonald corrected the malfunction as soon as ~~it was identified by~~ the technician *identified it.*

ESL TIPS FOR CHOOSING VOICE

Different languages place different values on active-voice and passive-voice constructions. In some languages, the passive is used frequently; in others, hardly at all. As a nonnative speaker of English, you may have a tendency to follow the pattern of your native language. But remember, even though technical writing may sometimes require the passive voice, active verbs are highly valued in English.

V

W

wait for / wait on

Wait on should be restricted in writing to the activities of hospitality and service employees; otherwise, use *wait for*. (See also **idioms.**)

- Be sure to *wait for* Ms. Sturgess to finish reading the report before asking her to make a decision.

- Joe's feet ached from *waiting on* tables at the restaurant.

Web page design

DIRECTORY
Overview   669
Web Page Navigation and Links   670
Graphics and Typography   673
Page and Site Layout   674
Home Pages   675

Overview

You can apply many of the technical writing principles covered throughout this handbook to designing **World Wide Web** pages. As you do when creating other documents, carefully consider your **purpose** and your **readers'** needs as you prepare your Web pages. Although some commercial and personal Web sites are intended for visitors with no specific purpose, most organizational sites have well-defined goals, such as reference, training, education, publicity, advocacy, or marketing. Before you begin building your site, create a clear statement of purpose that identifies your target audience.

EXTERNAL SITE   The purpose of this site is to enable our customers to locate product information, place online orders, and contact our customer service department with questions or comments.

INTERNAL SITE   The purpose of this site is to provide SN Security Corporation employees with a single, consistent, and up-to-date resource for materials about SN Security's Employee Benefits Package.

As those examples of purpose statements suggest, there are two general kinds of sites: external sites and internal sites. External sites target an audience from the entire Internet; internal sites are designed for audiences on *intranets*. An intranet is a computer network within an educational institution or company that is not accessible to audiences outside that institution or company.

Web Page Navigation and Links

Your foremost goal when you are designing a Web page is to establish a predictable environment in which users feel comfortable navigating your site and confident that they can easily find the information they need and move on. The information should be logically accessible to the user in the fewest possible steps (or clicks). To design an efficient navigation plan, it is especially helpful to draft a navigation chart of your site early in the process. Figure W–1 on page 671 shows an initial navigation plan for a Web site for a U.S. Department of Energy national laboratory. In Figure W–1, each higher-level page (for example, Basic Science) is linked to related subject areas (Biology, Chemistry, Materials, Mathematics, and so on). Under Scientific Facilities, related content areas are shown, with a specific site for the laboratory's Auto Shredder facility, which in turn is linked to a site with information about recovering plastics from scrapped autos. That page has an email link to the principal investigator for the Recovering Plastics project. The pattern is repeated, with greater or lesser detail, throughout the site.

The **hypertext** links that connect pages within a site as well as to pages at other sites make the Web site more useful. Links to customer service and online order forms are especially useful for cultivating potential customers. In short, to make the Web site more useful, create links to sites related to your visitors' organizational or professional needs and interests.

When you identify links, do not write out "click here for more

W

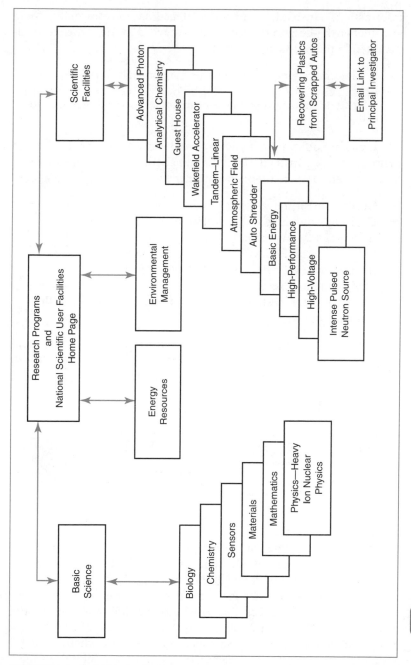

FIGURE W-1. Initial Navigation Plan

W

information." Instead, write the sentence as you normally would and anchor the link on the most relevant word in the sentence, as illustrated in the following example.

- For information about employment opportunities, visit <u>Human Resources</u>.
  [Hypertext links for text anchors are underlined and in a different color.]

If you are offering many links, avoid writing paragraphs that are dense with links. Instead, list links alphabetically in groups of about four to seven to make them easier for the viewer to see at a glance.

➠ Cost Engineering
➠ Decision Modeling and Simulation
➠ Emergency Management
➠ Environmental Policy Analysis and Planning
➠ Environmental Protection and Remediation
➠ Environmental Research
➠ Environmental Restoration, Compliance, and Pollution Prevention

In addition to identifying links with words, you can use icons and graphics, as shown in the preceding list; they are not only easy to use but also provide the visual cues Web users expect. Figure W–2 shows some icons that can be used for links.

FIGURE W–2. Icons That Can Be Used for Links

Include a link that allows visitors to send you email messages. Identify the email code with either text or a graphic like the mailbox icon shown in Figure W–2. When visitors click on the icon, an email dialog box opens in which they can type their message.

Because Web sites change constantly, you must maintain your site periodically and check that your links are still appropriate and working properly. Your Web pages should include the date they were created or last updated so users can determine whether the information is current. You might also put an icon (like the fourth one in Figure W–2) next to each new or updated item to alert users to changes.

Graphics and Typography

Your audience and your purpose should determine the graphic style and theme of your Web site. Graphic elements provide visual relief from dense text, which is difficult to read on a computer screen. However, do not overdo graphics, especially those with animation. Complex graphics and motion can backfire if they seem gratuitous or if they slow access to your site, especially if users depend on the site for accessible and visually clear information. Although it may seem like fun to use many different colors, avoid overly bold colors; opt instead for lighter colors, especially for backgrounds. Keep in mind that large or high-resolution graphics, like color photographs, can cause long delays as they download to the user's system. You might also consider giving visitors a graphics-free option for quicker access.

Writer's Checklist: Web Page Typography

Many of the guidelines on typography in **layout and design** are applicable to Web pages; however, not all apply, so keep the following in mind:

☑ Limit the number of typeface colors as well as styles.

☑ Make sure the typeface colors contrast sharply (but do not clash) with background colors for better readability.

☑ Do not use capital letters or boldface letters for all the text or large portions of it.

☑ Use generous blank space (the equivalent of white space) to separate text and graphics.

☑ Use only a few **heading** and subheading styles and use them consistently.

☑ Block-indent text sections that you expect viewers to read in detail.

When you have decided on fonts, interparagraph spacing, heading sizes, and so on, consider creating a style sheet that lists the specifications. A style sheet is especially helpful for large sites with numerous pages. In general, aim for consistency to establish a sense of unity and to provide visual cues that help visitors find

W

information. Consistency also encourages visitors to browse further and to return to your site.

Page and Site Layout

Take into account the viewing capabilities of visitors to your site. Many computer monitors cannot display more than about half of the typical Web page at any one time; on many screen settings, only the top four or five inches may be visible. For that reason, you should place important graphic elements and information in the upper half of the page. Then use short narrative passages or lists so viewers can scan and access information quickly. Organize material in a logical, predictable pattern; use general-to-specific, sequential, chronological, or other traditional **methods of development.** When no such pattern is appropriate, use an alphabetical or numerical sequence.

Because most users do not like scrolling down long pages, many Web designers recommend that Web pages (home pages in particular) contain no more than roughly one or two screens of information. If a larger amount of information is covered, include a concise table of contents at the top of the page linked to specific sections elsewhere on the page. Viewers can then click on the section they want without having to scroll through the entire page. You can also provide a link that returns to the top of the page.

When viewers access Web pages randomly, they often have no context for where they are. Therefore, incorporate the name of your company or organization on each page. You can do that with page footers—sections in a smaller type size that contain basic information about the origin of each page and when it was last updated—which are especially useful when they also provide links to other pages. Another way to orient users is to provide a set of links in graphics or words called a *tool bar* or *button bar,* as shown below.

| Home | Search | Order | What's New | Reference |

A tool bar at the top, bottom, or sides of each page serves as a table of contents and shows visitors the structure of your site. Tool bars on each page help you avoid sending users to a dead-end page (one with no link); in fact, every page should, if nothing else, link back to your home page.

Home Pages

A home page is the point of entry for all pages in your Web site. Many organizational home pages include an image map, with links to other pages at the site. Figure W–3 shows an image map for a federal agency, with icons that lead to content areas at the site. By clicking on a specific area, such as News & Information, viewers are linked to the relevant page or site. The image map also introduces visitors to the overall site design, identifies the purpose of the site, and provides an overview of major content areas. If you do not use an image map, at least display a small graphic banner, tool bar, or menu across the top of the home page.

FIGURE W–3. Image Map for a Home Page

An overly complex home page graphic can take a minute or longer to download. If large graphics are important to your site, consider using relatively small ones on your home page, gradually increasing the size of the graphics for pages deeper into your site. Users who go beyond a page or two into a site are more committed and therefore more willing to tolerate longer delays, especially if you offer them warnings that particular pages contain graphics that take a while to download. View your pages on a variety of browsers, systems, and monitors—colors, sizes, and other features can vary dramatically from one environment to another.

when and if

When and if (or *if and when*) is a colloquial expression that should not be used in writing.

- When ~~and if~~ your new position is approved, I will see that you receive adequate staff support.

- *If*
  ~~When and if~~ your new position is approved, I will see that you receive adequate staff support.

where . . . at

In phrases using the *where . . . at* construction, *at* is unnecessary and should be omitted.

- Where is his office ~~at~~?

where / that

Do not substitute *where* for *that* to anticipate an idea or fact to follow.

- *that*
  I read in the newspaper ~~where~~ computer chips will be used in the new process.

whether or not

When *whether or not* is used to indicate a choice between alternatives, omit *or not;* it is redundant, because *whether* communicates the notion of a choice.

- The project director asked whether ~~or not~~ the request for proposals had been issued.

(See also **as to whether.**)

while

While, meaning "during an interval of time," is sometimes substituted for connectives like *and, but, although,* and *whereas.* Used as a connective in that way, *while* often causes **ambiguity.**

- Ian Evans is sales manager, ~~while~~ *and* Joan Thomas is in charge of research.

Do not use *while* to mean *although* or *whereas.*

- *Although* ~~While~~ Ryan Patterson wants the job of engineering manager, he has not yet asked for it.

Restrict *while* to its meaning of "during the time that."

- I'll have to catch up on my reading *while* I am on vacation.

who / whom

Writers are often unsure whether to use *who* or *whom. Who* is the subjective **case** form, and *whom* is the objective case form. If you are in doubt about which form to use, substitute a personal **pronoun** to see which one fits. If *he, she,* or *they* fits, use *who.*

OPTIONS   *Who/whom* is the sales representative for our area? [*She* is the sales representative for our area.]
CORRECT   *Who* is the sales representative for our area?

If *him, her,* or *them* fits, use *whom.*

OPTIONS   It depends on *who/whom?* [It depends on *them.*]
CORRECT   It depends on *whom?*

who's / whose

Who's is the contraction of *who is*.

- *Who's* scheduled to attend the productivity seminar next month?

Whose is the possessive **case** of *who*.

- *Whose* department will be affected by the budget cuts?

Who's and *whose* are not interchangeable.

whose / of which

Normally, *whose* is used with persons, and *of which* is used with inanimate objects.

- The man *whose* car had been towed away was angry.
- The mantel clock, the parts *of which* work perfectly, is over one hundred years old.

If the use of *of which* causes a sentence to sound awkward, *whose* may be used with inanimate objects.

- There are added fields, for example, *whose* totals should never be zero.

-wise

Although the **suffix** *-wise* often seems to provide a tempting short-cut in writing, it leads more often to inept than to economical expression. It is better to rephrase the sentence.

- Our department ~~rates~~ high ~~efficiencywise.~~ *has a* *efficiency rating.*

The *-wise* suffix is appropriate, however, in instructions that indicate certain space or directional requirements.

- Fold the paper *lengthwise*.
- Turn the adjustment screw half a turn *clockwise*.

W

word choice   ⓔⓈⓁ

Mark Twain once said, "The difference between the right word and almost the right word is the difference between 'lightning' and 'lightning bug.'" The most important goal in choosing the right word in technical writing is the preciseness implied by Twain's comment. Vague words and abstract words defeat preciseness because they do not convey the writer's meaning directly and clearly.

VAGUE   It was a meaningful meeting.

PRECISE   The meeting helped both sides understand each other's positions.

In the first sentence, *meaningful* ironically conveys no meaning at all. See how the revised sentence says specifically what made the meeting "meaningful." Although abstract words may at times be appropriate to your topic, using them unnecessarily will make your writing dry and lifeless.

ABSTRACT   work, fast, food

CONCRETE   sawing, 110 mph, apple

Being aware of the connotations and denotations of words will help you anticipate the reader's reaction to the words you choose. Connotation is the suggested or implied meaning of a word beyond its dictionary definition. Denotation is the literal, or primary, dictionary meaning of a word. For example, the denotation of *Hollywood* is "a district of Los Angeles, consolidated with Los Angeles in 1910"; its connotation is "romance and glittering success."

Understanding **antonyms** and **synonyms** will increase your ability to choose the proper word. Antonyms are words with nearly opposite meanings (*fresh/stale*), and synonyms are words with nearly the same meaning (*notorious/infamous*).

Although many of the entries throughout this book will help you improve your word choices and avoid problems, the following ones should be particularly helpful:

abstract words/concrete words
affectation
biased language
clichés
conciseness/wordiness

W

connotation/denotation
euphemisms
idioms
jargon
malapropisms
trite language
vague words
vogue words

A key to choosing the correct and precise word is to keep current in your reading and to be aware of new words in your profession and in the language. In your quest for the right word, remember that there is no substitute for a good **dictionary.** (See also **English as a second language.**)

World Wide Web

The World Wide Web (or, simply, the Web) is a system that simplifies navigation around the **Internet** and combines the capabilities of all the other Internet access systems. The Web supports applications for email, discussion groups, and chat environments, as well as for searching, retrieving, and providing information. The Web is a multimedia environment: It brings together vast amounts of information in text, graphic, audio, and video form on a global scale. Links, called *hyperlinks,* to other Web pages are embedded in Web pages when the pages are created. A Web page is simply the information visible on your screen. A browser is a software program for accessing the Web, viewing Web documents, and navigating among Web sites. Netscape Navigator®ª and Microsoft®ᵇ Internet Explorer are two commonly used browsers. A home page, the first page of a Web site, identifies whose site it is and often contains a table of contents of the site. When you click on a hyperlinked feature, your browser jumps to the related Web document or image.

The ability to access and retrieve related documents makes hyperlinking a powerful tool for navigating from page to page and site to site. For instance, a typical university home page screen, like the

W

ªNetscape Navigator® is a registered trademark of Netscape Communications Corporation in the United States and other countries.

ᵇMicrosoft® Internet Explorer is either a registered trademark or trademark of Microsoft Corporation in the United States and/or other countries.

FIGURE W–4. The University of Wisconsin–Milwaukee Home Page

one shown in Figure W–4, may contain links to the individual colleges and professional schools on campus, the library system's on-line card catalog, and a brief history of the university and its admissions requirements. University sites also frequently contain home pages for individual professors and list their course schedules, syllabi, required texts, and office locations and hours.

Each page on the Web has a unique address known as a Uniform Resource Locator (URL). The URL is often made up of strings of letters and sometimes numbers. Fortunately, you do not have to remember or record lengthy URLs to retrieve most Web pages because their addresses are often automated as **hypertext** links. In fact, when you find a site that you want to revisit, you can bookmark the URL for future access. A *bookmark* is a browser feature that saves URLs in a file. To revisit a site, you open the bookmark list, which is displayed as a drop-down menu, and click on the site's title.

W

Despite their complexity, URLs make sense if you understand their components. For example, the URL for the Consumer Price Index (CPI) page at the U.S. Bureau of Labor Statistics is typical: (http://www.bls.gov/cpiovrvw.htm). The first element, *http://,* stands for *hypertext transfer protocol,* which tells you that the page is a Web site. Non-Web prefixes include *gopher, ftp,* and *telnet.* The next component is the host server name, *www.bls.gov,* which identifies the computer that stores the information. The suffix *gov* identifies the page as a government site. Other distinctive identifiers for Web page sponsors are *edu* (educational), *com* (commercial), and *org* (often a nonprofit organization). The next element, *cpiovrvw,* identifies the directory in which the page is stored and the file name associated with the page. The final suffix, *htm* (sometimes shown as *html*), indicates that the file is a hypertext document written in hypertext markup language. (See also **email, Internet research, search engines,** and **Web page design.**)

writing a draft

You are well prepared to write a rough draft when you have established your **purpose,** your **readers'** needs, and your **scope** and when you have done adequate **research** and have completed your outline. Writing a draft is simply transcribing and expanding the notes from your outline into **paragraphs,** without worrying about grammar, refinements of language, or spelling. Refinement comes with **revision** and **proofreading.**

Writing and revising are different activities. Write the rough draft as though you were casually explaining your subject to someone. Do not let worry about a good opening slow you down. There is no need in a rough draft to be concerned about an **opening** or **transitions** unless they come easily. Concentrate on getting your ideas on paper; do not try to polish or revise them.

Even with good preparation, writing a draft can be a chore and is an obstacle for many writers. Do not wait for inspiration to strike you—treat writing a rough draft as you would any other on-the-job task.

W

Writer's Checklist: Writing a Rough Draft

Experienced writers use the following tactics to get started and keep moving. Discover which tips are the most helpful to you.

☑ Set up your writing area with the equipment and materials (paper, dictionary, sourcebooks) you will need to keep going once you get started. Then hang out the "Do Not Disturb" sign.

☑ Use writing tools, separately or in combination, that are most comfortable for you: pencils, felt-tip pens, computer, and so on.

☑ The most effective way to start and keep going is to use a good outline as a springboard for your writing. (See also **outlining**.)

☑ As you write your rough draft, keep in mind your reader's knowledge of the subject. Not only will that help you write directly to your reader, it will also tell you which ideas need further development.

☑ When you are trying to write quickly and come to something difficult to explain, try to relate the new concept to something with which the reader is already familiar, as discussed in **figures of speech.**

☑ Start with the section that seems easiest to you to get the writing rolling. Your reader will never know that the middle section of the document was the first section you wrote.

☑ Give yourself a set time (10 or 15 minutes, for example) in which you write continuously, regardless of how good or bad your writing seems to be. The point is to keep moving.

☑ Do not let anything stop you when you are rolling along easily—if you stop and come back, you may lose momentum.

☑ Give yourself a small reward—a short walk, a soft drink, a short chat with a friend, an easy task—after you have finished a section.

☑ Reread what you have written when you return to your writing. Often, seeing what you have written returns you to a productive frame of mind.

W

X

X-ray

X-ray is usually capitalized and is always hyphenated as an **adjective** and a **verb** and usually hyphenated as a **noun.** If you hyphenate the noun, be sure to do so consistently.

ADJECTIVE   We ordered a portable *X-ray* unit.

VERB   The technician *X-rayed* the ankle.

NOUN   The device uses *x-rays* [or x rays] and gamma rays.

Y

"you" viewpoint

The "you" viewpoint (or "you" attitude) is a writing technique that places the **readers'** interest foremost. It is based on the principle that readers are naturally more concerned about their own needs than they are about the writer's (or the organization's) needs. The "you" viewpoint often, but not always, means using the words *you* and *your* rather than *we, our, I,* and *mine.* Consider the following sentence, which focuses on the writer's, rather than the readers', needs.

- *We must receive* your receipt with the merchandise before *we can process* your refund.

Even though the sentence uses *your* twice, the words in italics suggest that the **point of view** centers on the writer's need to receive the receipt in order to process the refund. You can make the request more politely, as follows:

- Please enclose the sales receipt with the merchandise so we can process your refund promptly.

The **tone** is now more polite and friendly (which is important in most **correspondence**), but the second version still emphasizes the writer's need to get the receipt. What is the reader's interest? The reader is not interested in helping the business process its paperwork. The reader simply wants the refund—by emphasizing that need, the writer can encourage the reader to act quickly. Consider the following revision, written with the "you" viewpoint.

- So you can receive your refund promptly, please enclose the sales receipt with the returned merchandise.

Because the reader's benefit is stressed as the reason the reader should enclose the receipt, the writer is more likely to accomplish the **purpose** of getting the reader to act.

As you can see, using the "you" viewpoint means more than just using the pronouns *you* and *your*. It means seeing a situation from the reader's viewpoint and writing to reflect that perspective. (See also **persuasion.**)

your/you're

Your is a personal pronoun denoting possession; *you're* is the contraction of *you are*.

- If *you're* going to the conference in Seattle, be sure to pack *your* umbrella.

Index

a. See also articles (adjectives)
  a/an usage, 1
  as an indefinite article, 20, 61, 63
  when to use, 1
a few/few, 230
a/an historic, 1
a lot/alot, 2
a while/awhile, 66
abbreviations, 2–8. *See also* acronyms
  and initialisms
  in abstracts, 13
  in addresses, 6–8, 138
  ampersands and, 5, 52
  apostrophes with, 55
  capitalization of, 83
  copy notations in letters, 143–44
  in correspondence headings,
    138–41
  in email, 201
  enclosure notations in letters,
    142–43
  in formal reports, 243
  forming, 3–8
  italics for, 327
  lists of, in formal reports, 243
  periods with, 446
  plurals of, 55
  proofreading for, 496
  symbols and, 243
  in tables, 618
  in titles, 633
  using, 2–3
above, 8
absolute phrases, 105
absolute words, 9, 23, 213
absolutely, 9
abstract nouns, 45, 409
abstract words/concrete words,
  9–10
  ethics in using, 215
  gobbledygook and, 255
  word choice and, 679
abstraction, levels of, 9–10
abstracts, 11–14
  in formal reports, 242
  library research using, 355–56
  trade journal articles with, 638–39

academic titles
  abbreviations for, 3, 4
  Ms./Miss/Mrs. usage with, 394
accent marks (diacriticals), 162, 238
accept/except, 14
acceptance letters, 14–15, 319
accident reports, 644
accumulative/cumulative, 16
accuracy of information
  in correspondence, 135–37, 144
  exposition and, 223
  library research using the Web and,
    353
  note taking and, 407
  proofreading for, 495
  revision and, 567
  testing instructions for, 296
accurate (absolute word), 23
accusative case, 85
acknowledgment
  acknowledgment letters for, 16
  adjustment letters for, 25
  cover letters and, 145–46
  documenting sources for, 173
  in formal reports, 242
  prefaces in formal reports and, 242
  in refusal letters, 546
  responding to inquiries and, 288
acknowledgment letters, 16–17
  response letters and, 288
acronyms and initialisms, 17–18. *See
  also* abbreviations
  apostrophes in, 18
  in formal reports, 240, 243
  new words with, 399
  proofreading for, 496
  in titles, 633
activate/actuate, 19
active listening, 359–61
active voice, 661, 664–65
  infinitives and, 655
  in titles, 633
activities section, on résumés, 561
activity reports, 484–87
actually
  as an intensifier, 543
  really and, 543

actuate/activate, 19
acute accent, 162
A.D., 5
ad hoc, 19
adapt/adept/adopt, 19
address, direct, 168
　commas with, 104
　emphasis with, 205
addresses, electronic
　for email, 200–201
　in résumés, 558, 559
　for Web, 600, 681–82
addresses, street
　abbreviations for, 7–8, 138
　commas with, 108
　in correspondence headings, 138
　numbers in, 416–17
　in résumés, 558–59
　slashes in, 600
adept/adapt/adopt, 18
adjective clauses, 208
adjective phrases, 451–53
adjectives, 19–25. *See also* modifiers
　centuries as, 151
　commas with, 23, 25, 107
　comparison of, 22–23
　complements as, 117
　conciseness and overuse of, 122
　converting verbs into, 594
　intensifiers with, 296–97, 390
　nouns used as, 23, 25, 410
　as a part of speech, 444
　phrases functioning as, 451–53
　placement of, 23
　possessive case and, 22
　prepositional phrases as, 453
　prepositions used with, 464
　types of, 19–22
　use of, 23–25
　verbals as, 653
　word order for, 620
adjustment letters, 25–29
　complaint letters and, 114–16
adopt/adapt/adept, 19
adverb clauses, 95
adverb connectives, 31, 584
adverb phrases, 451–52
adverbial conjunctions. *See* conjunctive
　adverbs
adverbial particles, 465
adverbs, 29–32, 390. *See also* modifiers
　as used as, 62
　commas with, 106, 108
　comparison of, 30–31
　conciseness and overuse of, 122
　intensifiers as, 296–97, 390
　nouns as, 410

as a part of speech, 444
phrases functioning as, 451–52
placement of, 31–32
prepositional phrases as, 451–52
prepositions used with, 464
semicolons with, 31, 591
types of, 29–30
verbals as, 653
word order for, 620
advertisements, in job searches, 335
advice/advise, 33
affect/effect, 33
affectation, 32–35
　acronyms and initialisms and, 18
　allusion and, 48
　causes of, 34
　elegant variation and, 194–95
　-ese words and, 213
　euphemisms and, 217
　foreign terms and, 238
　gobbledygook and, 255
　in terms of usage and, 280–81
　individual usage and, 286–87
　international correspondence and,
　　300
　jargon and, 330
　Latin terms as, 153
　long variants and, 366
　new words and, 399
　nominalizations and, 404
　revision for, 570
　technical writing style and, 623
　vogue words and, 664
　wordiness, 122
affinity, 36, 275
affirmative writing (positive writing),
　458–59
aforesaid, 36
afterthoughts, 149
age, and biased language, 71–73
agenda, 373–74
agree to/agree with, 37, 276
agreement, 37–38
　gender and, 249
　nor/or usage and, 405
　number and, 413
agreement of pronouns and an-
　tecedents, 38–43, 493
　gender and, 39–40, 492
　number and, 40–43, 492–93
agreement of subjects and verbs, 43–47
　collective nouns and, 410–11
　each and, 193
　everybody/everyone and, 217–18
　one of those . . . who and, 422
aids, visual
　flowcharts for, 234–37

global communication and, 251–52
global graphics and, 252–54
graphs as, 257–64
illustrations as, 277–79
maps as, 368–70
organizational charts as, 429
photographs as, 450
physical movement during delivery
    and, 477
practice with, 475–76
presentations using, 471–75, 476
tables as, 616–19
ain't, 256
aircraft, names of, 328
all
    agreement of pronouns and an-
        tecedents and, 43, 492
    agreement of subjects and verbs and,
        44
all around/all-around/all-round, 47
"all caps" (capital letters) usage, 204,
    346, 367
all ready/already, 49
all right/alright, 47
all-round/all around/all-around, 47
all together/altogether, 47
alleys (layout and design), 346
allude/elude/refer, 48
allusion, 48, 299
allusion/illusion, 49
almanacs, 354
almost/most, 49
alot/a lot, 2
alphabetical list of definitions
    dictionaries with, 163–67
    glossaries with, 254–55
    thesaurus with, 631
already/all ready, 49
alright/all right, 47
also, 49–50
altogether/all together, 47
a.m., 5, 415
ambiguity, 50–51
    clarity and, 93
    commas to prevent, 103, 107
    misplaced modifiers and, 391
    prepositional phrases and, 453
    pronoun reference and, 488
    split infinitives and, 655
    squinting modifiers and, 392
    telegraphic style and, 623
    that usage and, 630
    while usage and, 677
ambiguous pronoun reference
    agreement of pronouns and an-
        tecedents, 37–38
    hidden references and, 488

American English, 211–12
American Psychological Association
    (APA) style
    documenting sources using, 173,
        179–85
    quotation format in, 540–41
among/between, 69
amount/number, 51
ampersands, 5, 52
an. See also articles (adjectives)
    a/an usage, 1
    as an indefinite article, 20, 64
    when to use, 1
analogy, 232
    comparison method of development
        and, 113–14
    definition by, 156–57
    descriptions using, 159
    false, 365
analysis, in division-and-classification
    method of development, 169–72,
        384
and. See also conjunctions
    ampersand used in place of, 52
    both . . . and usage and, 74–75
and etc., 213–14
and/or, 52
and sign (ampersand), 5, 52
anecdotes, as openings, 426
Anglicized words, 237–38
annotated bibliography, 73–74
anonymous works, 178, 184
ante-/anti-, 52–53
antecedents, 489
    agreement of pronouns and, 38–42
    case of appositives and, 490–91
    first- and second-person pronouns
        and, 493
    hidden references and, 488
    pronoun gender and, 492
    pronoun reference and, 487–88
anti-/ante-, 52–53
anticipation, colons for, 584
anticipatory subjects (see expletives),
    222
anticlimax, in increasing-order-of-
    importance method of develop-
    ment, 281–82, 385
antithesis, 232
antonyms, 53
    thesaurus with, 631
    word choice and, 679
anxiety, and making presentations, 478
APA style
    documenting sources using, 173,
        179–85
    quotation format in, 540–41

apostrophes, 53–55, 527
  in acronyms and initialisms, 18
  in contractions, 128
  omission indicated with, 55
  plurals with, 53–54, 414
  possession with, 56–57, 459–61
  proofreaders' mark for, 494
appearance of a document. *See* format;
    layout and design
appendixes, 56
  in formal reports, 244
  illustrations placed in, 278
  in proposals, 497, 512
  in specifications, 606
application letters, 56–61
  acceptance letters and, 14–15
  job interviews and, 318
  job searches and, 336
  refusal letters and, 546–47
  résumés and, 557
appositive phrases
  colons with, 101
  commas with, 105
appositives, 61
  capitalization of, 82
  case of, 64, 86, 491
  commas with, 105
  dashes with, 149
  e.g./i.e. with, 194
  gerunds as, 653–54
  nouns as, 410
approximate numbers, 414
argumentation (persuasion), 448–50
arrangement of ideas
  methods of development and,
    382–85
  organization for, 428–29
  outlining and, 430–32
arrangement of papers. *See* format; lay-
    out and design; organization
arrangement of words
  paragraphs and, 434–38
  sentence construction and, 585–90
  syntax and, 614–15
articles (adjectives), 61, 63
  a/an usage and, 1
  as adjectives, 20
  capitalization of, in titles, 21, 63,
    82
  definite, 20, 61
  English as a second language (ESL)
    and, 206–07
  indefinite, 1, 20, 61
  omitting in writing, 19–20, 61, 63,
    623
  parallel structure repeating, 439
  telegraphic style of, 623

articles (journals), 172. *See also* trade
    journal articles
  citation formats for, 178, 179, 184
  formal writing style for, 608–09
  indexes of, in library research,
    354–55
  openings for, 423–27
  research using, 552
articles (newsletters), 399–403
  electronic, citation formats for, 179,
    185
  illustrations in, 349
articles (newspapers)
  citation formats for, 178, 184
  indexes of, in library research,
    354–55
as, 63
  because as a connective and, 68
  like/as usage and, 357–58
  subjective case after, 85, 87
as much as/more than, 63
as per, 444
as regards, 547
as such, 64
as to whether, 64
as well as, 64
ASCII files, for résumés, 563, 564
associations
  abbreviations for, 5
  capitalization of, 81–82
associative words (connotation/denota-
    tion), 127
assure/insure/ensure, 296
asterisks
  in email, 198
  for footnotes in tables, 618
  proofreaders' mark for, 494
astronomical bodies, 81
atlases, 356–57
attachments to email, 225
attitude
  correspondence and, 130–33
  point of view and, 456–58
  tone and, 633–34
  "you" viewpoint and, 685
attribute/contribute, 64–65
attribution, in documenting sources,
    173
attributive adjectives, 23
audience (readers), 542–43
  defining terms for, 153–55, 542–43
  for formal reports, 240
  global communication and, 251–52
  introductions to technical manuals
    and, 324
  jargon and, 344
  point of view and, 456–58

preparation and needs of, 462
presentations and analysis of, 467
purpose and, 527
reports and, 550
scope and, 575–76
segmenting documents for, 543
selecting the medium and, 580
technical manuals and, 621
telegraphic style and, 623
visualizing a typical member of, 543
"you" viewpoint and, 685–86
augment/supplement, 65
authors
    bibliographies and, 73–74
    citation formats for, 175, 176, 180
    copyright and, 128–30
    documenting sources for, 173, 180–81, 183
    of forewords, 242
    of formal reports, 240
    in library catalogs, 353–54
    quotations from, 538–41, 599
authors' rights
    copyright and, 128–30
    plagiarism and, 455–56
auxiliary verbs, 658
Ave., 2
average/median, 65
awhile/a while, 66
awkwardness, 66
    agreement of pronouns and antecedents and, 36–41
    agreement of subjects and verbs and, 42–45
    ambiguity and, 50–51
    and/or usage and, 52
    as usage and, 62
    clarity and, 93
    dangling modifiers and, 148–49
    expletives and, 222
    faulty parallelism and, 440–41
    faulty subordination and, 590–92
    garbled sentences and, 248
    intensifiers and, 296–97
    international correspondence and, 300
    mixed constructions and, 387–88
    pace and, 434
    passive voice and, 665–66
    point of view and, 466
    pronoun reference and, 488
    repetition and, 550
    revision for, 570
    sentence faults and, 590–92
    sentence fragments and, 592–93
    sentence length and, 594–95

shifts in tense and, 628
spelling out numbers and, 414
split infinitives and, 655
squinting modifiers and, 392
stacked modifiers and, 390–91
subordination and, 611–12
transition to prevent, 640

B.A., 4
back matter, 244–45
background, in openings, 426
backgrounds, in visual aids, 474, 475
bad/badly, 67
bad news letters
    adjustment letters, 25–29
    complaint letters, 114–15
    refusal letters, 546–48
    resignation letters or memos and, 553–55
bad news pattern, in correspondence, 133–35
balance
    lists and, 361
    outlining and, 432
    parallel structure and, 438–41
    writing style and, 610
balance/remainder, 67
bar graphs, 260–62, 264
bars (slashes), 600
B.C., 5
bcc: notation in letters, 143–44
B.C.E., 5
be, 393, 413
be sure and/be sure to, 68
because, 68, 544
because of/due to, 192
beginning reports
    abstracts and, 11
    executive summaries and, 219–20, 243, 550
    introductions for, 322
    openings for, 423–27
beginning sentences
    also used for, 48
    conjunctions for, 126
    ellipses and, 195–97
    expletives for, 222
    numbers and, 414
    sentence construction and, 590
    sentence variety and, 595
beginning-to-end approach
    chronological method of development, 91
    narration and, 395–96
    sequential method of development and, 596–97
being as/being that, 69

benefits to readers
  audience analysis and, 542–43
  correspondence and, 132
  "you" viewpoint and, 685
beside/besides, 69
between/among, 69
between you and me, 69
bi-/semi-, 70
biannual/biennial, 70
biased evidence, 365
biased language, 70–73. *See also* sexism
    in language
  email and, 200
  ethics in writing and, 216
  everybody/everyone usage and, 217–18
  female and, 230
  gender and, 249
  he/she usage and, 265
  pronoun reference and, 488
  revision for, 570
Bible, 83, 328
bibliographical guides, 355–56
bibliographies, 73–74
  abbreviations in, 5–6
  APA style for, 179–85
  commas in, 109–10
  documentation in, 172
  documenting sources in, 173–75
  in formal reports, 244
  library research using, 354–56
  literature reviews and, 363, 364
  MLA style for, 173–79
  periodicals listed in, 354–55
  style manuals for, 185–86
bids, invitation for (IFB), 512–20
biennial/biannual, 70
big words
  affectation and, 32–35
  gobbledygook and, 255
  long variants and, 366
bilingual dictionaries, 166, 167
bisexual men and women, and biased
    language, 73
blend words, 74, 399
blind copy notations in letters,
    143–44
block, writer's, while writing a draft,
    682–83
block-style formats, in letter, 138
blocked quotations
  format for, 541
  indentation of, 283
body language, 251
body of documents
  feasibility reports, 226
  formal reports, 243–44
  informal and short reports, 551
  letters, 141–42

presentations, 469–70
  proposals, 500
  sales proposals, 503–12
  test reports, 628
body of tables, 618
body positions, in global graphics,
    253–54
boilerplate, 456
boiling down ideas
  in note taking, 406–08
  paraphrasing and, 441–42
boldface
  in email, 198
  emphasis with, 204
  in headings, 267, 347
  layout and design with, 347
  in presentation visual aids, 475
  proofreaders' mark for, 494
  Web page design with, 673
bombast
  affectation and, 32–35
  gobbledygook and, 255
book guides, 355–56
book titles
  in bibliographies, 73–74
  capitalization of, 82
  citation formats for, 175, 182
  italics for, 327
  in library catalogs, 353–54
bookmarks, on the Web, 310, 681
books
  bibliographies of, 73–74
  documenting sources and, 173
  indexing of, 283–86
Boolean operators, 577
borrowed material
  copyright and, 128–30
  plagiarism and, 455–56
borrowed words, 237–38
both . . . and, 74–75, 125
boxed elements, in layout and design,
    346, 347
boxheads, 618
brackets, 75, 527
  ellipses with, 195
  inserting material into quotations
    with, 540
  parenthetical items in, 443
  proofreaders' mark for, 494
  sic placed in, 599
  slashes with, 600
brainstorming, 75–77
  job searches using, 333
  presentations with, 471
  research and, 551, 552
  résumé writing and, 557–58
breadth of coverage. *See* scope
breve, 162

brevity
  conciseness and, 119–22
  international correspondence and,
    298
broad references, 488
bromides (clichés), 95–96
browsers, 680
B.S., 4
building numbers (addresses), 416
bulleted lists, 361–62
  in email, 198–99
  in presentation visual aids, 472,
    474–75
bunch, 77
business letter writing
  acceptance letters, 14–15
  acknowledgment letters, 16
  adjustment letters, 23–28
  application letters, 56–61
  complaint letters, 114–15
  correspondence, 130–46
  cover letters, 145–46
  inquiries and responses, 287–88
  memos, 378–82
  parts of a letter and, 138–44
  reference letters, 544–46
  refusal letters, 546–47
  resignation letters, 553–54
business memos. *See* correspondence;
    memos
business names, abbreviations for, 5
business writing style
  formal writing style and, 608–09
  informal writing style and, 609–10
  international correspondence and,
    298–99
  tone and, 633–34
business writing types
  acceptance letters, 14–15
  acknowledgment letters, 16
  adjustment letters, 23–28
  application letters, 56–61
  complaint letters, 114–15
  correspondence, 130–46
  cover letters, 145–46
  description, 158–59
  feasibility reports, 225–27
  formal reports, 239–45
  inquiries and responses, 287–88
  instructions, 292–96
  investigative reports, 324–25
  job descriptions, 330–31
  memos, 378–82
  newsletter articles, 399–400
  press releases, 480–82
  progress and activity reports, 482–84
  proposals, 496–521
  reference letters, 544–46

refusal letters, 546–47
resignation letters or memos,
    553–54
résumés, 557–67
trip reports, 643
trouble reports, 644–45
businessese, and affectation, 34
but. See conjunctions
buzzwords (vogue words)
  affectation and, 34
  vogue words and, 664

callouts
  graphs with, 264
  photographs with, 450
  trade journal articles with, 639
cameras, 450
can, 659
can/may, 78
cannot/can not, 78
cannot help but, 78
cant (jargon), 330
canvas/canvass, 78
capital/capitol, 79
capital appropriations proposals, 499
capital letters, 79–84
  abbreviations and, 83
  acronyms and initialisms and, 18
  of articles in titles, 21, 62
  after colons, 101
  "all caps" usage and, 346, 367
  common nouns and, 80
  in email, 198, 201, 204
  emphasis with, 204
  events and concepts and, 82
  first word in a sentence, salutation,
    or close and, 80, 83–84, 141
  groups and, 80
  in headings, 267
  highlighting in layout and, 347
  individual letters of the alphabet and,
    83
  institutions and organizations and,
    81–82
  job titles and, 82–83
  layout and design and, 347
  personal titles and, 82–83
  places and, 81
  proofreaders' mark for, 494
  proofreading for, 496
  proper nouns and, 79
  titles of works and, 82, 126
  Web page design with, 673
capitol/capital, 79
captions
  illustrations with, 279
  layout and design and, 347
  in trade journal articles, 639

carbon copy notations in letters,
    143–44
card catalogs, 353–54
cardinal adjectives, 22
cardinal numbers, 414
case, 88
case (grammar), 84–88. *See also* posses-
    sive case
    of appositives, 61
    determining, 87, 492
    pronouns and, 490–92
case (letters). *See* capital letters; lower-
    case and uppercase letters
catalogs, in library research,
    353–54
causal relationships, 68
cause, definition by, 157
cause-and-effect method of develop-
    ment, 88–90, 384
cc: notation in letters, 143–44
CD-ROM resources
    dictionaries on, 165
    downloading notes from, 407
    hypertext on, 270
    library research using, 354
    search engines for, 577
C.E., 5
cedilla, 162
census data, 354
center on/center around, 90–91
centuries
    abbreviations for, 5
    as adjectives, 151
    date format for, 151
chair/chairperson, 91
chalkboards, 471–72
chapters
    capitalization of titles of, 84
    footers with titles of, 266
    headers with titles of, 265
    layout and design of titles of, 347
    number usage for, 416
charts
    flowcharts and, 234–37
    illustrations as, 277–79
    in lists of figures, 242
    organizational charts, 429, 521
    in presentations, 471–75
chat environment, 306, 307–08
Checklist of the Writing Process,
    xviii–xix
choppy writing
    garbled sentences and, 248
    pace and, 434
    sentence length and, 594–95
    subordination and, 611–12
    telegraphic style and, 623

transition and, 640–43
    word order and, 595
chronological method of development,
    91, 596–97
chronology, and narration, 395–96
circle (pie) charts, 262–63
circular definitions, 154
circumflex, 162
circumlocution
    affectation and, 34
    gobbledygook and, 255
    wordiness from, 120
citations
    APA style for, 179–85
    bibliographies with, 73–74
    MLA style for, 173–79
cite/sight/site, 91–92
claims
    adjustment letters responding to,
        23–28
    complaint letters for, 114–15
    refusal letters and, 546–47
clarity, 93–94
    ambiguity and, 50–51
    conciseness/wordiness and, 119–22
    defining terms and, 153–55
    emphasis for, 203
    figures of speech and, 231–33
    grammar and, 256–57
    hyphens for, 273
    idioms and, 275
    illustrations and, 278
    instructions and, 296
    in memos, 379
    in note taking, 406
    pace and, 434
    parallel structure and, 438–39
    passive voice and, 665
    positive writing and, 458–59
    semicolons for, 584
    sentence construction and, 589–90
    specifications and, 605
    subordination and, 610–12
    telegraphic style and lack of, 623
    transition and, 640–43
    unity and, 648
    word choice and, 679
classification, 170–72
clauses, 94–95
    comma splice and, 102, 573
    commas with, 105
    faulty subordination with, 591–92
    fused (run-on) sentences with,
        446–47, 573–74
    parallel structure of, 439
    semicolons with, 583, 584
    subjects missing in, 591

clearness. *See* clarity
clichés, 95–96, 232
  clarity and, 93
  in correspondence, 131
  revision and, 570
  trite language and, 643
climactic order of importance
  emphasis and, 204
  increasing-order-of-importance method of development and, 281–82
clip art, 188, 474
clipped forms of words, 96
clipped (telegraphic) style, 623
close-space mark, 494
closings. *See also* conclusions
  of meetings, 378
  paragraph length for, 436
  of presentations, 470
closings of letters
  in correspondence, 142
  of refusal letters, 547
clustering, 76–77
Co., 5
co-authored articles
  collaborative writing of, 97–100, 636
  in trade journal, 636
coherence, 96–97
  clarity and, 93
  conciseness/wordiness and, 119–22
  methods of development and, 384
  outlining and, 430
  pace and, 434
  paragraphs and, 437–38
  pronoun reference and, 488
  revision and, 567
  split infinitives and, 655
  subordination and, 610–12
  transition and, 640–43
  unity and, 648
cohesive writing. *See* coherence
coined words
  blend words and, 74
  new words and, 399
collaborative writing, 97–100
  hypertext and, 271
  trade journal articles and, 636
collective nouns, 410–11
  agreement of pronouns and antecedents, 40–41
  agreement of subjects and verbs, 46
  committee as, 111
college course title, 537
colloquialisms
  dictionaries listing, 165
  fine usage and, 233–34

quotation marks and, 537
varieties of English and, 211–12
colons, 100–102, 527
  anticipation or enumeration and, 584
  capitalization of the first words after, 80
  proofreaders' mark for, 494
  quotation marks and, 537–38
  in salutations, 101, 138
color
  global graphics and, 252, 254
  graphs with, 262, 263
  instructions with, 296
  layout and design with, 348
  maps with, 370
  photographs and, 450
  presentation visual aids and, 474
  warnings in instructions and, 295
  Web page design and, 673
column (bar) graphs, 260–62
column headings in tables, 618, 619
columns, in layout and design, 346, 348
combination of words (compound words), 118–19
combined words (blend words), 74
comma faults
  comma splice, 102–03, 573
  run-on (fused) sentences, 446–47, 573–74
  unnecessary commas, 110
comma splice, 102–03, 573
commands
  imperative mood, 392
  implied subject in, 586
  in instructions, 292–96
commas, 103–10, 527
  addresses with, 108
  adjectives with, 23, 25, 107
  adverbs with, 31
  avoiding unnecessary, 110
  dates with, 109, 150
  e.g./i.e. and, 194
  independent clauses with, 104, 107
  introducing elements with, 105–06
  names with, 109
  nonrestrictive clauses with, 104, 555–56
  numbers with, 109, 300, 416, 417
  omissions shown by, 107–08
  other punctuation used with, 108
  parenthetical elements with, 104–05, 149–50
  proofreaders' mark for, 494
  quotations with, 106, 445
commercial guides to library research, 355–56

commercial proposals. *See* proposals
committee, 111
committee meetings, minutes of, 385–87
common (mutual/in common), 394
common knowledge, 456
common nouns, 80, 409
commonly confused terms
  idioms and, 275–77
  usage and, 651–52
communication
  global audiences and, 251–52
  global graphics and, 252–54
  listening and, 358–61
  presentations and, 466–80
  selecting the medium for, 580–83
company (they/it), 411
company names
  abbreviations for, 5
  ampersands in, 5, 52
  capitalization of, 81
  in salutations, 141
company news
  newsletter articles on, 399–400
  press releases for, 480–82
company newsletter, 399–400
comparative degree
  absolute words and, 9, 23
  of adjectives, 21–22
  of adverbs, 30–31
compare/contrast, 111–12
comparison, 112
  absolute words in, 9, 23
  adjectives for, 22–23
  adverbs for, 30–31
  ambiguity and, 50–51
  analogy and, 232
  double, in the same sentence, 112
  figures of speech and, 231–33
  metaphor and, 233
  simile and, 233
comparison method of development,
  113–14, 384
complaint letters, 114–16
  adjustment letters responding to,
  23–28
  refusal letters replying to, 546–47
complement/compliment, 116–17
complements, 117–18
  linking verbs and, 657
  predicates with, 586
complete predicates, 586
completeness, and revision, 567
complex sentences
  construction of, 587
  emphasis and, 204
  sentence variety using, 594
compliment/complement, 116–17

complimentary closings
  capitalization of, 80
  in correspondence, 142
  in international correspondence, 299
components, definition by, 157
compose/comprise, 118
composing the report (writing a draft),
  121, 682–83
compound antecedents, 41
compound-complex sentences
  construction of, 587
  sentence variety using, 594
compound constructions, case in, 87
compound modifiers, 271–72
compound nouns
  determining case of, 87
  plurals of, 412
  possessive case of, 461
compound numbers, 271
compound predicates, 110, 586
compound pronouns, 87, 491–92
compound sentences
  construction of, 587
  emphasis and, 204
  sentence variety using, 594
compound subjects
  agreement of subjects and verbs
  and, 46
  commas with, 110
  determining case of, 87
  sentence construction and, 585
compound words, 118–19
  hyphens with, 271
  plurals of, 119
  possessives of, 119, 461
comprise/compose, 118
computer graphics. *See* graphics
computer searches. *See* Internet research; search engines
computer software
  citation formats for, 178, 184
  desktop publishing and, 343–49
  flowcharts of, 234, 237
  presentations with, 473–74
  reference manuals for, 620–21
  spell checker with, 606
concept/conception, 119
concepts
  capitalization of, 79, 81–82
  definitions of, 157
  nouns for, 409
conciseness/wordiness, 119–22
  affectation and, 34
  awkwardness and, 66
  causes of, 119–21
  clarity and, 93
  compound predicates for, 586

correspondence style and, 137
email and, 200
faulty parallelism and, 440–41
gobbledygook and, 255
illustrations and, 278
instructions and, 292
intensifiers and, 296–97
international correspondence and, 298
map information and, 369
memos and, 379
minutes of meetings and, 387
passive voice and, 666–67
redundant prepositions and, 464–66
revision for, 570
-size/-sized usage and, 599
subordination and, 610–12
telegraphic style and, 623
concluding a sentence
exclamation marks for, 218–19
parentheses and, 442
periods for, 445, 446, 538
question marks for, 529
quotation marks for, 538
conclusions, 123–25. See also summaries
abstracts on, 12
in correspondence, 131
in feasibility reports, 226
in formal reports, 243
in informal and short reports, 550, 551
in investigative reports, 325
to newsletter articles, 401
openings summarizing, 427
positive writing and, 458–59
in progress reports, 484
in proposals, 500, 503, 505
in refusal letters, 547
specific-to-general method of development and, 604
in test reports, 628
in trade journal articles, 638
in trouble reports, 645
concrete nouns, 409–10
concrete words
abstract words compared with, 9–10
figures of speech and, 231–33
nouns for, 409–10
word choice and, 679
concrete writing
abstract words/concrete words in, 9–10
connotation/denotation of words and, 127
description and, 158–59
figures of speech and, 231–33

nouns and, 409–10
word choice and, 679
conducting interviews
for information, 314–16
for a job, 316–21
conference calls, 582
conference proceedings and papers
citation formats for, 177–78, 179, 184
research using, 552
confidentiality, in email, 202
conflict, and meetings, 377–78
confusion, and ambiguity, 50–51
conjugation of verbs, 661
conjunctions, 125–26
adverbs as, 31
agreement of subjects and verbs and, 46
as used as, 62
beginning a sentence with, 126
capitalization of, 82, 126
commas with, 104
fused (run-on) sentences and, 447, 573–74
parallel structure and, 439, 440
as a part of speech, 444
sentence construction with, 587
telegraphic style and omission of, 623
in titles of works, 82, 126
conjunctive adverbs, 31, 126, 140
clauses with, 95
commas with, 106
compound sentences with, 587
fused (run-on) sentences and, 447
semicolons with, 584
sentence construction and, 587
connected with/in connection with, 126
connectives
adverbs and, 31
because as, 68
conjunctions as, 126
semicolons with, 583–84
connotation/denotation, 127
affectation and, 633
word choice and, 679
consensus, 127
consistency
figures of speech and, 232
illustrations and, 278
mixed constructions and, 387–88
in presentation visual aids and text, 475
revision for, 567
verbs and, 661
construction. See sentence construction
consumer complaint letters, 114–15

contact persons, on press releases,
    481
contents, tables of, 616
    appendixes listed in, 56
    in formal reports, 242
    headings in, 269
    parallel structure of, 440
continual/continuous, 127
continued tables, 618–19
continuity
    coherence and, 96
    unity and, 648
continuous/continual, 127
contractions, 128
    apostrophes in, 54, 55
    clipped forms of words and, 96
    in formal writing style, 608
    in informal writing style, 609
contrary to fact (subjunctive mood),
    392–93
contrast
    antithesis and, 232
    comparison method of development
        and, 113–14
    sentence length and, 594–95
contrast/compare, 111–12
contrasting thoughts, and commas, 107
contribute/attribute, 64–65
conversational English and style
    informal writing style and, 609–10
    nonstandard English and, 211–12
    tone and, 633–34
conversations, in quotation marks, 536
convincing your reader (persuasion),
    448–50
coordinate nouns, 460
coordinating conjunctions, 125
    clauses with, 95
    commas with, 104, 110
    compound sentences with, 587
    fused (run-on) sentences and, 447,
        573–74
    subordination using, 610–12
coordination
    conjunctions and, 125
    parallel structure and, 438–41
copy marking, 495
copy notations in letters, 143–44
copying
    copyright and, 128–30
    documenting sources and, 173–75
    plagiarism and, 455–56
    quotations and, 538–41
copyright, 128–30
    permissions and, 128, 130, 243,
        279
    plagiarism and, 455–56

    publication date with, 182
    quotations and, 539
    trade journal articles and, 640
Copyright Office, 129
Corp., 5
corporate authors, 175, 177, 183
corporate names
    abbreviations for, 5
    in salutations, 141
corrections
    proofreaders' marks for, 495
    revision and, 567–70
correlative conjunctions, 125–26
    double negatives and, 187
    parallel structure of, 439
correspondence, 130–46
    audience and, 130–33
    format of, 137–44, 245
    informal writing style of, 609–10
    international, 298–305
    letterhead for, 138, 581
    openings in, 424
    parts of, 138–44
    point of view in, 458
    presenting good and bad news in, 133–35
    style and accuracy of, 135–37
    tone of, 130–33, 633
    "you" viewpoint in, 685
correspondence types
    acceptance letters, 14–15
    acknowledgment letters, 16–17
    adjustment letters, 25–29
    application letters, 56–61
    complaint letters, 114–16
    cover letters, 144–47
    email, 197–202
    inquiries and responses, 287–91
    memos, 378–83
    reference letters, 544–46
    refusal letters, 546–48
    resignation letters or memos, 553–55
cost analysis, in proposals, 505–12
could, 659
count nouns, 410
    adjectives with, 22
    English as a second language (ESL)
        and, 205–06
    fewer/less usage and, 230
cover letters and memos, 144–47
    for formal reports, 239
    meeting agendas with, 374–75
    proposals with, 501, 504, 505
credible/creditable, 147
credit
    bibliographies and, 73–74
    copyright and, 128–30
    documenting sources for, 173

for illustrations, 279
note taking and, 407
paraphrasing and, 441
plagiarism and, 455–56
quotations and, 539
creditable/credible, 147
criterion/criteria, 147
critique, 147
cross-cultural differences
　correspondence style and, 133
　global communication and, 251–52
　global graphics and, 252–54
　international correspondence and,
　　300–305
　meetings and, 377, 378
　presentations and, 479–80
cross-references, 285–86
crowd, 411
cultural differences
　correspondence style and, 133
　global communication and, 251–52
　global graphics and, 252–54
　international correspondence and,
　　298–99
　meetings and, 377, 378
　quotation mark usage and, 538
cumulative/accumulative, 16
curriculum vitae (résumés), 557–67
customer complaint letters, 114–15

-*d* ending
　with past participles, 656
　with past tense, 625
daggers, proofreaders' mark for, 494
dangling modifiers, 148–49
　ambiguity and, 50–51
　infinitive phrases as, 454
　participial phrases and, 452–53
　passive voice and, 665
　testing for, 148–49
dashes, 149–50, 527. *See also* em
　dashes; en dashes; hyphens
　gaps in tables signified by, 618
data/datum, 150
data sheets (résumés), 557–67
databases, online
　search engines for, 577
　Telnet for connecting to, 308–09
dates, 150–51
　abbreviation for, 5
　apostrophes with, 55
　commas with, 109, 150
　in correspondence headings, 138
　in international correspondence,
　　151, 300, 416
　numbers in, 415
　plurals of, 55

on press releases, 480
　slashes in, 600
dates of publication
　citation formats for, 176, 181–82
　footers with, 265
　on formal reports, 241
datum/data, 150
days
　capitalization of, 82
　date format for, 150–51
de facto/de jure, 153
deadwood
　gobbledygook and, 255
　wordiness and, 119–22
decided/decisive, 151
decimal numbering system for head-
　ings, 269
decimals, 415
　commas in, 109, 300
　in international correspondence, 300
　periods in, 416, 417, 446
decisive/decided, 151
declarative (indicative) mood, 122,
　392, 393
declarative sentences
　construction of, 588
　sentence variety using, 594
　word order for, 619
decreasing-order-of-importance
　method of development, 151–52,
　385
　press releases and, 480
　reference letters and, 545
deduction, in general-to-specific
　method of development, 384–85
defective/deficient, 153
defining terms, 153–55
　connotation/denotation of words
　　and, 127
　definition method of development
　　and, 155–58
　dictionaries and, 163–67
　glossaries for, 254–55
　for readers, 542–43
definite/definitive, 155
definite articles
　as adjectives, 20, 61–62
　using, 63
definition method of development,
　155–58
definitions
　connotation/denotation of words
　　and, 127
　defining terms using, 153–55
　definition method of development
　　and, 155–58
　descriptions using, 159

definitions (*cont.*)
dictionaries with, 163–67
glossaries with, 254–55
introductions to technical manuals and, 324
openings providing, 425
readers' need for, 542–43
definitive/definite, 155
degree, adverbs of, 30
degree of comparison
adjectives and, 21–22
adverbs and, 30–31
degrees, academic, 4
deity, 83
delete mark, 494
deletions
apostrophes showing, 53, 55
contractions and, 128
ellipses for, 195–97, 540
proofreading for, 496
quotations and, 540
delivering speeches, 475–79
delivery schedule, in proposals, 505
demonstrative adjectives, 20–21, 39–40
demonstrative pronouns, 489
denotations of words, 127, 679
dictionaries and, 163–67
glossaries and, 254–55
department names
capitalization of, 81
in salutations, 141
dependent clauses, 94
commas with, 107
complex sentences with, 587
compound-complex sentences with, 588
conjunctions with, 126
faulty subordination with, 591–92
relative pronouns and, 489
subordination and, 610–12
depiction (description), 158–59
depth of coverage. *See* scope
derivations of words, 157, 164
description, 158–59
forms of discourse and, 245–46
of mechanisms, 158–59
of mechanisms in operation (process explanation), 234–37
point of view and, 456–58
process explanation and, 482
spatial method of development for, 602–03
descriptions, job, 330–31
descriptive abstracts, 11–12
descriptive adjectives, 23
descriptive (subordinate) clauses, 94
descriptive grammar, 256

descriptive words. *See* adjectives; adverbs; modifiers
descriptive writing, 158–59
design, 343–49. *See also* format; Web page design
ethics in using, 215
format and, 245
highlighting devices for, 346–48
of illustrations, 279
organization and, 429
page design and, 348–49
persuasion and, 450
placement of illustrations and, 278
presentation visual aids and, 474
proofreading for, 495
typography in, 343–44
warnings in instructions and, 295
desk dictionaries, 165, 352
desktop publishing
format and, 245
layout and design and, 343–49
despite/in spite of, 161
details, in openings, 425–26
developing ideas
in memos, 379
methods of development and, 382–85
organization and, 428–29
outlining and, 430–32
paragraphs and, 434–38
sentence construction and, 589–90
device, descriptions of, 158–59, 482
diacritical marks
in dictionaries, 164
for foreign words, 161
list of, 162
diagnosis/prognosis, 162
diagrams
flowcharts and, 234–37
instructions with, 293–94
schematic, 575, 621
dialect
dictionaries listing, 165
writing style and, 608, 610
dialectal English, 212
dialogue, in quotation marks, 536
diction, 162
dictionaries, 163–67
defining terms and, 153–55
denotations of words in, 127
elements of an entry in, 164–65
foreign words used in English in, 238, 327
glossaries and, 254–55
library research using, 352, 354
noun usage described in, 412
plurals given in, 660

spelling given in, 606
synonyms given in, 613–14
types of, 165–67
usage in, 652
word choice and, 680
dieresis, 162, 462
differ from/differ with, 168, 276
differences, in comparison method of
development, 113–14
different from/different than, 168, 276
digests of writing
abstracts as, 11–14
executive summaries as, 219–20
digital cameras, 450
direct address, 168
commas with, 104
emphasis with, 205
direct communication, in letters,
133–35
direct objects, 418
complements as, 117
gerunds as, 653–54
nouns as, 410
direct quotations, 539
commas with, 106
copyright and, 129
documenting sources for, 173, 180
plagiarism and, 455–56
direct statement, for emphasis, 205
direct word meanings (connotation/de-
notation), 127
directions
instructions, 292–96
procedures, 468–69
directions, geographical
capitalization of, 81
on maps, 370
disabilities, people with, and biased
language, 72
discourse, forms of, 245–46
description, 158–59
exposition, 223
narration, 395–96
persuasion, 448–50
discreet/discrete, 168–69
discussion groups, 306
dishonesty, and plagiarism, 455–56
disinterested/uninterested, 169
dissertation abstracts, 11
dissertations, 177, 183
ditto marks, 538
dividing time periods
months, using *bi-* or *semi-,* 70
years, using *biannual/biennial,* 70
dividing words
dictionaries for, 164
hyphens for, 274

division (organization), 81
division-and-classification method of
development, 169–72, 384
document design
format and, 245
layout and design and, 343–49
documentation, 172
library research using, 352–53
numbering chapters and figures in,
416
style, 173, 185–86
technical, 172
documenting sources, 172–86
abbreviations in, 5–6
ampersands used in, 52
APA style for, 179–85
bibliographies for, 73–74
colons used in, 101
commas used in, 109–10
copyright and, 129–30
in formal reports, 243–44
MLA style for, 173–79
note taking and, 406
plagiarism and lack of, 455–56
quotations and, 539
research and, 552
style manuals for, 185–86
documents, government
citation formats for, 175, 181
library research using, 356–57
dots, spaced
ellipses, 195–97
periods, 445–46, 527, 616, 618
double comparison, 112
double daggers, proofreaders' mark for,
494
double negatives, 186–87
irregardless/regardless as, 325
positive writing and, 458–59
double possessives
Dr., 4, 138
drafts
Checklist of the Writing Process on,
xviii–xix
collaborative writing of, 97, 98
in email, 197–98
wordiness in, 121
writing, 682–83
drawings, 188–90
instructions with, 293–94
layout and design of, 349
in lists of figures, 242
photograph use compared with,
450
schematic diagrams with, 575
in technical manuals, 621
in trade journal articles, 639

due to/because of, 192
dummy, in layout and design, 348

each, 193
  agreement of pronouns and an-
    tecedents and, 39, 43, 492
  agreement of subjects and verbs and,
    45, 47
each and every, 193
each other, 490
earth, 81
east, 81
economic/economical, 193–94
economic groups, 81
economical writing
  conciseness and, 119–22
  parallel structure and, 438–41
-ed ending
  part participles with, 656
  past tense with, 625
editing. *See also* revision
  proofreaders' marks for, 494,
    495
  proofreading and, 494
editions, 177, 183
editorial guidelines, 279
editors, 175, 177, 183, 216
educational record, on résumés,
    559–60
effect
  affect/effect usage, 33
  cause-and-effect method of develop-
    ment and, 88–90
e.g./i.e., 194
either . . . or, 125, 439
electronic documents. *See* CD-ROM
  resources; Internet research; Web
  page design
electronic mail. *See* email
electronic résumé, 563–64
elegant variation, 194–95
  affectation and, 34
  repetition and, 549
ellipses, 195–97
  colons and, 102
  for deletions or omissions, 195–97,
    539–40
  periods as, 445
elliptical constructions
  commas with, 107–08
  dangling modifiers as, 148
  in informal writing style, 609
  sentence variety using, 594
elude/refer/allude, 48
em dashes
  emphasis using, 204
  proofreaders' mark for, 494

email, 197–202, 301
  acknowledgment letters sent via, 16
  attachments to, 225
  business writing style of, 86
  citation formats for, 179, 182
  colloquial English in, 211–12
  confidentiality considerations with,
    202
  design considerations for, 198–99
  direct address in, 168
  discussion groups and, 306
  informal writing style for, 608, 609–10
  inquiries using, 287–88
  international correspondence using,
    298
  meeting agenda sent via, 374
  netiquette in using, 200–202
  openings in, 424–25
  outlining, 431
  persuasion and, 448
  point of view in, 457
  questionnaires sent out in, 530, 532
  readers of, 542
  reasons for selecting, 581
  research using, 310
  résumés sent via, 563–64
  signature blocks with, 199–200
  tone of, 633–34
  Web page design with links for, 672
  writing style for, 198
email addresses
  in résumés, 558, 559
  using, 200–202
eminent/imminent, 202
emoticons, 201
emotional associations (denotations) of
    words, 127
emphasis, 203–05
  "all caps" (capital letters) and, 367
  clarity and, 93
  commas and, 103–04
  conjunctions and, 126
  dashes for, 149–50
  direct address for, 205
  in email, 198
  highlighting devices for, 346–48
  intensifiers and, 296–97
  inverted sentence order for, 589
  italics for, 327
  in newsletter articles, 400
  pace and, 434
  parallel structure and, 438
  quotation marks for, 536
  really for, 543
  repetition for, 549
  sentence length and, 594–95
  subordination and, 611

topic sentences and, 436
voice and, 665
writing style and, 610
employee newsletters, 399–400
employment
  acceptance letters in, 14–15
  application letters for, 56–61
  description of, on résumés, 560
  interviewing for, 316–21
  job descriptions and, 330–31
  job searches and, 333–37
  refusal letters and, 546–47
  resignation letters or memos and,
    553–54
  résumés for getting, 557–71
employment agencies, 336
en dashes, proofreaders' mark for, 494
-en ending, 656
enclosure notations in letters,
  142–43
encyclopedias
  citation formats for articles in, 177, 184
  library research using, 352, 354
ending a report. See closings
ending a sentence
  abbreviations and, 446
  ellipses and, 196
  exclamation marks for, 218–19
  periods for, 445, 446
  prepositions and, 466
  question marks for, 529
endings. See closings
endnotes/footnotes
  commas in, 109–10
  in formal reports, 244
  on graphs, 260
  parentheses and, 443
  secondary information placed in,
    396
  in tables, 618
English, varieties of, 211–12, 651
English as a second language (ESL),
  205–11
  dictionaries for, 166
  list of entries explaining usage for,
    210–11
ensure/assure/insure, 296
entries, in indexes, 283–86
enumeration
  colons for, 584
  transition with, 641
epithets, 83
equal/unique/perfect, 213
equations, 370–71, 639
-er ending
  adjectives with, 21–22
  adverbs with, 30–31

-es ending
  plurals of noun with, 411
  plurals of numbers with, 414
  singular verbs with, 413
-ese ending, 213
ESL. See English as a second language
-est ending
  adjectives with, 22–23
  adverbs with, 30–31
et al., 180
etc., 213–14
ethics in writing, 214–17
  biased language and, 70–73
  logic errors and, 365
  passive voice and, 666
  persuasion and, 449
  plagiarism and, 455–56
  positive writing and, 458–59
  résumés and, 561
  revision and, 570
  scales on graphs and, 258
  sexism in language and, 598–99
  technical writing style and, 622
  trouble reports and, 645
ethnic groups
  biased language and, 72–73
  capitalization of, 80
etymology, 163, 164
euphemisms, 217
  affectation and, 34
  connotation/denotation of words
    and, 127
  ethics in using, 215
  gobbledygook and, 255
events, 81–82
every
  agreement of pronouns and an-
    tecedents and, 40, 43
  agreement of subjects and verbs and,
    47
everybody/everyone, 40, 217–18
everyone, 39
evidence
  cause-and-effect method of develop-
    ment and, 88–90
  logic errors from biased or sup-
    pressed, 365
-ex prefix, 273
exact (absolute word), 9, 23
exact writing
  abstract words/concrete words in, 9–10
  connotation/denotation of words
    and, 127
  description and, 158–59
  positive writing and, 458–59
  word choice and, 679–80
exaggeration, in hyperbole, 233

except/accept, 14
excess words
 revision for, 570
 wordiness from, 119–22
exclamation marks, 218–19, 527
exclamations
 interjections with, 297–98
 inverted sentence order for, 589
exclamatory sentences
 construction of, 588
 interjections and, 297–98
 sentence variety using, 594
executive summaries, 219–20
 in formal reports, 243
 in informal and short reports, 550
 in proposals, 497, 504, 505
explaining a process, 482
explanatory footnotes. *See* foot-
  notes/endnotes
explanatory phrases, and sentence fragments, 592–93
explanatory series, 149
expletives, 222–23
 awkwardness from overuse of, 66
 it used as, 325
 passive voice and, 665
 sentence construction with, 586
 there used as, 631
 wordiness and, 120
explicit/implicit, 223
exploration of origin, definition by, 157–58
exposition, 223, 245
extended definitions, 156
extent of coverage. *See* scope
eye contact, in presentations, 476–77,
  479

facsimile (fax), 96, 224–25, 581
fact, 224
fact sheets (résumés), 557–67
facts, and logic errors, 366
factual writing, in exposition, 223
fad words, 664
fair use, and copyright, 129, 130
false analogies, 365
false elegance (affection), 32–35
family relationships, 83
fancy words
 affectation using, 32–35
 elegant variation and, 194–95
 long variants and, 366
fancy writing
 affectation and, 32–35
 gobbledygook and, 255
 jargon and, 330
fashionable words, 664
faults, comma
 comma splice, 102–03, 573

run-on (fused) sentences, 446–47,
  573–74
 unnecessary commas, 110
faults, period, 446–47
faults, sentence, 590–92
 faulty subordination and, 591–92
 missing subjects in clauses and, 591
 missing verbs and, 592
 rambling sentences and, 591–92
faulty parallelism, 440–41
 lists and, 361
 outlining and, 432
faulty pronoun reference
 agreement of pronouns and antecedents and, 36–41
 ambiguity and, 50–51
 hidden references and, 488
faulty subordination, 591–92, 612
fax (facsimile), 96, 224–25, 581
feasibility reports, 225–27, 633
federal government proposals, 512–21
federal government publications
 citation formats for, 175, 181
 library research using, 356–57
federal government specifications, 604,
  605–06
feel bad/feel badly, 67
female, 230
 feminine pronouns and titles agreement and, 37–38
 agreement of pronouns and antecedents and, 38–39
 gender and, 249
 he/she usage and, 265
 Ms./Miss/Mrs. usage and, 394
 sexist language and, 70–72
 in salutations, 138
few/a few, 230
fewer/less, 230
figurative language, 231–33
figuratively/literally, 231
figure numbers
 for illustrations, 278–79
 for maps, 369
 number usage in, 416
 for photographs, 450
figures
 as illustrations, 277–79
 in lists of figures, 242
figures, lists
 in formal reports, 242
 illustrations listed in, 279
figures of speech, 231–33
 in international correspondence, 299
 slang with, 212
 technical writing style and, 622
 types of, 232–33
file name notations in letters, 142

file transfer protocol (FTP), 309
film titles
  capitalization of, 82
  italics for, 328
final drafts
  revision of, 567–70
  writing, 682–83
finalize/finish, 366
finding a job
  acceptance letters, 14–15
  application letters, 56–61
  interviewing for a job, 316–21
  job search, 333–37
  resignation letters or memos, 553–54
  résumés, 557–67
finding information
  inquiry letters for, 287–88
  interviews for, 314–16
  library research and, 351–57
  presentations and, 468
  questionnaires and, 530–41
  research and, 551–53
findings. *See* conclusions; recommendations
fine, 233–34
fine writing (affection), 32–35
finish/finalize, 366
finite verbs, 658
first
  as an adjective, 22
  as an adverb, 22
first/firstly, 234
first names in salutations, 141
  spelling out, 3
first person
  antecedents and, 493
  case and, 490–92
  narration with, 395
  person and, 447
  point of view using, 456–58
first words, capitalization of
  in closings of letters, 141
  after colons, 80
  in sentences, 80, 83–84
firstly/first, 234
"Five Steps to Successful Writing,"
  x–xvii
flammable/inflammable/nonflammable,
  234
flip charts, 471
flow of writing
  clarity and, 93–94
  coherence and, 96
  conciseness/wordiness and,
    119–22
  emphasis and, 203–05
  pace and, 434
  parallel structure and, 438–41

sentence variety and, 594–96
subordination and, 610–12
transition and, 640–43
unity and, 648
writing a draft and, 682–83
flowcharts, 234–37
  forms of, 234–35
  as illustrations, 277–79
  in technical manuals, 621
flowery writing
  affectation and, 32–35
  gobbledygook and, 255
fonts (type)
  electronic résumés and, 564
  instructions with, 296
  layout and design and, 343–45
  proofreaders' marks for, 494, 495
  warnings in instructions and, 295
footers, 265–66, 347
footnotes/endnotes
  commas in, 109–10
  in formal reports, 244
  on graphs, 260
  parentheses and, 443
  secondary information placed in, 396
  in tables, 618
for, in causal relationships, 68
for example
  e.g./i.e. and, 194
  sentence fragments and, 593
forceful/forcible, 237
forecasts, in openings, 436
foreign customers
  global communication and, 251–52
  global graphics and, 252–54
  international correspondence and,
    298–301
foreign languages
  new words borrowed from, 399
  translations of international corre-
    spondence into, 298
foreign punctuation (diacritical marks),
    161, 162
foreign speakers, and idioms, 275
foreign words in English, 237–38
  ad hoc, 19
  affectation and, 34
  diacritical marks and, 161
  dictionary listings of, 238, 327
  italics for, 328
  new words and, 399
  quid pro quo, 536
foreword/forward, 238–39
forewords, 242
formal definitions
  defining terms with, 153–54
  definition method of development
    and, 155–58

formal English
    usage and, 651
    varieties of English and, 211
    writing style and, 608–09
formal letters, contractions in, 128
formal memos, 381
formal reports, 239–45
    abstracts for, 11–14
    appendixes in, 56
    cover letters for, 144–47
    elements of, 239–45
    format of, 245
    glossaries in, 244, 254, 255
    table of contents in, 616
formal writing style, 135, 608–09
format, 245. *See also* layout and design;
    Web page design
    of correspondence, 137–44
    of documentation citations,
        173–75
    of electronic résumés, 562–64
    of formal reports, 240–45
    headers and footers and, 265–66
    headings and, 267–70
    of illustrations, 278–79
    indentation and, 283
    of indexes, 283
    of mathematical equations, 370–71
    of memos, 383
    organization and, 429
    of presentation visual aids, 473–74
    of press releases, 480–82
    of procedures, 469–70
    of progress reports, 484
    proofreading for, 495
    of proposals, 497
    of quotations in text, 540–41
    of trade journal articles, 639
formation of words
    blend words and, 74
    new words and, 399
former/latter, 245
forms of address
    abbreviations for, 3–4
    capitalization of, 82–83
    chair/chairperson and, 91
    Ms./Miss/Mrs. usage with, 394
    in salutations, 141
    sexist language and, 71
forms of discourse, 245–46
    description, 158–59
    exposition, 223
    narration, 395–96
    persuasion, 448–50
formula/formulae/formulas, 246
fortuitous/fortunate, 246
forward/foreword, 238–39

forwarding letters, 288
fractions
    numbers in, 415
    slashes with, 600
fragments, permissible sentence (minor
    sentences), 588
fragments, sentence, 592–93
free writing
    brainstorming using, 75–77
    writing a draft using, 683
French words, 238. *See also* foreign
    words in English
front matter, 240–43
FTP (file transfer protocol), 309
full-block letter style, 138
functional résumés, 560
functional shift, 246–47
fused sentences, 446–47
fused words (blend words), 74
future perfect tense, 625, 626–28,
    660
future progressive tense, 624–25, 628
future tense, 624, 627, 660

garbled sentences, 248
    awkwardness and, 66
    sentence fragments and, 594
gay men, and biased language, 73
gender, 249
    agreement of pronouns and an-
        tecedents and, 38–39, 492, 493
    female usage and, 230
    he/she usage and, 265
    male usage and, 368
    Ms./Miss/Mrs. usage and, 394
    pronouns and, 492, 493
    salutations and, 141
general encyclopedias, 352
general-to-specific method of develop-
    ment, 250, 384–85
general truths, and present tense, 625
general types of writing
    description, 158–59
    exposition, 223
    forms of discourse, 245–46
    instructions, 292–96
    narration and, 395–96
    persuasion and, 448–50
    process explanation, 482
generalizations, and logic errors,
    364–65
genitive case (possessive), 86,
    459–61
    of adjectives, 22
    apostrophes for, 56–57, 459–61
    of compound words, 119
    determining, 87, 492

of *it,* 329
pronouns and, 490–91
geographic features
capitalization of, 81
maps and, 368–70
geographical directions
capitalization of, 81
on maps, 370
geographical names
abbreviations for, 6–8
capitalization of, 81
commas with, 109
possessive case of, 460
German words, 237. *See also* foreign
words in English
gerund phrases, 455
dangling modifiers and, 148–49
gerunds, 653–54, 658
English as a second language (ESL)
and, 207–08
objective case of the pronoun objects
of, 85
possessive form of a pronoun with,
461
as substantives, 612
gestures
global communication and,
251–52
presentations with, 477, 479
getting ready (preparation), 462–63
global communication, 251–52
international correspondence and,
298–99
meetings and, 377
presentations and, 479–80
global graphics, 252–54
graphs and, 264
international correspondence and,
298–99
gloss overs (euphemisms), 217
glossaries, 254–55
in formal reports, 244, 255
global graphics with, 254
in proposals, 497
goals of a writer, and purpose, 527
gobbledygook, 255
affectation and, 34
blend words and, 74
long variants and, 366
wordiness and, 119–22
God, 83
good/well, 255–56
good news pattern, in correspondence,
133–35
goodwill, establishing
acknowledgment letters for, 16–17
adjustment letters for, 25–29

correspondence style and, 131–32,
133–35, 137
Gopher search system, 179, 185,
309
government agencies as authors, 175,
181
government proposals, 520–25
government publications
citation formats for, 175, 181
copyright for, 130
library research using, 356–57
government specifications, 604,
605–06
grammar, 256–57
mixed constructions and errors of,
389
proofreading for, 496
revision for, 570
grammar books, 167
grammatical agreement, 37–38
agreement of pronouns and antecedents, 36–41
agreement of subjects and verbs,
42–45
graphics
flowcharts as, 234–37
global business environment and,
252–54
graphs as, 257–64
illustrations as, 277–79
in lists of figures, 242
maps as, 368–70
organizational charts as, 429
outlining and planning for, 432
photographs as, 450
in presentations, 471–75
Web page design with, 673–74,
675–76
graphs, 257–64
as illustrations, 277–79
in laboratory reports, 339
in lists of figures, 242
in presentations, 471–75
grave accent, 162
Greek words, 237. *See also* foreign
words in English
greetings. *See* salutations
grid lines, in graphs, 257–60
group (collective) nouns, 410–11
grouping of words
clauses and, 94–95
phrases and, 451–55
sentence construction and, 585–90
groups
capitalization of names of, 80
collective nouns for, 409
speaking before. *See* presentations

guides to reference books, 355–56
gutters (layout), 346

habitual action, and present tense, 626
hackneyed expressions
    clichés as, 95
    trite language and, 643
hand gestures, 251
handbooks, in library research, 352–53
hasty generalizations, and logic errors,
    364–65
have, 626
he/she, 265
    case and, 491
    determining case of, 87, 492
    female and, 230
    gender and, 249
    male and, 368
    person and, 447
    sexist language and, 71–72
headers and footers, 265–66, 347
headings, 267–70
    in correspondence, 138
    decimal numbering system for, 269
    in formal reports, 242
    general style for, 267
    in instructions, 295
    italics for, 328–29
    layout and design and, 347
    in memos, 381–82
    in outlining, 432
    parallel structure of, 440
    in process explanations, 482
    in résumés, 558–59, 563
    in tables of contents, 242, 616
    in trade journal articles, 639
    in trip reports, 643
    Web page design and, 673
helping verbs, 455, 658
her, 447, 491
herewith, 33
hidden references, 488
highlighting, 346–48
highway numbers, 416
him, 447, 491
his, 447
historical events, 82
historical present, 626
history of a topic, as an opening, 426
history of words, 164
holidays, 82
home pages, 675–76, 680
honorable (as title), 4
horizontal axis, 257–60
hours, 415
house numbers, 417
house organ articles, 399–400

however, 31
humor, and malapropisms, 368
hyperbole, 233
hypertext links, 270–71, 354, 670–73,
    680
hyphenated nouns, plurals of, 412
hyphens, 271–74, 527
    in compound words, 118–19
    dashes and, 149–50
    with prefixes, 462
    proofreaders' mark for, 494

I
    antecedents and, 493
    between you and me usage and, 69
    case and, 491
    person and, 447
    point of view and, 466–67
    subjective case and, 85
icons
    global graphics and, 254
    Web page design and, 347–48, 672
idea development
    brainstorming for, 75–77
    methods of development and,
        382–85
    organization and, 428–29
    outlining and, 430–32
idioms, 275–77
    ambiguity and, 50–51
    clarity and, 93
    common pairings causing problems,
        275–77
    dictionaries with, 163, 166
    in international correspondence, 299
    phrasal verbs as, 465
    varieties of English and, 212
i.e./e.g., 194
-ies, 414, 459
if, 393
if and when, 676
IFB (invitation for bids), 512–20
illegal/illicit, 277
illogical constructions (sentence faults),
    590–92
illogical shifts (mixed constructions),
    387–88
illusion/allusion, 49
illustrations, 277–79
    descriptions with, 159
    drawings and, 188–90
    flowcharts and, 234–37
    global graphics and, 252–54
    graphs and, 257–64
    indexing of, 284
    instructions using, 293–94
    layout and design of, 349

in laboratory reports, 339
in lists of figures, 242
maps as, 368–70
newsletter articles with, 400
organization and planning for, 429
organizational charts as, 429
outlining and, 432
photographs as, 450
placement of, in text, 278
in presentations, 471–75
in process explanations, 482
schematic diagrams and, 575
tables and, 616
in technical manuals, 621
in trade journal articles, 638, 639
illustrations, lists of
in formal reports, 242
illustrations in, 279
image maps, 675
imaginative comparison (figures of
speech), 231
imminent/eminent, 202
imperative mood, 392
conciseness/wordiness and, 122
in instructions, 292
imperative sentences
construction of, 588
sentence variety using, 594
impersonal point of view, 466–67
implicit/explicit, 223
implied meanings (connotation/denota-
tion), 127
implied subjects
mood and, 392
sentence construction and, 586
imply/infer, 276, 279–80
importance, relative
emphasis and, 203–05
subordination and, 610–12
imprecise words
ambiguity and, 50–51
vague words and, 653
in/into, 280
in common/mutual, 394
in connection with/connected with, 126
in order to, 280
in re, 542
in regard to, 547–49
in spite of/despite, 161
in terms of, 280–81
inanimate objects, 460
Inc., 5
incident (trouble) reports, 644–45
incomparable (absolute) words, 9, 23
incomplete comparisons
ambiguity and, 50–51
comparison usage and, 112

incomplete sentences
missing subjects and, 591
missing verbs and, 592
sentence fragments and, 592–93
inconsistency
in logic errors, 364–65
in mixed constructions, 387–88
increasing-order-of-importance method
of development, 281–82, 385
indeed, 297
indefinite adjectives, 22
indefinite articles
a/an usage, 1
as adjectives, 20, 61–62
using, 63
indefinite pronouns, 490
agreement of subjects and verbs and,
44
number and, 492
one as, 421
possessive case of, 461
indentation, 283
of mathematical equations, 370–71
of quotations, 540–41
in Web page design, 673
independent clauses, 94
comma splice and, 102–03, 573
commas linking, 104, 107, 108
complex sentences with, 587
compound-complex sentences with,
588
compound sentences with, 587
conjunctions joining, 126
dangling modifiers and, 148–49
e.g./i.e. with, 194
run-on sentences with, 573
semicolons with, 583, 584
sentence length and use of, 594–95
simple sentence with, 587
index cards, in note taking, 408
indexes, 283–86
compiling, 284
cross-references in, 285–86
in formal reports, 245
Internet research using, 310
periodical, in library research,
354–55
in technical manuals, 621
wording entries in, 284–85
indexing, 283–86
indicative mood, 122, 392, 393
indirect communication, in letters,
133
indirect objects, 419
complements as, 117
nouns as, 410
indirect questions, 529

indirect quotations, 539
  commas with, 106
  paraphrasing and, 441
  quotation marks and, 536
indirect references, in allusion, 48
indirect word meanings
  (connotation/denotation), 127
indiscreet/indiscrete, 286
individual, 286–87
induction (specific-to-general method
  of development), 604
inductive (indirect) communication, in
  letters, 133
industrial specifications, 604–05
infer/imply, 276, 279–80
inferior type, proofreaders' mark for,
  494
infinitive phrases, 454
infinitives, 654–55, 660
  English as a second language (ESL)
    and, 207–08
  objective case of the pronoun sub-
    jects of, 85
  parallel structure repeating, 439, 440
  split, 655
  as substantives, 612
inflammable/flammable/nonflammable, 234
inflated words
  affectation using, 32–35
  long variants and, 366
inflection, 477, 479
informal definitions, 154
informal English
  informal writing style and, 609–10
  usage and, 651
informal memos, 381
informal reports, 550. *See also* memos
informal tables, 619
informal writing style
  defining terms in, 155
  using, 609–10
information, locating
  inquiry letters for, 287–88
  interviews for, 314–16
  library research and, 351–57
  presentations and, 468
  research and, 551–53
information, requesting
  inquiry letters for, 287–88
  interviews for, 314–16
  questionnaires for, 530–41
information-processing flowcharts,
  234, 237
informative abstracts, 12–13
-ing ending
  gerunds with, 653, 658
  present participles with, 656
  progressive tense with, 628

ingenious/ingenuous, 287
initialisms and acronyms, 17–18. *See
  also* abbreviations
  apostrophes in, 18
  in formal reports, 240, 243
  new words with, 399
  proofreading for, 495
  in titles, 633
initials
  in correspondence, 142
  in memos, 382
  periods with, 446
inoffensive words (euphemisms), 217
inquiries and responses, 287–91
  categories of, 287
  job searches and, 335
  responding to inquiries, 291
  writing inquiries, 287–89
insert marks, 494
inserted clauses or phrases, for emphasis, 595
inset text, 283
inside/inside of, 291
inside address, 138
insoluble/insolvable, 291
institutions, names of
  abbreviations for, 5
  ampersands in, 5
  capitalization of, 81
instructions, 292–96
  illustrations used in, 293–94
  lists for, 361
  passive voice in, 665
  point of view in, 466
  procedures as, 468–69
  sequential method of development
    for, 384, 596–97
  spatial method of development for,
    602–03
  testing, 296
  usability testing of, 649
  user manuals with, 620
  warnings used in, 295
  writing, 292–93
insure/ensure/assure, 296
intellectual property. *See* copyright
intensifiers, 296–97, 390
  conciseness/wordiness and, 122
  emphasis and, 204
  very and, 663
intensive pronouns, 490
intention, and sentence construction, 588
intention of a writer
  point of view and, 456–58
  purpose and, 527
intercultural communication, 299. *See
  also* global communication; international correspondence

interesting details, in openings, 425–26
interface, 297, 664
interjections, 297–98
  commas with, 105
  exclamation marks with, 218
  as a part of speech, 444
interlibrary loans, 353
internal correspondence. *See* memos
internal proposals, 497–501
international audiences, xv
international correspondence, 298–305
  cross-cultural examples of, 300–304
  culture and writing style and,
    298–99
  date format in, 151, 300, 416
  diacritical marks for, 161
  email conventions and, 201
  global communication and, 251–52
  global graphics and, 252–54
  language used in, 299–300
  number usage in, 416
  presenting good news and bad news
    and, 133
International Organization for Stan-
  dardization (ISO) symbols, on
  flowcharts, 235, 237
Internet, 305–09. *See also* World Wide
  Web
  citing sources on, 174, 176, 179
  communications using, 306–08
  connecting with other systems using,
    308–09
  copyrighted materials on, 129–30
  email on, 197–202
  netiquette in using, 200–202
Internet relay chat (IRC), 307–08
Internet research, 309–14. *See also* research
  evaluating sources in, 313
  indexes (subject directories) in,
    310–11
  job searches using, 333–34
  library research and, 353–54
  note taking during, 407
  search engines for, 310, 577–80
  technical information on the Web
    and, 312
interoffice communication. *See* memos
interpolation within quotations, 75
interpretations, multiple, and ambiguity, 50–51
interrogation
  interviewing for information and,
    314–16
  interviewing for a job and, 316–21
interrogative adverbs, 31
interrogative clauses, 529
interrogative pronouns, 489

interrogative sentences
  construction of, 588
  question marks for, 529
  sentence variety using, 594
  word order for, 620
interviewing for information, 314–16
  listening and, 359–61
  questionnaires compared with, 530,
    531
  research and, 552
  usability testing with, 651
interviewing for a job, 316–21
  conducting the interview, 318–19
  follow-up communications after,
    319–21
  preparing for the interview, 316–18
  résumés and, 557
interviews, citation formats for, 179
into/in, 280
intransitive verbs, 657
introductions, 322–25
  of activity reports, 486
  conclusions related to, 124
  in feasibility reports, 225–26
  in formal reports, 243
  in informal and short reports, 550
  in instructions, 293
  in literature reviews and, 363
  openings and, 423
  presentations with, 468–69
  purpose of, 322
  in progress reports, 484
  in proposals, 500, 503
  tone of, 634
  in trade journal articles, 322–25, 638
introductory clauses, 105, 108
introductory words and phrases
  commas with, 105, 108
  expletives as, 222
  infinitive phrases as, 454
  introductions with, 322
  transition with, 642
  wordiness and, 122
inverted commas. *See* quotation
  marks
inverted sentences, 589, 595
investigation
  interviewing for information and,
    314–16
  investigative reports for, 324–25
  library research and, 351–57
  questionnaires and, 530–41
  research and, 551–53
investigative reports, 325–26
  documenting sources in, 173
  laboratory reports as, 339
  trouble reports and, 644–45
invitation for bids (IFB), 512–20

IRC (Internet relay chat), 307–08
irregardless/regardless, 327
irregular verbs, 164
ISO symbols, on flowcharts, 235, 237
"is when" (or "is where") definitions,
    154–55
it, 325
    case and, 491
    as an expletive, 325
    person and, 447
    as a pronoun, 325
it is, 222, 586
italics, 327–29
    in email, 198
    emphasis with, 204, 327
    foreign words in English set in, 238
    in headings, 267, 329
    layout and design using, 347
    proofreaders' mark for, 494
    quotation marks used instead of, 328
    titles with, 328, 537
items in a series
    commas for, 106–07, 110
    etc. with, 213–14
    lists for, 361–62
    semicolons for, 108, 584
its/it's, 54, 128, 329
    case and, 491
    person and, 447

jammed (stacked) modifiers, 390–91
    awkwardness and, 66
    gobbledygook and, 255
    nouns used as adjectives and, 24
jargon, 330
    affectation and, 34
    blend words and, 74
    -ese suffix for, 213
    ethics in using, 215
    functional shifts and, 247
    international correspondence and,
        300
    revision for, 570
    spin-off as, 607
    stacked modifiers and, 390–91
job descriptions, 330–31
job interviews, 316–21
job offers, refusing, 321, 546
job search, 333–37
    acceptance letters and, 14–15
    application letters for, 56–61
    interviewing for a job and, 316–21
    résumés for, 557–67
job titles
    abbreviations for, 3–4
    application letters mentioning, 56
    capitalization of, 82–83

chair/chairperson and, 91
job descriptions for, 330–31
Ms./Miss/Mrs. usage with, 394
    in salutations, 141
    sexist language and, 71
jobs, finding
    acceptance letters, 14–15
    application letters, 56–61
    interviewing for a job, 316–21
    job search, 333–37
    resignation letters or memos, 553–54
    résumés, 557–71
joining words
    blend words for, 74
    compound words for, 118–19
    conjunctions for, 126
joint possession, 460
jokes, in email, 200
journal articles, 172. *See also* trade
    journal articles
    citation formats for, 178, 179, 184
    formal writing style for, 608–09
    indexes of, in library research, 354–55
    introductions for, 322–25
    openings for, 423–27
    research using, 552
journalism, 399
journals
    abstracts of articles in, 11–14
    citation formats for, 178, 179, 184
    indexes of, 354–55
Jr., 4
judicial/judicious, 337
    justification of margins, 346

key words
    abstracts using, 11–14
    electronic résumés with, 564
    indexing of, 284
    library research using, 352, 353–54
    repetition using, 549
    transition by repeating, 641
keys
    on drawings, 190
    on graphs, 258, 262
    on maps, 370
kind
    agreement of pronouns and an-
        tecedents and, 39–40
    demonstrative adjective with, 20–21
kind of/sort of, 338
know-how, 338

labels
    for drawings, 190
    for flowcharts, 235–37
    for graphs, 258, 260, 262, 264

for illustrations, 159
for maps, 369
in trade journal articles, 639
laboratory reports, 91, 339
lack of reason (logic error), 364
language
    absolute words, 9
    abstract words/concrete words,
        9–10
    affectation, 32–35
    antonyms, 53
    blend words, 74
    clichés, 95 96
    clipped forms of words, 96
    compound words, 118–19
    connotation/denotation, 127
    contractions, 128
    defining terms, 153–55
    diction, 162
    dictionaries, 163–67
    direct address, 168
    double negatives, 186–87
    elegant variation and, 194–95
    English, varieties of, 211–12
    euphemisms, 217
    foreign words in English, 237–38
    functional shift, 246–47
    glossaries, 254–55
    gobbledygook, 255
    idioms, 275–77
    intensifiers, 296–97
    jargon, 330
    long variants, 366
    malapropisms, 368
    new words, 399
    prefixes, 462
    suffixes, 613
    synonyms, 613–14
    thesaurus, 631
    trite language, 644
    usage, 651
    vague words, 653
    vogue words, 664
    word choice, 679–80
Latin words, 237–38
    abbreviations using, 5–6
    ad hoc, 19
    de facto/de jure, 153
    e.g./i.e., 194
    et al., 180
    etc., 213–14
    formula/formulae/formulas, 246
    quid pro quo, 536
    sic, 599
    via, 663–64
latter/former, 245
lay/lie, 339–43

layout and design, 343–49. See also for-
    mat; Web page design
    ethics in using, 215
    format and, 245
    highlighting devices for, 346–48
    of illustrations, 279
    organization and, 429
    page design and, 348–49
    persuasion and, 450
    placement of illustrations and, 278
    presentation visual aids and, 473–74
    proofreading for, 495
    technical writing style and, 622
    typography in, 343–45
    warnings in instructions and, 295
lead-ins (openings), 423–27
lead paragraphs, 400
lead (topic) sentences, 435–36, 438
leaders, 616
least, 23, 32
leave/let, 350
left-justified margins, 346
legal rights of an author
    copyright and, 128–30
    plagiarism and, 455–56
legal terms
    ad hoc, 19
    de facto/de jure, 153
    libel/liable usage and, 350–51
    party, 444
    quid pro quo, 536
    re, 542
legalese
    affectation using, 34
    aforesaid and, 36
    gobbledygook and, 255
legends
    on graphs, 258, 260
    on maps, 370
legislative documents, titles of, 328
lend/loan, 350
length
    of paragraphs, 436
    of press releases, 482
    of sentences, and emphasis, 203
    of sentences, for variety, 594–95
lesbians, and biased language, 73
less
    adjectives with, 23
    adverbs with, 32
    fewer/less usage and, 230
let/leave, 350
let-stand mark, 494
letter reports. See letters; memos; re-
    ports
letter types
    acceptance letters, 14–15

letter types (*cont.*)
  acknowledgment letters, 16
  adjustment letters, 23–28
  application letters, 56–61
  complaint letters, 114–15
  cover letters, 145–46
  email, 197–202
  inquiries and responses, 287–88
  memos, 378–82
  reference letters, 544–46
  refusal letters, 546–47
  resignation letters or memos, 553–54
letter writing. *See* correspondence;
  memos
letterhead, in correspondence, 138,
  581
letters, 130–46
  abbreviations for personal titles in, 4
  audience and, 130–33
  citation formats for, 178
  format of, 245
  formats and parts of, 137–44, 245
  informal writing style for, 609–10
  international correspondence and,
    298–301
  letterhead for, 138, 581
  memos, 378–82
  openings in, 424
  parts of, 138–44
  point of view in, 458
  presenting good news and bad news
    in, 133–35
  style and accuracy of, 135–37
  tone of, 130–33, 633
  "you" viewpoint in, 685
letters of acceptance, 14–15
letters of the alphabet
  in acronyms and initialisms, 15–17
  apostrophes for plurals of, 55
  capitalization of, 83
  italics for, 328
letters of application, 56–61
  job searches and, 336
  résumés and, 557
letters of inquiry
  application letters as, 56–61
  job searches and, 335
  writing, 287–88
letters of response, 288
letters of transmittal (cover letters),
  145–46
levels of heads
  in headings, 267
  in outlines, 432
lexicons. *See* dictionaries
liable/libel/likely, 350–51
libel/liable/likely, 350–51
librarians, reference, 351

library research, 351–57
  abstracts in, 11
  copyright and, 129–30
  literature reviews and, 363
  note taking and, 406–08
  paraphrasing and, 441–42
  plagiarism and, 455–56
lie/lay, 339–43
lightface type, 494
-like, 357
like/as, 357–58
like and unlike (comparison method of
  development), 113–14
likely/libel/liable, 350–51
limiting adjectives, 19–20
line graphs, 257–60
lingo (jargon), 330
linking verbs, 657–58
linking words
  conjunctions for, 126
  prepositions and, 463–66
links, with hypertext, 270, 670–73, 680
listening, 358–61
  interviewing for information and,
    318
  interviewing for a job and, 318–19
  meetings and, 377
lists, 361–62
  bulleted, 198–99, 361–62, 472,
    474–75
  colons with, 100
  etc. with, 213–14
  layout and design of, 349
  in memos, 381–82
  numbered, 446
  parallel structure of, 440–41
  in presentation visual aids, 472,
    474–75
lists of abbreviations and symbols, 243
lists of books and articles
  bibliographies for, 73–74
  documenting sources and, 173–75
  in formal reports, 243–44
  library research using, 353–54
lists of definitions
  dictionaries with, 163–67
  glossaries with, 254–55
lists of figures or illustrations
  in formal reports, 242
  illustrations listed in, 279
lists of items. *See* series of items
lists of references, 243–44
lists of symbols, in formal reports, 243
lists of tables, 242
lists of terms, 254–55
lists of topics
  indexes and, 283–86
  table of contents as, 616

lists of words
  dictionaries, 163–67
  glossaries, 254–55
  thesaurus, 631
literal meanings
  connotation/denotation of words
    and, 127
  defining terms and, 153–55
literally/figuratively, 231
literature, 362
literature reviews, 362–64, 637
literature searches
  Internet research for, 310
  library research using, 353–54
  research with, 551–53
litotes, 233
little, 23
loaded arguments (logic error), 366
loan/lend, 350
loan words, 237
localisms, 212
locating information
  Internet research for, 310
  library research for, 351–57
  research for, 551–53
logic errors, 364–66
  awkwardness and, 66
  double negatives and, 187
  ethics in writing and, 215
  mixed constructions and, 389
  outlining and detection of, 430
logical thinking
  coherence and, 96–97
  comparison and, 112
  defining terms and, 153–55
  faulty subordination and,
    591–92
  methods of development and,
    382–85
  mixed constructions and, 389
  organization and, 428–29
  outlining and, 430–32
long dashes. See em dashes
long sentences, and variety, 594–95
long variants, 366
  affectation and, 34
  gobbledygook and, 255
loose/lose, 367
loose sentences, 588, 595–96
lose/loose, 367
lowercase and uppercase letters, 367.
  See also capital letters
  in acronyms and initialisms, 18
  proofreaders' mark for, 494
lucid writing
  clarity and, 93–94
  coherence and, 96–97
-ly ending in adverbs, 32, 234, 272

M.A., 4
macron, 162
magazines. See also trade journal ar-
    ticles
  citation formats for articles in, 178,
    184
  indexes of, in library research,
    354–55
  informal writing style for, 609–10
  titles of, 328
magnification (hyperbole), 233
mailing lists, 306
main clause, 590
main idea statement (stating the subject)
  introductions and, 322
  outlining and, 431
  purpose and, 527
  topic sentences and, 435–36
main verbs, 455, 658
malapropisms, 368
male, 368
man
  he/she usage and, 265
  male usage and, 368
management communications
  informal and short reports, 549–50
  memos, 378–82
  policies and procedures, 456–58
  progress and activity reports, 482–84
manner, adverbs of, 30
manuals
  as documentation, 172
  library research using, 352–53
  style, 173, 185–86
  technical, 172
manuscript preparation
  format and, 245
  mathematical equations and, 370–71
  proofreaders' marks for, 495
  proofreading and, 495–96
  revision and, 567–70
  for trade journal articles, 639
  writing a draft, 682–83
maps, 368–70
  as illustrations, 277–79
  library research using, 356–57
  in lists of figures, 242
margins
  in correspondence, 138
  in layout and design, 346
mark of anticipation. See colons
mark of possession. See apostrophes
masculine nouns and pronouns
  agreement and, 37–38
  agreement of pronouns and an-
    tecedents and, 38–39
  gender and, 249
  he/she usage and, 265

masculine nouns and pronouns (*cont.*)
  male usage and, 368
  sexist language and, 70–72, 488
mass nouns, 410
  amount/number usage and, 51
  English as a second language (ESL)
    and, 205–06
  fewer/less usage and, 230
mathematical equations, 370–71, 639
may, 659
may/can, 78
maybe/may be, 372
M.D., 5
me
  between you and me usage and, 69
  objective case and, 85–86
  person and, 447
mean (average), 65
meaning
  ambiguity and, 50–51
  connotation/denotation of words
    and, 127
  defining terms and, 153–55
  dictionaries and, 163–67
  ethics in writing and, 214–15
  glossaries and, 254–55
  listening and, 359
  prefixes and changes in, 462
  symbols and, 613
  thesaurus and, 631
  vague words and, 653
measurements
  abbreviations for, 3
  in international correspondence, 300
  numbers for, 414–15
  in tables, 618
mechanical arrangement. *See* format
mechanisms, descriptions of, 158–59
mechanisms in operation (process ex-
    planation), 234–37
media/medium, 372
median/average, 65
medical articles, 363
medical terms, 157
medium/media, 372
meeting minutes, 385–89
  assigning task for, 375
  chronological method of develop-
    ment for, 91
meetings, 372–78
  conducting, 376–78
  conference calls as, 582
  planning, 372–75
  selecting, as a medium for communi-
    cation, 582
memo, 96
memos, 96, 378–83
  exposition in, 223

  with formal reports, 239
  format of, 383
  informal and short reports as, 550
  informal writing style for, 609–10
  international correspondence and,
    298–301
  letterhead for, 138, 581
  meeting agenda sent using, 374
  openings of, 423–27
  outlining, 431
  persuasion and, 448
  proposals as, 497
  reasons for selecting, 581
  subject lines of, 82, 383
  tone of, 633
  trip reports as, 643
  trouble reports as, 645
  writing, 379–83
Messrs., 4
metaphor, 233
method of development
  clarity and, 93
  "Five Steps to Successful Writing"
    on, xv–xvi
  organization and, 428
  outlining and, 431, 432
  Web page design and, 674
methods of development, 383–85
  cause-and-effect, 88–90
  chronological, 91
  comparison, 113–14
  decreasing-order-of-importance,
    151–52
  definition, 155–58
  division-and-classification,
    169–72
  general-to-specific, 250
  increasing-order-of-importance,
    281–82
  sequential, 596–97
  spatial, 602–03
  specific-to-general, 604
methods of discourse
  description and, 158–59
  exposition and, 223
  forms of discourse and, 245–46
  narration and, 395–98
  persuasion and, 448–50
metonymy, 233
metric system, 109, 300
might, 659
military format of dates, 150
military titles
  abbreviations for, 3–4
  capitalization of, 82–83
mind mapping, 76–77
mine, 447, 491
minor sentences, 588

minutes (time), 415
minutes of meetings, 385–89
chronological method of development for, 91
note taking during, 389
misplaced modifiers, 391
dangling modifiers and, 148–49
participial phrases and, 454
Miss
Ms./Miss/Mrs. usage and, 394
in salutations, 138–41
missing subjects in clauses, 590
missing words, ellipses for, 195–97
misused words (malapropisms), 368
mixed constructions, 389
mixed words
blend words and, 74
malapropisms and, 368
MLA style
documenting sources using, 173–79
ellipses in, 195
quotations in, 540–41
modal auxiliary verbs, 659
Modern Language Association (MLA) style
documenting sources using, 173–79
ellipses in, 195
quotations in, 540–41
modified-block letter style, 138
modifiers, 390–92. *See also* adjectives; adverbs
ambiguity and, 50–51
conciseness and overuse of, 122
dangling, 148–49, 665
jammed, 24
misplaced, 391
periodic sentences and placement of, 588, 595–96
restrictive and nonrestrictive, 555–56
squinting, 392
stacked (jammed), 24, 255, 390–91
wordiness from repetition of, 120
monthly, bi- or *semi-,* 70
monthly progress reports, 482–84
months
abbreviations for, 5
capitalization of, 82
commas with, 109
in correspondence headings, 138
date format for, 150–51
numbers for, 415
mood, 392–93
conciseness/wordiness and, 122
helping verbs and, 658
shifts in, 661
moon, 81
MOOs (Multi-User Dungeons Object Oriented), 308

more
absolute word comparison with, 213
as an adjective, 23
as an adverb, 32
more than/as much as, 62
most
absolute word comparison with, 213
as an adjective, 23
as an adverb, 32
almost/most usage and, 49
emphasis with, 204
movement, in presentations, 476–77
movie titles
capitalization of, 82
italics for, 328
Mr., 5
Mrs.
as an abbreviation, 5
Ms./Miss/Mrs. usage and, 394
in salutations, 138–41
Ms./Miss/Mrs., 5, 394
ms/mss, 6
much, 23
MUDs (Multi-User Dungeons), 308
multiauthor books, 176, 181, 183
multimedia dictionaries, 165
multiple interpretations, and ambiguity, 50–51
multiple words (compound words), 118–19
Multi-User Dungeons (MUDs), 308
Multi-User Dungeons Object Oriented (MOOs), 308
multivolume works
citation formats for, 176, 177, 182, 183
titles of, 632
musical works, titles of, 328
must, 659
mutual/in common, 394
my, 491
myself, 490

-n ending, 656
names
abbreviations for, 3–4
capitalization of, 79, 81–82
commas with, 109
different words (appositives) for, 61
italics for, 328
names, first
abbreviating, 3
in salutations, 141
naming (direct address), 168

narration, 395–98
  chronological method of develop-
    ment and, 91
  forms of discourse and, 246
  point of view and, 456–58
nationalities
  biased language and, 72–73
  capitalization of, 80
nature, 398
n.d., 176
needless to say, 398
negative definitions, 158
negative information
  positive presentation of, 458–59
  refusal letters and, 546–47
  resignation letters or memos and,
    553–54
negatives, double, 186–87, 325
neither, double negatives with, 187
neither . . . nor, 125
  agreement of pronouns and an-
    tecedents and, 41
  parallel structure of, 439
neo-, 399
neologisms, 399
nervousness, and presentations,
  478
netiquette, 200–202
networking, in job searches, 334
new words, 399
news releases, 480–82
newsgroups, 306–07
newsletter articles, 399–403
  electronic, citation formats for, 179,
    185
  illustrations in, 349
newspaper articles
  citation formats for, 178, 184
  indexes of, in library research, 354–55
newspaper titles, 327
no, 104
no doubt but, 403
no one, 404–05
nodes, with hypertext, 270
nominalizations, 404
nominative case, 85
nominatives, predicate
  complements and, 117
  nouns and, 410
  sentence construction and, 587
non sequiturs, 365
none, 404–05
nonfinite verbs, 654, 658–60
nonflammable/flammable/inflammable,
  234
nonliteral use
  figuratively/literally usage and, 231
  idioms and, 275

nonnative speakers, and idioms, 275
nonrestrictive clauses, 555–56
  commas for, 104–05, 555–56
  which used with, 556, 630
nonspecific (vague) words, 653
nonstandard English, 211–12
nor, double negatives with, 187
nor/or, 405
normative grammar, 256
north, 81
not
  double negatives with, 186–87
  in verb phrases, 455
not one, 404–05
not only . . . but also, 125, 439
notable/noticeable, 405
note cards
  note taking using, 408
  outlining using, 431–32
note taking, 406–08
  copyright and, 129–30
  interviewing for information and,
    317
  listening and, 360
  minutes of meetings and, 387
  paraphrasing and, 441–42
  plagiarism and, 455–56
  quotations and, 539
  research and, 553
notes, format of, in documenting
  sources, 173
notes (reminders), for presentation
  delivery, 474, 479
noticeable/notable, 405
noun phrases, 285, 451–52, 455
noun predicates, 117
noun stacks (jammed modifiers), 24,
  390–91
noun substitutes
  pronouns as, 489–93
  substantives as, 612
nouns, 409–12
  abstract words/concrete words as,
    9–10
  as adjectives, 23, 25, 410
  amount/number usage and, 51
  case of, 84–88
  clauses with, 94–95
  as a part of speech, 444
  phrases functioning as, 451–52, 455
  plurals of, 411–12
  possessive case of, 459–61
  pronouns modifying, 491–92
  as substantives, 612
  types of, 409–10
  usage of, 410
  verbals as, 653
nowhere near, 412

n.p., 176
number, 413, 660
  agreement and, 37–38
  agreement of pronouns and an-
    tecedents and, 39–40, 492, 493
  agreement of subjects and verbs and,
    46
  collective nouns and, 410–11
  pronouns and, 492, 493
  number/amount, 51
numbered lists, 446
numbering
  of figures and tables in formal re-
    ports, 242
  in outlines, 431–32
  in presentation visual aids, 475
  in procedures, 469–70
  of tables, 618
numbers, 413–17
  addresses and, 416–17
  average/median usage and, 65
  beginning sentences with, 414
  capitalization of units identified by, 84
  commas with, 109, 416, 417
  dates and, 415
  document structure and, 417
  few/a few usage and, 230
  fewer/less usage and, 230
  fractions and, 415
  italics for, 328
  mathematical equations with, 370–71
  measurements and, 414–15
  in parentheses, 416, 443
  in percentages, 415, 445
  plurals of, 414
  punctuating, 109, 416, 417
  spelling out, 414, 416, 443
  time and, 415
  using numerals or words for, 414
numeral adjectives, 22
numerical data
  graphs for, 257–64
  tables for, 616–19

object-verb-subject sentence order
  sentence construction and, 589–90
  sentence variety and, 594–96
objective case, 85–86
  determining, 87, 492
  pronouns and, 87, 490–91
objective complements, 117, 410
objective (impersonal) point of view,
  466–67
objectives of writers
  openings stating, 427
  preparation and, 462
  purpose and, 527
  on résumés, 559

objects, 418–19
  commas with, 110
  complements as, 117
  gerunds as, 653–54
  nouns as, 410
  prepositions and, 463
  observance/observation, 419
  observation, in research, 552
  o'clock, 415
of
  agreement of pronouns and an-
    tecedents and, 39–40
  demonstrative adjective with, 20–21
  possessive case using, 460, 461
  redundant use of, 288, 432,
    464–66
of which/whose, 678
officialese
  affectation and, 34
  gobbledygook and, 255
oh, 297
OK/okay, 420
omissions
  apostrophes for, 55
  commas showing, 107–08
  ellipses for, 195–97, 539–40
  proofreading for, 496
  quotations and, 540
  slashes for, 600
  telegraphic style and, 623
omitted letters
  apostrophes for, 53, 55
  contractions with, 128
  slashes for, 600
omnibus words, 653
on/onto/upon, 420
on account of, 420
on the grounds that/on the grounds of, 421
on to (on/onto/upon), 420
one, 421
  agreement of pronouns and an-
    tecedents and, 39
  agreement of subjects and verbs and,
    45
  point of view and use of, 466
one another, 490
one of the..., 43
one of those...who, 422
online card catalogs, 353–54
online documents. *See* CD-ROM resources; Internet research; Web
  page design
online searches. *See also* Internet research
  Gopher system for, 309
  library research using, 353–54
  search engines for, 310, 577–80
  Telnet used with, 308–09

online sources
    citation formats for, 176, 182–83,
        185
    Telnet for connecting to, 308–09
only, 31–32, 423
onto/on/upon, 420
openings, 423–27
    of correspondence, 130–31,
        133–35
    introductions and, 322
    of memos, 379–80
    paragraph length for, 436
    of presentations, 468–69
    of process explanations, 482
    of proposals, 497
    rhetorical questions as, 468, 573
    of test reports, 628
    tone of, 634
    writing a draft and, 682
operation of mechanisms (process ex-
    planation), 234–37
operators' manuals, 620
opinions, and logic errors, 366
opposite meanings (antonyms), 53
or. See also conjunctions
    agreement of pronouns and an-
        tecedents and, 41
    and/or usage and, 52
    nor/or usage and, 405
oral/verbal, 428
oral presentations. *See also* presenta-
    tions
    global communication and, 251–52
    global graphics and, 252–54
    increasing-order-of-importance
        method of development and,
        281–82
order of importance, 385
    in decreasing-order-of-importance
        method of development, 151–52
    in increasing-order-of-importance
        method of development, 281–82
order of words
    of modal auxiliary verbs, 659
    sentence construction and, 585–90
    sentence variety and, 595
    subject-verb-object pattern and, 586,
        589–90, 614
    syntax and, 614–15
ordered writing
    methods of development and,
        382–85
    organization and, 428–29
    outlining and, 430–33
ordering ideas
    organization and, 428–29
    outlining and, 430–33
ordinal adjectives, 22

ordinal numbers
    adding *-ly* suffix to, 234
    adjectives and, 22
    using numerals or words for, 415
organization, 428–29
    awkwardness and, 66
    Checklist of the Writing Process on,
        xviii
    drawings and, 188
    "Five Steps to Successful Writing"
        on, xv–xvi
    illustrations and, 278
    methods of development and,
        382–85
    outlining and, 431
    of paragraphs, 436
    of presentations, 468–71
    of résumés, 558–61
    of technical manuals, 621
    of trade journal articles, 637
    transition and, 642
organization names
    ampersands in, 52
    apostrophes in, 54
    capitalization of, 81–82
    in salutations, 141
    as singular nouns, 411
organizational charts, 429, 521
organizations
    abbreviations for, 5
    capitalization of names of, 81–82
    formal reports and, 242
    organizational charts for, 429
    press releases from, 480–82
organizing ideas
    methods of development and,
        382–85
    organization and, 428–29
    outlining and, 430–33
organizing reports
    methods of development for,
        382–85
    organization and, 428–29
    outlining and, 430–32
orient/orientate, 366, 430
origins of words, 164
orphans, in columns, 348
our, 447, 491
ours, 447, 491
outlining, 430–33
    brainstorming and, 76
    clarity and, 93
    coherence and, 96
    creating, 431–33
    drawings and, 188
    "Five Steps to Successful Writing"
        on, xvi
    method of development and, 385

note taking and, 408
organization and, 428
of paragraphs, 436–37
parallel structure and, 440
technical manuals and, 621
trade journal articles and, 637
transition and, 642
types of, 430–31
unity and, 648
writing a draft using, 683
outside of, 433
over with, 433
overhaul manuals, 621
overhead transparencies, 472
overloaded sentences
garbled sentences and, 248
pace and, 434
sentence construction and, 589–90
squinting modifiers and, 392
stacked modifiers and, 390–91
overstatement
affectation and, 32–35
gobbledygook and, 255
hyperbole and, 233
long variants and, 366
overtones (connotation/denotation), 127
ownership
adjectives and, 22
apostrophes for, 53–54
possessive case showing, 459–61
pronouns and, 491
ownership of written material
copyright and, 128–30
documenting sources and, 173
plagiarism and, 455–56

p., 180
pace, 434
clarity and, 93
coherence and, 96
presentations and, 477
revision for, 570
that usage and, 630
variety in sentence beginnings and, 595
packaging of technical manuals, 619
page design (online Web format), 669–76. *See also* format; layout and design
electronic résumés and, 562–64
graphics and typography in, 673
home pages and, 675–76
hypertext and, 270–71
navigation and links and, 670–73
page and site layout in, 674
photographs and, 450
typography and, 345

page design (paper format)
correspondence, 137–44
headers and footers and, 265–66, 347
layout and design of, 348–49
page margins, in correspondence format, 138
page numbers
citation format for, 174–75
footers with, 266
in formal reports, 241
number usage in, 416, 417
pamphlets, 175
papers (conference proceedings). *See also* trade journal articles
citation formats for, 177–78, 179, 184
research using, 552
titles of, 631–32
topic selection for, 634
paragraphs, 434–38
ellipses for omission of, 196–97
hypertext links in, 672
indentation of the beginning of, 283
lead, in newsletter articles, 400
length of, 436
outlining and, 431–32
proofreaders' marks for, 494
topic sentences in, 435
transition between, 641–43
transition within, 640–41
unity and coherence in, 437–38
writing, 436–37
parallel structure, 438–41
in abstracts, 13
both... and usage and, 74–75
faulty parallelism and, 440–41
headings and, 270
lists with, 361
outlining with, 432
sentence construction and, 590
writing style and, 610
paraphrasing, 441–42
copyright and, 129–30
documenting sources for, 173
listening and, 360
meeting closings and, 378
note taking and, 406–08
plagiarism and, 455–56
parentheses, 527
brackets used within, 75
commas after, 108
dashes compared with, 149–50
numbers in, 416, 443
periods with, 445–46
proofreaders' mark for, 494
punctuation with, 442–43
secondary information placed in, 396
symbols in, 613

parenthetical elements
  brackets enclosing, 443
  commas enclosing, 104–05
  dashes with, 149
  nonrestrictive phrase or clause as,
    555–56
  parentheses enclosing, 442–43, 584
parenthetical method of documentation, 173–75, 179–80
part-by-part method of comparison,
  113–14
participial phrases, 452–53
participles, 655–56, 660
  dangling modifiers and, 148–49
  sentence fragments and, 655–56
particles, 465
particular (specific)-to-general method
  of development, 604
parts (mechanisms), descriptions of, 158–59
parts of a book, capitalization of, 84
parts of a business letter, 138–44
parts of speech, 443–44
  adjectives, 19–23
  adverbs, 29–32
  conjunctions, 126
  dictionaries with, 163, 164
  functional shifts and, 246–47
  interjections, 297–98
  nouns, 409–12
  parallel structure and, 440–41
  prepositions, 463–66
  pronouns, 489–93
  verbs, 656–61
party, 444
passive voice, 664–68
  awkwardness from overuse of, 66
  conciseness and, 122
  ethics in using, 214
  infinitives and, 655
  problems with, 665–66
  revision for, 570
  technical writing style and, 623
  using, 666–67
past participles, 656
past perfect tense, 625, 660
past progressive tense, 624–25, 628
past tense, 625, 660
patterns, sentence
  sentence construction and, 585–90
  sentence variety and, 594–96
pedantic writing
  affectation and, 32–35
  gobbledygook and, 255
people, 411
people
  capitalization of names of, 79
  personal/personnel usage and, 447–48
  persons/people usage and, 448

per, 444, 600
percent/percentage, 415, 445
perfect
  as an absolute word, 9
  equal/unique/perfect usage and, 213
perfect participles, 656
performance standards, 520
period faults, 446–47
periodic progress reports, 482–84
periodic sentences, 588, 595–96
periodical indexes and guides, 354–56
periodicals
  bibliography citation format for, 182
  library research using, 354–56
periods, 445–47, 527
  after abbreviations, 3
  in decimals, 416, 417, 446
  ellipses with, 196
  gaps in tables signified by, 618
  incorrect use of (period faults),
    446–47
  leaders with, 616
  in numbers in global communica-
    tion, 417
  with parentheses, 442–43, 445–46
  proofreaders' mark for, 494
  quotation marks with, 445, 538
  quotations with, 445
periods, spaced
  ellipses, 195–97
  periods, 445–46, 527, 616, 618
periods of time, 82
permissible sentence fragments (minor
  sentences), 588
permissions
  copyright and, 128, 130
  for illustrations, 279
  in prefaces of formal reports, 243
  for trade journal articles, 640
person, 447, 660
  agreement and, 37–38
  gender and, 249
  he/she usage and, 265
  mixed constructions with, 387–88
  narration and, 395
  number and, 413
  point of view and, 456–58
  pronouns and, 493
personal/personnel, 447–48
personal data sheets (résumés),
  557–67
personal names
  abbreviations for, 4
  in salutations, 141
personal (subjective) point of view
  connotation/denotation of words
    and, 127
  writer's point of view and, 46–67

personal pronouns, 489
case of, 84–88
person of, 447
point of view and, 456–58
sexual bias and, 488
personal titles
abbreviations for, 4
capitalization of, 82–83
personality, in informal writing style, 609
personification, 233
personnel/personal, 447–48
persons/people, 448
perspectives of a writer, 456–58
persuasion, 448–50
forms of discourse and, 246
inquiry letters and, 287–89
logic errors and, 364
memos and, 379–80
presentations and, 470
proposals and, 496
"you" viewpoint and, 685–86
Ph.D., 5
phenomenon/phenomena, 450
phone, 96
phone calls, selecting as the medium for communication, 582
phonic marking (diacritical marks), 161, 162
photocopying
copyright and, 128–30
note taking and, 407
photographs, 450
layout and design and, 349
in lists of figures, 242
in newsletter articles, 400
in technical manuals, 621
in trade journal articles, 639
phrasal verbs, 465
phrases, 451–55
functions of, 451–52
introductory, with commas, 105
parallel structure of, 439
physical appearance. *See* format
physical movement, in presentations, 477
pictograms, 264
pictorial illustrations
icons, 254, 347–48
illustrations, 277–79
photographs, 450
picture graphs, 264
pie graphs, 262–63
place, adverbs of, 30
place names, 81
placement services, 334–35

plagiarism, 455–56
copyright and, 129
documenting sources to avoid, 173
listing references in formal reports to avoid, 244
note taking and, 406
paraphrasing and, 442
quotations and, 539
research and, 552
technical writing and, 216
plain English
tone and, 633–34
varieties of English and, 211–12
planets, 81
planning documents
collaborative writing and, 98, 635–36
drawings and, 188
organization and, 428–29
outlining and, 430–32
preparation and, 462–63
research and, 551
scope and, 576
trade journal articles and, 636–37
planning meetings, 372–75
play titles, 82
pluperfect tense, 625
plurals
of acronyms and initialisms, 18
agreement and, 37–38
agreement of pronouns and antecedents and, 38–39
apostrophes for, 54–55
of compound words, 119
data/datum usage and, 150
dictionaries with, 164
foreign words in English and, 238
gender and, 492
none usage and, 404
of nouns, 411–12
number and, 413
of numbers, 414
of personal pronouns, 447
possessive case of, 459
of pronouns, 413, 492
sexist language and use of, 70–72, 488
some usage and, 601
verbs and, 660
p.m., 5, 415
poem titles, 328
point sizes, in layout and design, 343–45
point of view, 456–58. *See also* "you" viewpoint
clarity and, 93

point of view (*cont.*)
  in correspondence, 130–31, 133
  in description, 158–59
  in drawings, 190
  narration and, 395
  paragraphs and, 437
  person and, 447
  pronouns and, 489
  readers and, 542–43
  tone and, 633–34
politeness in letters, 133
political divisions
  abbreviations for, 6–7
  capitalization of, 81
political titles
  abbreviations for, 4
  capitalization of, 82
pompous style
  affectation and, 32–35
  gobbledygook and, 255
popular words, 664
portmanteau words (blend words),
    74
position descriptions, 330–31
position of modifiers
  of adjectives, 23
  of adverbs, 31–32
  of dangling modifiers, 148–49
  squinting modifiers and, 392
position of words, and emphasis, 203
positive degree
  of adjectives, 21–22
  of adverbs, 30–31
positive writing, 458–59
  double negatives rephrased for, 187
  persuasion with, 449
  style and, 610
possession, marks of (apostrophes),
    56–57
possessive adjectives, 22
possessive case, 86, 459–61
  of adjectives, 22
  apostrophes for, 56–57, 459–61
  of compound words, 119
  determining, 87, 492
  of *it*, 328
  pronouns and, 490–92
possessive pronouns, 43, 460, 493
post hoc, ergo propter hoc, 365
postal abbreviations, 7–8, 138
pp., 180
practicable/practical, 462
practicing presentations, 475–76, 479
precis (abstracts), 11–14
preciseness
  abstract words/concrete words and,
    9–10

vague words and, 653
word choice and, 679–80
predicate adjectives, 23
predicate complements, 117
predicate nominatives
  complements and, 117
  nouns and, 410
  sentence construction and, 587
predicate nouns
  complements as, 117
  sentence construction and, 586–88
predicates
  complements in, 117
  nouns in, 410
  sentence construction and,
    586–88
prefaces, 242
prefixes, 462
  hyphens with, 272–73
  long variants with, 366
preliminaries (preparation), 462–63
preparation, 462–63
  Checklist of the Writing Process on,
    xviii
  "Five Steps to Successful Writing"
    on, xi–xiv
  purpose and, 527
  readers/audience and, 542
  research and, 551
  scope and, 576
  specifications and, 605
  topic selection and, 634
prepositional phrases, 451–52
  commas with, 110
  passive voice and, 665–66
prepositional verb, 465
prepositions, 463–66
  as used as, 62
  capitalization of, in titles of works,
    82
  common errors using, 464–66
  functions of, 463–64
  idioms and, 275–77
  objects of, 410, 419
  parallel structure repeating, 439
  as a part of speech, 444
  sentences ending with, 466
prescriptive grammar, 256
present participles, 656
present perfect tense, 626, 661
  English as a second language (ESL)
    and, 208–09
  infinitives and, 654–55
  progressive form of, 624–25
present progressive tense, 209, 624–25,
    628
present tense, 624, 625–26, 654–55, 661

presentations, 466–80
    anxiety and, 478
    audience analysis for, 467
    delivering, 475–79
    global communication and, 251–52,
        479–80
    global graphics and, 252–54
    information gathering in, 468
    purpose of, 467
    structuring, 468–71
    visual aids used in, 471–74
president, 82
press releases, 479–81
pretentious writing
    affectation and, 32–35
    gobbledygook and, 255
    jargon and, 330
    long variants and, 366
    revision for, 570
primary research, 552
principal/principle, 482
printing. *See also* typography
    proofreaders' marks and, 495
    typeface and type size choices and,
        344
priority/prioritization, 366
problem reporting
    activity reports for, 484, 486
    complaint letters for, 114–15
    trouble reports for, 644–45
problem reports
    chronological method of develop-
        ment for, 91
    trouble reports as, 644–45
problem-solution development (cause-
        and-effect method of develop-
        ment), 88–90
problem-solution pattern
    for presentations, 470
    for proposals, 500
problems statements in openings, 425
procedures
    process explanations and, 482
    writing, 468–69
proceedings
    citation formats for, 177–78, 179, 184
    research using, 552
process descriptions
    flowcharts for, 234–37
    process explanation for, 482
process explanation, 482
product descriptions
    in proposals, 505
profanity (expletives), 223
professional groups
    jargon used by, 330
    job searches and networking in, 334

professional journals. *See also* trade
        journal articles
    job searches and, 336
professional titles
    abbreviations for, 4
    capitalization of, 82–83
    chair/chairperson usage and, 91
    Ms./Miss/Mrs. usage with, 394
prognosis/diagnosis, 162
program reference manuals, 620–21
programs (computer software)
    citation formats for, 178, 184
    desktop publishing and, 245, 343–49
    flowcharts of, 234, 237
    presentations with, 473–74
    reference manuals for, 620–21
    spell checker with, 606
progress report letters, 424
progress reports, 424, 482–87
progressive tense, 624–25, 628
projection, in presentations, 477
projects
    activity reports on, 484–86
    progress reports on, 482–84
    specifications for, 605
prolixity (wordiness), 119–22
pronoun-antecedent agreement,
        36–41
    ambiguity and, 50–51
    pronoun reference and, 486–88
pronoun reference, 487–88
    agreement of pronouns and an-
        tecedents and, 38–43
    ambiguity and, 50–51
    revision for unclear, 570
pronouns, 488–93
    agreement of antecedents and,
        38–43
    appositives and, 61
    as used as, 62
    case of, 84–88, 490–92
    gender and, 249, 492
    nouns modified by, 492
    number and, 413, 492–93
    parallel structure repeating, 439
    as a part of speech, 444
    person and, 447, 493
    plurals of, 413
    point of view and, 456–58
    possessive case of, 459, 461
    as substantives, 612
    telegraphic style and omission of, 623
    types of, 489–90
pronunciation
    diacritical marks and, 161
    dictionaries with, 163, 164
proofreaders' marks, 494, 495

proofreading, 494–96, 570
    international correspondence and,
        300
    proofreaders' marks for, 494, 495
    spelling checked during, 606–07
proper names
    capitalization of, 79
    italics for, 328
proper nouns, 79, 409
proportion, in drawings, 190
proposals, 496–525
    appendixes in, 56
    cover letters for, 147
    documenting sources in, 173
    executive summaries for, 219–20
    government proposals, 512, 520–25
    persuasion and, 448
protectorates, abbreviations for, 6–7
protocols, 651
proved/proven, 526
provinces, abbreviations for, 8
pseudo-/quasi-, 526
public documents (government publi-
        cations)
    citation formats for, 175, 181
    copyright and, 130
    library research using, 356–57
public domain, and copyright, 129
public speaking. *See* presentations
publication of articles, 639–40
publication clearances, 639–40
publication information
    citation formats for, 176, 182
    copyright in, 130
    for formal reports, 240–41
    parentheses for, 443
punctuation
    for abbreviations, 3
    for acronyms and initialisms, 16, 18
    for adjectives, 23
    for adverbs, 31
    comma splice and, 102–03, 573
    for conjunctions, 104
    in date formats, 150–51
    e.g./i.e. and, 194
    global graphics and, 254
    for items in a series, 106–07, 108,
        110, 584
    with mathematical equations, 371
    numbers, 109, 416, 417
    parentheses ending a sentence and,
        442, 445
    sentences within parentheses and,
        442–43, 446
punctuation marks, 526–27
    apostrophes, 53–55
    brackets, 75

colons, 100–102
commas, 103–10
dashes, 149–50
ellipses, 195–97
exclamation marks, 218–19
hyphens, 271–74
parentheses, 442–43
periods, 445–47
question marks, 529–30
quotation marks, 536–48
semicolons, 583–85
slashes, 600
purple prose
    affectation and, 32–35
    gobbledygook and, 255
purpose, 527–28
    abstracts on, 11
    clarity and, 93
    coherence and, 96
    conclusions and, 123
    feasibility reports and, 225
    "Five Steps to Successful Writing"
        on, xi–xii
    of formal reports, 240
    illustration and, 277–78
    of informal and short reports, 550
    interviewing for information and,
        314
    introductions and, 322
    level of abstraction and, 9
    listening and, 359
    of meetings, 372–73
    memos and, 381
    preparation and, 462
    of presentations, 467
    scope and, 575
    selecting the medium and, 580
    titles and, 631
    tone and, 633
    unity and, 638
    usability testing and, 651
    Web page design and, 669–70

qualifiers. *See* modifiers
qualities, capitalization of, 79
quasi-/pseudo-, 526
query letters, 287–88
question marks, 254, 527, 529–30
questioning as research
    interviewing for information and,
        314–16
    research and, 552
questionnaires, 530–41
    interviewing for information and,
        314
    logic errors and, 365
    questions used in, 531–32

research using, 551
  selecting recipients for, 532–36
questions
  adverbs for, 31
  inquiry letters with, 287–88
  interviewing for information and,
    314–16
  interviewing for a job and, 316–21
  inverted sentence order for, 589
  listening and, 360
  mood and, 392–93
  pronouns and, 489
  in questionnaires, 531–32
  verb phrases beginning, 455
questions, rhetorical, 573
  as openings, 468, 573
  as titles, 400, 573
quid pro quo, 536
quite, 296–97
quitting a job, 553–54
quotation marks, 527, 536–38
  citation format with, 174, 180
  colon placement and, 101–02
  commas with, 108
  other punctuation used with,
    537–38
  periods with, 445, 538
  proofreaders' mark for, 494
  question marks with, 530
  semicolons with, 537–38, 585
  titles in, 327
quotations, 538–41
  capitalization of the first word of,
    80
  citation formats for, 180
  commas with, 106, 445
  copyright and, 129–30
  documenting sources for, 173, 180
  ellipses in, 195–97
  indentation of, 283
  as openings, 426, 469
  paraphrasing and, 441–42
  periods with, 446–47
  plagiarism and use of, 455–56
  question marks with, 530
  quotation marks for, 536–37
  sic in, 599
quoting copyrighted material, 129–30
  note taking and, 406
  plagiarism and, 455–56
  quotations and, 539

races
  biased language and, 72–73
  capitalization of, 80
raise/rise, 542
rambling sentences, 119–22, 591–92

rank (military)
  abbreviations for, 4
  capitalization of, 82–83
rather, 296
re, 542
readers, 542–43
  defining terms for, 153–55
  "Five Steps to Successful Writing"
    on, xii
  for formal reports, 240
  global communication and, 251–52
  introductions to technical manuals
    and, 324
  jargon and, 344
  point of view and, 456–58
  preparation and needs of, 462
  presentations and analysis of, 467
  purpose and, 527
  reports and, 550
  scope and, 575–76
  segmenting documents for, 543
  technical manuals and, 621
  telegraphic style and, 623
  visualizing a typical member of, 543
  "you" viewpoint and, 685–86
really, 204, 543
reason, lack of, and logic errors, 364
reason is because, 544
reciprocal pronouns, 490
recommendations
  conclusions and, 123
  in feasibility reports, 226
  in formal reports, 243
  in informal and short reports, 550,
    551
  in investigative reports, 325
  openings summarizing, 427
  in proposals, 512
  reference letters for, 544–46
recording information. See note taking
recording minutes of meetings. See
    meeting minutes
reducing ideas
  in note taking, 406–08
  paraphrasing and, 441–42
redundancy
  conciseness/wordiness and, 120, 122
  of usage and, 288, 432, 464–66
  one usage and, 421
  prepositions and, 464–66
  repetition and, 549–50
  -size/-sized usage and, 599
  in titles, 632
  up with verbs and, 651
refer/allude/elude, 48
reference, indirect (allusion), 48, 299
reference, and pronouns, 486–88

reference books
  dictionaries, 163–67
  library research using, 351–53, 355–56
reference letters, 544–46
reference librarians, 351
reference lists, 180–82
reference works
  library research using, 351–53,
    355–56
  scholarly abbreviations in, 6–7
references
  commas in, 109–10
  in formal reports, 243–44
  résumés with, 561
  in trade journal articles, 639
references cited
  bibliographies with, 73–74
  citation format for, 180–85
reflexive pronouns, 490
refusal letters, 546–48
  international correspondence and,
    299
  interviewing for a job and, 321
regarding/with regard to, 548–49
regardless/irregardless, 327
regards, 547–49
rehearsing presentations, 475–76,
  479
rejection letters, 133–34
relationship of words
  grammar and, 256–57
  syntax and, 614–15
relationships, family, 83
relative adjectives, 120
relative importance
  emphasis and, 203–05
  subordination and, 610–12
relative pronouns, 489
  agreement of pronouns and an-
    tecedents and, 43
  agreement of subjects and verbs and,
    43–44
  clauses with, 95
  in restrictive clauses, 630
  sentence fragments and, 592–93
  subordination using, 612
  wordiness and, 120
religious groups
  biased language and, 72–73
  capitalization of, 80
religious titles
  abbreviations for, 3–4
  capitalization of, 82–83
remainder/balance, 67
remote pronoun reference, 488
rendering (illustrations), 277–79
repeated material in columns, 543

repetition, 549–50
  awkwardness and, 66
  dashes and, 149
  emphasis and, 203–04
  of that, 630
  transition using, 641
  wordiness from, 120
rephrasing
  paraphrasing and, 441–42
  revision and, 570
replying to email, 199, 202
report elements, 239–45
  abstracts, 11–14, 242
  appendixes, 56, 244
  bibliography, 73–74, 244
  conclusions, 123–25, 243
  cover letters, 144–47
  executive summaries, 219–20, 243
  forewords, 242
  glossaries, 244, 254–55
  headings, 267–70
  illustrations, 277–79
  indexes, 245
  introductions, 243, 322
  lists of abbreviations and symbols,
    243
  lists of figures, 242
  lists of tables, 242
  prefaces, 242
  recommendations, 243
  references, 243–44
  tables of contents, 242
  text, 243
  title pages, 240–41
  titles, 631–32
report titles, 631–32
  italics for, 327
  quotation marks for, 537
report types
  abstracts, 11–14
  activity reports, 484
  executive summaries, 219–20
  feasibility reports, 225–27
  formal reports, 239–45
  investigative reports, 325–26
  laboratory reports, 339
  memos, 378–82
  presentations, 466–80
  progress reports, 482–87
  proposals, 496–521
  trip reports, 643
  trouble reports, 644–45
reports, 549–50. See also formal reports
  citation formats for, 177, 183
  as documentation, 172
  elements of, 239–45
  executive summaries for, 219–20

e.g./i.e. and, 194
fused (run-on) sentences and, 447,
     573–74
proofreaders' mark for, 494
quotation marks and, 537–38, 585
sentence beginnings
     also used for, 48
     capitalization of first words in, 80,
          83–84
     conjunctions in, 126
     ellipses and, 195–97
     expletives for, 222
     numbers and, 414
     sentence variety and, 595
sentence construction, 585–90
     awkwardness and, 66
     comma usage and, 104
     faulty subordination and, 591–92
     garbled sentences and, 248
     hypertext links and, 670–73
     in international correspondence,
          299–300
     mixed, 389
     parallel structure and, 438–41
     predicates and, 586–88
     rambling sentences and, 591–92
     sentence fragments and, 592–93
     sentence types classified by, 587–88
     sentence variety and, 594–96
     style and, 607–10
     subjects and, 585, 586
sentence endings
     abbreviations and, 446
     exclamation marks for, 218–19
     periods for, 445, 446
     prepositions and, 466
     question marks for, 529
sentence faults, 590–92
     faulty subordination and, 591–92
     missing subjects in clauses and, 591
     missing verbs and, 592
     rambling sentences and, 591–92
sentence fragments, 592–93
     clauses compared with, 94
     missing verbs and, 592
     participles and, 655–56
     period faults and, 446–47
sentence length
     emphasis and, 203
     variety and, 594–96
sentence openings
     also used for, 48
     capitalization of first words in, 80,
          83–84
     conjunctions in, 126
     ellipses and, 195–97
     expletives for, 222

numbers and, 414
sentence variety and, 594–95
sentence outlines, 430–31
sentence structure
     garbled sentences and, 248
     sentence classification and, 587
sentence types
     emphasis and selection of, 204
     sentence construction and, 587–88
sentence variety, 594–96
     inverted sentences and, 589, 595
     loose sentences and, 588, 595–96
     periodic sentences and, 588, 595–96
     sentence construction and, 589–90
     sentence length and, 594–95
     subordination and, 611
     style and, 607–10
sentence word order
     adjective placement and, 23, 25
     inverted sentences and, 589, 595
     modal auxiliary verbs and, 659
     sentence construction and, 585–90
     sentence variety and, 595
     subject-verb-object pattern and, 586,
          589–90, 614
     syntax and, 614–15
sentences
     classification of, 587–88
     inverted, 589, 595
     length of, 594–95
     loose, 588, 595–96
     periodic, 588, 595–96
     transition between, 641
     without subjects, 590–92
     without verbs, 592
     word order in, 595
separating items in a series
     commas for, 106–07, 110
     semicolons for, 108, 584
separating words
     dictionaries for, 163–67
     hyphens for, 274
sequence
     in drawings, 190
     ordinal adjectives and, 22
sequence of events, 91
sequence of ideas
     coherence and, 96
     sequential method of development
          and, 384, 596–97
sequence of tenses, 624
sequential method of development,
          384, 596–97
series, 44
series (publishing), 176, 177, 182, 183
series of items
     commas in, 106–07, 110

series of items (*cont.*)
  etc. with, 213–14
  question marks with, 529
  semicolons in, 108, 584
serif typefaces, 345
service, 597
service manuals, 620
set/sit, 598
set in text, 283
setup manuals, 621
sexism in language, 598–99. *See also* biased language
  agreement of pronouns and antecedents, 38–39
  chair/chairperson and, 91
  ethics in writing and, 215
  everybody/everyone usage and, 217–18
  female and, 230
  gender and, 249
  he/she usage and, 265
  male and, 368
  masculine personal pronouns and, 488
  Ms./Miss/Mrs. usage and, 394
  revision for, 570
  salutations and, 141
sexual orientation, and biased language, 73
shading
  on graphs, 258, 262, 263
  in layout and design, 348
  on maps, 370
shall/will, 599, 627
she
  case and, 87, 491
  determining case of, 87, 492
  gender and, 249
  he/she usage and, 265
  person and, 447
  sentence variety with, 596
  sexist language and, 71–72
shift
  functional shift, 246–47
  mixed constructions and, 387–88
  in mood, 661
  in person, 447
  in tense, 628, 661
  in voice, 661
shilling signs (slashes), 600
ships, names of, 328
shop talk, 330
short sentences, and variety, 594–95
shortened word forms. *See also* abbreviations
  clipped forms of words and, 96
  contractions and, 128
  functional shifts and, 246–47

shorthand, in note taking, 407
should, 659
sic, 75, 599
sight/cite/site, 91–92
signature blocks, in email, 199–200
signatures
  on correspondence, 142
  on memos, 382
  on minutes of meetings, 386
similar meanings (synonyms), 613–14
similarities, in comparison method of development, 113–14
simile, 233
simple predicates, 586
simple sentences
  construction of, 587
  emphasis and, 204
  length of, 594–95
  sentence variety using, 594
since, 68
single quotation marks, 537
singular
  agreement and, 37, 43
  data/datum usage and, 150
  number and, 413
  of personal pronouns, 447
  some usage and, 601
sit/set, 598
site/cite/sight, 91–92
site preparation descriptions, 512
-size/-sized, 599
skills section, on résumés, 561
slang, 212
  dictionaries listing, 165
  jargon and, 330
  quotation marks and, 537
  writing style and, 608, 610
slant lines (slashes), 600
slashes, 415, 527
slides, in presentations, 472–73
small capital letters, 494
small letters (lowercase letters), 367
so/so that/such, 600–601
so-called, 537
social groups, 80
social varieties of English, 211–12
software
  citation formats for, 178, 184
  desktop publishing and, 245, 343–49
  flowcharts of, 234, 237
  presentations with, 473–74
  reference manuals for, 620–21
  spell checker with, 606
solicited proposals, 501–03
solidus (slashes), 600
solutions (cause-and-effect method of development), 88–90

some, 601
some/somewhat, 601
some time/sometime/sometimes, 602
somewhat/some, 601
song titles, 328
sort (singular and plural forms)
    agreement of pronouns and an-
        tecedents and, 40, 41
    demonstrative adjective with, 20–21
sort of/kind of, 338
sound values for letters (diacritical
    marks), 161
source lines
    in graphs, 260
    in tables, 618
sources, citing
    bibliographies for, 73–74
    documenting sources and, 173–75
sources of information
    interviews for, 314–16
    in library research, 351–57
    questionnaires as, 530–41
    research and, 551–53
    technical, on the Web, 312
south, 81
space, inserting, 494
spaced periods
    ellipses, 195–97
    periods, 445–46, 527, 616, 618
spacing
    correspondence format and, 138
    layout and design using, 348
    Web page design using, 673
spatial method of development, 384,
    602–03
speaking. *See* presentations
special events, 81–82
special-purpose manuals, 620–21
special-purpose (investigative) reports,
    324–25
specie/species, 603
specific words
    abstract words/concrete words and,
        9–10
    word choice and, 679–80
specific-to-general method of develop-
    ment, 385, 604
specifications, 604–06
    introductions to, 322–23
    invitation for bids (IFB) and, 520
speech, figures of, 231–33
    in international correspondence, 299
    slang with, 212
    technical writing style and, 622
    types of, 232–33
speech, parts of, 443–44
    adjectives, 19–23

adverbs, 29–32
    conjunctions, 126
    dictionaries with, 163, 164
    functional shifts and, 246–47
    interjections, 297–98
    nouns, 409–12
    parallel structure and, 440–41
    prepositions, 463–66
    pronouns, 489–93
    verbs, 656–61
speeches. *See* presentations
speed of presenting ideas, 434, 478
spell-checker software, 606
spell-out mark, 494
spelling, 606–07
    clipped forms of, 96
    dictionaries for, 163, 164
    hyphens between letters for, 273
    proofreading for, 494–96
    spell checker software for, 606
spelling out numbers, 414, 415, 416,
    443
spin-off, 607
splice, comma, 102–03, 573
split infinitives, 655
spoken English, varieties of, 211–12
squinting modifiers, 392
Sr., 5
stacked (jammed) modifiers, 390–91
    awkwardness and, 66
    gobbledygook and, 255
    nouns used as adjectives and, 23
staff (they/it), 411
stale expressions
    clichés and, 95–96
    trite language and, 643
standard English, 211
start-to-finish (sequential) approach
    chronological method of develop-
        ment and, 91
    narration and, 395–96
    sequential method of development
        and, 596–97
starting documents (openings), 423–27
    of correspondence, 130–31
    introductions and, 322
    of memos, 379–80
    paragraph length for, 436
    preparation and, 462–63
    of presentations, 468–69
    of process explanations, 482
    of proposals, 497
    rhetorical questions as, 468, 573
    tone of, 634
    writing a draft and, 682
starting reports
    abstracts and, 11

starting reports (*cont.*)
  executive summaries and, 219–20,
    243, 550
  introductions for, 322
  openings for, 423–27
starting sentences
  also used for, 48
  conjunctions for, 126
  ellipses and, 195–97
  expletives for, 222
  numbers and, 414
  sentence construction and, 590
  sentence variety and, 594–95
statements of finance, 53
statements of the problem, 425
statements of responsibilities, 512
statements to stockholders, 52
states
  abbreviations for, 7
  capitalization of, 81
  in correspondence headings, 138
statistical information
  graphs for, 257–64
  library research using, 356–57
  in proposals, 512
  in tables, 616–19
status reports
  activity reports, 484–86
  progress reports, 482–84
step-by-step (sequential) method of development, 384, 596–97
steps, in instructions, 292
stereotypes, and biased language,
  70–72
stet (let stand proofreaders' mark),
  494
stockholders, statements to, 52
stories, and narration, 395–96
strata/stratum, 607
street numbers, 416
structure of sentences, 587–88
structure of writing
  organization and, 428–29
  outlining and, 430–32
  parallel structure and, 438–41
  presentations and, 468–71
  sentence construction and, 585–90
stubs, in tables, 618
style, 607–10
  correspondence and, 135–37
  email and, 198
  formal, 608–09
  informal, 609–10
  international correspondence and,
    298–99
  memos and, 379–81
  newsletter articles and, 400
  sentence length and, 594–95

telegraphic style and, 623
  tone and, 633–34
style guides and manuals
  abbreviations in, 3
  documenting sources using, 173,
    185–86
  trade journal articles and, 639–40
style sheets
  for trade journal articles, 639–40
  Web page design using, 673
subentries, in indexes, 283–86
subheadings
  italics for, 329
  Web page design and, 673
subject dictionaries, 352
subject encyclopedias, 352
subject lines
  capitalization of, 82
  in correspondence, 141
  in email, 199
  in memos, 383
subject searches
  in Internet research, 310
  in library research, 353–54
  search engines for, 310, 577–80
subject-verb agreement, 43–45
subject-verb-object pattern, 586,
  589–90, 614
subjective case, 85
  determining, 87, 492
  pronouns and, 87, 490–92
subjective complements, 85, 118
  agreement of subjects and verbs and,
    44–45
  gerunds as, 653–54
  nouns as, 410
subjective overtones (connotation/
  denotation), 127
subjective (personal) point of view
  connotation/denotation of words
    and, 127
  writer's point of view and,
    46–67
subjects (topics) of reports, 634
subjects of sentences
  agreement of verbs and, 42–45
  anticipatory (expletives), 222–23
  dangling modifiers and, 148–49
  missing, in clauses, 591
  missing, in sentence faults, 592
  nouns as, 410
  sentence construction and, 585, 586,
    589–90
  sentence fragments with missing,
    592–93
subjunctive mood, 392–93
submitting an article for publication,
  173, 639–40

subordinate clauses, 94–95
  commas with, 108
  dangling modifiers and, 148–49
  loose sentences with, 588, 595–96
  sentence fragments and, 592–93
subordinate phrases
  loose sentences with, 588, 595–96
  sentence fragments, 592–93
subordinating conjunctions, 126
  clauses with, 95
  sentence fragments and, 592–93
  subordination using, 612
subordination, 611–12
  in abstracts, 13
  awkwardness and, 66
  clarity and, 93
  clauses and, 94
  conciseness using, 121
  faulty, 591–92
  sentence construction and,
    589–90
  sentence length and, 594–95
  writing style and, 608, 610
substandard English, 211–12
substantives, 612
substitute words (euphemisms), 217
subtitles
  in formal reports, 240
  in heading structures, 267–70
such/so/so that, 600–601
such as, 593
suffixes, 613
  -ese, 213
  hyphens with, 273
  -like, 357
  long variants with, 366
  -wise, 678
summaries. See also conclusions
  abstracts as, 11–14
  executive summaries as, 219–20
  openings providing, 427
summarizing information
  abstracts for, 11–14
  conclusions and, 123–24
  executive summaries for, 219–20
  minutes of meetings and, 387
  note taking and, 406–08
  paraphrasing and, 441–42
sun, 81
superfluous details
  revision for, 570
  wordiness from, 119–21, 122
superior type, proofreaders' mark for,
  494
superlatives
  adjectives, 23
  adverbs, 30–31
supplement/augment, 65

supplementary material, in formal reports, 244
suppressed evidence, 365
surveys
  investigative reports and, 324–25
  questionnaires for, 530–41
  research using, 551
  usability testing with, 651
sweeping generalizations, 364–65
symbols, 613
  diacritical marks and, 161
  flowcharts with, 234–37
  fonts containing, 344, 345
  global graphics and, 252–54
  maps with, 370
  picture graphs with, 264
  proofreaders' marks using, 495
  punctuation as, 526–27
  in tables, 618
  in technical manuals, 621
symmetry, and parallel structure,
  438–41
symposium papers, 178, 184
synonyms, 613–14
  dictionaries with, 163, 164
  elegant variation with, 194
  euphemisms and, 217
  informal definitions with, 154
  lists of, in formal reports, 243
  thesaurus with, 631
  word choice and, 679
synopses
  abstracts as, 11–14
  executive summaries as, 219–20
syntax, 614–15

-t ending, 656
table of contents, 616
  appendixes listed in, 56
  in formal reports, 242
  headings in, 269
  parallel structure of, 440
table numbers
  illustrations with, 278–79
  number usage in, 416, 618
tables, 616–19
  elements of, 621–24
  in email, 198–99
  graphs compared with, 257, 263
  as illustrations, 277–79
  indexing of, 284
  informal, 619
  in laboratory reports, 339
  in lists of tables, 242
  in technical manuals, 621
  in trade journal articles, 638
tables, lists of
  in formal reports, 242

tables, lists of (*cont.*)
  illustrations listed in, 279
tabular illustrations
  illustrations, 277–79
  tables, 616–19
taking notes, 406–08
  copyright and, 129–30
  interviewing for information and, 317
  listening and, 360
  minutes of meetings and, 387
  paraphrasing and, 441–42
  plagiarism and, 455–56
  quotations and, 539
  research and, 553
tales (anecdotes), as openings, 426
technical manuals, 172, 619–21
  headers and footers in, 265–66
  introductions for, 322–23
  technical writing style and, 622
  types of, 620–21
  usability testing of, 649
technical writing style, 621–23
  contractions and, 128
  correspondence style compared with, 130
  specifications and, 604–06
technology, workplace
  email, 197–202
  fax, 224–25
  hypertext, 270–71
  Internet, 301–09
  Internet research, 309–14
  Web page design, 669–76
  World Wide Web, 309, 680–82
telegraphic style, 623
  articles and, 19–20, 61–62
  conciseness and, 121
telephone calls, selecting as the
    medium for communication, 582
telling a story (narration)
  introducing presentations by, 469
  narration and, 395–96
Telnet, 308–09
templates, for form letters, 137
tenant/tenet, 624
tense, 624–30, 660
  agreement of subjects and verbs and, 46
  helping verbs and, 658
  mixed constructions and, 387–88
  narration and, 395–96
  progressive form of, 624–25, 628
  shift in, 628, 658
terms
  connotation/denotation of, 127
  defining, 154–55

dictionary definitions of, 163–67
  glossary definitions of, 254–55
territories, abbreviations for, 7
test reports, 216, 339, 628
testing
  instructions, 296
  usability, 621, 649–51
text
  indentation of, 283
  in formal reports, 243
texts, reference, in library research, 351–53, 355–56
than, subjective case after, 85, 98
thank-you letters, after job interviews, 319
that, 628–29
  as an adjective, 21
  agreement with antecedent of, 41, 43
  because usage and, 544
  restrictive clauses introduced by, 556
that/which/who, 630
that is (i.e.), 194
the, 22. *See also* articles
their
  case and, 491
  person and, 447
  their/they're/there usage and, 631
theirs, 447, 491
them, 447, 491
there/their/they're, 631
there are (there is, there will be)
  as an expletive, 222–23
  sentence construction and, 586
thesaurus, 631. *See also* dictionaries
  synonyms in, 614
these
  as an adjective, 21
  agreement of pronouns and antecedents and, 39
theses, 177, 183, 362
they
  case and, 491
  determining case of, 87, 492
  person and, 447
they're/there/their, 631
third person
  case and, 490–92
  narration with, 395
  person and, 447
  point of view using, 466–67
this
  as an adjective, 21
  agreement of pronouns and antecedents and, 39
  pronoun reference and, 486–88
those
  as an adjective, 21

agreement of pronouns and antecedents and, 39
thumbnail sketches, 348
tightening up
conciseness from, 121–22
revision for, 567–70
time
abbreviations for, 5
adverbs of, 30
capitalization of, 82
commas with, 109
in international correspondence,
300
numbers in, 415
slashes in, 600
tense shift and change in, 624–25,
628
time method of development
chronological method of development and, 91
instructions and, 292–93
narration and, 395
time periods
abbreviations for, 5
capitalization of, 82
dividing, using *bi-* or *semi-*, 70
tired expressions
clichés and, 95–96
trite language and, 643
title pages
in formal reports, 240–41
in sales proposals, 504, 505
title searches
Internet research for, 310
in library research, 353–54
titles (document titles), 631–32
articles in, 21, 61–62
capitalization of, 82, 126
citation formats for, 176, 182
conjunctions in, 82, 126
of formal reports, 240
of illustrations, 278
italics for, 328, 537
of memos, 382
of newsletter articles, 400
of organizational charts, 429
prepositions in, 464
of proposals, 505
quotation marks for, 327, 537
rhetorical questions as, 400, 573
of tables, 618
tone of, 634
titles (personal, professional, and job
titles)
abbreviations for, 4
capitalization of, 82–83
chair/chairperson and, 91

Ms./Miss/Mrs. usage with, 394
in salutations, 141
sexist language and, 71
titles of sections
headings for, 267–70
outlining and, 432
quotations for, 327
to
compare used with, 112
helping verbs and, 658
infinitives with, 654, 660
prepositional phrases with, 454
as a sign of the infinitive, 454
to/too/two usage and, 633–34
tone, 633–34
abstracts and, 11
adjustment letters and, 25
correspondence and, 130–33
email and, 200
memos and, 379–81
newsletter articles and, 400
point of view and, 456–58
style and, 607–10
technical writing style and, 621
trouble reports and, 645
"you" viewpoint and, 685
too/to/two, 633–34
topic lists (indexes), 283–86
topic outlines, 430–31, 445
topic sentences, 435–36, 438
topics, 634
toward/towards, 634–35
trade journal articles, 635–40
citation formats for, 178, 179, 184
as documentation, 172
documenting sources in, 173
formal writing style for, 608–09
gathering data for, 636–37
illustrations in, 279
introductions for, 322–25
literature reviews and, 362–63
obtaining publication clearances for,
640
organizing, 637
planning, 636–37
preparing the manuscript for,
639–40
research using, 552
sections of, 637–40
titles of, 631–32
trade journals
job searches using, 336
submitting articles to, 173, 639–40
tradition, and business writing styles,
298–99
trailing (loose) sentence construction,
588

train names, 328
training section, in proposals, 512
transition, 640–42
  clarity and, 93
  coherence and, 96
  conjunctions and, 126
  in lists, 361, 362
  in newsletter articles, 400
  pace and, 434
  paragraphs and, 438
  in presentations, 470–71
  in process explanations, 482
  quotations in text with, 541
  repetition and, 549
  revision for, 567
  unity and, 648
  writing a draft and, 682
transitional words
  commas with, 105
  narration with, 395
  telegraphic style and omission of,
    623
transitive verbs, 418, 657
translations
  citation formats for, 177, 183
  international correspondence and,
    298
transmittal (cover) letters, 144–47
  for formal reports, 239
  meeting agendas with, 374
  proposals with, 501, 504, 505
transparencies, 472
transpose mark, 494
transposition (inverted sentences), 589,
  595
travel (trip) reports, 91, 643
trendy words, 664
trip reports, 643
trite language, 24, 95, 644
trouble reports, 91, 644–45
truth, and ethics in writing, 214–17
try and, 647
tutorials, 620
two/to/too, 633–34
two-way indicator, in transition, 640
two-word verb, 465
type (singular and plural forms)
  agreement of pronouns and an-
    tecedents and, 40, 41
  agreement of subjects and verbs and,
    44
  demonstrative adjective with,
    20–21
types of business writing
  abstracts, 11–14
  acceptance letters, 14–15
  acknowledgment letters, 16

adjustment letters, 23–28
application letters, 56–61
complaint letters, 114–15
correspondence, 130–46
executive summaries, 219–20
feasibility reports, 225–27
formal reports, 239–45
inquiries and responses, 287–88
instructions, 292–96
investigative reports, 324–25
memos, 378–82
newsletter articles, 399–400
press releases, 480–82
process explanation, 482
progress and activity reports, 482–84
proposals, 496–521
reference letters, 544–46
refusal letters, 546–47
resignation letters or memos, 553–54
résumés, 557–67
trade journal articles, 635–40
trip reports, 643
trouble reports, 644–45
types of words
  abstract words/concrete words, 9–10
  blend words, 74
  clipped forms of words, 96
  compound words, 118–19
  euphemisms, 217
  foreign words in English, 237–38
  idioms, 275–77
  new words, 399
  vague words, 653
  vogue words, 664
typist's initials in letters, 142
typography
  electronic résumés and, 564
  email limitations with, 198–200
  layout and design and, 343–45, 357
  presentation visual aids and, 474,
    475
  proofreading for errors of, 496
  Web page design and, 673–74

umlaut, 162
unabridged dictionaries, 165–66, 352
unattached (dangling) modifiers,
  148–49
unclearness
  ambiguity and, 50–51
  vague words and, 653
underlining
  in email, 198
  emphasis with, 204
unequal importance
  emphasis and, 203–05
  subordination and, 610–12

Uniform Resource Locator (URL)
  documenting online sources with, 176, 182
  Web address with, 681–82
uninterested/disinterested, 169
unique
  as an absolute word, 9, 23
  equal/unique/perfect usage and, 213
United States, 8
unity, 648
  clarity and, 93
  coherence and, 96
  paragraphs and, 437–38
  point of view and, 456–58
  in process explanations, 482
  transition and, 643
unlike and like (comparison method of development), 113–14
unsolicited proposals, 501–03
unsolvable (insoluble/insolvable), 290
unspecific (vague) words, 653
up, 649
upon/on/onto, 420
uppercase letters, 367. *See also* capital letters
URL (Uniform Resource Locator)
  documenting online sources with, 176, 182
  Web address with, 681–82
U.S., 8
U.S. government proposals, 512–21
U.S. government publications
  citation formats for, 175, 181
  library research using, 356–57
U.S. government specifications, 604, 605–06
us
  case and, 491
  determining case of, 87, 492
  person and, 447
usability engineering, 649
usability testing, 621, 649–51
usage, 651–52
  dictionaries on, 165, 166
  intensifiers and, 296–97
  international correspondence and, 299–300
  proofreading for, 496
use, 366, 652
Usenet newsgroups, 306–07
user-centered design (UCD), 649
user manuals, 620
user testing, 651
utilitarian writing
  exposition and, 223
  instructions and, 292–96

utilization, 33, 366
utilize, 652

vague pronoun reference, 488
vague words, 653
  clarity and, 94
  clichés and, 95–96
  pronoun reference and, 486–88
  revision for, 570
vagueness
  ambiguity and, 50–51
  vague words and, 653
vantage point (point of view), 456–58
variables, in graphs, 257–64
variety, in sentences, 594–96
variety of vocabulary
  synonyms and, 613–14
  thesaurus and, 631
vendor descriptions, in proposals, 512
verb phrases, 455, 657
verbal/oral, 428
verbals, 592, 653–56
verbatim usage
  copyright and, 129–30
  note taking and, 406–08
  plagiarism and, 455–56
  quotation marks and, 536–38
  quotations and, 538–41
verbs, 656–61
  agreement of subjects and, 43–45
  conjugation of, 661
  converting into adjectives, 594
  forms of, 658–60
  missing, in sentences, 592
  mood and, 392–93
  nominalizations and, 401–04
  parallel structure repeating, 439
  as a part of speech, 444
  phrases functioning as, 451–52, 455
  properties of, 660–61
  sentence construction with, 586–88, 589–90
  sentence fragments with missing, 592–93
  types of, 657–58
  up used with, 649
  used as nouns (verbals), 653–56
  voice of, 664–68
vernacular English, 211–12
vertical axis, 257–60
very, 663
  emphasis with, 204
  as an intensifier, 296, 663
via, 663–64
video conferencing, 582–83
videotaping
  of practice presentations, 476

videotaping (*cont.*)
  research with, 552
viewpoints of writers
  point of view and, 456–58
  "you" viewpoint and, 685–86
vigorous writing
  positive writing and, 458–59
  style and, 607–10
  voice and, 665–68
virgules (slashes), 600
visit/visitation, 366
visual aids
  flowcharts for, 234–37
  global communication and,
    251–52
  global graphics and, 252–54
  graphs as, 257–64
  illustrations as, 277–79
  maps as, 368–70
  organizational charts as, 429
  photographs as, 450
  physical movement during delivery
    and using, 477
  practice with, 475–76
  presentations using, 471–75, 476
  tables as, 616–19
vitae (résumés), 557–67
vivid writing
  descriptions and, 158–59
  figures of speech and, 231–33
vocabulary
  dictionaries and, 163–67
  glossaries and, 254–55
  thesaurus and, 631
  word choice and, 679–80
vocative expression (direct address),
    168
vogue words, 664
  stacked modifiers and, 390–91
  word choice and, 679
voice (grammar), 660–61, 664–68
  helping verbs and, 658
  mixed constructions and, 387–88
  shift in, 661
  in titles, 632
  technical writing style and, 623
voice (sound), in presentations,
    477–78
voice mail messages, 582
volume numbers, 416

wait for/wait on, 277, 669
warnings, in instructions, 295, 621
we
  antecedents and, 493
  case and, 491
  determining case of, 87

person and, 447
point of view using, 458
weak verbs, and nominalizations,
    401–04
Web. *See* World Wide Web
Web forums, 308
Web page design, 669–76. *See also* for-
    mat; layout and design
  electronic résumés and, 562–64
  graphics and typography in, 673
  home pages and, 675–76
  hypertext and, 270–71
  navigation and links and, 670–73
  page and site layout in, 674
  photographs and, 450
  typography and, 345
Web sites. *See also* World Wide Web
  for job searches, 333, 336
  for library research, 354
  policies posted on, 468
  search engine sites, 579–80
  slashes in addresses for, 600
weekly, bi- or *semi-*, 70
weekly progress reports, 482–84
well, 297
well/good, 255–56
were, 392–93
west, 81
when and if, 676
where . . . at, 676
where/that, 676
whether (as to whether), 64
whether or not, 676
which
  agreement with antecedent of, 43
  as an interrogative pronoun, 489
  nonrestrictive clauses introduced by,
    556
  that/which/who usage and, 630
while, 677
white space, 349, 673
whiteboards, 471–72
who
  agreement with antecedent of, 43
  case of, 86–87, 491
  as an interrogative pronoun, 489
  one of those . . . who, 422
  that/which/who usage and, 630
who/whom, 677
  case of, 98–99, 491
whole-by-whole method of compari-
    son, 113–14
whom
  case of, 86–87
  as an interrogative pronoun, 489
  who/whom usage and, 677
who's/whose, 678

whose/of which, 678
widows, in columns, 348
will, 659
will/shall, 599, 626
-wise, 678
wish, verb forms used with, 393
with
  compare used with, 112
  contrast used with, 112
with regard to/regarding, 548–49
without (w/o), 600
word associations (connotation/denotation), 127
word beginnings (prefixes), 462
word books
  dictionaries, 163–67
  glossaries, 254–55
  thesaurus, 631
word choice, 679–80
  abstract words/concrete words, 9–10
  ambiguity and, 50–51
  antonyms and, 53
  blend words and, 74
  clarity and, 93
  clipped forms of words and, 96
  compound words and, 118–19
  connotation/denotation of words
    and, 127
  dialectal English and, 212
  diction and, 162
  dictionaries for, 163–67
  ethics in writing and, 214–15
  euphemisms and, 217
  foreign words in English, 237–38
  gobbledygook and, 255
  grammar and, 256–57
  idioms and, 275–77
  international correspondence and,
    299–300
  jargon and, 330
  long variants and, 366
  malapropisms and, 368
  new words, 399
  policies and procedures and, 455–56
  pronoun reference and, 486–88
  repetition and, 549–50
  revision and, 570
  technical writing style and, 622–23
  telegraphic style and, 623
  thesaurus and, 631
  trite language and, 643
  vague words and, 653
  vogue words and, 664
word definitions
  connotation/denotation of words
    and, 127
  defining terms using, 153–55
definition method of development
  and, 155–58
  descriptions using, 159
  dictionaries with, 163–67
  glossaries with, 254–55
  openings providing, 425
  readers' need for, 542–43
word division
  dictionaries for, 163–67
  hyphens for, 274
word endings
  dictionaries showing, 164
  suffixes and, 613
word-for-word usage
  plagiarism and, 455–56
  quotations in, 538–41
word groups
  clauses as, 94–95
  phrases as, 451–55
  sentence construction and,
    585–90
word lists
  dictionaries, 163–67
  glossaries, 254–55
  thesaurus, 631
word meanings
  connotation/denotation of words
    and, 127
  defining terms and, 153–55
  dictionaries for, 163–67
  glossaries and, 254–55
  openings providing, 434
  readers' need for, 542–43
word order
  adjective placement and, 23
  inverted sentences and, 589, 595
  modal auxiliary verbs and, 659
  sentence construction and,
    585–90
  sentence variety and, 594–95
  subject-verb-object pattern and, 586,
    589–90, 614
  syntax and, 614–15
word processing, and letter formats, 137
word types
  abstract words/concrete words, 9–10
  blend words, 74
  clipped forms of words, 96
  compound words, 118–19
  euphemisms, 217
  foreign words in English, 237–38
  idioms, 275–77
  new words, 399
  vague words, 653
  vogue words, 664
word usage, in dictionaries, 165, 166,
    412

wordiness. *See also* conciseness/
wordiness
faulty subordination and, 591–92
rambling sentences and, 591–92
squinting modifiers, 392
stacked (jammed) modifiers, 24,
255, 390–91
repetition and, 549–50
words, defining, 153–55
connotation/denotation of words
and, 127
definition method of development
and, 155–58
descriptions and, 159
dictionaries and, 163–67
glossaries and, 254–55
openings and, 425
readers' needs and, 542–43
words as words, italics for, 328
words commonly confused
idioms and, 275–77
usage and, 651–52
workplace technology
email, 197–202
fax, 224–25
hypertext, 270–71
Internet, 301–09
Internet research, 309–14
Web page design, 669–76
World Wide Web, 309, 680–82
works cited
APA style for, 179–85
bibliographies of, 73–74
MLA style for, 173–79
in trade journal articles, 639
World Wide Web (WWW), 309,
680–82. *See also* Internet; Web
page design
citing sources on, 174, 176, 179,
182, 185
copyrighted materials on, 129–30
dictionaries available on, 165,
167
downloading notes from, 407
evaluating sites on, 313
forums on, 308
hypertext on, 270–71
job searches using, 333–34
library research using, 353–54
research on, 309–14
search engines and, 577–80
slashes for addresses on, 600
technical information on, 312
worn figures of speech
clichés as, 95–96
trite language and, 643
would, 659

writer's block, while writing a draft,
682–83
writers' objectives
openings stating, 427
preparation and, 462
purpose and, 527
on résumés, 559
writing a draft, 682–83
Checklist of the Writing Process on,
xviii–xix
collaborative approach to, 98
conclusions and, 123–24
"Five Steps to Successful Writing"
on, xvi–xvii
introductions and, 322
openings and, 423–27
point of view and, 456–58
titles and, 631–32
wordiness during, 121
writing letters. *See* correspondence; memos
Writing Process, Checklist of,
xviii–xix
writing styles
for email, 198
formal, 608–09
informal, 609–10
for memos, 379–81
writing types, general
description, 158–59
exposition, 223
forms of discourse, 245–46
narration, 395–96
persuasion, 448–50
wrong words
idioms and, 275–77
malapropisms and, 368
WWW. *See* World Wide Web

xeroxing (photocopying)
copyright and, 128–30
note taking and, 407
X-ray, 684

years
abbreviations for, 5
biannual/biennial usage and, 70
commas with, 109
date format for, 150–51
numbers for, 415, 417
yes, 104
you, 421, 447
antecedents and, 493
case and, 491
"you" viewpoint, 685–86. *See also*
point of view
correspondence and, 132
persuasion and, 448–50

your, 447
  case and, 491
  your/you're usage and, 686
yours, 447, 491

zero points, in line graphs, 260
zeros, 417
zip codes, 108, 417